AF517406

Biopesticides in Sustainable Agriculture
Progress and Potential

About the Editors

Opender Koul, Fellow of the National Academy of Agricultural Sciences and the Indian Academy of Entomology, is an insect toxicologist/physiologist/chemical ecologist and currently the Director of the Insect Biopesticide Research Centre, Jalandhar, India and Adjunct Professor, Kasetsart University, Bangkok, Thailand. He has been a Visiting Scientist at the University of Kanazawa, Japan (1985-1986), University of British Columbia, Canada (1988-1992), and Institute of Plant Protection, Poznan, Poland (2001). His extensive research experience concerns insect-plant interactions, spanning toxicological, physiological and agricultural aspects. He has been honoured with an Indian National Science Academy Medal (INSA), the Kothari Scientific Research Institute Award, KEC Science Society Award, Recognition Award of National Academy of Agricultural Sciences of India (2005); Prof. R.C. Saxena Memorial Medal (2009); and Late P.P. Singhal Memorial Award (2010) for outstanding contributions in the field of Insect Toxicology/ Physiology and Plant Protection. Dr. Koul was also nominated for the Recognition Award of ESA (USA) in 2006. Dr. Koul is on panel of experts in many committees and leading international and national journals. He has also been an informal consultant to BOSTID, NRC of USA at ICIPE, Nairobi, UNIDO Expert on Biopesticides in China (2006) and Nigeria (2007), Biopesticides Expert, Tsukuba, Japan (2009) and Member External Advisory Board (ADAPPT Net Work for African, Caribbean and Pacific Group of States, European Development Fund), UK, 2009. He has authored over 200 research papers and articles, and is the author/editor of 20 books on different aspects of biopesticides.

G. S. Dhaliwal is currently Visiting Professor, Insect Biopesticide Research Centre, Jalandhar, India. He obtained Ph.D. from the Indian Agricultural Research Institute, New Delhi in 1972 and remained Post-doctoral Fellow at the International Rice Research Institute, Los Banos, Philippines, for two years. A member of the World Food Prize Nominating Academy, he is Fellow of the National Environmental Science Academy (NESA), Society of Plant Protection Sciences, Society of Pesticide Sciences India, Plant Protection Association of India, Indian Society for the Advancement of Insect Science and Entomological Society of India. Honoured with Best Scientist Award of NESA and Lifetime Achievement Award of International Allelopathy Foun-dation, he has been Founder President of the Indian Society for the Advancement of Insect Science and Society of Biopesticide Sciences India, and President of the Indian Ecological Society, He has authored more than 50 book chapters, 250 research papers and review articles, and is author/ editor of more than 40 books on different aspects of pest management, environment and sustainable agriculture.

Sucheta Khokhar is currently Dean, College of Agriculture, and served as Head, Department of Entomology (until June 2010) at CCS Haryana Agricultural University, Hisar, She obtained her Ph.D. from CCS Haryana Agricultural University, Hisar, in 1979 and was a FAO Post-doctoral Training Fellow at Ohio State University, USA in 1999 in the Diagnostic Clinics for Insects and Other Pathogens. A Fellow of the Applied Zoological Research Association (AZRA), CRRI, Cuttack, she is the recipient of Dr. Mrs. Jagadeswari Rao Women Scientist Award 1999 and 2000 bestowed by AZRA. She has established the new identity of 35 insects from Haryana and adjoining areas, and discovered 10 new species of insects. She has published more than 70 research papers in national and international journals of scientific repute and contributed 10 chapters in the books and 3 review papers. She is a co-editor of two books/compendia. She is life member of several scientific societies. She has guided more than 12 post-graduate students and taught taxonomy, morphology, classification and biosystematics courses of Entomology to undergraduate and post-graduate students at CCS Haryana Agricultural University, Hisar and Indian Agricultural Research institute, New Delhi.

Ram Singh did his Ph.D. from the Indian Agricultural Research Institute, New Delhi, in 1985. He was awarded the European Commission Post-doctoral Fellowship (1991-92) and FAO Post-doctoral Training Fellowship (1998) when he worked on various aspects of insect-plant interactions and integrated pest management at the Horticulture Research International, Wellesbourne, United Kingdom. A Fellow of the Royal Entomological Society, London, he is the recipient of University Gold Medal and Aspee Gold Medal, from HAU, Hisar and ICAR award from New Delhi. He has published more than 150 research papers in national and international journals of scientific repute. He has also contributed 17 chapters in the books and 11 review papers. Dr. Singh has co-edited/authored four books and four compendia. He is life member of several scientific societies and also the Fellow of Plant Protection Association of India. He has presented his work at several international conferences in India, Netherlands, Scotland and United Kingdom.

Biopesticides in Sustainable Agriculture

Progress and Potential

Editors

Opender Koul G.S. Dhaliwal
Sucheta Khokhar Ram Singh

Published by:

Scientific Publishers (India)
5 A, New Pali Road, P.O. Box 91
Jodhpur 342 001 (India)

E-mail: info@scientificpub.com
Website: www.scientificpub.com

Branch Office
Scientific Publishers (India)
4806/24, Ansari Road, Daryaganj
New Delhi - 110 002 (India)

Print 2026

ISBN: 978-81-7233-865-7

eISBN: 978-93-86102-01-0

Printed in India

PREFACE

Sustainable agriculture may be described as any system that provides for the profitable production of crops while minimizing adverse environmental impacts and utilizing natural resources to ensure agricultural production indefinitely. In other words, sustainable agriculture is a system that can evolve indefinitely towards greater human utility, greater efficiency of resource use, and a balance with the environment that is favourable to the humans and other species. To achieve the ecological goals of sustainable agriculture, crops and pest management practices need to be directed at suppressing the incidence and intensity of a wide range of crop pests. The basic approach of pest regulation is consistent with the primary goals of sustainability. Integrated pest management (IPM) includes ecologically based management processes developed with an understanding of natural cycles and natural regulators of those species that compete with humans for resources in agriculture production systems. Biopesticides are an integral part of IPM systems and play a pivotal role in sustainable agriculture. The present book gives the state-of-art of the role of biopesticides in sustainable production of various agricultural crops and outlines the future strategies for efficient utilization of biopesticides in agricultural production systems.

In all, the book has been divided into 16 chapters. The first chapter gives an overview of biopesticide use on global scale and describes salient features of the role of biopesticides in Indian agriculture. The current status and future potential of biopesticides in management of pests of grain crops, viz. rice, maize, pulses and oilseeds has been described in the next four chapters. The management of pests of commercial crops, viz. cotton and sugarcane using biopesticides has been outlined in chapter 6 and 7, respectively. The next two chapters enumerate the role of biopesticides in management of pests of vegetables and fruits. The biopesticide strategies for management of pests of tobacco, spices and tuber crops have been described in chapters 10, 11 and 12, respectively. The management of diseases of coconut with biopesticides, with special reference to coconut root wilt, has been discussed in chapter 13. The next chapter focuses on the management of pests of forest trees with biopesticides. The penultimate chapter describes the role of biopesticides in tea from Japanese perspective. The last chapter is devoted to management of stored product pests with bipesticides.

We are thankful to Dr K.S. Khokhar, Vice-Chancellor, CCS Haryana Agricultural University, Hisar; Dr Arvind Kumar, Deputy Director General (Education), Indian Council of Agricultural Research, New Delhi, for their encouragement and support. We are grateful to Dr O.P. Tokey, Dean, Postgraduate Studies and Dr R.K. Saini, Head, Department of Entomology, CCS Haryana Agricultural University, Hisar, for their help from time to time. Thanks are also due to Dr Viaks Jindal and Ms Bharathi Mohindru, Assistant Entomologists, Punjab Agricultural University, Ludhiana and Rajwinder Singh and Dharamvir Kanda, Insect Biopesticide Research Centre, Jalandhar, for their help at various stages of preparation of the book.

We express our gratitude to the Department of Biotechnology, Government of India; Council of Scientific and Industrial Research, Agricultural and Processed Food Products Export Development Authority and the Indian National Science Academy, for the financial support. We are also thankful to all the contributors for submitting their manuscript within the stipulated period. Finally, our special thanks are due to Mr. Tanay Sharma, Scientific Publishers (India), Jodhpur, for his keen interest in bringing out this volume.

Opender Koul
G. S. Dhaliwal
Sucheta Khokhar
Ram Singh

CONTRIBUTORS

R. Aravind
Plant pathology and Microbiology (PLPM), Texas A & M University, 2132 TAMU College Station, Texas-77843, USA

H. Basappa
Directorate of Oilseeds Research (ICAR) Rajendranagar, Hyderabad-500030, India

T. Bharathimeena
Department of Agricultural Entomology, Centre for Plant Protection Studies, TNA University, Coimbatore-641003, India

S. Devasahayam
Indian Institute of Spices Research, Marikunnu P.O., Kozhikode-673012, Kerala, India

G. S. Dhaliwal
Insect Biopesticide Research Centre, 30-Parkash Nagar, Model Town, Jalandhar-144003, India

A. K. Dhawan
Department of Entomology
Punjab Agricultural University
Ludhiana-141004, India

P. Duraimurugan
Division of Crop Protection
Indian Institute of Pulses Research, Kanpur- 208024, India

C. Durairaj
Department of Agricultural Entomology, Centre for Plant Protection Studies, Tamil Nadu Agricultural University, Coimbatore-641003, India

E. R. Harish
Central Tuber Crops Research Institute, Sreekariyam-695017, Thiruvananthapuram, India

Saroj Jaipal
CCS Haryana Agricultural University, Regional Research Station, Uchani, Karnal-132001(Haryana) India

C. A. Jayaprakas
Central Tuber Crops Research Institute
Sreekariyam-695017, Thiruvananthapuram, India

M. Kalyanasundaram
Department of Agricultural Entomology, Centre for Plant Protection Studies, Tamil Nadu Agricultural University, Coimbatore-641003, India

Gururaj Katti
Directorate of Rice Research
Hyderabad - 500030, India

Sucheta Khokhar
College of Agriculture
CCS Haryana Agricultural University, Hisar-125004, India

Opender Koul
Insect Biopesticide Research Centre, 30-Parkash Nagar, Model Town, Jalandhar-144003, India

V. Krishnamurthy
Central Tobacco Research Inst.
Rajahmundry- 533105, India

Nitin Kulkarni
Tropical Forest Research Inst.
Jabalpur- 482021, India

A. Kumar
Indian Agricultural Research Institute
New Delhi-110012, India

Pradyumn Kumar
Directorate of Maize Research
Pusa Campus
New Delhi-110012, India

Vijay Kumar
Department of Entomology
Punjab Agricultural University
Ludhiana-141004, India

M. Mani
Division of Entomology and Nematology, Indian Institute of Horticultural Research
Bangaluru -560089, India

Madoka Nakai
Division of Bioregulation and Biointeraction Institute of Agriculture, Tokyo University of Agriculture and Technology,
Saiwai, Fuchu,
Tokyo 183-8509, Japan

R. K. Patil
Department of Entomology
University of Agricultural Sciences,
Dharwad-580005, India

L. Ragesh
Central Tuber Crops Research Institute, Sreekariyam-695017,
Thiruvananthapuram, India

S. Gunneswara Rao
Central Tobacco Research Inst.
Rajahmundry- 533105, India

Y.R. Sarma
Aramam, M 10/5 KSHB Colony
Malaparamba, Kozhikode
Calicut-673009, Kerala, India

Hem Saxena
Division of Crop Protection
Indian Institute of Pulses Research
Kanpur- 208024, India

D.R. Sharma
Department of Horticulture
Punjab Agricultural University
Ludhiana-141004, India

P. S. Shera
Department of Plant Breeding
Punjab Agricultural University
Ludhiana-141004, India

Ram Singh
College of Agriculture
CCS Haryana Agricultural University
Hisar-125004, India

U. Sreedhar
Central Tobacco Research Institute
Rajahmundry- 533105, India

R. S. Sreerag
Central Tuber Crops Research Institute, Sreekariyam-695017,
Thiruvananthapuram, India

N. Srinivasan
Division of Crop Protection
Central Plantation Crops Research Inst., Regional Station (ICAR),
Kayangulam
Krishnapuram- 690533,
Kerala, India

S. B. Suby
Directorate of Maize Research
Pusa Campus
New Delhi-110012, India

CONTENTS

1

Biopesticide Based Integrated Pest Management for Sustainable Agriculture

G. S. DHALIWAL[1]*, OPENDER KOUL[1], SUCHETA KHOKHAR[2] AND RAM SINGH[2]

[1]Insect Biopesticide Research Centre, 30-Parkash Nagar, Model Town, Jalandhar-144003, India
[2] Department of Entomology, CCS Haryana Agricultural University, Hisar-125004, India
*E-mail: gurmail_d@yahoo.in

INTRODUCTION

The green revolution technology has accentuated insect pest problems in agriculture. The intensive use, misuse and abuse of pesticides have led to the problems of pesticide resistance, resurgence and contamination of different components of the environment. In addition, there have been outbreaks of secondary insect pests in several regions of the world. Consequently, there has been an increasing trend in the magnitude of crop losses due to insect pest problems. According to recent estimates, insect pests inflict a loss of 10.8% to principal agricultural crops at global level (Oerke, 2006), whereas the losses in Indian agriculture are to the tune of 16.4% valued at ₹ 900 billion (Dhaliwal *et al.*, 2013). The losses due to insect damage are likely to increase as a result of changes in crop diversity and increase in the incidence of insect pests due to climate change (Sharma, 2010).

Pesticides have remained the principal weapon to tackle the menace of insect-pests. The world's pesticide consumption is estimated to be about 3.0 million tones, which includes 1500 active ingredients and 50,000 or more commercial products (Pimentel, 2007). The current consumption of pesticides in India is about 44 thousand tones which includes 228 registered pesticides. The overuse and misuse of pesticides have caused several adverse effects on nature and natural resources like environmental pollution, pesticide resistance, pest resurgence, and emergence of secondary pest problems. The increasing awareness of health related effects of chemical pesticides has accelerated the demand for organic food. Therefore, the quest for shifting the focus on to more reliable, sustainable and environment friendly agents of pest control has recently started gaining momentum. Biopesticides are expected to fill in this slot and play a prominent role in future integrated pest management (IPM) programmes (Dhaliwal and Koul, 2007, 2010).

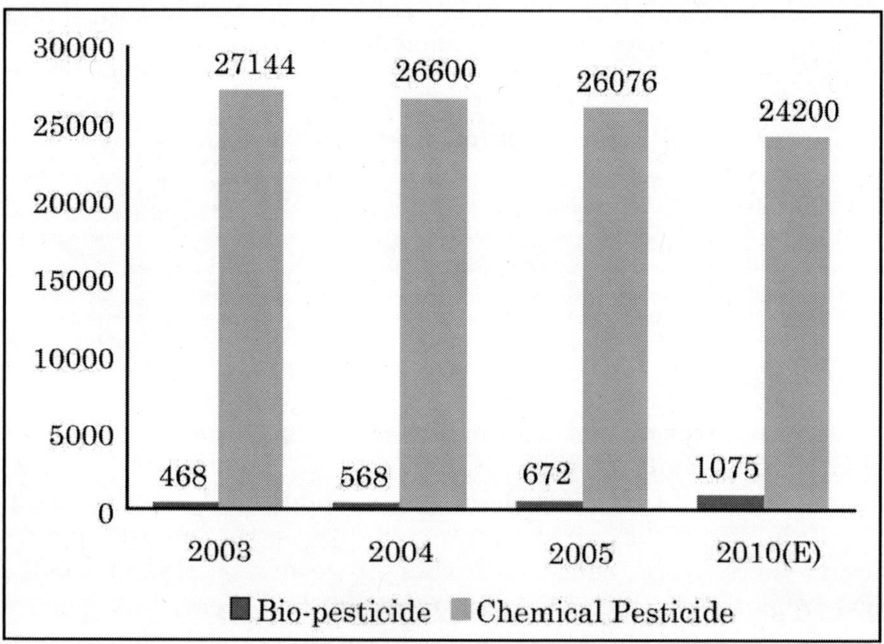

Fig. 1.1. Trends in global chemical pesticide and bio-pesticide market ($ million) (After BCCI, 2010).

Biopesticides are fast emerging as an alternative to the synthetic chemical pesticides for management of insect pests, pathogens and weeds. The global weighted average consumption level of biopesticides is approximately 1 kg/ha. With the world organic farming area comprising about 24 million ha, global biopesticide consumption is thus estimated at about 24 million kg. In terms of use, orchards claim the largest share (55%) of the total biopesticides used. The current consumption of organically produced fruits and vegetables at the global level is valued at US$ 27 billion. India produces around 4,00,000 metric tones of certified organic products. Regionwise, North America consumes the largest share (40%) of the global biopesticides production followed by Europe and Oceanic countries accounting for 20% each. While the overall market for pesticides is showing a decline in average annual growth rate of 1.1%, biopesticide market is growing rapidly, increasing from $ 672 million in 2005 to over $ 1 billion in 2010 (Fig. 1.1) at an annual growth rate of 9.9%. At present, about 1000 commercial biopesticide formulations with more than 225 registered biopesticide active ingredients are known in the global market. Biopesticides accounted for 2.5% of the total global pesticide market in 2005, which increased to 4.5% in 2010 (BCCI, 2010).

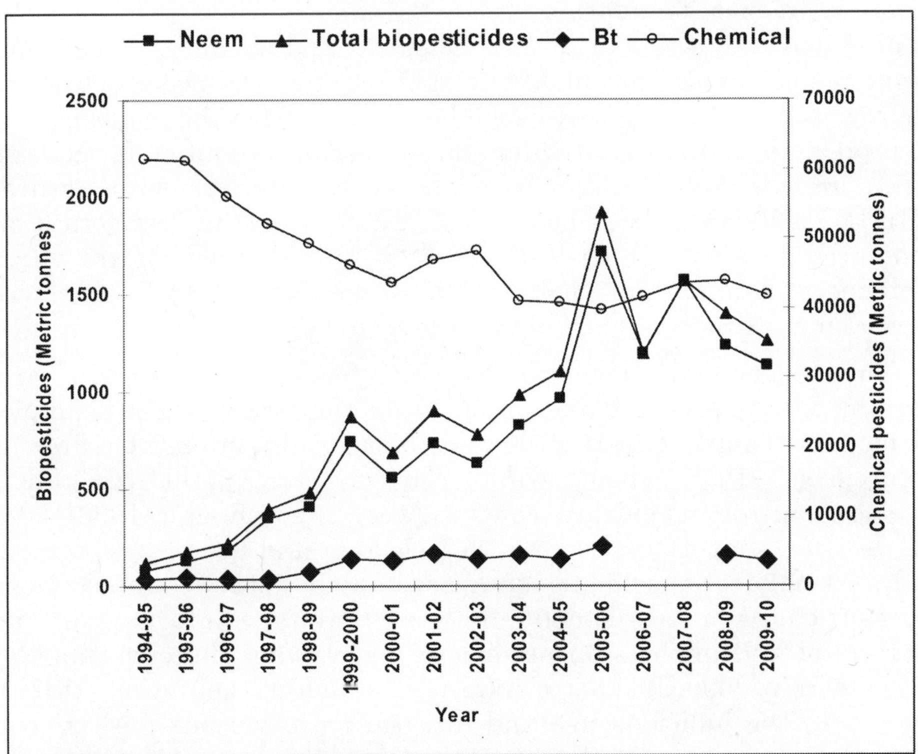

Fig. 1.2. Trends in the use of chemical pesticides and biopesticides in India, 1994-95 to 2009-10. (Data for Bt for 2006-07 and 2007-08 not available) (www.ppqs.gov.in).

Biopesticides represent only 3% of the overall pesticide market in India (Fig. 1.2) and are expected to exhibit an annual growth rate of

2.5% in the coming years. So far, only 12 types of biopesticides have been registered in India, prominent among them being neem-based products and *Bacillus thuringiensis* (Berliner) (Bt). The overall growth rate of biopesticides at global level is estimated to be about 20% per annum or even more by the middle of twenty-first century (Sinha and Biswas, 2008; Dhaliwal *et al.*, 2012)

CONCEPT OF BIOPESTICIDE BASED IPM

Biopesticide based integrated pest management (BBIPM) may be defined as the manipulation of natural enemies (predators, parasitoids, microbes, etc.), biorational compounds (semiochemicals, insect growth regulators, plant-derived materials, microbe-derived materials, etc.) and transgenic crops in such a manner so as to keep pest populations below damaging levels and maintain ecosystem harmony. In fact, foundation of BBIPM was laid by Hoskins *et al.* (1939) when they said "............ biological and chemical control are considered as supplementing to one another or as the two edges of the same sword .... nature's own balance provides the major part of the protection that is required for the successful pursuit of agriculture .... insecticides should be used so as to interfere with natural control of pests as little as possible ....". Even the first definition of integrated control (Stern *et al.*, 1959) as "applied pest control which combines and integrates biological and chemical control" considers biopesticides as integral part of the integrated approach.

The modern definitions of IPM, which focus on the ecological and environmental issues, lay due emphasis on biopesticides. For example, Frisbie and Smith (1991) proposed 'biologically intensive IPM' or 'biointensive IPM', which would rely on host plant resistance, biological control and cultural control. According to Kogan (1998), IPM is a decision support system for the selection and use of pest control tactics, singly or harmoniously coordinated into a management strategy, based on cost/benefit analyses that take into account the interests of and impacts on producers, society and the environment. The essence of biopesticides is reflected in another definition (USDA, 2000), *viz.*, the judicious use and integration of various pest control tactics in the context of the associated environment of the pests in ways that complement and facilitate the biological and other natural controls of the pests to match economic, public health, and environmental goals.

A special committee of the National Research Council's Board of Agriculture (NRC, 1996), proposed 'ecologically based pest management (EBPM), also called 'ecologically based integrated pest management (EBIPM) (Kennedy and Sutton, 2000), which lays emphasis to shift the IPM paradigm from focusing on pest management strategies relying on pesticide management to a system approach relying primarily on biological knowledge of pests and on their interaction with the crops. This suggests that EBIPM programmes should represent a sustainable approach to manage pests combining biological, chemical, physical and cultural tools to ensure favourable economic, ecological and sociological consequences (Koul and Cuperus, 2007). It signifies that biopesticides would remain the core of future IPM programmes which should be environmentally sound, economically viable and socially acceptable. Therefore, biopesticide based integrated pest management would open new vistas and provide a road map to sustainable crop protection in twenty first century (Koul *et al.*, 2009).

BIOPESTICIDES IN INDIAN AGRICULTURE

India is very rich in biodiversity of biocontrol agents. The maximum number (120) of natural enemies have been recorded from *Helicoverpa armigera* (Hubner), followed by *Scirpophaga incertulas* (Walker) (72), *S. excerptalis* (Walker) (61), *Chilo infuscatellus* Snellen (50), *Oryctes rhinoceros* Linnaeus (45), *Spodoptera litura* (Fabricius) (42), *Opisina arenosella* Walker (41), *Bemisia tabaci* (Gennadius) (34), *Chilo partellus* (Swinhoe) (32), *Aproaerema modicella* (Deventer) (30), *Plutella xylostella* (Linnaeus) (29), *Achaea janata* (Linnaeus) (27), *Idioscopus clypealis* Lethiery (23), *Rastrococcus iceryoides* (Green) (23) and *Coccus viridis* (Green) (23). In addition, many species of spiders have been recorded from different crop ecosystems; maximum from rice (191), followed by cotton (106), sugarcane (58), maize (39), coconut (26),tobacco (23), legumes (19), wheat (14) and mango (11). Of the 218 species of predatory mites recorded in India, 150 species are from family Phytoseiidae. More than 450 species of coccinellids, 60 species of chrysopids and 26 species of trichogrammatids and anthocorid bugs have been described from India (Singh, 2003).

The earliest attempt at introduction of a natural enemy in India was made in 1898 when a coccinellid, *Cryptolaemus montrouzieri* Mulsant, from Australia was released against coffee green scale, *Coccus viridis* (Green). It established but failed to control the scale. It is now a common mealybug predator in South India. However, concerted efforts at setting an insect to kill an insect began in 1920s and soon many successes were achieved. The notable among these are the introduction of *Rodolia cardinalis* (Mulsant) against *Icerya purchasi* Maskell (1926–48), *Aphelinus mali* (Haldeman) (1930–37) against *Eriosoma lanigerum* (Hausmann), *Encarsia perniciosi* (Tower) against *Quadraspidiotus perniciosus* (Comstock) (1959-66), and *Leptomastix dactylopii* Howard against mealybugs (1980) (Ramani, 2004). The establishment of Regional Station of Commonwealth Institute of Biological Control at Bangalore in 1957 gave a fillip to biopesticide work in India. The biopesticide research and development got a further boost with the establishment of All India Coordinated Research Project on Biological Control of Crop Pests and Weeds (AICRPBC) in 1977. The AICRPBC was upgraded to the level of Project Directorate of Biological Control (PDBC) in 1993. Recently, PDBC has been further upgraded as National Bureau of Agriculturally Important Insects (NBAII) in 2008 (Dhaliwal and Koul, 2010).

So far 15 biopesticides (fungi 8, bacteria 3, viruses 2 and botanicals 2) with various formulations have been registered under the Insecticides Act 1968 in India. Neem-based products with 0.03, 0.15 and 1% azadirachtin content have been registered. The registration committee has also granted regular registration to *Bacillus thuringiensis* and *B. subtilis* with a view to commercialize formulations based on bacilli for control of various crop pests. Various fungi like *Ampelomyces quisqualis, Beauveria bassiana, Metarhizium anisopliae, Paecilomyces lilacinus, Trichoderma harzianum, Trichoderma viride, Verticillium chlamydosporium* and *Verticillium lecanii* have been registered. Nuclear polyhedroviruses (NPVs) of *Helicoverpa armigera* (Hubner) and *Spodoptera litura* (Fabricius) have also been registered. An antagonistic fungus *Trichoderma viride* has been registered provisionally under Section 9 (3B) for groundnut seedling wilt (Wahab, 2010). Consumption of biopesticides has increased from 219 metric tones in 1996-97 to 683 metric tones in 2000-01, and about 85% of the biopesticides used are neem-based products (Table 1.1).

In India, about 16 commercial preparations of *B. thuringiensis,* 38 fungal formulations based on *Trichoderma, Metarhizium* and *Beauveria,* and about 45 baculovirus based formulations of *Helicoverpa* and *Spodoptera* are available. Microbials are expected to replace at least 20% of chemical pesticides. About 128 units in the country (80 private companies) are supplying biopesticides. Besides, ICAR Institutes (8), State Agricultural Universities (10) and Central Integrated Pest Management Centres (30) and 4 parasitoid producing laboratories are also supplying natural enemies. The first private insectary, Biocontrol Research Laboratory, was established at Bangalore in 1981. There are as many as 42 parasitoid breeding centres in the State Department of Agriculture, particularly in peninsular India. Many sugar factories have established laboratories of their own for mass production of the egg parasitoids. The commercial concerns also supply pheromone traps for lepidopteran monitoring and mating disruption, and plant products such as neem based pesticides (Wahab, 2010).

Table 1.1. Annual availability of biopesticides in India

Sr. No.	Biopesticide	Quantity/annum
1	Neem (300 ppm)	1,000,000 l
2	Neem (1500 ppm)	250,000 l
3	*Bacillus thuringiensis*	50,000 kg
4	Nuclear polyhedrosis virus (liquid)	500,000 LE
5	*Beauveria*	Meager
6	Pheromone traps	500,000
7	Lures	2 million
8	*Trichogramma*	1 million
9	*Chrysoperla* and other natural enemies	Meager
10	*Trichoderma*	500 T

Source: Sinha and Biswas (2008)

Standardized cost effective and commercially viable mass production technologies of various biopesticides have been developed. These include **(i) Microbials.** NPVs of *H. armigera, S. litura, Hyblaea puera* (Cramer) and *Spilarctia obliqua* (Walker); granulosis viruses (GVs) of *C. infuscatellus* and *P. xylostella* and non-occluded virus of *O. rhinoceros* and *Trichoderma viride, T. harzianum, T. koningii, T. virens, Myrothecium verrucaria, Aspergillus niger, Bacillus thuriengiensis, B. subtilis, Pasteuria penetrans, Pseudomonas fluorescens, Beauveria bassiana, Verticillium chlamydosporium, Verticillium lecanii, Paecilomyces lilacinus, Nomuraea rileyi* and *Debaromyces henseni* **(ii) Entomopathogenic nematodes.** *Steinernema carpocapsae, S. bicornutum, S. feltiae and Heterorhabditis indicus* **(iii) Parasitoids and predators.** *Trichogramma chilonis* Ishii, *T. japonicum* (Ashmead), *Cryptolaemus montrouzieri* (Mulsant) and *Chrysoperla carnea* (Stephens) **(iv) Plant products.** *Azadirachtin* from neem, *Azadirachta indica* (A. Juss.) **(v) Pheromones.** Insect sex pheromones for *Chilo sacchariphagus indicus* (Kapur), *C. infuscatellus, S.*

incertulas, Pectinophora gossypiella (Saunders), *H. armigera* and aggregation pheromones for *Rhyncophorus ferrugineous* (Olivier).

Field efficacy of these biopesticides has been established in multi-location tests on various crops, *viz.*, cotton, rice, groundnut, sunflower, sesamum, soybean, chickpea, black gram, green gram, pigeonpea, lentil and other pulses, cabbage, cauliflower, chilli, tomato and other vegetables, tobacco, sugarcane, spices, coconut, tea, coffee, etc. (Dhaliwal and Koul, 2007, 2010; Rabindra and Venkatesan, 2010; Wahab, 2010).

BIOPESTICIDE BASED IPM: SUCCESS STORIES

Biopesticides form the core of integrated pest management programmes in many agricultural crops. A few selected examples are described herein.

Cotton

The cotton production systems of the world illustrate well the ecological problems associated with intensive pesticide use. A landmark in BBIPM has been the validation of the system at village Ashta, Distt. Nanded (Maharashtra), which is a representative of more than 2 million ha rainfed cotton in Central India. The baseline information indicated less than a quintal average seed cotton yield per ha in *H. armigera* epidemic year, when the farmers had sprayed more than 12–13 chemical spray rounds. All the farmers of the village were involved and the BBIPM approach covered over 200 ha cotton area. The prominent management practices adopted in BBIPM module were as follows (Bambawale *et al.*, 2004).

- Seeds were treated with imidacloprid @ 7g a.i. per kg.
- Synchronized sowing was completed within a week in the entire village, soon after the onset of monsoon rains.
- Planting of maize as border crop interfaced with cowpea was done to provide substrate for build up of ladybird beetles and their migration in cotton in due course.
- A row of *Setaria* was planted between every 9th and 10th row of cotton to enhance the activity of predatory birds by serving as food source and acting as live perch.
- Pheromone traps were installed in the fields @5/ha to monitor the arrival of *H. armigera* and timing the action to be taken.

- *T. chilonis* was released in cotton fields @ 1,50,000 per ha, when 2–8 adults of *H. armigera* per pheromone trap were captured continuously for 3–4 days a week.
- Neem seed kernel extract (5%) was sprayed a week after the release of *T. chilonis.*
- Application of HaNPV @ 250 LE per ha was made based on the economic threshold level of *H. armigera* .
- Need based application of fungicides was done to control grey mildew.

The BBIPM module resulted in substantial reduction in pesticide use, conserved natural fauna and created a congenial atmosphere for the natural forces of defence to act. The cost of plant protection was reduced by 31.3%, chemical pesticide consumption reduced by 94.3% and seed cotton yield increased by 56.9%. The average cost: benefit ratio in IPM was 1:1.473 as compared to 1:1.018 in non-IPM. The system has become self-sustainable as the farmers have themselves become decision makers and have started adopting many IPM practices on their own.

The introduction of gene technology has added a new dimension to pest management in cotton. Transgenic insect protective cotton with an insecticidal gene (*cry1Ac*) from *B. thuringiensis* was first developed in 1996 and has since been cultivated in many countries of the world. India is the largest Bt cotton growing country with its ever expanding area; increasing from 0.05 million ha in 2002 to 10.6 million ha in 2011. The increase in Bt cotton area over the years has become possible with the development of new Bt cotton hybrids; 3 in 2002, 4 in 2004, 20 in 2005, 62 in 2006, 135 in 2007, 274 in 208, and 522 in 2009. In 2006, Monsanto developed stacked Bt cotton (Bollgard II) with two *cry1Ac* and *cry2Ab* genes and since then dual stacked Bt cotton area is expanding all over the country. Bollard II occupied 4% of Bt cotton in 2006, 8% in 2007, 27% in 2008 and 57% in 2009 (Gujar *et al.*, 2010; James, 2011).

Amidst the challenges in agriculture, India's Bt cotton success story stands out even in the ongoing controversy over the genetically modified crops. Bt cotton has revived the dwindling cotton production in India and rekindled the hopes of cotton farmers. In 2011, 7 million small and marginal resource-poor farmers planted Bt cotton over 88% of hectarage under this crop. The deployment of Bt cotton has resulted in India becoming the number one exporter of cotton globally as well as the second largest cotton producer in the world. Bt cotton has literally revolutionized cotton production in India. In a short span of 10 years (2002–11), Bt cotton has generated economic benefits for farmers valued at US$ 9.4 billion, reduced the pesticide load by almost 50% and contributed to the doubling of yield, and transformed India from a cotton importer to a major exporter. India celebrated a decade of Bt cotton in 2011, which has been a boon to cotton, Indian agriculture and the country. Bt cotton is expected to achieve new heights in the future due to the availability of a wide variety of hybrids with stacked genes (James, 2011).

In spite of the economic and environmental benefits of Bt cotton, its ecological impact on various components in general and arthropod diversity and resistance potential in particular, remains a point of concern. Some of the studies have shown no significant differences between Bt and non-Bt cottons for populations of major non-target insect pests (leafhoppers, whiteflies, ash weevils, aphids, dusky bug, red cotton bug and green bug), and the generalist predators (lady-birds, chrysopids and spiders) (Dhillon and Sharma, 2009). However, some new pests like the mealybug, *Phenacoccus solenopsis* Tinsley, and mirid bug, *Creontiodes biseratense* (Distant) have started appearing in serious proportions on Bt cotton in certain regions. The need for applying pesticides to manage these pests has offset the advantages of reduced insecticidal sprays that Bt cottons offered. Moreover, there are also unconfirmed reports of survival of *H. armigera* and *P. gossypiella* on Bt cotton, which should not be ignored. Increased use of pesticides in Bt cotton in recent years will need to be viewed from the threat of pesticide resistance evolution in non-target pests and destruction of sustainable natural control agents (Gujar *et al.*, 2010).

Rice

The potential of BBIPM in rice remained subdued due to the initial success of pesticides for the control of plant hoppers and leafhoppers. It was only after the occurrence of widespread outbreaks of these insects which led to the withdrawal of several pesticides in many Asian countries that paved the way for the revival of biopesticide approach (Jahn *et al.*, 2007). The Food and Agriculture Organization (FAO)'s Inter-country programme for the Development and Application of Integrated Pest Control in Rice in South-east Asia remains one of the best examples of IPM implementation in the tropical region. The hallmarks of the Asian Rice IPM were a reliance on conservation of biological control and the use of Farmers Field Schools (FFSs) to investigate, implement and extend IPM technologies. FFSs are organized at the village level, which involve farmers in the development and implementation of pest management practices, because rice IPM is built upon a foundation of conserving endemic natural enemies (mainly predatory bugs, egg parasitoids and spiders), farmers are involved in observing natural enemy attacks on rice pests, and investigating relationships between production practices and natural enemy diversity, dynamics and effectiveness. The success of this approach is evident not only through its implementation in the rice production system, but also in its application to other crop-pest systems in Asia and other parts of the world (O'Neil *et al.*, 2003).

BBIPM in rice essentially comprises of the following inputs (Dhaliwal and Koul, 2007):

- Installation of pheromone traps @ 20 per ha to monitor the incidence of stem borer and leaf folder.
- Release of *T. chilonis* and *T. japonicum* @ 1.0 to 1.5 lakh per ha at weekly intervals.
- Spray of neem seed kernel extract or commercial formulations of neem against planthoppers.
- Need based application of chemical pesticides.

The large scale validation of IPM approach has been carried out on Basmati rice (Pubjab Basmati-1) in village Shikohpur of Baghpat district in western Uttar Pradesh, where farmers solely depended on the use of intensive pesticidal applications. In IPM fields, more abundance of natural enemies was recorded as compared to non-IPM fields due to least pesticidal interventions. The common predators like spiders, dragonflies, damselflies, carabids and mirid bug, *Cyrtorhinus lividipennis* Reuter, were more abundant in IPM fields, while among parasitoids, *Cotesia, Telenomus, Tetrastichus, Stenobracon* and *Xanthopimpla* spp. were quite common. By following the IPM tactics, the pesticide use was drastically cut, other input costs were reduced and farmers could secure residue-free produce with increased yield. The cost : benefit ratios were 1:3.18 (IPM) and 1:2.28 (non-IPM) in 2000, 1:3.16 (IPM) and 1:2.12 (non-IPM) in 2001, and 1:2.21 (IPM) and 1:1.64 (non-IPM) in 2002 (Singh *et al.*, 2005).

Vegetable Crops

Several insect-pests and their natural enemies co-exist in the vegetable cropping system. The delicate balance between these two groups is often disrupted largely due to misuse of pesticides and monocropping of vegetables in specialized production centres. This has resulted in the emergence of several insect pests in serious proportions on different groups of vegetable crops. Among these, the diamondback moth (DBM), *P. xylostella,* on cabbage and fruit borer, *H. armigera* on tomato are quite serious.

Cabbage. BBIPM system has been developed by the Indian Institute of Horticultural Research (IIHR), Bangalore and it has become increasingly popular among the farmers. The BBIPM is based on the following components.

- Sow paired rows of bold-seeded mustard as a trap crop at the beginning and after every 25 cabbage rows. Among the paired rows, the first is sown 15 days prior to cabbage planting while the second is sown 25 days after planting.
- Spray *B. thuringiensis* formulation or neem seed kernel extract (NSKE) (4%) or neem soap (1%) or pongamia soap (1%) to control DBM and other pests at 10 days after transplanting (DAT).
- Release, *Trichogrammatoidea bactrae* Nagaraja at 0.50–0.75 lakh eggs/ha (5 releases at 10 days interval).
- At 20–25 DAT, spray NSKE or neem soap or pongamia soap and repeat 2-3 times at 10–15 days interval.
- Remove lower and diseased leaves periodically and apply pesticides on need basis.

The above BBIPM package has been refined and validated at Bangalore and Varanasi. The total number of sprays with chemical pesticides in IPM plots was 0.8 as compared to 14.4 in non-IPM plots at Bangalore. The corresponding number of sprays at Varanasi was 3.50 and 11.17. The mustard crop attracted 80% of DBM for colonization. Spraying of NSKE was safe to *Cotesia plutellae* Kurdjumov, the dominant parasitoid of DBM. The cost:benefit ratio was 1:2.89 in IPM and 1:2.60 in non-IPM plots at Bangalore, and 1:3.70 in IPM and 1:1.89 in non-IPM plots at Varanasi (Sardana *et al.*, 2004).

Tomato. Tomato is another vegetable crop in which successful BBIPM technology has been developed at IIHR for the management of fruit borer, *H. armigera.* The important components of BBIPM package are as follows:

- Transplant 25-day old tomato and 50-day old African marigold seedlings simultaneously in a pattern of one row of marigold for every 16 rows of tomato. It must be ensured that the first and the last row of plots should be of marigold. However, if the area is small, marigold can be planted all round the tomato crop.

- Apply neem cake @ 250 kg/ha at 20 DAT.
- Spray HaNPV @ 250 larval equivalents (LE)/ha at 28 DAT and repeat 2 more times at weekly interval.
- Install pheromone traps @ 12.5/ha.
- Release *T. chilonis* @ 1 lakh/ha at weekly interval.
- Spray NPV on marigold flowers and buds to kill the fruit borer in them.
- Collect tomato fruits infested with borers and destroy them.
- Apply pesticides on need basis to control mites and diseases.

The above technology has been refined and validated at Bangalore and Varanasi. The total number of chemical pesticide sprays was 4.7 in IPM plots and 28.0 in non-IPM plots at Bangalore. The corresponding number of sprays was 3.83 and 14.17 at Varanasi. Most of the components of BBIPM favoured the population of larval parasitoid, *Campoletis chloridae* Uchida. The cost: benefit ratio was 1:4.82 for IPM and 1:0.61 for non-IPM plots at Bangalore. The corresponding ratios were 1:3.30 and 1:2.02 for Varanasi (Sardana *et al.*, 2004).

Fruit Crops

The pest scenario in fruit crops is changing because of increase in area and change in agronomic practices like close planting, drip irrigation, regular pruning, etc. Biopesticide based IPM programme is gaining importance in the major fruit crops because of its environment friendly nature (Reddy, 2010).

Citrus. BBIPM system has been developed for the management of various pests explained as under.

(i) Citrus leaf miner, *Phyllocnistis citrella* Stainton

- Under low to moderate infestation, spray 4% neem seed kernel extract during new flush emergence.
- Under heavy infestation, give sandwich sprays involving 4% neem seed kernel extract- cypermethrin 25 EC at 0.5 ml/l- neem seed kernel extract 4% at fortnightly intervals.

- Combined release of *Mallada boninensis* (Okamoto) (30 larvae/tree) and *Tamarixia radiata* (Waterston) (40 adults/ tree) resulted in 23–26% reduction in leaf miner population.
- Spraying with fish oil resin soap and nicotine sulphate (1:1) + 50 parts of water in the early stages of infestation help in reducing the pest population.
- The parasitoid, *Tetrastichus phyllocnistoides* (Narayanan) and *Ageniaspis* sp. cause upto 80% parasitism in Punjab.

(ii) Blackfly, *Aleurocanthus woglumi* Ashley

- Avoid close planting and water logging.
- The affected shoots should be clipped off and destroyed.
- Avoid excessive nitrogen and irrigation.
- Spray dimethoate 30 EC at 2ml/l during new flush and repeat at 10–12 days interval.
- Spray neem seed kernel extract at 4% which does not affect natural enemies.
- The pest is regulated by parasitoides like *Amitus hesperidium* Silvestri and *Encarsia clypealis* (Silvestri) up to 80–90% in Karnataka and Maharashtra. At Gawahati, *Prospaltella divergence* and *Encarsia merceti* (Silvestri) are dominant parasitoides accounting for 90% of the total parasitoides.
- *Chrysoperla carnea* (Stephens) or *M. boninensis* can also be released at 10–15 eggs/ Ist instar larvae/plant. Combined release of *M. boninensis* (40 adults/tree) resulted in 28–30, 42–45 and 23–26% reduction in black fly, psylla and leaf miner population, respectively.

(iii) Lemon butterflies, *Papilio* spp.

- Collect and destroy larvae which look like bird droppings in the early stage.
- Spray endosulfan 35 EC at 2ml/l on a new flush
- Release *Trichogramma chilonis* Ishii at 500 adults per tree.
- Conservation of indigenous braconid larval parasitoid, *Distatrix papilionis* (Viereck) keeps the pest under check. The parasitoids, *T. chilonis* and *D. papilionis* account for a cumulative parasitism of 88%.

- Application of *Bacillus thuringiensis* var. *kurstaki* at 1 ml/l + spreader is very effective for the control of citrus butterflies. The susceptibility of citrus caterpillar to Bactospeine has been reported.

(iv) Psylla, *Diaphorina citri* Kuwayana

- Prunning of affected and died shoots
- The vegetative growth should be regulated by judicious application of irrigation.
- The eulopid nymphal parasitoid, *Tamarixia radiata* (Waterston) causes upto 90% parasitism in Punjab and 30–90% in Maharashtra. Among predators the chrysipids *M. boninensis, Brumus suturalis* (Fabricius) and the coccinellid *Coccinella rependox* prey upon nymphal psylla and can be successfully used for the biological control of the pest.

Mango

Among important insect pests of mango, hoppers and mealybugs have become the focus of BBIPM.

(i) Hoppers, *Amritodus* spp., *Idioscopus* spp.

- Orchard sanitation, regular ploughing and removal of weeds.
- If the hopper population is low to moderate, two sprays of azadirachtin at 3000 ppm at 2ml/l or azadirachtin at 10000 ppm at 1 ml/l first spray before flower opening and second spray at pea-size fruit stage gave effective control of hoppers. Sprays of neem/pongamia soaps were also highly effective.
- Biological suppression of hoppers was achieved by release of predator, *M. boninensis* at 10–12 larvae/tree. The reduviid predator, *Irantha armiper* Stal was found preying voraciously on hoppers. *Verticillium lecanii* has been observed wiping out of hoppers.

(ii) Mealybug, *Drosicha mangiferae* (Green)

- Alkathene sheet (30 cm wide, 400 gauges) should be put around the tree trunk at 0.5 to 1.0 m above the ground level and fastened by twine with a grease barrier at the lower end.
- Sanitary conditions should be maintained near the plants.

- Application of karanj based biodynamic preparation resulted in 100% mortalty of mealybugs within 7 days, while neem based biodynamic preparation took 9 days for 100% mortality.
- *Menochilus sexmaculata* (Fabricius), *Chrysopa scelestes* Banks and *Beauveria bassiana* are effective against mealybugs.
- Against crawled nymphs of mango mealybug garlic oil 0.1% was at par with carbaryl 0.1%.
- Soil drenching with 5% NSKE/ *Beauveria bassiana* around the tree trunk to prevent migration of mealybugs on tree.

Apple

There are several insect pests of this temperate fruit where the pest management is dominated by biological based strategies. Beside others, San Jose Scale and Wooly aphid are the major insect pests. Suggested BBIPM is summarized (Reddy, 2010) below:

- Spraying 5% NSKE/Neemarin at blossom stage to control blossom thrips.
- Spraying of horticultural mineral oil at 2 litres/200 litres of water in June-July
- Releasing of chrysopids like *Chrysoperla carnea* (Stephens) and *C. scelestes* for aphid control.
- Release of predatory mites like *Amblyseus fallacies* (Garman) and *Zitzellia mali* (Ewing) against European red spider mite.
- Soil application of neem and mustard cake at 250 kg/ha for controlling the soil insect pests, collar rot and root rot pathogens.
- Spraying of *Bacillus thuringiensis* (Bilep) at 1 kg/ha against mites and fruit scrapper pests

PLANTATION CROPS

BBIPM considerations are especially important in tropical plantation crops where, unlike annual crops, biological interactions have a chance to reach equilibrium or near equilibrium conditions. Adoption of BBIPM is feasible for the effective management of rhinoceros beetle, *O. rhinoceros*, one of the major pests of coconut palm. The beetle suffers from natural epizootics produced by the fungus, *Metarhizium anisopliae,* and a baculovirus (OBV). The rlease of baculovirus infested beetles has been successfully used for the suppression of the beetle in several countries including India, South Pacific Islands, Fiji, Mauritius, Seychellus and Papua New Guinea. In India, the damage by *O. rhinoceros* was substantially checked when the OBV infected beetles were released in Minicoy and Andioth islands of Lakshadweep, Chhittilappily in Trichur and Sipighat of Andaman islands. The release of *O. rhinoceros* beetle infected with Kerala isolate of the virus resulted in over 99% reduction in palm damage within three years of augmentation in Andaman islands (Prasad *et al.*, 2008).

There have been reports of resurgence of the beetle in some other regions including India (Kerala) and Western Samoa. There is also the possibility of development of resistance to baculovirus in Indonesa, East Java and South Sulawese. In order to overcome these problems, BBIPM has been proposed for the management of *O. rhinoceros* (Nair *et al.* 2005).

- The potential breeding sites of the beetle can be concealed by planting legumes as cover crops.
- The adult beetles can be extracted by means of a metal rod (about 0.5 m long) with a hook at one end during peak period of pest abundance (June-September) from the infested sites.
- The entomopathogen, *M. anisopliae*, could be mass multiplied using cheaper substrates (both solid and liquid), the spores could be harvested and breeding places of the beetle be treated at the rate of 10^{11} spores/m^3.
- Release of virus-infected beetle (10–15/ha), preferably during dusk, effectively reduces the pest population. The spread of the virus can be promoted by leaving some dead standing palms.
- The commercially available pheromone (Rhinolure) traps could be employed to remove the active floating population in a given area to reduce the population load of the pest. These beetles could be made infective by inoculating OBV and released for dissemination of the viral pathogen in the field.
- The breeding sites of the beetle could be treated with carbaryl (0.01%) where biological control is not possible.

CONCLUSIONS

Biopesticide based integrated pest management (BBIPM) plays a pivotal role in conservation of ecosystem leading to sustainable agriculture. Many of the components of BBIPM are self-perpetuating, ecofriendly and target-specific, and hence do not cause disruption to ecosystem. Therefore, national strategies should be developed and implemented with a mission mode zeal to make BBIPM more effective and feasible. Intensified efforts are required to develop more efficient products and technologies, and enhanced capability and capacity for production, availability, access and refinement of BBIPM technology. Increased emphasis on formulation research and quality products with improved shelf and field life is necessary to make BBIPM successful. A countrywide programme should be launched for promotion, adoption and assessment of BBIPM through participatory mode involving public and private sectors, self help groups, farmers and other stake holders. There is a growing demand for organically produced fresh and processed food as a result of increasing health consciousness among the people. The area under organic cultivation in India is estimated to be around 1,00,000 ha, which is likely to increase substantially in the coming years. Further, some of the states like Sikkim and Uttrakhand have been declared as organic by their respective governments. These developments would further stimulate the promotion of BBIPM into a holistic and sustainable crop management strategy.

REFERENCES

Bambawale, O.M., Singh, A., Sharma, O.P., Tanwar, R.K., Kanwar, V., Dhandapani, A., Sanle, U.R., Jafri, A.A., Vennila, S., Patil, S.B., Bhosle, B.B. *et al.* (2004) Cotton. In: A. Singh, H.R. Sardana and N. Sabir (eds.), *Validated IPM Technologies for Selected Crops.* National Centre for Integrated Pest Management, New Delhi, pp.1–32.

BCCI (2010) *Technical Market Research Report, RC-204R.* Business Communication Company Inc., Gillette, Wyoming, USA/Chesham, Bucks, UK.

Dhaliwal, G.S. and Koul, O. (2007) *Biopesticides and Pest Management: Conventional and Biotechnological Approaches.* Kalyani Publishers, New Delhi.

Dhaliwal, G.S. and Koul, O. (2010) *Quest for Pest Management: From Green Revolution to Gene Revolution.* Kalyani Publishers, New Delhi.

Dhaliwal, G.S., Singh, R. and Jindal, V. (2013) *A Textbook of Integrated Pest Management.* Kalyani Publishers, New Delhi.

Dhaliwal, G.S.; Koul, O., Khokhar, S. and Singh, R. (2012) Biopesticides: springboard to food and environment security. In: O. Koul, G.S. Dhaliwal, S. Khokhar and R. Singh (eds.), *Biopesticides in Food and Environment Security: Issues and Strategies.* Scientific Publishing (India), Udaipur, pp. 1–11.

Dhillon, M.K. and Sharma, H.C. (2009) Impact of Bt engineered cotton on target and non-target arthropods, toxin flow through different trophic levels, and seed cotton yield. *Karnataka J. Agric. Sci.*, **22,** 462–466.

Frisbie, R.E. and Smith, J.W. Jr. (1991) Biologically intensive integrated pest management: The future, In: J.J. Menn and A.I. Steinhauer (eds.), *Progress and Perspectives for the 21st Century.* Entomological Society of America, Lanham, Maryland, pp. 151–164.

Gujar, G.T., Kalia, V. and Dhillon, M.K. (2010) Bt cotton sustainability in India-Environmental concerns, In: Souvir.Natn. Symp. *Perspectives and Challenges of Integrated Pest Management for Sustainable Agriculture,* November 19–21, 2010, Nauni, Solan, Himachal Pradesh, pp. 64–69.

Hoskins, W.M., Borden, A.D. and Michellbacher, A.E. (1939) Recommendations for a more discriminating use of insecticides. *Proc. 6th Paci. Sci. Congr.*, **5**, 119–123.

Jahn, G.C., Litsinger, J.A., Chen, Y. and Barrion, A.T. (2007) Integrated pest management of rice. In: O.Koul and G.W. Cuperus (eds), *Ecologically Based Integrated Pest Management*, CAB International, Wallingford, UK, pp. 315–366.

James, C. (2011) *Global Status of Commercialized Biotech/GM Crops: 2011.* ISAAA Brief No. 43. International Service for the Acquisition of Agribiotech Applications, Ithaca, New York.

Kennedy, G.G. and Sutton, T.B. (2000) *Emerging Technologies for Integrated Pest Management: Concepts, Research and Implementation.* APS Press, St. Paul, Minnesota, USA.

Kogan, M. (1998) Integrated pest management: Historical perspectives and contemporary developments. *Annu. Rev. Entomol.*, **43**, 243–270.

Koul, O. and Cuperus, G.W. (2007) *Ecologically Based Integrated Pest Management.* CAB International, Wallingford, U.K.

Koul, O., Dhaliwal, G.S. and Kaul, V.K. (2009) *Sustainable Crop Protection : Biopesticide Strategies.* Kalyani Publishers, New Delhi.

Nair, C.P.R., Mohan, C. and Rajan, P. (2005) Advances in integrated pest management in coconut. In: B.S. Chhillar, V.K. Kalra, S.S. Sharma and R. Singh (eds), *Advances in the Integrated Pest Management of Horticultural, Spices and Plantation Crops*, CCS Haryana Agricultural University, Hisar, pp. 179–184.

NRC (1996) *Ecologically Based Pest Management: New Solutions for a New Century.* National Academy Press, Washington, DC.

Oerke, E.C. (2006) Crop losses to pests. *J. Agric. Sci.*, **144**, 31–43.

O'Neil, R.J., Yaninek, J.S., Landis, D.A. and Orr, D.B. (2003) Biological control and integrated pest management. In: K.M. Maredia, D. Dakous and D. Mota-Sanchez (eds), *Integrated Pest Management in the Global Arena*, CAB International, Wallingford, UK, pp.19–30.

Pimentel, D. (2007) Area-wide pest management: Environmental, economic and food issues. In: M.J.B. Vreysen, A.S. Robinson and J. Hendrichs (eds), *Area-wide Control of Insect Pests: From Research to Field Implementation,* Springer, Dordrecht, The Netherlands, pp. 35–47.

Prasad, G.S., Jayakumar, V. and Sharma, T.V.R.S. (2008) Management of coconut rhinoceros beetle (*Oryctes rhinoceros*) by augmentation of *Oryctes baculovirus* (Kerala isolate) in little Andaman Islands. *Indian J. Agric. Sci.*, **78**, 962–965.

Rabindra, R.J. and Venkatesan, T. (2010) Prospects of biological control of pests in organic farming. In: *Souvn. Natn. Symp. Perspectives and Challenges of* Integrated Pest Management for Sustainable Agriculture, November 19–21, 2010, Nauni, Solan, Himachal Pradesh, pp. 61–63.

Ramani, S. (2004) Classical biological control attempts in India. In: S. Ramani, B.S. Bhumannavar and R.J. Rabindra (eds.), *Quarantine Procedures and Facilities for Biological Control Agents*, Project Directorate of Biological Control, Bangalore, pp. 5–15.

Reddy, P.P. (2010) *Insect, Mite and Vertebrate Pests and their Management in Horticultural Crops.* Scientific Publishers (India), Jodhpur.

Sardana, H.R., Trivedi, T.P., Bambawale, O.M., Sabir, N., Singh, R.V., Arora, S., Krishnamoorthy, P.N., Pandey, P.K., Pandey, K.K. and Sengupta, A. (2004) Vegetable crops. In: A. Singh, H.R. Sardana and N. Sabir (eds), *Validated IPM Technologies for Selected Crops,* National Centre for Integrated Pest Management, New Delhi, pp. 119–147.

Sharma, H.C. (2010) Global warming and climate change: Impact on arthropod biodiversity, pest management, and food security. In: *Souvn. Natn. Symp. Perspectives and Challenges of Integrated Pest Management for Sustainable Agriculture,* November 19–21, 2010, Nauni, Solan, Himachal Pradesh, pp.1–14.

Singh, A., Garg, D.K., Jeswani, M.D. and Pathak, M. (2005) *Resource Manual on IPM in Basmati Rice.* National Centre for Integrated Pest Management, New Delhi.

Singh, S.P. (2003) Role of predators and parasitoids in biological control of crop pests. In: O. Koul, G.S. Dhaliwal, S.S. Marwaha and J.K. Arora (eds), *Biopesticides and Pest Management,* Vol. 1, Campus Books International, New Delhi, pp. 196–221.

Sinha, B. and Biswas, I. (2008) Potential of biopesticide in Indian agriculture vis-à-vis rural development (http://www.nistads.res.in/ indiasnt2008/ t6rural/t6rur17.htm).

Stern, V.M., Smith, R.F., Van den Bosch, R. and Hagen, K.S. (1959) The integration of chemical and biological control of the spotted alfalfa aphid, I. The integrated control concept. *Hilgardia,* **29**, 81–101.

USDA (2000) Definition for integrated pest management. United States Department of Agriculture CSREES, Washington, DC (http://www.reeusda.gov).

Wahab, S. (2010) Biotechnological approaches in the management of plant pests, diseases and weeds for sustainable agriculture. In: O. Koul, G.S. Dhaliwal, S. Khokhar and R. Singh (eds), *Biopesticides in Food and Environment Security,* Souvenir Third Conference on Biopesticides: Emerging Trends (BET 2010), October 20-22, 2010, Hisar, India, pp.15–40.

2

Current Status and Potential of Biopesticides in Pest Management in Rice

GURURAJ KATTI

Directorate of Rice Research, Hyderabad - 500030, India
E-mail: gururajkatti@yahoo.com

INTRODUCTION

The five countries namely India, China, Indonesia, Bangladesh, and Thailand account for 68.1% of the world's total rice area harvested and a total of 3.4 million hectares will be added by India, Pakistan, Thailand and Bangladesh in next year which more than compensates for the total decline of 1.5 million hectares in China, and 0.5 million hectares in the U.S., Japan, South Korea, and Vietnam. There is a global challenge to increase rice production due to constraints in land and water, farm demographics, climate change, and rice carbon footprint issues. Milled rice producers are China, India, Indonesia, Bangladesh, and Vietnam. In unison they account for 72.3% of global output. Overall, milled rice production has expanded by 51.3 mmt, with 62.2% coming from India, Indonesia, Bangladesh, Vietnam, and Thailand combined; and 24.7% comes from China, Egypt, Cambodia, Myanmar, Pakistan, the Philippines, and Nigeria combined (Wailes and Chavez, 2012). Global production of rice in 2010/11 was 451.385 thousand metric tonnes which touched 464.625 thousand metric tonnes in 2011/12 (Table 2.1).

Table 2.1. World's rice supply and utilization in 2010/11 and 2011/12 (*Source*: Wailes and Chavez, 2012)

	10/11	**11/12**
	Thousand ha	
Area Harvested	157,278	160,206
	Metric tonnes/ha	
Yield	2.87	2.90
	Thousand metric tonnes	
Production	451,385	464,625
Beginning stocks	94,200	97,938
Domestic Supply	545,585	562,563
Consumption	446,222	460,016
Ending stocks	97,938	102,797
Domestic Use	544.160	562,813
Trade	33,690	33,037

In India, rice is the staple food for 65% of the population. It is the most important source for meeting the caloric and dietary protein needs of the people as well as for generating employment and income particularly for low income groups in rural areas. There has been a marked improvement in rice production in the last 30 years primarily due to the improved technologies such as high yielding varieties, increased fertilizer and other input use efficiency, improved irrigation facilities and adoption of improved crop and pest management practices. In India, for example, Gulati *et al.* (2011) characterized India's historical rice trade policy as "stop-go" which contributed to the uncertainty in the international rice market. The government sets minimum support prices and provides input subsidies at the farm level; coupled with policies on procurement, stocks, and distribution. In fact, India has been record exporter of rice in 2012 and globally India, Pakistan, Thailand, United States and Vietnam account for 80% global rice exports (Fig. 2.1). During 2010/11 the production of rice in India was 95.3 thousand metric tonnes that touched to 102.0 thousand metric tonnes in 2011/12 (Table 2.2). At the present population growth, India needs to add not less than 2.5 million tonnes of milled rice every year to sustain the present level of self sufficiency. In order to achieve this, adoption of modern and intensive agricultural practices by the farmers is inevitable. However, along with the practice of intensive agriculture, there is a constant battle to overcome biotic constraints like insect pests, diseases, nematodes and weeds.

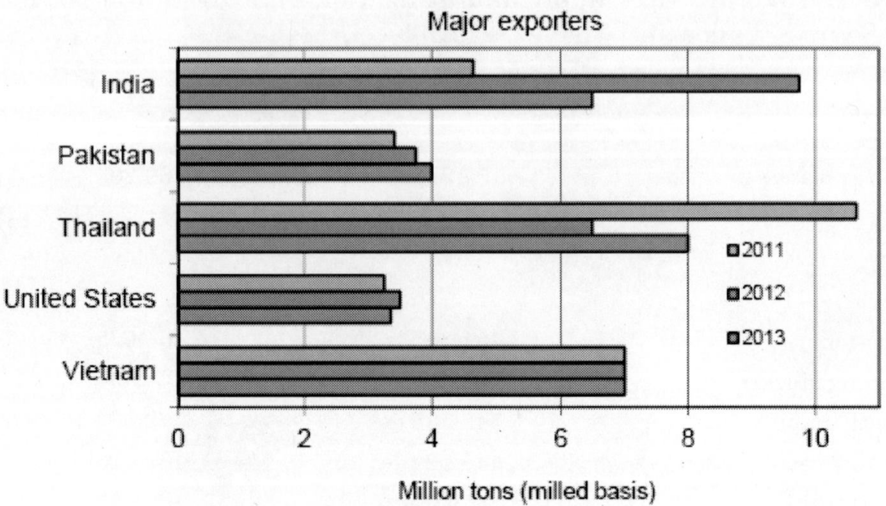

Fig. 2.1. Rice export forecasts for 2012 & 2013. These five countries account for morethan 80% of global rice exports. (*Source*: http://www.fas.usda.gov/psd).

Table 2.2. India's rice supply and utilization in 2010/11 and 2011/12 (*Source*: Wailes and Chavez, 2012)

	10/11	**11/12**
	Thousand ha	
Area Harvested	42,700	45,000
	Metric tonnes/ha	
Yield	2.23	2.27
	Thousand metric tonnes	
Production	95,300	102,000
Beginning stocks	20,500	23,000
Domestic Supply	115,800	125,000
Consumption	90,000	94,000
Ending stocks	23,000	24,494
Domestic Use	113,000	118,500
Net trade	2,800	6,500

INSECT PESTS – MAJOR BIOTIC STRESSES IN RICE

About 128 species of insects have been reported to ravage the rice fields. Out of this only 15 to 20 insect are regarded as economically obnoxious species (Kalode, 2005). About a dozen are reported to cause economic losses to rice crop in India.

Major Pests

Five insect pests viz., rice yellow stem borer, *Scirpophaga incertulas* (Walker), gall midge, *Orseolia oryzae* (Wood-Mason), *Cnaphalocrocis medinalis* (Guenee), brown planthopper (BPH), *Nilaparvata lugens* (Stal) and whitebacked planthopper (WBPH), *Sogatella furcifera* (Horvath), are of national importance as their incidence has significant impact on rice yields across the diverse rice ecosystems.

Minor Pests

Other pests viz., whorl maggot, *Hydrellia* spp. in all major rice growing areas, rice hispa, *Dicladispa armigera* (Olivier) occurring in Andhra Pradesh, Himachal Pradesh, Bihar, West Bengal, Orissa and north eastern region, green leafhopper, (GLH) *Nephotettix virescens* (Distant) prevalent in Bihar, West Bengal, Assam, Orissa, Madhya Pradesh, Andhra Pradesh and Tamil Nadu, gundhi bug, *Leptocorisa* spp. in Uttar Pradesh, Bihar, West Bengal, Orissa, Madhya Pradesh, Manipur and parts of Andhra Pradesh, climbing cutworm, *Mythimna separata* Walker in coastal upland rice growing areas, swarming caterpillar, *Spodoptera mauritia* Boisduval in low lying rice in Bihar, Gujarat, West Bengal, Assam and Orissa and thrips in several southern and eastern parts of India, have regional significance. Some pests such as rice mealy bug, termites, and case worm are pests of growing concern particularly in eastern region. Recently, leaf and panicle mites, black bugs, blue beetle have also started causing serious concern in some parts of the country.

LOSSES CAUSED ACROSS ECO-SYTEMS

At the global level, Oerke *et al.* (1994) reported losses of 51.4% of the attainable yields in rice of which 20.7% was due to animal pests including insect pests, 15.6% due to weeds and 15.1% due to pathogens. In India, assessment of losses through analysis of 135 multi-location trials under the All India Coordinated Rice Improvement Project (AICRIP) has revealed that the avoidable losses due to insect pests averaged 28.8% in rice crop across the different ecosystems.

BIOINTENSIVE MANAGEMENT OF INSECT PESTS

Biointensive approach of managing pests is an ecologically based strategy that focuses on long term solution of pests through a combination of techniques such as use of resistant varieties, biological control, modification of agronomic practices and habitat manipulation. There are several factors that favour biological control of pests in tropical rice. The major ones are the stable water supply, moderate host-plant resistnce and the potential of plants to compensate the growth after pest attack, a short fallow period, proximity to non-crop vegetation on rice bunds and withholding of the application of insecticides in early season (Way and Heong, 1994). Although some of these factors are relatively intrinsic to the system, others are conditions that are strongly dependent on farmers adopting appropriate management tactics. Asynchronous rice plantings are now more widely accepted which prevents pest buildup provided that insecticide use is moderated, allowing natural enemies to build up (Way and Heong, 1994, Ives and Settle, 1997). The presence of rice crops for much of the year and over large areas makes this system quasi-perennial in nature, thus approaching in terms of stability the woody perennial systems referred to above in which conservation biological control efforts have enjoyed success. This effect has contributed to a broader trend whereby rice farmers have shifted from a field-by-field approach to rice pest management toward an area-wide management approach that may extend over large regions (Matteson, 2000). An ecological engineering is another approach based on applying selected herbivore-induced plant volatiles (HIPV) elicitors to rice to promote their sink status for natural enemy populations could be especially powerful if linked with manipulation of the nearby vegetation to make it a more effective source habitat for predators and parasitoids (Gurr, 2009). Use of biopesticides mainly, microbial insecticides and botanical products will have to be an integral component of such an approach to minimize the risks to the human health, beneficial and non target organisms and environment.

Natural Regulation of Insect Pests in Rice Ecosystem by Entomopathogens

In fact, paddy ecosystems are richly endowed with natural regulatory mechanisms to take care of the insect pests and entomo-pathogens are one of the major groups among the natural enemies (Table 2.3). Of the eleven insect pathogens reported to attack stem borer of rice, *Bacillus thuringiensis* is the major one. Against leaf folder,

Table 2.3. List of insect pathogens of major insect pests in rice

S.N.	Insect Pest	Pathogen reported	Reference
1.	Stem borer	**Bacteria**	
		Bacillus subtilis Ehrenberg	Annonymous (1977)
		Bacillus thuringiensis var. *kurstaki* Berliner	Nayak *et al* (1978)
		Serratia marcescens Bizio	Annonymous (1977)
		Fungi	
		Mucor hiemalis, Fusarium moniliforme [*Gibberella fujikuroi*] and *Scopulariopsis* sp.	Yasodha and Narayanasamy (2004)
		Beauveria bassiana (Balsamo) Vuillemin	Annonymous (1977)
		Nomurea pracinae	Rao and Israel (1977)
		Virus	
		Nuclear polyhedrosis virus	Nayak and Srivastava (1979a) Godse and Nayak (1983)
		Entomopathogenic nematodes (EPN)	
		Hexamermis sp.	Bhatnagar *et al* (1985)
		Parasitorhabditis sp.	Rao *et al* (1977)
		Neoaplectana carpocapsae Weiser	Rao and Manjunath (1966)
2.	Leaf folder	**Bacteria**	
		Bacillus thuringiensis and *Serratia marcescens*	Joshi *et al* (1987)
		Fungi	
		Beauveria bassiana, Metarhizium anisopliae, M. fluvoviride, Nomuraea rileyi, Paecilomyces sp., *Aspergillus* spp., *Fusarium* spp., *Zoophthora radicans* and *Entomophthora* sp.	Joshi *et al.* (1987), Ambethgar (1995, 1996, 1997), Parathasarathy and Narayanasamy (1998), Padmaja and Kaur (2001), Neelam Joshi *et al.* (2004), Ambethgar *et al.* (2007) Padmavathi *et al.* (2008)
		Pseudomonas sp.	Dangar (2008)
		Syncephalastrum racemosum and *Penicillium oxalicum*	Joshi *et al.* (1987)
		Mucor hiemalis, Fusarium moniliforme [*Gibberella fujikuroi*] and *Scopulariopsis* sp.	Yasodha and Narayanasamy (2004)
		Aspergillus parasiticus, A. niger, Penicillium sp.	Neelam Joshi *et al.* (2004)

S.N.	Insect Pest	Pathogen reported	Reference
		Virus	
		Nuclear polyhedrosis virus	Nayak and Srivastava (1979a) Godse and Nayak (1983)
		Granulosis virus	Joshi *et al.* (1987)
		Entomopathogenic nematodes (EPN)	
		DD-136 nematode	Prasad and Srinivas (1991)
		Neoaplectana carpocapsae	Joshi *et al.* (1987)
		Steinernema carpocapsae	Srinivas and Prasad (1991)
3	Plant-hoppers	**Fungi**	
		Fusarium pallidoroseum (Cooke) Sacc.	Rao *et al.* (2008)
		Beauveria bassiana Meturhizium anisopliae Nomuraea rileyi (four)	Rachappa *et al.* (2007)
		Mucor hiemalis Wehme	Ambethgar (2000)
		Pandora delphacis	Narayanasamy (1995, 1998), Ambethgar (1996), Narayanasamy *et al* (1992)
		Absidia corymbifera	Gunathilagaraj *et al* (1987)
		Entomopathogenic nematodes (EPN)	
		Hexamermis sp.	Satpathi (1999; 2009), Satpathi *et al* (2007, 2008), Satpathi (1999)
4	Hispa	*Beauveria bassiana*	Rupesh *et al* (2005), Hazarika *et al* (1998), Hazarika and Pujari (1995), Pujari *et al* (1994), Pujari and Hazarika (1991), Hazarika & Pujari (1990)
		Fusarium pallidoroseum (Cooke) Sacc.	Rao *et al* (2008)
		Metarhizium anisopliae Nomuraea rileyi (Four),	Rachappa *et al* (2007)
		Mucor hiemalis Wehme	Ambethgar (2000)
		Pandora delphacis	Narayanasamy (1995, 1998), Ambethgar (1996) , Narayanasamy *et al* (1992)
		Absidia corymbifera	Gunathilagaraj *et al* (1987)
5	Ear head bug	*Metarhizium anisopliae* and *Beauveria bassiana*	Vietnam (Nguyen *et a*l, 2005)
6	Case-worm	Nuclear Polyhedrosis virus	Litsinger

S.N.	Insect Pest	Pathogen reported	Reference
7	Cutworm	*M. anisopliae*, *F. lateritium* and *L. muscarium*	Rajesh Anand *et al* (2009)
		Nomuraea rileyi (Farlow) Samson	Ambethgar and Kumaran (1998)
8	Swarming caterpillar	Nuclear Polyhedrosis virus	Nair *et al.* (1985)

eighteen fungal, two bacterial, two viral pathogens and as well as two entomopathogenic nematodes have been reported. Among the entomopathogens reported on plant and leafhoppers, the fungi, *Pandora delphacis, Metarrhizium flavoviridae, Beauveria bassiana, Erynia radicans, Entomophthora, Entomophaga aulicae* and *Fusarium* sp. have shown promise. Though the natural action of entomopathogens against the pests has been well documented and reported but their inherent ability to naturally regulate the pest populations below the economically damaging levels, has not been evident. Recently, bioformulations combining entomopathogens like *Pseudomonas* and *Beauveria* under glass house and field conditions have shown significant promise (Karthiba *et al.*, 2010). These studies on rice plants do emphasize the possibility of controlling both pest and diseases using single formulation.

BIOPESTICIDE STANDARDIZATION LEADING TO TRANSFORMATION IN THEIR POTENTIAL

Pest control methods have been evolving and diversifying in response to public awareness of environmental and health impacts of synthetic chemical pesticides and resulting legislation. In this process, biopesticides were developed as key components of integrated pest management (IPM) programs, mainly as a means to reduce the load of synthetic chemical products being used for control of pests. The other focus was on to standardize the contents of active principles for reliable, better and consistent results following their use.

Biopesticides are certain types of pesticides derived from such natural materials as animals, plants, bacteria, and certain minerals. The biopesticides used so far fall into two major catgories viz., Microbial pesticides and botanical pesticides (Ranga Rao, 2007).

Microbial Insecticides

Microbial pesticides contain a microorganism (bacterium, fungus, virus, protozoan or alga) as the active ingredient which is relatively specific for its target pest(s). The most widely known microbial pesticides are derivatives of the bacterium *Bacillus thuringiensis*, or Bt, which produces a toxin protein that is harmful mainly to lepidopterans. In the last decade, Bt transgenics have been developed which are plant-pesticides where in the Bt toxin gene is introduced into the plants own genetic material so that the plant, itself manufactures the toxin that destroys the pest.

Evaluation of Bt formulations in rice

Several Commercial formulations of Bt developed by Private companies have been evaluated for their effectiveness against mainly leaf folder and stem borer in rice.

Under the All India Coordinated Rice Improvement Programme (AICRIP), three commercially available formulations viz., Delfin 85%, Dipel 3.5% and BTK II were evaluated focusing on their efficacy against leaf folder, but concomitant effectiveness against stem borer was also investigated. It was evident that all the three formulations performed moderately against leaf folder, while the effect was marginal in case of stem borer, across ecosytems. The Bt formulations registered better effectiveness at higher doses (Fig. 2.2), however the check insecticide chlorpyriphos showed consistent superiority both in terms of less pest incidence as well as higher yield (Fig. 2.3). Rath (1997) found that BTK II and Dipel 3.5% were more effective than Delfin 85% and Biolep. Panda *et al* (1999) observed that Bioasp, Biolep, Biotox, dipel and Delfin @ 2000 g a.i./ha were more effective against leaf folder than stem borer and their performance was on par with that of recommended monocrotophos treatment. Roshan Lal (2001) also reported that the four Bt formulations were as effective as standard chlorpyriphos @ 250 g a.i./ha. Recently, studies carried out on bio-efficacy of an indigenously developed DOR Bt formulation (Ramandeep Kaur *et al* 2008; DRR, 2007-08) revealed that the formulation at a higher dose of @ 2.0 kg/ha was effective in controlling the rice leaf folder (Fig. 2.4) and increasing the grain yield of rice (Fig. 2.5).

Evaluation of Bt formulations in combination with other components of IPM

Research efforts have also been aimed at evaluation of Bt formulations as one of the eco-friendly components of IPM modules for adoption across rice ecosystems along with other components such as botanical insecticides, insect growth regulators, biocontrol agents, conventional chemical insecticides as well.

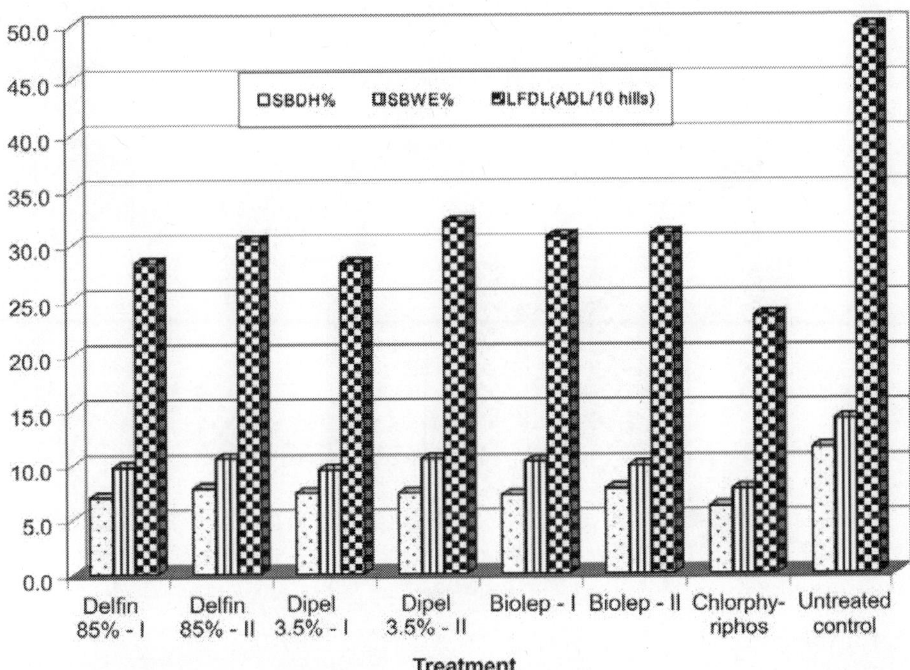

Fig. 2.2. Efficacy of microbial insecticides against stem borer and leaf folder in rice (AICRIP 1995-1997).

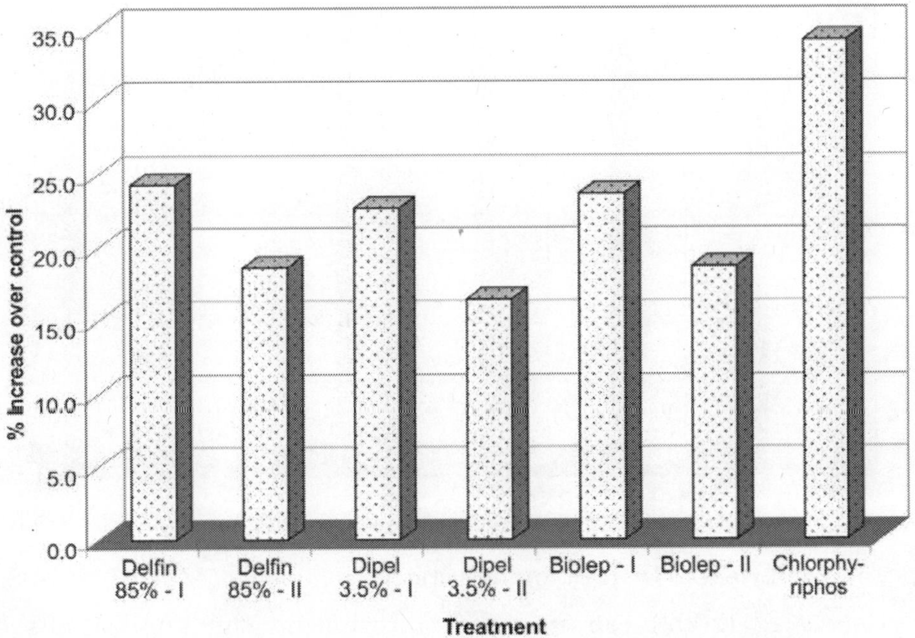

Fig 2.3. Effect of microbial insecticide treatments on yield (AICRIP 1995-97).

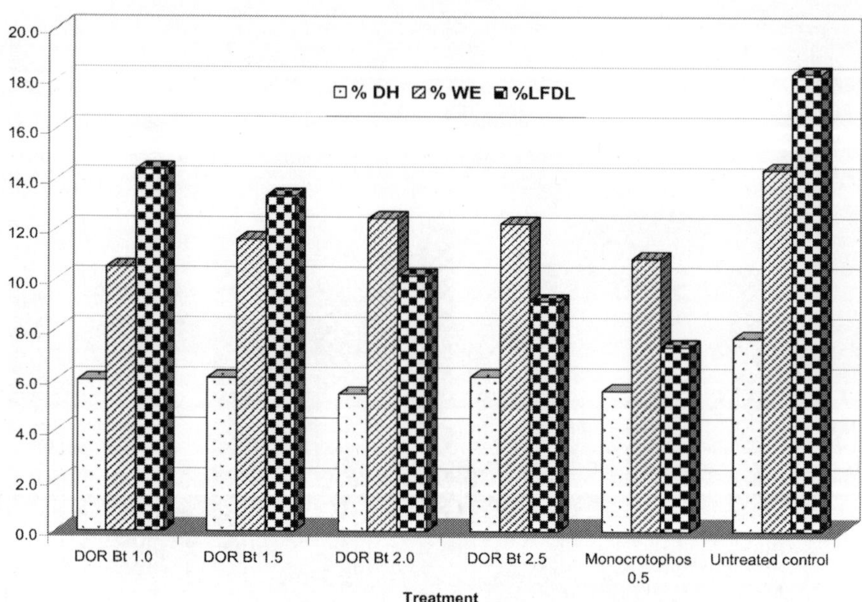

Fig 2.4. Efficacy of DOR Bt formulation (AICRIP, 2007-08).

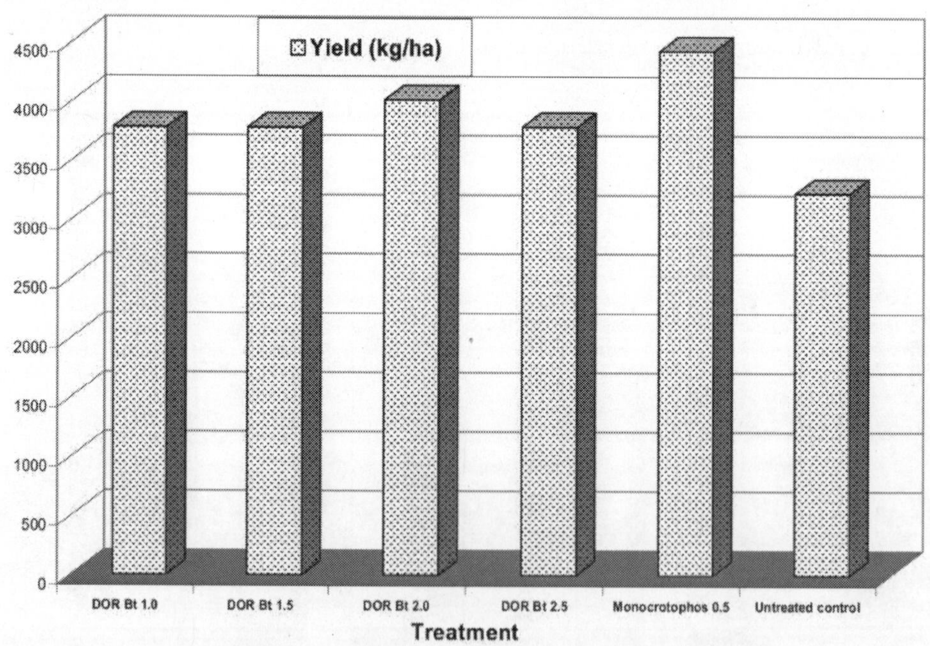

Fig 2.5. Grain yield in DOR Bt and insecticide treatments (AICRIP 2007-08).

Rao *et al* (2002) reported that alternating the applications of insecticides with Biobit or Dispel sprays were more effective than sole biopesticidal treatments in controlling leaf folder and increasing rice grain yield over control. The combination of biobit with systemic insecticides was also found to be economical as well as eco-friendly as it resulted in the best control of the pest and also conserved the natural enemies such as coccinellids and spiders in the rice field (Rao and Singh, 2003,; Rao *et al.,* 2006). In another study, CAMB Bt (*Bacillus thuringiensis*) at 250 g/acre and CAMB fungi (*Metarrhizium anisopliae*) at 250 g/acre applied alone or in combination resulted in significant reduction in leaf folder and stem borer incidence (Shahid *et al.*, 2003).

Overall, the utility of Bt formulations in paddy ecosystems has been mainly limited by the inability of externally applied sprays to actually reach the target pest stages which are mostly hidden in case of stem borer which completes most of its life cycle within the plant system, while in case of leaf folder, the larvae feed remaining within the leaf folds thereby escaping exposure from direct spray. So, the timing of application is crucial for the effectiveness of Bt formulations. However, in recent times, the versatile biotechnology tool has provided a novel option of incorporating Bt genes like *cry IA (b)* and *cry IA (c)* which can trigger continuous production of insecticidal toxins in the plant system itself to overcome this problem. Already Bt transgenic rice varieties for resistance to yellow stem borer are in advanced stage of testing in India (Manju Sharma *et al.*, 2003).

Studies on efficacy of other microbial insecticides in rice

In India , so far, eight microbial pesticides have been registered in India which include five of bacterial origin (four *Bacillus* species and one *Pseudomonas fluorescens*), three fungal origin (two *Trichoderma* species and one *Beauveria bassiana*) and one viral i.e. Nuclear Polyhedrosis virus (NPV). In India, there are very few reports on evaluation of other microbial insecticides in rice, that too restricted to *Beauveria bassiana* based products, against leaf folder but without much success (Rao *et al* 1999 & 2003; Sher Singh *et al.,* 2005).

Botanical Insecticides

Botanical insecticides are synthetic derivatives of the naturally occurring secondary metabolites synthesised by plants species, which act on the insect growth and survival. They have long been advertised as attractive substitutes to synthetic chemical-insecticides, for controlling many insect pests because botanicals reputedly pose little threat to the environment or to the human health. Although, there is enormous scientific literature documenting bioactivity of plant derivatives to arthropod pests yet only pyrethrum and neem are well established commercially (Isman, 2006). In India, a wide variety of commercial neem formulations have been tested and sold and newer ones continue being marketed by local formulators. Four major types of botanical products have been used for insect pest management (pyrethrum, rotenone, neem and essential oils), along with three others in limited use (ryania, nicotine and Sabadilla). Additional plant extracts and oils (eg. Garlic oil, *Capsicum* oleoresin etc.) have also been or being used in a limited way, specific to regions (Anand Prakash *et al.*, 2008). However, neem and its products have more potential and investigated better than other products. Two types of crude botanical insecticides can be obtained from the seeds of neem tree. Neem oil, obtained by cold pressing seeds and seed residue (cake) after removal of oil, contains the major active principle, azadiractin equal to 0.2 to 0.6% by wt. The active ingredient is concentrated to the level of 10 to 50% in the technical grade material used to produce commercial products. Neem oil and neem cake have been extensively tested for their efficacy against various pests of rice. There is a plethora of such reports on their utilization in rice pest management (Table 2.4). However, these botanicals were relegated to a trivial position in the market place because of their moderate efficacy and inconsistent results due to variation in composition of naturally active principles *vis- a- vis* superiority of chemical insecticides in terms of their curative effect, easy application and availability.

Evaluation of efficacy of neem formulations in rice

Wide ranging greenhouse and field studies were carried out at the Directorate of Rice Research (DRR), Hyderabad to evaluate the efficacy of ready to use neem formulations against the insect pests of rice (Krishnaiah *et al.*, 2008). Evaluation studies were carried out on two types of neem formulations *viz.,* (i) Oil based formulations with 300 ppm of azadirachtin content and (ii) Solvent based formulations with 1500 ppm or more of azadirachtin content in the formulation. The studies focused on antifeedant, growth regulating, development-modifying and insecticidal effects.

Under glasshouse conditions, studies on feeding deterrent effects revealed that Rakshak and Neemgold 4 were better than Neem Azal T/S, in case of BPH and leaf folder (Tables 2.5 and 2.6). Mayabini Jena (2005) also reported that antifeedant and oviposition deterrent activities were more prominent than the knock down effects. In case of

Table 2.4. Pesticidal activity of neem products tested against insect pests of rice

Neem product & dose/ conc.	Type of application	Target insects	Biological activity	References
Neem oil	Spraying	*Hydrellia phillipina*	Antifeedant	Murthy (1975)
Neem oil	Spraying	BPH	Strong repellency	Balasubramanian (1979)
Neem oil 10%	Spraying	LF and GLH	Antifeedant Reduced life span	Mariappan *et al.* (1982a)
Neem oil	Spraying	GLH & RTV	Population reduction	Mariappan and Saxena (1983)
Neem oil	Neem oil coated urea	*Hydrellia philippina, N. virescens* and BPH	Reduced incidence	Krishnaiah and Kalode (1984)
Neem oil		*L. oratorius*	Deformity	Saxena *et al.* (1985)
Neem coated urea	Soil application	GM, GLH & RLF	Reduced incidence	David (1986)
Neem Seed Kernel Extract (NSKE)	Spray application using ULV sprayer	BPH, RLF	Checked incidences of the test insects	Rajasekaran *et al.* (1987)
Neem oil		BPH	Insecticidal activity	Velusamy *et al.* (1987)
Neem oil (1%), neem cake extract (5%)	Spraying	BPH and WBPH	Reduced emergence	Ramaraju & Sundarababu (1989)
NSKE 5%+ 0.16% teepol	Spraying	LF	Reduced population significantly	Mohan and Gopalan, (1990)
3% neem oil	Spraying	GM	Reduced infestation	Samalo *et al.* (1990)
1-4% neem oil	Spraying	LF and YSB	Reduced incidences	Singh *et al.*, (1990)
5% neem oil	Spraying	*L. acuta*	Reduce population	Gupta *et al.* (1990)
3% Neem kernel powder	Spraying	*N. virescencs*	Inhibited nymphal growth	Krishnaiah and Kalode (1990)
Neem cake @150 kg/ha + 3% neem oil spray	Soil application and Spraying	LF	Effectively checked insect infestation	Krishnaiah *et al.*, 1990, Krishnaiah and Kalode (1990)
Neem cake with urea	Soil application	GLH and WBPH	Reduce population	Viswanathan & Kandiannan, (1990)

Neem product & dose/ conc.	Type of application	Target insects	Biological activity	References
2% neem oil	Spraying	LF, *Hieroglyphus banian*	Reduced infestation	Mohan *et al.* (1991)
Neemax	Spraying	WBPH	Reduction in pest incidence	Shukla *et al.* (1991)
Welgro	Spraying	WBPH	Reduction in pest incidence	Shukla *et al.* (1991)
Neem oil @7.5 kg/ha	Spraying	BPH and WBPH	Reduced infestation	Sontakke (1993)
Welgro2%	Spraying	GM and YSB	Reduced incidence	Nanda *et al.* (1993)
Nemidin 1000 ppm	–	WBPH	Inhibit the development of larvae	Nelson *et al.* (1993)
Neem oil	Spraying	Gall midge, stem borer, leaf folder and WBPH	Reduction in damage and effect on predators	Sontakke (1993)
Neem oil (3%) & neem seed kernel extract(5%)	Spraying	WBPH and GLH	Reduction in populations	Shukla and Kaushik (1994)
Neem oil	Spraying	WBPH	Reduction in population	Sontakke *et al.* (1994)
Neem seed kernel extract & neem cake extract	Seedling root dip – Greenhouse study	GLH	Reduction in population and effect on growth and emergence	Dash and Senapati (1994)
Neem cake @150 kg/ha	Soil application	*Hydrellia philippina*	Reduce damage of whorl maggots	Bhatia *et al.* (1994)
0.5 and 1.0% Achook	Spraying	*L. acuta*	Effectively controlled the pest	Prakash and Rao (1994)
Margoside CK and Margoside OK 1%	Spraying	5th instar BPH	57-80% mortality	Jena and Dani (1994)
Neem seed kernel extract(5%)	Spraying	Leaf folder	Reduction in damage	Latha *et al* (1994)
Neem seed kernel extract(5%)	Spraying	Leaf folder	Reduction in damage	Latha *et al* (1994)
Neem cake	Pot experiments with pellets	WBPH	Persistent toxicity	Logiswaran and Venugopal (1995)

Neem product & dose/ conc.	Type of application	Target insects	Biological activity	References
Neem cake, neem seed kernel extract (5%), neem leaf decoction & neem oil (3%)	Soil application and spraying	Leaf folder	Reduction in leaf damage	Ambethgar (1996)
Neem oil and neem cake	Seedling root dip, soil application and spray	GLH, BPH and predators of BPH	Reduction in incidence of GLH and BPH. Little effect on predators	Babu *et al.* (1998)
Nimbecidine, Neemax, Neem Gold, Neem Azal T/S and Fortune Aza	Spraying	Leaf folder, WBPH and stem borer	Reduction in pest incidence	Korat *et al.* (1999)
Neem oil based formulations	Spraying	GLH and rice yellow dwarf disease	Reduction in GLH survival and incidence of disease	Rajappan *et al.* (1999)
NSKE 5%, Neem azal F 5%, Neem Azal T/S	Spraying	LF and egg parasitoid, *Trichogramma chilonis*	Contact toxicity	Prabal Saikia and Parameswaran, 2001
Neemazal	Spraying	Hispa	Reduction in damage	Sharma and Kaul (2003)
Neem seed kernel extract	Laboratory studies	Leaf folder	Effect on gut enzymes	Nathan *et al.* (2004)
Neem oil, Neem seed kernel extract & neem seed kernel poeder	Spraying	Gundhi bug	Reduction in bug population	Singh (2006)
Neem limonoids	Laboratory studies	Leaf folder	Toxicity and behavioural effects	Nathan *et al.* (2006)
NSKE 5%, Neem Oil 3%, Neem Leaf Extract 3%	Seed Treatment, Seedling root dip and Foliar spray	WBPH	Effects on survival and growth index	Sujeetha, (2008)
Multiplex	Spraying	Gundhi bug	Reduction in grain damage	Singh *et al.* (2009)

leafhoppers and planthoppers, disruption of growth resulted in reduction in size and weight of insects after feeding on plants treated with crude or commercial neem formulations. Consequently the proportion of nymphs becoming adults was also affected. However, in lepidopterous insects larval pupal intermediaries were observed (Krishnaiah and Kalode, 1988). Although there are reports that oviposition by BPH, WBPH and GLH are affected when confined to plants treated with neem oil or neem formulations, there was no consistency in such effects (Kalode and Krishnaiah, 1991). Further, studies with neem formulations (Krishnaiah, 2000) revealed that the oil based neem formulations were more effective in oviposition deterrency than solvent based neem formulations in spray formulation. As regards direct toxicity effects, the studies have revealed that constituents other than azadirachtin also play a role in exercising toxic effect against BPH. Some neem formulations with high azadirachtin content like Neem Azal T/S have exhibited exhibited some systemic activity when given as a seedling root dip adversely affecting the growth and development of BPH and GLH nymphs confined to treated plants (Krishnaiah 2000). Neem formulations as spray also adversely affected the survival of BPH through toxic effects. Saikia and Parameswaran (2001) also reported more than fifty percent mortality of leaf folder larvae after direct exposure to Neem Azal-F 5% treatment.

Table 2.5. Effect of Neem formulations on feeding of BPH

Treatment	Concentration (%)	Area (mm^2)*
NG4	0.5	355.00 c
NG4	1.0	413.33 de
NG4	2.0	233.67 ba
Rakshak	0.3	358.67 c
Rakshak	0.5	224.67 ab
Rakshak	1.0	105.00 a
Neem Azal TIS	0.1	309.00 bc
N eem Azal TIS	0.3	356.00 c
Neem Azal TIS	1.0	353.33 c
Control	-	470.0 e

N = 10 fifth instar nymphs; * = Mean of honey dew area from 3 replicates
Data in column followed by same letter are not significantly different at $P = 0.05$
Source: Krishnaiah *et al.*, 2008

Table 2.6. Effect of Neem formulations on feeding of leaf folder larvae

Treatment	Azadirachtin content (ppm)	Cone. of test solution (%)	Leaf area consumed (mm²)* 24 HAT
NG4	300	1.0	438.33 d
NG4	300	2.0	278.33 bcd
NG4	300	4.0	120.00 ab
Rakshak	1500	0.1	233.33 abc
Rakshak	1500	0.5	130.00 abc
Rakshak	1500	1.0	116.67 a
Neem Azal TIS	10000	0.1	288.33 cd
Neem Azal TIS	10000	0.3	171.67 abc
N eern Azal TIS	10000	1.0	130.00 abc
Control	NIL	-	396.67 d

N=2, third instar larvae; * = Mean of 3 replications, HAT = hours after treatment Data in column followed by same alphabets are not significantly different at *P*=0.05. *Source*: Krishnaiah *et al.*, 2008

The extensive research on the field efficacy of neem products has included evaluation of several neem based formulations for the control of brown planthopper, yellow stem borer and leaf folder. Field experiments at DRR as well as multi location trials under the All India Coordinated Rice Improvement Programme (DRR, 1995-97) revealed that neem formulations viz., Achook, Nimbecidine, Neemax, Neemgold and Econeem at recommended concentrations (2% in oil based formulation) were moderately effective against stem borer (6.5 to 7.1% dead hearts-DH and 10.2 to 11.6% white ears-WE) and leaf folder (17.0 to 26.0 average damaged leaves - ADL per 10 hills) compared to standard insecticide check (5.4% DH, 8.0% WE and 19.2% ADL) but were significantly superior to control (11.3% DH, 14.8% WE and 42.4% ADL) (Fig. 2.6).

The neem formulations were not effective against rice gall midge. It is also evident from other studies that the neem formulations effectively controlled BPH and WBPH, moderately suppressed stem borers but were less effective against gall midge compared to recommended insecticides In all these studies, standard check insecticide treatments yielded significantly higher than neem formulations (Korat *et al.,* 1999; Dash *et al.*, 2001; Multani *et al.,* 2002). However, there are also few studies reporting the superiority or parity of commercial neem formulations in their performance compared to the recommended insecticides both in terms of reducing pest incidence and resulting in higher yields (Krishnaiah, *et al.*, 1993; Kaul and Sharma, 1999; Singh *et al.*, 1999; Prasad *et al.*, 2004) (Fig. 2.7).

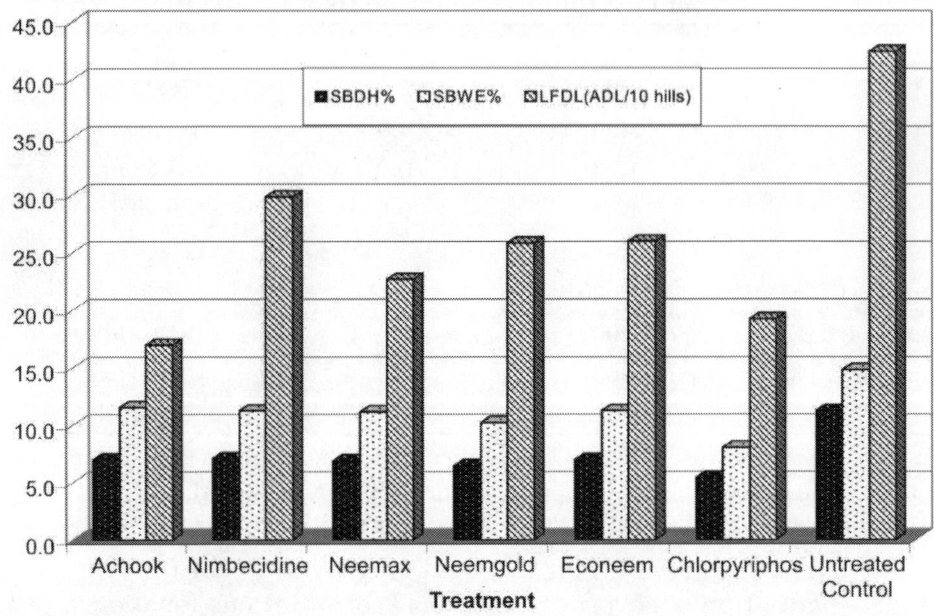

Fig. 2.6. Efficacy of botanical insecticides against leaf folder and stem borer in rice (AICRIP 1995-97).

Studies on safety of neem formulations to natural enemies

There has been a general impression that neem and other plant products are safe to non target organisms. But, studies on impact of neem formulations on natural enemies (i.e., beneficial predators and parasitoids that attack pests) have documented effects ranging from harmless to adverse (Lim Guan Soon and Bottrell, 1994). Crude formulations of neem such as neem oil, neem cake and other non-edible oils and cakes have been reported to be safer to natural enemies compared to synthetic insecticides (Krishnaiah and Kalode, 1984; Rajashekhran 1996; Babu *et al.*, 1998; Dash *et al.*, 2001). Investigations carried out at DRR, Hyderabad have revealed that, commercial neem formulations such as, Neemax, Rakshak, and Fortune Aza, were also safer to planthopper predators like velid bug, *Microvelia douglasi atrolineata* and mirid bug, *Cyrtorhinus lividipennis* and egg parasitoid, *Trichogramma japonicum* (Jhansilakshmi *et al.*, 1997 & 1998). However, Neem Gold, Neem Azal and NG4 resulted in high mortality of velid predator (Jhansilakshmi *et al.*, 1997). Other workers have also reported the safety of Fortune Aza to egg parasitoids (Borah *et al.*, 2001; Srinivasan *et al.*, 2001).

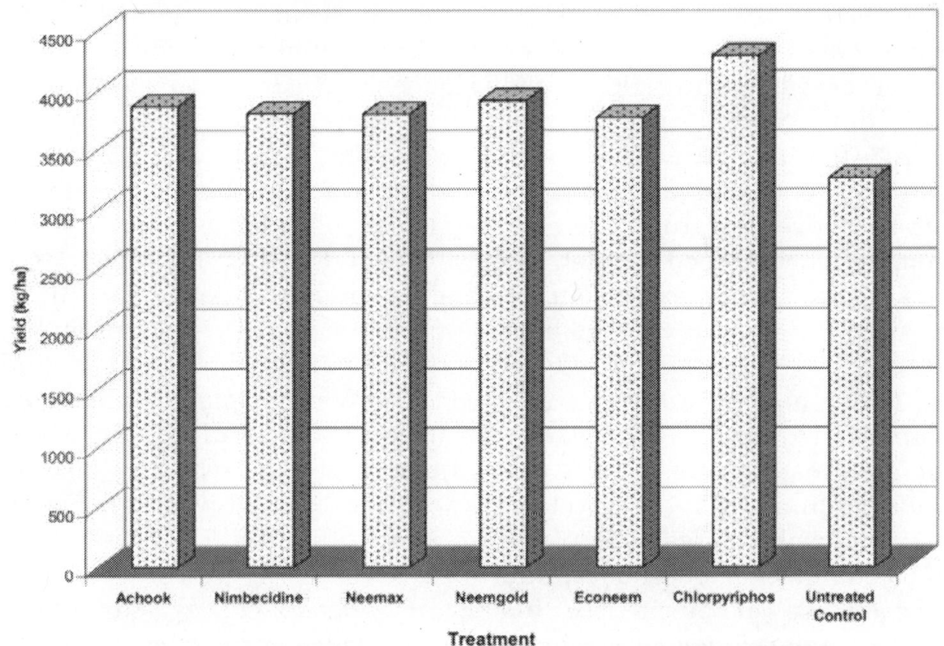

Fig. 2.7. Effect of botanical insecticides on yield (AICRIP 1995-97).

Evaluation of neem formulations in combination with other components of IPM

In order to make best use of the neem formulations in IPM there is a need to optimize the number as well as timing of their applications to derive maximum benefits. Trials conducted at DRR, Hyderabad revealed that carbofuran 3G @ 0.75 kg a.i./ha applied at 25 DAT followed by two sprays of NG 4 (2%) at 50 and 70 DAT reduced pest incidence and increased grain yield similar to three sprays of monocrotophos (0.4 kg a.i./ha/ application) at 25, 50 and 70 DAT revealing the possibility of reducing environmental contamination without lowering either pest control or grain yield. Replacement of monocrotophos at 50 DAT with NG 4 maintained similar pest control and grain yield. However, three NG 4 applications at 25, 50 and 70 DAT resulted in lower insect pest control and grain yield (Krishnaiah, 2000). Increase in the effectiveness of neem products when combined with insecticides has also been reported by other workers (Sudhakar 2000; Sharma and Kaul, 2003). In deep water rice also, integrated treatments with neem components plus one or two synthetic chemical applications were found very effective in controlling the pest population build up superior to chemical controls (Chakraborti, 2003). Recently extent of suppression of rice ear head bug (gundhi bug), *Leptocoryza acuta* Th. population by six selected neem formulations was observed in the field of paddy cultivar *Swarna mashuri* (MTU 7029) during four consecutive kharif crop seasons of 2005-2008 in India. All the treatments were effective significantly to suppress *L. acuta* incidence and accordingly the extent of grain damage. Yield also increased over control after commercial formulation of nimbecidine, neem oil, neem seed kernel extract, neem leaf extract, neem root extract, neem bark extract in descending order (Chakraborty, 2011). In another field study, combination of botanicals with egg parasitoid, *Trichogramma japonicum* reduced populations of both stem borer and leaf folder as well as resulted in conservation of spider population, compared to insecticidal treatments (Sher Singh *et al*, 2005). Neem formulations have also been found quite useful in reducing disease incidence in addition to insect pests when integrated with other non-pesticidal components of IPM. Dodan and Roshan Lal (1999) reported that combination of nimbecidine application with pre-transplanting incorporation of burnt rice husk and release of egg parasitoid, *Trichogramma japonicum* reduced the incidence of neck blast disease as well as stem borer damage in rice on par with that of recommended pesticide combinations. Evaluation of different IPM modules experiments in farmers fields at Karakkad village, Pattambi, Kerala over three years showed that IPM module comprising of alternate spraying of econeem formulation with ecofriendly insecticides coupled with release of egg parasitoids against leaf folder and monitoring of yellow stem borer with sex pheromone traps resulted in significant reduction of stem borer and leaf folder, highest yield and most optimum cost benefit ratio (Karthikeyan *et al.*, 2010).

FURURISITIC SCENARIO OF BIOPESTICIDES IN RICE IPM

Over the past 25 years, the research on biopesticides has evolved towards being more ecologically holistic and more oriented towards both production systems and industry's concerns. In the rice IPM scenario, biopesticides provide environmentally friendly options in many ways. The relatively shorter duration of persistence of the available botanicals can be suitably exploited to prevent secondary pest out breaks resulting from misuse of synthetic insecticide application. Synthetic pesticides with single active principle are likely to induce the development of resistance in insects. Botanicals on the other hand contain complex array of compounds with multiple effects and there is less likelihood of development of resistance. Therefore, wherever possible, botanicals can be alternated with synthetic pesticides to hinder the development of insecticide resistance. Concurrently, with advances in development of latest and more efficient analytical techniques, research on identification of newer active principles of these biopesticides may lead to synthesizing newer molecules with better efficacy and enhanced persistence under field conditions.

On the other hand, the biopesticides face a number of constraints in their development, manufacture and utilization. Lack of multidisciplinary research, inadequate public private partnerships and poor understanding of their quality aspects, are the critical bottlenecks. Generally, farmers are accustomed to quick knock-down effects of pesticides. Therefore, they may not be satisfied with the slower action of biopesticides. There is, thus, a need to educate farmers about the special behavioural effects of these products and also create awareness among extension specialists and policy makers of the potential utilization of biopesticides. More focused research efforts in production, formulation and development of effective delivery systems are needed to effectively harness their potential and convince the farmers about their role as equally efficient and eco-friendly alternatives to conventional chemical pesticides.

In the recent years, there has been a spurt in efforts to develop organic pest management methods involving use of biopesticides in view of the strong influence and growth of the organic foods market in the developed countries. In our country also, there has been a distinct trend in the decreased use of conventional chemical insecticides with a concomitant but gradual increase in consumption of biopesticides (See Chapter 1).

There are also reports of rice pests already developing resistance to even newly introduced agrochemicals leading to synthetic chemicals being registered at a slower rate than in the past. This situation has helped to reopen the market for a new generation of biopesticides. With fast paced changes in development of effective delivery systems and possibility of identifying newer potential biomolecules, a relook at the utility of biopesticides may be worthwhile in future.

References

AICRIP (1995-97) *Progress Reports of All India Co-ordinated Rice Improvement Project, Entomology and Pathology,* Vol. 2. Directorate of Rice Research, Rajendranagar, Hyderabad, India.

Chakraborti, S. (2003) Management of insect pests in deep water rice using ecofriendly approach. *J. Appl. Zool. Res.*, **14**, 65–67.

Chakraborty, K. (2011) Extent of suppression of rice ear head bug, *Leptocoryza acuta.* The population by different neem formulation in field of paddy cultivar *swarna mashuri* (mtu 7029) during kharif season. *ARPN J. Agric. Biol. Sci.*, 6, 37–41.

Dash, A.N., Mukherjee, S.K. and Sontakke, B.K. (2001) Efficacy of some commercial neem formulations against major pests of rice and their safety to natural enemies. *Pest Manage. Econ. Zool.*, **9**, 59–64.

Directorate of Rice Research (DRR) (2007-2008) *Crop Protection (Entomology, Plant Pathology).* Progress Report, All India Coordinated Rice Improvement Programme (ICAR), Directorate of Rice Research, Rajendranagar, Hyderanad.

Dodan, D.S. and Lal, R. (1999) Integrated management of neck blast and stem borer in scented rice. *Haryana Agric. Univ. J. Res.*, **29**, 47–49.

Gulati, Ashok, Jain, S. and Sharma, U. (2011) *Indian Rice Landscape: Trade, Production and Government Intervention in Marketing.* Kyushu University, Fukuoka, Japan.

Gurr, G.M. (2009) Prospects for ecological engineering for planthoppers and other arthropod pests in rice. In: K.L. Heong and B. hardy (eds.), *Planthoppers: New Threats to the Sustainability of Intensive Rice Production Systems in Asia*, International Rice Research Institute, Los Baños, Philippines, pp. 371–388.

Karthiba, L., Saveetha, K., Suresh, S., Raguchander, T., Saravanakumar, D. and Samiyappan, R. (2010) PGPR and entomopathogenic fungus bioformulation for the synchronous management of leaffolder pest and sheath blight disease of rice. *Pest Manage. Sci.*, **66**, 555–564.

Katti, G., Krishnaiah, N.V., Pasalu, I.C., Jhansilakshmi, V., Padmakumari, A.P. and Padmavathi, Ch. (2005) Scope of botanicals in rice pest management with special reference to neem formulations. In: *Proceedings of National Workshop on Promotion of Biopesticides and Biofertilizers in Agriculture*, National Institute of Rural Development, Hyderabad, pp. 158–163.

Isman, M.B. (2006) Botanical insecticides, deterrents, and repellents in modern agriculture and an increasingly regulated world. *Annu. Rev. Entomol.*, **51**, 45–66.

Ives, A.R. and Settle, W.H. (1997) Metapopulation dynamics and pest control in agricultural systems. *Am. Nat.*, **149**, 220–246.

Jhansilakshmi, V., Katti, G., Krishnaiah, N.V. and Mahesh Kumar, K. (1998) Safety of neem formulations vis-a-vis insecticides to *Cyrtorhinus lividipennis* Reuter (Hemiptera : Miridae), a predator of brown planthopper, *Nilaparvata lugens* (Stal) in rice crop. *J. Biol. Control*, **12**, 119–122.

Jhansilakshmi, V., Katti, G. and Krishnaiah, N.V. (1997) Safety of neem formulations and insecticides to *Microvelia douglasi atrolineata* Bergroth (Heteroptera: Veliidae), a predator of planthoppers in rice ecosystem. *J. Biol. Control*, **11**, 33–36.

Jhansilakshmi, V., Katti, G., Krishnaiah, N.V. and Lingaiah, T. (1997) Laboratory evaluation of commercial neem formulations vis-a-vis insecticides against egg parasitoid, *Trichogramma japonicum* Ashmea (Hymenoptera: Trichogrammatidae). *J. Biol. Control*, **11**, 29–32.

Kalode, M.B., 2005, Insect pest of rice and their management. In: S.D. Sharma and B.C. Nayak (eds.), *Rice in Indian Perspective*, Today and Tomorrow Printers and Publishers, New Delhi (India), pp. 819–854.

Karthikeyan, K., Sosamma Jacob, Pethummal Beevi and Purushothaman, S.M. (2010) Evaluation of different integrated pest management modules for the management of major pests of rice (*Oryza sativa*). *Indian J. Agric. Sci.*, **80**, 59–62.

Kaul, B.K. and Sharma, P.K. (1999) Effect of neem based insecticides against the major insect pests of rice in the hills of Himachal Pradesh (India). *J. Entomol. Res.*, **23**, 377–379.

Kaur, Ramandeep, Virk, J.S., Joshi, Neelam (2008) Bio-efficacy of DOR Bt, a *Bacillus thuringiensis* formulation against rice leaf folder, *Cnaphalocrosis medinalis* (Guenee) in *Punjab J. Biol. Control*, **22**, 475–477

Korat, D.M., Dodia, J.F., Patel, M.C. and Pathak, A.R. (1999) Evaluation of some neem formulation against insect pests of paddy. *Gujarat Agric. Univ. Res. J.*, **25**, 112–116.

Krishnaiah, N.V., Jhansi Lakshmi, V., Pasalu, I.C., Katti, G.R. and Padmavathi, Ch. (2008) *Insecticides in Rice IPM – Past, Present and Future.* Technical Bulletin No. 30. Directorate of Rice Research, Rajendranagar, Hyderabad.

Krishnaiah, N.V., Katti, G., Krishnaiah, K., Pasalu, I.C., Mahesh Kumar, K. and Lingaiah, T. (1998) Studies of systemic action of neem formulations in rice. *Pestic. Res. J.*, **10**, 155–159.

Krishnaiah, N.V., Katti, G. and Mahesh Kumar, K. (1997) A greenhouse study on systemic action of neem formulations in rice. *Indian J. Plant Prot.*, **25**, 24–26.

Krishnaiah, N.V., Pasalu, I.C., Katti, G., Kondala Rao, Y., Mahesh Kumar, K. and Lingaiah, T. (2000) Effect of neem formulations on resurgence of brown planthopper, *Nilaparvata lugens* (Stal) under field conditions. *Indian J. Plant Prot.*, **28**, 143–147.

Lingaiah, T., Krishnaiah, N.V., Mahesh Kumar, K., Katti, G., Krishnaiah, K. and Pasalu, I.C. (1999) Antifeedant effect of neem formulations against rice leaf folder, *Cnaphalocrocis medinalis* and hispa, *Dicladispa armigera*. *Pestic. Res. J.*, **11**, 93–96.

Matteson, P.C. (2000) Insect pest management in tropical Asian irrigated rice. *Annu. Rev. Entomol.*, **45**, 549–574.

Mayabini, Jena (2005) Integrated pest management with botanical pesticides in rice with emphasis on neem products. *Oryza*, **42**, 124–128.

Multani, J.S., Dhaliwal, G.S., Singh, Sandeep, Kaur, Gagandeep and Dilawari, V.K. (2002) Efficacy of four commercial neem formulations against whitebacked planthopper, *Sogatella furcifera* (Horvath) on rice. *Pestic. Res. J.*, **14**, 248–250.

Prakash, A., Rao, J. and Nandagopal, V. (2008) Future of botanical pesticides in rice, wheat, pulses and vegetables pest management. *J. Biopestic.*, **1**, 154–169.

Prasad, S.S., Gupta, P.K., Singh, R.B. and Kanaujia, B.L. (2004) Evaluation of neem products against yellow stem borer, *Scirpophaga incertulas* on deep water rice. *Ann. Plant Prot. Sci.*, **12**, 426–428.

Ranga Rao, G.V., Rupela, O.P., Rameshwar Rao, V. and Reddy, Y.V.R. (2007) Role of biopesticides in crop protection: Current status and future prospects. *Indian J. Plant Prot.*, **35**, 1–9.

Rao, N.B.V.C. and Singh, V.S. (2003) Eco-friendly management of rice leaf folder *Cnaphalocrocis medinalis* (Guenee). *Indian J. Plant Prot.*, **31**, 17–19.

Rao, N.B.V.C., Singh, V.S. and Subash Chander (2003) Studies on efficacy of various insecticides and biopesticides on rice leaf folder, *Cnaphalocrocis medinalis* (Guenee). *J. Entomol. Res.*, **27**, 297–303.

Rao, N.B.V.C., Singh, V.S. and Subash Chander (2006) Effect of various combinations on rice leaf folder and its coleopteran predators. *J. Entomol. Res.*, **30**, 219–223.

Rath, P.C. (1999) Evaluation of *Bacillus thuringiensis* and neem formulations against yellow stem borer of rice under rainfed lowlands. *Oryza*, **36**, 398–399.

Roshan Lal (2001) Field evaluation of some biopesticides for management of rice leaf folder, *Cnaphalocrocis medinalis* (Guenee). *Pestic. Res. J.*, **13**, 62–65.

Shahid, A.A., Nasir, I.A., Zafar, A.U., Sumrin, Aleena, Chaudhry, B. and Sheikh Riazuddin (2003) The use of CAMB biopesticides to control pests of rice, *Oryza sativa*. *Asian J. Plant Sci.*, **2**, 1079–1082.

Sharma, M., Charak, K.S. and Ramanaiah, T.V. (2003) Agricultural biotechnology research in India : Status and policies. *Curr. Sci.*, **84**, 297–302.

Sharma, P.K. and Kaul, B.K. (2003) Management of *Dicladispa armigera* (Linn.) on rice with combination of insecticide and ecofriendly neem formulations. *Crop Res.* (Hisar), **25**, 248–250.

Singh, Sher, Prasad, C.S., Lok Nath and Tiwari, G.N. (2008) Eco-friendly management of *Scirpophaga incertulas* (Walk.) and *Cnaphalocrocis medinalis* (Guen.) in basmati rice. *Ann. Plant Prot. Sci.*, **16**, 11–16

Singh, S.P. (1994) *Fifteen Years of AICRP on Biological Control.* Project Directorate of Biological Control, Bangaluru, India.

Soon, L.G. and Bottrell, D.G. (1994) *Neem Pesticides in Rice: Potential and Limitations.* International Rice Research Institute, Philippines.

Wailes, E.J. and Chavez, E.C. (2012) *International Rice Baseline with Deterministic and Stochastic Projections, 2012-2021.* Department of Agricultural Economics and Agribusiness, University of Arkansas, Fayetteville, AR.

Way, M.J. and Heong, K.L. (1994) The role of biodiversity in the dynamics and management of insect pests of tropical irrigated rice- A review. *Bull. Entomol. Res.*, **84**, 567–587.

3

Biopesticides in Maize Pest Management: Progress and Potential

Pradyumn Kumar and S. B. Suby

Directorate of Maize Research, Pusa Campus, New Delhi-110012, India
E-mail: pradyumn.kumar@gmail.com

INTRODUCTION

The global corn production has gone up to 849 million tonnes in 2012-13, mainly due to a larger crop in countries such as China, Russia and Canada as indicated in the latest monthly report released by United States Department of Agriculture (USDA). Overall global corn production has increased by 9.4 million tonnes in 2012/13 and China contributing maximum to the output by 8.0 million tonnes. Corn production for Canada has gone up by 1.5 million tones. Similarly, for Russia increase of 1.0 million tonnes is a new record. Corn production has increased for North Korea and Chile too by 0.4 million tonnes and 0.3 million tonnes, respectively. However, corn production in Argentina did decline by 0.50 million tonnes due to poor crop yield in 2011/12 (Fig. 3.1). India, however, contributes to about 2.6% of the global production (Fig. 3.2).

Global corn consumption for 2012/13 has reached to about 925 million tonnes (Fig. 3.3) mostly with increases of 6.0 million tonnes and 1.5 million tonnes, respectively, for China and European Union. Global corn exports have gone up by 0.5 million tonnes each for Canada and Russia. Increases in local marketing year 2011/12 exports for Argentina, Paraguay, and South Africa support higher 2012/13 consumption for Northern Hemisphere countries (http://www.indiainfoline.com). USDA estimate of the global corn ending stocks for 2012/13 is projected as 0.4 million tonnes lower at 117.60 million tonnes, mostly on reductions in Argentina, Colombia, and Ukraine. Small increases in a number of countries, including Canada, are partly offsetting.

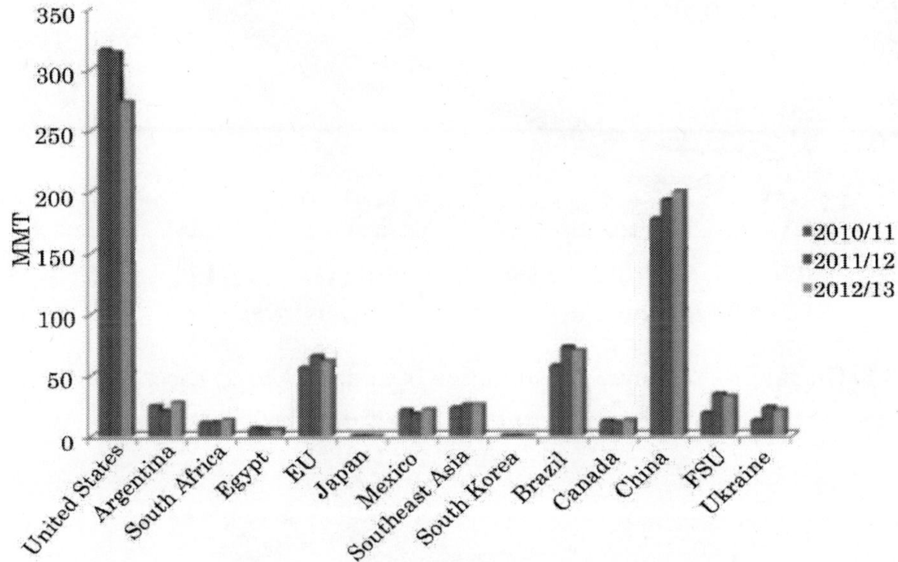

Fig. 3.1. World corn production. *Source*: US Grains Council

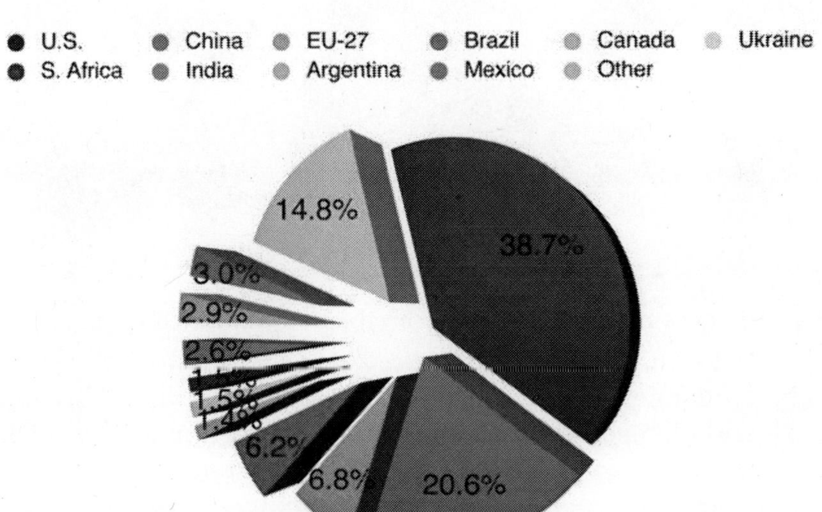

Fig. 3.2. World corn production. *Source:* World Markets and Trade (2011).

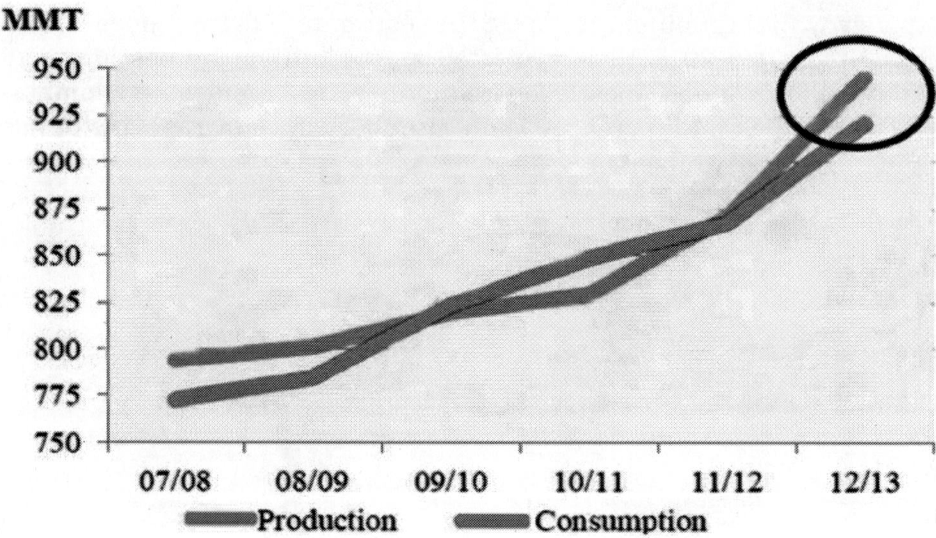

Fig. 3.3. Global production & consumption of maize. *Source:* thecropsite.com

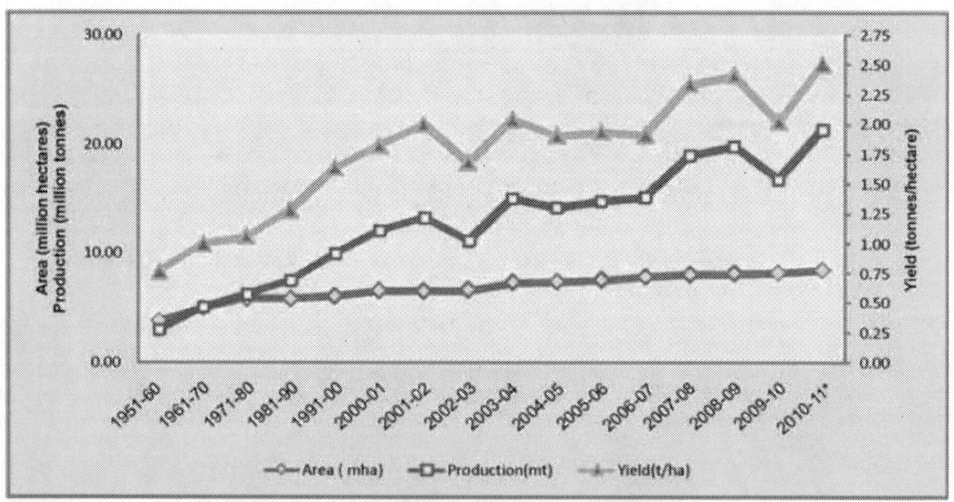

Fig. 3.4. All India area wide growth, production and yield of maize. *Source*: DACNET

India ranks seventh in maize production (Fig. 3.2). The demand for maize and of specialty corn such as sweet corn, pop corn and baby corn is growing day by day with change in food habits and increase in purchasing power of the people. All India area wide growth, production and yield of maize is shown in Fig. 3.4. Maize grown in *Rabi* is known to be free from the attack of major insect pests which usually attack *Kharif* crop. However, pink stem borer, *Sesamia inferens* is a major pest. The larvae attack almost all parts of the plant such as leaf, stem tassel, and ear. Due to larval feeding, the grown up plants show many slit-like oval elongated holes on the leaf blades. They also form tunnel inside the stem and exit holes at the surface.The decaying shoots in grown up plants cause cob to rot, thus causing complete loss of grain. The crop suffers 15-20% loss in field due to infestation of insect pests thus plant protection is unavoidable in economic maize production. Changes in conventional cultivation such as adoption of zero tillage can change the pest scenario in maize. Maintaining permanent bed and leaving the crop residues in the field will make the ecosystem favorable for soil pests like chaffer beetle, cut worms, etc. Chemical pesticides are largely used to protect maize from pest ravage.

Increase in awareness of pesticide residues in food among consumers and growing concern for environmental hazards driving agriculturists to find eco-friendly and economic means of pest management strategy. Conservation of naturally occurring biological control agents and applied biological control in maize ecosystem is usually adequate in bringing the insect pest populations within economic threshold levels. It works on the simple principle of ecological balance where the out proportion of one biological entity is balanced by another. Insects attain pest status when this balance is lost. Maize is an ideal crop for the adoption of biopesticides due to its characteristics like comparatively longer duration, partial shade and high humidity in the canopy which are conducive for the multiplication of natural enemies. Further, the production of copious pollen grain support large number of predators such as lady bird beetles, green lacewing, hoverflies, etc. Its widely spaced and semi-erect leaves provide excellent hiding space for natural enemies in its whorls during high temperature. It is perhaps for this reason that though over 130 insect pests have been reported in maize, but only a few require regular human intervention for their management while most of them are kept below economic injury level by adequate number and types of natural enemies (Sarup *et al.*, 1978). The spotted stem borer, *Chilo partellus* (Swinhoe) is a key pest throughout India mainly in 'kharif' and the pink stem borer, *Sesamia inferens* (Walker) causes serious damage in 'rabi' and spring. Shoot fly, *Atherigona* spp. cause considerable crop loss in spring sown maize in north India (Kumar *et al.*, 2005). The chaffer beetle, *Chiloloba acuta* (Wiedemann) adult is becoming a serious problem in north-west India. The adult is gregarious and eat the pollen of maize, thus seriously affects seed setting while larvae remain in soil and feed on organic matter.

Helicoverpa armigera (Hubner) is an emerging threat in peninsular India, especially in coastal Andhra Pradesh, as the pest shifted from cotton to maize because of the wide adoption of Bt-cotton in this area. The pest lays eggs on silk and larvae causes ear damage. Pesticide spraying is not feasible due to the height of the plant at that stage.

PREDATORS AND PARASITOIDS

Biological control of maize pests is known from decades. Ashley (1979) reported 53 species of parasitoids of the maize earworm belonging to 43 genera and 10 families, of which the family Tachinidae comprised 53% of the species. Molina-Ochoa *et al.* (2003) reported 150 species of parasitoids of *S. frugiperda* in the Americas and the Caribbean basin belonging to 14 families, nine in Hymenoptera and four in Diptera. The family Tachinidae is not only the most diverse of the Diptera, but overall as well (55 species). To date none of these species of parasitoids has had a controlling effect on the maize earworm, because their reproduction in the laboratory is not easy. However, *Telenomus remus* Nixon (Hymenoptera: Scelionidae) can be grown relatively easily under laboratory conditions (MartAnez *et al.*, 2012). The specialist larval parasitoid *M. grandii* is the only introduced parasitoid that seems to be an important biological mortality factor of European corn borers in USA (Siegel *et al.*, 1987). The parasitic hymenopterans from family Encyrtidae infests insects belonging to Homoptera, Heteroptera, Lepidoptera, Coleoptera, Diptera, Hymenoptera, Neuroptera, Orthoptera, Blattoidea and Megaloptera. These small parasitic wasps effectively suppress many agricultural pests in the tropics and subtropics (Trjapitzin 2008) and more so the maize crop pests (Krawczyk *et al.*, 2011). The hoverfly species, *Sphaerophoria scripta* (L.), *S. rueppelli* (Wied.), *S. menthastri* (L.) and *Episyrphus balteatus* (Deg.) are seen to be aphidivorous and maximum of parasitism of 84.4% has been recorded. Parasitoids of syrphid pupae were identified as *Pachyneuron grande* Thomson, *Syrphophagus aeruginosus* (Dalman), *Aphidencyrtus aphidivorus* (Mayr) and *Dendrocerus pupparum* (Graham), as well as ichneumonid wasps. The predominant parasitoid species that attacked the syrphid pupae of all types to the greatest extent was *P. grande* (Krawczyk *et al.*, 2011). Nocturnal predators like the carabid beetle, *Cyclotrachelus alternans*, have a large appetite for corn rootworms, the most costly pest of corn in the world. On the other hand *Poecilus chalcites* attacks corn root worms during the day (http://phys.org/news176195831.html).

Habitat management has been shown to benefit natural enemies by the simultaneous operation of more than one mechanism. Work with *Eriborus terebrans* (Gravenhorst), the primary parasitoid of *Ostrinia nubilalis* (Hubner) in maize in Michigan showed that females were most frequently captured in maize fields close to wooded field edges and that parasitism of *O. nubilalis* was higher in these areas (Landis and Haas, 1992). During the growing season, high temperature and low humidity also may constrain natural enemy populations or the activity of individuals. Orr *et al.* (1997) interplanted ryegrass (*Lolium multiflorum* Lambert) in seed maize (*Zea mays* L.) fields to reduce the temperature of the soil surface, increasing survival of augmentatively released *Trichogramma brassicae.*

In India, integration of augmentative releases of *Trichogramm chilonis* (Ishii) at the rate of 150 thousand parasitized eggs/ha during the oviposition period of stem borers in IPM of maize has been widely and successfully demonstrated. Conscious introduction of biological control agents brings this equilibrium back by reducing the pest population in a density responsive manner. Conservation of naturally occurring biological control agents and applied biological control in maize ecosystem is usually adequate in containing the insect pest problem. *Cotesia flavipes* Cameron, a gregarious larval parasitoid is one of the most potential biological control agents of stem borers occurring in maize ecosystem and is used in biological control against tropical stem borers (Potting *et al.*, 1997; Matama-Kauma *et al.*, 2001). The female wasp lays 20-40 eggs in a single host larva. The eggs on hatching produce maggots which feed inside the larva. The larvae continue to feed for 3–4 days and then stop feeding. The maggots devour the body content and come out to spin cocoon thus finally leaving only head capsule. A cluster of white cocoons inside the stem, tassel, silk and sometime even on leaf is common sight at flowering stage of the crop. *C. falvipes* parasitize the larvae in early instars while the late instars are parasitized by a tachnid fly. The fly lays 1–2 eggs inside the larva. The maggots feed on haemolymph and then viscera. The full grown maggot comes out of the body and form a puparium. *Campoletis chlorideae* Uchida, a naturally occurring hymenopterous parasitiod of *H. armigera* can be augmented for controlling this pest as it is proved that the parasitoid is strongly attracted to herbivore-induced maize volatiles (Yan and Wang, 2006).

Jalali and Singh (2003) recorded forty two natural enemy spp. attacking *C. partellus* in India including one egg parasitoid *Trichogramma chilonis*, thirty seven larval parasitoids including *Cotesia flavipes* and *Myosoma chinensis*, and three pupal parasitoids like *Xanthopimpla stemmator* and *Tetrastichus howardi* being the major ones. Predators consist mainly the anthocorid bug, *Orius tantillus* which was found active throughout the year. Other potential larval predators were reduviid bugs (2 species) and spiders (17 species). In the pupal stage only potential predator recorded has been earwig, *Euborellia annulipes.* Egg parasitoid *T. chilonis* was found active in the field in all months except in December. Larval parasitoid *C. flavipes* was found active in all months except between November and February when only one generation is possible and pupal parasitoid *T. howardi* can complete two generations between November and February. The combination of *T. chilonis* when released with the gap of three days at the rate of 100,000/ha five times during 1st generation and three times during 2nd generation followed by a spray of *B. thuringiensis* controls pest infestation and highest yield of maize is obtained. Studies on the effectiveness of *Trichogramma chilonis* Ishii in the suppression of *Chilo partellus* in summer maize during 2006 and 2007 at Ludhiana in India reduced incidence of maize stem borer in maize plot with single release of *T. chilonis* at the rate of 100,000/ha on 12-day-old crop (14.3%) followed by 15-dayold crop (15.3%) which was significantly lower as compared to untreated check (33.0%). The maize yield in these crops also increased significantly (Vijay Kumar and Uma Kanta, 2011).

Aphid colony in tassel often harbors variety of lady bird beetles, *Syrphid* maggots, larvae of *Chrysoperla carnea* (Stephens), spiders, etc. Sometime *Pyrilla perpusilla* Walker migrate from sorghum and infest the maize during flowering season but its population is arrested by *Epiricania melanoleuca* (Fletcher) before causing any economic loss. The larvae of *E. melanoleuca* stick to the abdomen of nymphs and adults of *P. perpusilla* and suck the haemolymph of the host. This external parasite debilitates the host and renders it sterile. It is advised not to burn the trash having the viable cocoons of *E. melanoleuca.* These may be removed and released in *P. perpusilla* infested fields (Gangwar *et al.*, 2008). It is thus obvious that natural enemy guild of maize pests and the nature of crop, beside seed treatment with some insecticide and fungicide, maize cultivation requires little chemical pesticides provided the natural enemies are conserved and 'Tricho-cards' are released in synchrony with stem borer egg laying.

MICROBIAL BIOPESTICIDES

Microbial biopesticides contribute to major class of biopesticides that are commercially available and bacteria contribute to about 60% of the products (Fig. 3.5). *Bacillus thuringiensis* (Bt) is a species of bacteria that produces proteins that are toxic to insects. Due to this toxicity Bt has been used as a safe microbial insecticide for over 50 years to control pest caterpillars. Bt has specific pH levels, enzymes, and midgut receptors are required to activate and bind a given Cry toxin to midgut cells, which leads to pore formation in the insect's intestine and death (Federici 2002). The specificity of these toxins implies a "lock and key" analogy. Midgut receptor can be considerd as the "lock" and the Cry protein as the "key" then insect death only occurs when the two make a match match. In case of maize pests, the relative susceptibility of 4 tropical maize pests, *Spodoptera frugiperda* (J. E. Smith), *Diatraea grandiosella* Dyar, *D. saccharalis* (F.), and *Helicoverpa zea* (Boddie), to the lepidopteran-specific CryI-type proteins produced by Bt is well known. The toxin with the highest potency against *H. zea* larvae has been reported as CryIAc toxin. *S. frugiperda* larvae were susceptible to CryID and CryIF toxins. The CryIB toxin showed to be highly toxic against *D. grandiosella* and *D. saccharalis* (http://oar.icrisat.org/).

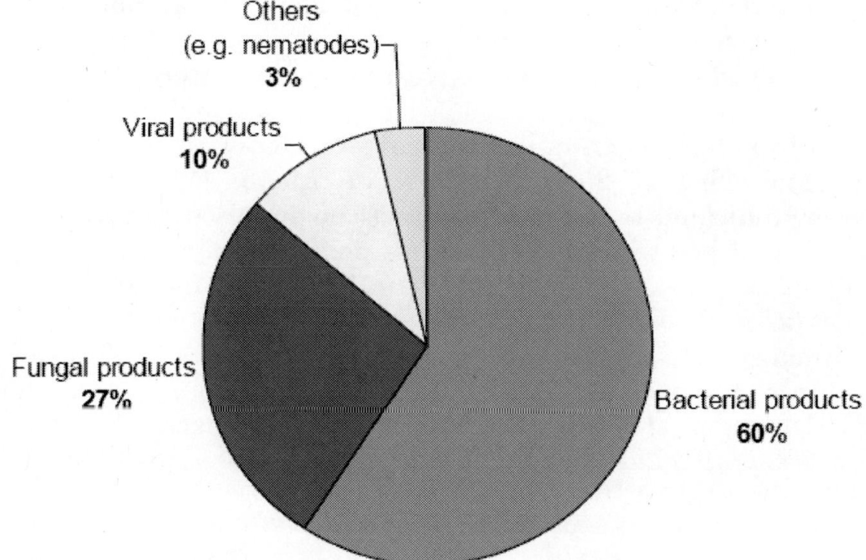

Fig. 3.5. Worldwide microbial biopesticides products (Source: Kabaluk *et al.*, 2010)

However, in recent past substantial emphasis has been given to Bt corn. There are a number of Cry toxins that are categorized by their spectrum of activity. For maize pests, primary Cry proteins are Cry1 and Cry2 for Lepidoptera and Cry3 proteins for Coleoptera (Schnepf *et al.* 1998). Maize can be genetically engineered to produce these specific Cry toxins. Growers are attracted to Bt corn because of the yield protection and improved grain quality commonly found with Bt maize. Interestingly, due to the introduction of Bt maize, recent research suggests there has been an areawide suppression of European corn borer populations (Hutchison *et al.* 2010). This is beneficial to both Bt and non-Bt maize growers. Another advantage with Bt corn is no effects on non-target organisms, which confirms the specificity of the Bt proteins. Most studies suggest Bt maize has little if any impact on predators and parasitoids and, when compared with maize treated with chemical insecticides, Bt maize often results in increased biodiversity (O'Callaghan *et al.* 2005, Romeis *et al.* 2008). Specialist insects that depend on target pests are the exception to the generalization that Bt maize does not impact non-target organisms. This is particularly true for some parasitoids, which may become less abundant along with their herbivorous hosts (Pilcher *et al.* 2005, Romeis *et al.* 2008, Storer *et al.* 2008). Also, fewer saprophagous dipterans (i.e., fly larvae that feed on decaying organic matter) have been observed in Bt maize fields, which has been attributed to the indirect effect of reduced lepidopteran plant injury (Candolfi *et al.* 2004, Dively *et al.* 2004). Studies with monarch butterfly caterpillars, *Danaus plexippus*, suggested monarch butterfly populations would be reduced from feeding on milkweed leaves coated with Bt maize pollen (Jesse and Obrycki 2000, Losey *et al.* 1999). Follow-up studies, however, indicated that the impact was negligible because of limited exposure and low toxicity of Bt maize pollen to monarch caterpillars (Dively *et al.* 2004, Hellmich *et al.* 2001, Sears *et al.* 2001, Stanley-Horn *et al.* 2001). Gene pyramiding is also shown to play a significant role in maize crop prortection. Sweet corn, *Zea mays* L., transformed to express a novel vegetative insecticidal protein, Vip3A (event MIR162, Syngenta Seeds, Inc.) produced by the bacterium, *Bacillus thuringiensis* (Bt) has been evaluated in USA in various field seasons for efficacy against the corn earworm, *Helicoverpa zea* (Boddie). Hybrids expressing the Vip3A protein and pyramided in hybrids also expressing the Cry1Ab Bt protein (event Bt11, ATTRIBUTE®, Syngenta Seeds, Inc.) were compared to hybrids expressing only Cry1Ab or to genetically similar non-Bt hybrids each year. These studies have been extended to *Ostrinia nubilalis* (Hübner) and

Spodoptera frugiperda (J.E. Smith) as well. It has been pbserved that the non-Bt hybrids, without insecticide protection, averaged between 43 and 100% ears infested with a range of 0.24 to 1.74 *H. zea* larvae per ear. By comparison, in the pyramided Vip3A × Cry1Ab hybrids, no larvae were found and only minimal kernel damage was recorded. Hybrids expressing only Cry1Ab incurred a moderate level of *H. zea* feeding damage, with surviving larvae mostly limited to the first or second instar as a result of previously documented growth inhibition from Cry1Ab (Burkness *et al.*, 2010). These results suggest that the Vip3A protein, pyramided with Cry1Ab, provide a valuable resistance management strategy under field conditions.

SmartStax is another genetically modified (GM) maize that has eight GM traits combined or 'stacked' together, six for insect resistance (Bt) and two for herbicide tolerance. Current stacked GM trait crops on the market only have up to three traits each. SmartStax was created through collaboration between Monsanto and Dow AgroSciences (http://www.i-sis.org.uk/SmartStaxCornCorporateWarOnBees.php), allowing the two corporations to share GM traits. The traits are combined together using crosses between existing transgenic corn lines rather than using genetic transformation of a single maize strain. USDA/APHIS (United States Department of Agriculture/ Animal and Plant Health Inspection Service) the usual regulator of GM crops approved the crops without an approval process, because the transgenic traits had been granted unregulated status previously, and those traits were combined using conventional breeding. Nevertheless, USEPA (United States Environment Protection Agency) were obligated to regulate the stacked crop varieties containing plant-incorporated protectants (PIPS). EPA invited public comment prior to registration of stacked varieties of corn but they approved SmartStax corn without allowing public comment prior to its registration.

Western corn rootworm (WCR) is the most destructive maize pest in North America has become a serious invasive pest in Europe (Miller *et al.*, 2005; Vidal *et al.*, 2005). Most of the yield losses attributed to this pest are the result of damage to the maize roots caused by the soil-dwelling WCR larvae. One of the approaches to control these pests in maize has been the release of mass-produced entomopathogenic nematodes (EPN) (Kuhlmann and van der Brugt, 1998). These nematodes are obligate insect parasites in symbiosis with bacteria. After entering the host, the infective juvenile (the infective stage of EPN) releases its symbiotic bacteria, which multiply and kill the insect by septicaemia and serve as food for EPN. When nutrients are consumed and space is exhausted, a new generation of infective juveniles leave the cadaver and search for a new host. Among the species available on the market, *Heterorhabditis megidis* Poinar (Rhabditida: Heterorhabditidae) exhibits behavioural traits highly interesting for biological control. Recently it has been established that some plant metabolites play a significant role in attracting infective juveniles of *H. megidis*. (*E*)-ß-caryophyllene, a sesquiterpene released by WCR-damaged maize roots is one of them (Rasmann *et al.*, 2005; Rasmann and Turlings, 2007; Köllner *et al.*, 2008), and the presence of this signal is essential for this nematode to be effective as a control agent. Another nematode *H. bacteriophora*, is also highly virulent EPN against all WCR larval stages (Toepfer *et al.*, 2005; Kurzt *et al.*, 2009; Hiltpold *et al.*, 2009), however, it has been shown that it does not respond to (*E*)-ß-caryophyllene (Rasmann and Turlings, 2008; Hiltpold *et al.*, 2009). However, some recent studies have explored the feasibility of enhancing this responsiveness through artificial selection. Because of their short generation time, small genome size and ease of culture, EPNs are ideal subjects for genetic improvement (Hiltpold *et al.*, 2010). *Chiloloba acuta* control has also been achieved by using soil application of six types of entomopathogenic nematodes *viz. Heterorhabditis indica, H. bacteriophora, Steinernema glaseri, S. thermophilum, S. riobrave* and *Steinernema* sp. strain kr-1. Mortality of grubs was observed in all the species except in *H. indica* while in *S. thermophilum* it was maximum followed by *H. bacteriophora, Steinermema* sp. strain kr-1, *S. glaseri* and S. *riobrave* (Dhanya, 2009).

Fungi constitute a large group of more than 500 species which can parasitize insects. Most of the taxonomic groups contain entomopathogenic genera, such as *Metarhizium, Beauveria, Verticillium, Nomuraea, Entomophthora, Neozygites,* to name a few (Desphande, 1999). Fungal biocontrol agents are promising because they act by contact and do not need ingestation, can be mass produced very easily and are quite host specific. Field trials of entomopathogenic fungi are known from 1980s to determine their efficacy in soil applications using *Metarhizium anisopliae* and *Beauveria bassiana* against the southern corn rootworm and *Diabrotica undecimpunctata* in corn (Krueger and Roberts, 1997). *Beauveria bassiana* (Balsamo) Vuillemin has been reported as an endophyte in maize (Cherry *et al.,* 2004; Wagner and Lewis, 2000; Arnold and Lewis, 2005). The pioneering work on entomopathogenic endophytes has been conducted using maize (*Zea mays* L.) and the European corn borer, *Ostrinia nubilalis* (Hu¨bner) as a model system. Season-long suppression of insects, measured as reduced tunneling of *O. nubilalis* has been achieved by applying *B.*

bassiana in an aqueous (Lewis and Cossentine, 1986) or granular (Lewis and Bing, 1991) formulation on corn plants. It has also been established that local strains of entomopathogenic fungi are usually more toxic than the imported isolates. For example, control. In an isolate selection program of fungal isolates against ECB, ten isolates consisting of eight isolates from Iran and two from other countries were assayed using the dipping method on third instar larvae of ECB. Results showed that BEH isolate which was isolated from the soil of insects living in the field, caused the highest mortality in larvae in comparison with other isolates (Safavi *et al.*, 2010)

Viruses are also another group of microbial biopesticides that effect insects significantly. A microbial insecticide (SFNPV) against *Spodoptera frugiperda* on maize was developed in late 1980s (Valicente *et al.*, 1989). In field trials different concentrations of this nuclear polyhedrosis virus were applied as aqueous suspensions using a tractor mounted or a backpack-manual sprayer to control *Spodoptera frugiperda* (Smith) larvae on maize *(Zea mays* L.). A higher NPV persistence was obtained with doses over 1.25×10^{12} P.I.B./ha and 93.4% mortality was obtained (Cruz *et al.*, 1997). In China, *Helicoverpa armigera* SNPV (HaSNPV), which has been produced and applied to maize and other crops was first authorized as a commercial microbial insecticides in 1993 (Zhang *et al.*, 1995; Sun and Peng, 2007). About 20,000 ha of maize infested with pests is controlled by SfNPV in Brazil annually (Moscardi, 1999). Symptoms of bacterial infection in *C. partellus* and NPV in *H. armigera* have been observed in maize ecosystem. Farrar *et al.* (2009) evaluated biologically-based spray treatments, including NPV, neem and spinosad as supplemental controls for the fall armyworm *S. frugiperda*, and corn earworm *H. zea* on transgenic sweet corn expressing a *Bacillus thuringiensis* toxin Cry1Ab. Even though the Control of the fall armyworm was improved in both whorl-stage and tassel-stage corn by the use of either a nucleopolyhedrovirus or neem, only spinosad consistently reduced damage to ears, which was caused by both pest species.

BOTANICALS

Botanicals can be a suitable alternative to chemical pesticides as the fear of ecological back lashing will be at the bay hence it contain slow acting, biologically active substances. Use of botanicals for pest control has become an intrinsic part of local tradition and folklore all over the world. Mihale *et al.* (2009) documents that in Africa, indigenous knowledge (IK) based control methods used by farmers ranged from animal by-products (cow's urine and dung), plant parts *Azadirachta indica* (Meliaceae), *Tephrosia vogelii* (Fabaceae), *Tamarindus indica* (Fabaceae), *Aloe* spp. (Asphodelaceae), red pepper, *Capsicum* spp. (Solanaceae), *Nicotiana tabacum* (Solanaceae) to ash (general and specific) in the field. They also use neem, *Chenopodium opulifolium* (Chenopodiaceae), *Ocimum suave* (Labiatae), *Senna siamea* (Fabaceae or Caesalpinioideae), tobacco and *Eucalyptus* spp. (Myrtaceae) and plant by-products (rice husks, ash from rice husks and red maize cobs and general ash) to control storage pests. Fruit and leaf extracts of three locally available plant species in Ethiopia viz., chinaberry (*Melia azedarach* L.), endod (*Phytolacca dodecandra* L.) and pepper tree (*Schinus molle* L.) were tested for insecticidal activity against the maize stalk borer *Busseola fusca*. Both leaves and fruits of chinaberry and endod and fruits of pepper tree extracts significantly reduced the number of larvae and levels of infestation in field (Gebreamlak, 1999). Toxic effects of six indigenous plants, namely papaya (*Carica papaya*), Ivy gourd (*Coccinia indica*), bitter gourd (*Momordica charantia*), curry leaf (*Murraya koenigii*), water hyacinth (*Eicchornia crassipes*) and betel leaf (*Piper betle*) were evaluated against the major pest of stored maize, *Sitophilus zeamais* in India. Crude acetone extracts of *Eicchornia crassipes* caused 100% mortality in *S. zeamais* after 72 h of exposure followed by *P. betel* in grain surface coating method at 0.5% (w/w) without affecting germination. Further, the same plant extract had significantly reduced the development of F1 progeny by 97%. The crude extracts of *C. papaya* were found to be moderately potent as contact toxicants. Leaf extract of *M. koenigii* was superior over other leaf extracts tested for vapor toxicity, while extracts of *E. crassipes* was moderate in action (Chakradhar *et al.*, 2010).

The crude acetone extracts of chilli plant *Capsicum annuum,* bitter gourd *Momordica charantia,* brinjal *Solanum melongena,* papaya *Carica papaya* and curry leaf *Murraya koenigii* caused significant mortality to rice weevil, *Sitophilus oryzae* and red flour beetle, *Tribolium castaneum* in fumigation method and the extracts of *C. annuum* and *M. charantia* exhibited a delayed toxicity (i.e., 72 h) in contact form. Among them, *C. annuum* extract was the best at median lethal concentrations (as mg/20g diet) of 21.2 and 24.8 against *S. oryzae* and *T. castaneum* respectively. Ethyl acetate eluted fractions of *C. annuum, S. melongena* and *M. charantia* showed 100% mortality to *T. castaneum* and *S. oryzae* within 24 h of treatment in vapor form (Usha Rani and Devanand, 2011). Sharma (2009) has reported 50%

mortality in *Acrida exaltata* (Acrididae) adults feeding on maize leaves treated with 1.0% leaf extract of Tulsi, *Ocimum sanctum.*

Neem is the most potential botanical known to control variety of insects pests and diseases. The seed extract of neem, *Azadirachta indica* A. Juss is rich in azadirachtin that acts as a potent antifeedant and insect growth regulator (Koul, 1992). Maize kernels soaked for 20 min in a 1% solution of neem oil extractive was resistant to attack by *Sitophilus oryzae* (Attri and Prasad 1980). In Togo, Adhikary (1981) found that neem treatment of corn stored in sacks or unpeeled corn cobs held in bins was quite simple and effective against *S. zeamais*, *Tribolium* spp., *Rhyzopertha dominica*, and *Cathartus* spp. Jalali and Singh (2004) determined the efficacy of Various neem products (Nimbecidine 1000 ml/ha, Neemguard 2500 ml/ha, Neemgold 500 ml/ha, Nemark 2500 ml/ha, Neemzal 1000 ml/ha, Neemrich I 1500 ml/ha, Neemrich II 2500 ml/ha and Achook 2500 g/ha) against, *Chilo partellus* (Swinhoe) on maize and found a reduction in egg laying by 64.6 to 97.6% and egg hatching by 18.0 to 37.0%. Nimbecidine caused maximum ovipositional deterrence, whereas Achook and Neemrich-I showed significantly higher number of eggs than control in no-choice test. Different neem products caused 56.0 to 68.0% larval mortality.

Calneem oil, a commericial neem oil formulation was found effective in significantly reduceding the oviposition rate, development of eggs and immature stages, and emergence rate of the tropical warehouse moth *Ephestia cautella* (Walker) progeny in maize grains. Grains treated with Calneem oil had significantly reduced grain damage and the oil had a repellency of 53-70% compared with the negative control (Shehu *et al.*, 2010). It was found neem oil added to the diet of *S. frugiperda* caused total mortality at a dose of 0.4% whilst still in the first instars, prolonged the larval and pupal stages, and reduced the pupal weight. Histo-physiological alterations such as degeneration of the epithelial lining of the midgut and in the peritrophic matrix were found at all concentrations of neem oil (Roel *et al.*, 2010). Gutierrez-Garcia *et al.* (2010) recommends weekly application of 2% neem oil to get best protection against fall armyworm *Spodoptera frugiperda* in Mexico. Viana *et al.* (2010) reported that aqueous extracts of green ground neem leaves (5000 ppm) applied three times are effective to protect maize from fall armyworm *Spodoptera frugiperda* damage in Brazil and the time of application does not affect its efficiency for controlling larvae. Moyin-Jesu (2010) reported that the modified neem leaf extract (1:1 ratio 600 ml wood ash + 600 ml neem leaf extract) applied at 1200 l/ha (3 l/25 m^2) was most effective for pest control and seed treatment compared to neem leaf and wood ash alone in Nigeria.

Neem cake is used as a soil amendment which has soil nurturing, insecticidal as well as nematicidal properties. Shaukat *et al.*, reported that Fertinemakil Plus, a phytonematicide containing neem cake and a fungicide at 200 ml/row gave maximum reduction in the population of the plant parasitic nematode *Pratylenchus zeae* while both Fertinemakil Plus and Carbofuran (Furadan, at 15 kg/ha) were on par for controlling *Helicotylenchus indicus* in maize. Extracts of *Azadirachta indica* and *Nicotiana tabacum* were found effective for post-harvest preservation of stored maize against infestation by larger grain borer, *Prostephanus truncatus* (Osipitan *et al.*, 2010).

Complete oviposition deterrent effects of neem (2.5 and 5%), *Jatropha curcas* (5%) seed oils and *Hyptis suaveolens* (20%) leaf extracts have been reported (Agboka *et al.*, 2009). Oil emulsions of *A. indica* and *J. curcas* oils are ovicidal with LD_{50} values of 1.3 and 0.8%, respectively. Oviposition inhibition in case of *Sitophilus zeamais* has been demonstrated after the application of *J. caracas* A dose-dependent insectidical and antifeedant effects were also observed (Jide-Ojoa and Ojo, 2011).

Essential oils from plants are also toxic to maize pests. Cinnamon, Clove, Marjoram and Ginger oils as well as Eugenol reduce oviposition of the corn borer, *Sesamia cretica.* The yield of maize ears shows increase (El-Hosary, 2011). The essential oil of *Illicium simonsii* fruit was tested against *S. zeamais* adults and found to have strong fumigant toxicity with a LC_{50} value of 14.95 mg/l air as well as contact toxicity with LD_{50} value of 112.74 µg/adult. Overall, 40 components have been isolated, of which β-caryophyllene (10.30%), δ-cadinene (9.52%), methyl eugenol (8.94), β-elemene (5.84%), and α-amorphene (5.20%) have been reported as active (Chu ShaSha *et al.*, 2010). Similarly, essential oil of *Ostericum sieboldii* flowering aerial parts was found to possess strong contact toxicity against *S. zeamais* and *T. castaneum* adults with LD_{50} values of 13.82 µg/adult and 8.47 µg/adult, respectively. It also showed fumigant toxicity against *S. zeamais* and *T. castaneum* adults with LC_{50} values of 27.39 mg/l air and 20.92 mg/l air, respectively (Liu *et al.*, 2010). Suthisut (2011) has evaluated the fumigant toxicity of essential oils from rhizomes of *Alpinia conchigera, Zingiber zerumbet, Curcuma zedoaria* and their major compounds; camphene, camphor, 1,8-cineole, α-humulene, isoborneol, α-pinene, β-pinene and terpinen-4-ol against adults of *Sitophilus zeamais, Tribolium castaneum, Anisopteromalus calandrae* and *Trichogramma deion* larvae. Adults of *S. zeamais* and *T. castaneum* were more susceptible to the treatments. *Zingiber zerumbet* oils (LC_{50} = 26 µl/l in air) and *C. zedoaria* oils (LC_{50} = 25 µl/l in air) were significantly more toxic to adults of *A. calandrae* than *A. conchigera* oils (LC_{50} = 37 µl/l in air) whereas *T. deion* larvae were more sensitive to *A. conchigera* oils (LC_{50} = 62 µl/l in air) than *Z. zerumbet* and *C. zedoaria* oils (LC_{50} > 593 µ/l in air). *T. castaneum* was more susceptible than *S. zeamais* to the eight pure compounds. Terpinen-4-ol was highly toxic to both insects. Insecticidal formulations based on the essential oil of *Xylopia aethiopica* (Annonaceae) and kaolinite-clay (particles size less than 50 µm) were tested against *S. zeamais* and found that 4.35% (w/w) of essential oil and clay is required to cause 50% mortality (LC_{50}). Depending up on storage conditions and duration, efficacy was found decreasing from 91 to 37% for formulation stored in closed boxes and from 91 to 17% for formulation stored in open boxes, after 1 and 15 days, respectively (Nguemtchouin *et al.*, 2010).

INERT MATERIALS

The use of chemically inert materials, such as wood ashes, sand or other minerals, powders or plant products in large quantities to fill up the interstitial space in grain bulks and provide a barrier to insect movement, is quite widespread (FAO, 1999). The effectiveness of the diatomaceous earth Silicosec, a mineral industrial filter cake and domestic wood ash for the control of the maize weevil, *Sitophilus zeamais*, on three maize genotypes has been determined and found Silicosec was the most effective followed by filter cake and wood ash. The treatments reduced progeny emergence, percentage grain damage and grain weight losses, but did not affect percentage seed germination (Girma *et al.*, 2008). Coitinho *et al.* (2006) have evaluated the residual effect of natural insecticides on adults of *Sitophilus zeamais* on stored maize grains. Clove (*Syzygium aromaticum* L.), white pepper (*Piper nigrum* L.), neem (*Azadirachta indica* A. Juss) and silica dioxide powders as well as neem, eucalyptus (*Eucalyptus globulus* Labill and *Eucalyptus citriodora* Hook.), eugenol, andiroba (*Carapa guianensis* Aubl.), rosemary (*Lippia gracillis* HBK.), cedar (*Cedrela fissilis* Vell.) and pequi (*Caryocar brasiliense* Camb.) Powders were tested at concentration of 1 kg/t in 50 g-corn-plot and the oils at dosage of 50 µl in 20 g-corn-plot. Among powders, silica dioxide was the more efficient in causing mortality and reducing adults' emergence under different storage periods. All the oils caused mortality above 79% at the initial storage period, however, mortality was lower than

2.5% after 60 and 120 days. The adult emergence increased as the period of storage increased, although at 120 days the best performance was obtained using neem, followed by andiroba, cedar, rosemary and *E. globulus* oils.

CONCLUSIONS

Corn is the largest cultivated crop grown in the U.S., Asia and Africa in terms of acreage planted and net value. Corn rootworm (CRW, *Diabrotica* Spp.) is one of a spectrum of insect pests that a farmer may choose to control. Other insect pests include cutworms, wireworms, white grubs, flea beetles, seedcorn maggots, black cutworms, and cornborer species. However, CRW is one of the most damaging insect pests of corn and is responsible for economic damages worth >1 billion US$. One of the major methods to control these pests has been soil-applied chemical insecticide. In terms of the ecofriendly approaches in recent past, the use of biopesticides has made some headway, which includes the use of microbial and botanical pesticides. Amongst microbials, use of Bt formulations make a major component of these products. In fact, Cry3Bb1 protein from Bt has been reported as the major toxic protein against corn worms and has least risk to the environment. The Cry3Bb1 protein is expressed by the corn plant; thus, reducing the exposure to non-target organisms. The major pest controlled by Bt corn is the European corn borer. Other important insects controlled are corn earworm, southwestern corn borer, and other stalk boring insects. Bt corn use has exceeded 20% of field corn acreage. However, there are some constraints too. GM Corn seems to lose some edge because corn rootworms in the US may have developed resistance to a protective chemical made by a genetically modified corn. Corn rootworms are one of the most damaging agricultural pests in the US. They impair corn's ability to absorb water and nutrients from the soil. The economic impact of the insects, in terms of crop damage and the cost of pesticides, used to be around $1 billion a year, until the biotechnology firm Monsanto introduced its rootworm-killing variety of corn in 2003. However, it has recently been shown that western corn rootworms on two Illinois farms have developed resistance to Cry3Bb1, an insecticide produced by Monsanto's genetically modified YieldGard corn (*PLoS One*, doi.org/d5mqwn). Therefore, resistant rootworms could further damage corn crops. YieldGard was planted on more than 150,000 square kilometres of US farmland recently and apparently the resistant rootworms may spread, which cannot be summarily ignored.

Among botanicals, significant emphasis these days is on essential oil compounds. Some essential oils express bioactivity against stored product pests, such as oils of basil, citrus peel, eucalyptus, various mint species, lavender, and rosemary. Nutmeg oil has been determined to significantly impact maize weevil, *Sitophilus zeamais*. Neem has been used with variable results to manage corn ear worms and corn borers. NPV, neem and spinosad as supplemental controls for the fall armyworm *S. frugiperda*, and corn earworm *H. zea* have been shown to have potential. Such combinations do control pests, for example both whorl-stage and tassel-stage corn damage can be prevented by the use of either a nucleopolyhedrovirus or neem and at the same time spinosad consistently can reduce damage to ears caused by lepidoteran pests of maize.

REFERENCES

Adhikary, S. (1981) Togo experience in moving from neem research to its practical application for plant protection, In: H. Schmutterer, K.R.S. Ascher, and H. Rembold (eds.), *Natural Pesticides from the Neem Tree* (*Azadirachta indica* A. Juss), GTZ, Germany, pp. 215–222.

Agboka, K., Mawufe, A.K., Tamo, M. and Vidal, S. (2009) Effects of plant extracts and oil emulsions on the maize cob borer, *Mussidia nigrivenella* (Lepidoptera: Pyralidae) in laboratory and field experiments. *Int. J. Trop. Insect Sci.*, **29**, 185–194.

Aly Khan Shaukat, S.S., Bushra Nawab Khanzada, K.A. and Solangi, M.S. (2009) Management of nematodes associated with maize (*Zea mays* L.) using organic and inorganic amendments. *Pak. J. Agric. Res.*, **22**, 165–167.

Arnold, A.E. and Lewis, L.C. (2005) Ecology and evolution of fungal endophytes, and their roles against insects. In: F.E. Vega and M. Blackwell (eds.), *Insect-Fungal Associations: Ecology and Evolution*. Oxford University Press, New York, pp. 74–96.

Ashley, T.R. (1979) Classification and distribution of fall armyworm parasites. *Florida Entomol.*, **62**, 114–123.

Attri, B.S. and Prasad, G.R. (1980) Studies on the pesticide value of neem oil byproducts. *Pestology*, **4**, 16–20.

Burkness, E.C., Dively, G., Patton, T., Morey, A.C. and Hutchison, W.D. (2010) Novel Vip3A *Bacillus thuringiensis* (Bt) maize approaches high-dose efficacy against *Helicoverpa zea* (Lepidoptera: Noctuidae) under field conditions: Implications for resistance management. *GM Crops*, **1**, 337–343.

Candolfi, M. P., Brown, K., Grimm, C., Reber, B. and Schmidli, H.(2004) A faunistic approach to assess potential side-effects of genetically modified

Bt-corn on non-target arthropods under field conditions. *Biocontrol Sci. Technol.*, **14**, 129–170.

Chakradhar, D. and Ajay Narendra Nagaraju, K. (2010) Laboratory evaluation of grain protectant efficacy of different foliar extracts against maize weevil - *Sitophilus zeamais* (Motschulsky). *Arch. Phytopathol. Plant Prot.*, **43**, 1809–1818.

Cherry, A.J., Lomer, C.J., Djegui, D. and Shulthess, F. (1999) Pathogen incidence and their potential as microbial control agents in IPM of maize stem borers in West Africa. *BioControl*, **44**, 301–327.

Chu ShaSha, Liu ShaoLiang , Jiang GuoHua, Liu ZhiLong (2010) Composition and toxicity of essential oil of Illicium simonsii Maxim (Illiciaceae) fruit against the maize weevils. *Records Nat. Prod.*, **4**, 205–210.

Coitinho, R.L.B., de C. Oliveira, J.V., de Gondim Junior, M.G.C. and Camara, C.A.G. da (2006) Residual effect of natural insecticides in the control of *Sitophilus zeamais* Mots. on stored corn. *Caatinga*, **19**, 183–191.

Cruz, I., Figueiredo, M.L.C., Valicente, F.H. and Oliveira, A.C. (1997) Application rate trials with a nuclear polyhedrosis virus to control *Spodoptera frugiperda* (smith) on maize. *An. Soc. Entomol. Brasil*, **26**, 145–152.

Deshpande, M.V. (1999) Mycopesticide production by fermentation: Potential and challenges. *Crit. Rev. Microbiol.*, **25**, 229–243.

Dhanya, K.M. (2009) Potential of biopesticides for the management of *Chiloloba acuta* (Wiedemann) (Coleoptera: Scarabaeidae). MSc. Thesis, Indian Agricultural Institute, New Delhi, India.

Dively, G.P., Rose, R., Sears, M.K., Hellmich, R.L., Stanley-Horn, D.E., Calvin, D.D., Russo, J.M. and Anderson, P.L. (2004) Effects on monarch butterfly larvae (Lepidoptera : Danaidae) after continuous exposure to Cry1Ab-expressing corn during anthesis. *Environ. Entomol.*, **33**, 1116–1125.

El-Hosary, R.A. (2011) Evaluation of some essential oils against *Sesamia cretica* Led. under field conditions. *J. Am. Sci.*, **7**, 563–568.

FAO (1999) *The Use of Spices and Medicinals as Bioactive Protectants for grains.* Food and Agriculture Organization, Agricultural Services Bulletin No. 137, Rome.

Farrar, R.R. Jr., Shepard, B.M., Shapiro, M., Hassell, R.L., Schaffer, M.L. and Smith, C.M. (2009) Supplemental control of lepidopterous pests on Bt transgenic sweet corn with biologically-based spray treatments. *J. Insect Sci.*, **9**, 8.

Federici, B.A. (2002) Case study: Bt crops-A novel mode of insect control. In: K.T. Atherton (ed.), *Genetically Modified Crops: Assessing Safety*, Taylor & Francis, UK, pp. 164–200.

Gangwar, S.K., Srivastava, D.C., Tewari, R.K., Singh, M.R. and Rajak, D.C. (2008) Management of *Pyrilla perpusilla* Walker in sugarcane with ectoparasitoid *Epiricania melanoleuca* Fletcher during epidemics in sub-tropical India. *Sugar Technol.*, **10**, 162–165.

Gebre-Amlak (1999) Insecticidal activity of chinaberry, endod and pepper tree against the maize stalk borer (Lepidoptera: Noctuidae) in Southern Ethiopia. *Int. J. Pest Manage.*, **45**, 9–13.

Girma, D., Tadele, T. and Abraham, T. (2008) Efficacy of silicosec, filter cake and wood ash against the maize weevil, *Sitophilus zeamais* Motschulsky (Coleoptera: Curculionidae) on three maize genotypes *J. Stored Prod. Res.*, **44**, 227–232.

Gutierrez-Garcia, S. del C., Sanchez-Escudero, J., Perez-Dominguez, J.F., Carballo-Carballo, A., Bergvinson, D. and Aguilera-Pena, M.M. (2010) Effect of neem on damage caused by fall armyworm *Spodoptera frugiperda* (Smith) (Lepidoptera: Noctuidae) and three agricultural variables on resistant and susceptible maize. *Acta Zool. Mex.*, **26**, 1–6.

Hellmich, R.L., Siegfried, B.D., Sears, M.K., Stanley-Horn, D.E., Daniels, M.J., Mattila, H.R., Spencer, T., Bidne, K.G. and Lewis, L.C. (2001) Monarch larvae sensitivity to *Bacillus thuringiensis* purified proteins and pollen. *Proc. Nat. Acad. Sci. USA*, **98**, 11925–11936.

Hiltpold, I., Toepfer, S., Kuhlmann, U. and Turlings, T.C.J. (2009) How maize root volatiles influence the efficacy of entomopathogenic nematodes against the western corn rootworm. *Chemoecology*, **20**, 155–162.

Hiltpold, I., Baroni, M., Toepfer, S., Kuhlmann, U. and Turlings, T.C.J. (2010) Selection of entomopathogenic nematodes for enhanced responsiveness to a volatile root signal helps to control a major root pest. *J. Expt. Biol.*, **213**, 2417–2423.

Hutchison, W.D., Burkness, E.C., Mitchell, P.D., Moon, R.D., Leslie, T.W., Fleischer, S.J., Abrahamson, M. and Hamilton, K.L. (2010) Areawide suppression of European corn borer with Bt maize reaps savings to non-Bt maize growers. *Science*, **330**, 222–225.

Jalali S.K. and Singh S.P. (2003) Bio-ecology of *Chilo partellus* (Swinhoe) (Lepidoptera : Pyralidae) and evaluation of its natural enemies – A review. *Agric. Rev.*, 24, 79–100.

Jalali, S.K. and Singh, S.P. (2004) Effect of various neem products on survival and feeding capacity of *Chilo partellus* (Swinhoe) on maize . *J. Entomol. Res.*, **28**, 329–336

James, C. (2003) *Global Review of Commercialized Transgenic Crops: 2002 Feature: Bt Maize.* ISAAA Briefs No. 29.

Jesse, L.C.H. and Obrycki, J.J. (2000) Field deposition of Bt transgenic corn pollen: Lethal effects on the monarch butterfly. *Oecologia*, **125**, 241–248.

Jide-Ojoa, C.C. and Ojo, O.O. (2011) Evaluation of the biological effects of leaf extracts of *Jatropha curcas* against *Sitophilus zeamais* (Coleoptera: Curculionidae). *eJ. Environ. Agric. Food Chem.*, **10**, 2166–2172.

Köllner, T., Held, M., Lenk, C., Hiltpold, I., Turlings, T.C.J., Gershenzon, J. and Degenhardt, J. (2008) A maize (E)-β-caryophyllene synthase implicated in indirect defense responses against herbivores is not expressed in most American maize varieties. *Plant Cell*, **20**, 482–494.

Koul O. (1992): Neem allelochemicals and insect control. In: S.J.H. Rizvi and B. Rizvi (eds.), *Allelopathy, Basic and Applied Aspects*, Chapman and Hall, London, 389–412.

Krawczyk, A., Hurej, M. and Jackowski, J. (2011) Syrphids and their parasitoids from maize crop. *J. Plant Prot. Res.*, **51**, 93–97.

Krueger, S.R. and Roberts, D.W. (1997) Soil treatment with entomopathogenic fungi for corn root worm (*Diabrotica* spp.) larval control. *Biol. Control,* **9**, 67–74.

Kuhlmann, U. and van der Brugt, W.A.C.M. (1998) Possibilities for biological control of the western corn rootworm, *Diabrotica virgifera virgifera* LeConte, in Central Europe. *Biocontrol*, **19**, 59–68.

Kumar, P., Sekhar, J.C. and Chaudhary, R. (2005) Management of insect pests of maize in tropics. In: P.H. Zaidi and N.N. Singh (eds.), *Stresses on Maize in Tropics*, Directorate of Maize Research, New Delhi, India, pp. 1–12.

Kurzt, B., Hiltpold, I., Turlings, T.C.J., Kuhlmann, U. and Toepfer, S. (2009) Comparative susceptibility of larval instars and pupae of the western corn rootworm to infection by three entomopathogenic nematodes. *Biocontrol*, **54**, 255–262.

Lewis, L.C. and Cossentine, J.E. (1986) Season long intraplant epizootics of entomopathogens, *Beauveria bassiana* and *Nosema pyrausta*, in a corn agroecosystem. *Entomophaga*, **31**, 36–69.

Lewis, L.C. and Bing, L.A. (1991) *Bacillus thuringiensis* Berliner and *Beauveria bassiana* (Balsamo) Vuillemin for European corn borer control: Program for immediate and season long suppression. *Can. Entomol.*, **123**, 387–393.

Liu Zhi Long, Chu ShaSha and Jiang GuoHua (2011) Insecticidal activity and composition of essential oil of *Ostericum sieboldii* (Apiaceae) against *Sitophilus zeamais* and *Tribolium castaneum*. *Records Nat. Prod.*, **5**, 74–81.

Losey, J.E., Rayor, L.S. and Carter, M.E. (1999) Transgenic pollen harms monarch larvae. *Nature*, **399**, 214-215.

MartAnez, A.G., Tolon-Becerra, A. and Lastra-Bravo, X.B. (2012) Biological control of *Spodoptera frugiperda* eggs using *Telenomus remus* nixon in maize-bean-squash polyculture. *Am. J. Agric. Biol. Sci.*, **7**, 285-292.

Matama-Kauma T., Kyamanywa S., Ogwang J.A., Omwega C.O., Wilson, H.R. (2001) Cereal stemborer species complex and establishment of *Cotesia flavipes* Cameron in eastern Uganda. *Insect Sci. Appl.* 21, 317–326.

Messeguer, J. (2003) Gene flow assessment in transgenic plants. *Plant Cell Tiss. Organ Cult.*, **73**, 201-212.

Mihale, M.J. Deng, A.L. Selemani, H.O. Mugisha-Kamatenesi, M. Kidukuli, A.W. Ogendo, J.O. (2009) Use of indigenous knowledge in the management of field and storage pests around Lake Victoria basin in Tanzania. *Afr. J. Environ. Sci. Technol.*, **3**, 251–259.

Miller, N., Estoup, A., Toepfer, S., Bourguet, D., Lapchin, L., Derridj, S., Kim, K.S., Reynaud, P., Furlan, L. and Guillemaud, T. (2005) Multiple transatlantic introductions of the western corn rootworm. *Science*, **310**, 992.

Molina-Ochoa, J., Carpenter, J.E., Heinrichs, E.A. and Foster, J.E. (2003) Parasitoids and parasites of *Spodoptera frugiperda* (Lepidoptera: Noctuidae) in the Americas and Caribbean basin: An inventory. *Florida Entomol.*, **86**, 254–289.

Moscardi, F. (1999) Assessment of the application of baculoviruses for control of Lepidoptera. *Annu. Rev. Entomol.*, **44**, 257–289

Moyin-Jesu, E.I. (2010) Comparative evaluation of modified neem leaf, wood ash and neem leaf extracts for seed treatment and pest control in maize (*Zea mays* L.). *Emirates J. Food Agric.*, **22**, 37–45.

Nguemtchouin MMG, Ngassoum MB, Ngamo LST, Gaudu X, Cretin M (2010) Insecticidal formulation based on Xylopia aethiopica essential oil and kaolinite clay for maize protection. *Crop Prot.*, **29**, 985–991.

O'Callaghan, M., Glare, T.R., Burgess, E.P.J. and Malone, L.A. (2005) Effects of plants genetically modified for insect resistance on nontarget organisms. *Annu. Rev. Entomol.*, **50**, 271–292

Osipitan, A.A., Odebiyi, J.A., Lawal, O.A. and Somade, A.A. (2010) Evaluation of the extracts of some tropical plants in the management of *Prostephanus truncatus* (Horn) infestation in maize (*Zea mays* L.). *Arch. Phytopathol. Plant Prot.*, **43**, 1001–1010.

Pilcher, C.D., Rice, M.E. and Obrycki, J.J. (2005) Impact of transgenic *Bacillus thuringiensis* corn and crop phenology on five nontarget arthropods. *Environ. Entomol.*, **34**, 1302–1316.

Potting, R.P.J., Vet, L.E.M. and Overholt, W.A. (1997) Geographic variation in host selection behaviour and reproductive success in the stemborer parasitoid *Cotesia flavipes* (Hymenoptera: Braconidae). *Bull. Entomol. Res.*, **87**, 515–524.

Rasmann, S. and Turlings, T.C.J. (2007) Simultaneous feeding by aboveground and below ground herbivores attenuates plant-mediated attraction of their respective natural enemies. *Ecol. Lett.*, **10**, 926–936.

Rasmann, S. and Turlings, T.C.J. (2008) First insights into specificity of below ground tritrophic interactions. *Oikos*, **117**, 362–369.

Rasmann, S., Köllner, T.G., Degenhardt, J., Hiltpold, I., Toepfer, S., Kuhlmann, U., Gershenzon, J. and Turlings, T.C.J. (2005) Recruitment of entomopathogenic nematodes by insect-damaged maize roots. *Nature*, **434**, 732–737.

Roel, A.R., Dourado, D.M., Matias, R., Porto, K.R.A., Bednaski, A.V. and Costa, R.B. da (2010) The effect of sub-lethal doses of *Azadirachta indica* (Meliaceae) oil on the midgut of *Spodoptera frugiperda* (Lepidoptera, Noctuidae). *Rev. Brasil. Entomol.*, **54**, 505–510.

Romeis, J., Bartsch, D., Bigler, F., Candolfi, M.P., Gielkens, M.M.C., Hartley, S.E., Hellmich, R.L., Huesing, J.E., Jepson, P.C., Layton, R., Quemada, H., Raybould, A., Rose, R.I., Schiemann, J., Sears, M.K., Shelton, A.M., Sweet, J., Vaituzis, Z. and Wolt, J.D. (2008) Assessment of risk of insect-resistant transgenic crops to nontarget arthropods. *Nat. Biotechnol.*, **26**, 203–208.

Safavi, S.A., Kharrazi, A., Rasoulian, G.R. and Bandani, A.R. (2010) virulence of some isolates of entomopathogenic fungus, *Beauveria bassiana* on *Ostrinia nubilalis* (Lepidoptera: Pyralidae) larvae. *J. Agric. Sci. Technol.*, **12**, 13–21.

Sarup, P., Panwar, V.P.S., Marwaha, K.K. and Siddiqui, K.H. (1978) Management of maize pests with particular reference to the stalk borer, *Chilo partellus* (Swinhoe) under resource constraints. *J. Entomol. Res.*, **2**, 5–14.

Schnepf, E., Crickmore, N., Van Rie, J., Lereclus, D., Baum, J. and Feitelson, J. (1998) *Bacillus thuringiensis* and its pesticidal crystal proteins. *Microbiol. Mol. Biol. Rev.*, **62**, 775–806.

Sears, M.K., Hellmich, R.L., Stanley-Horn, D.E., Oberhauser, K.S., Pleasants, J.M., Mattila, H.R., Siegfried, B.D. and Dively, G.P. (2001) Impact of Bt corn pollen on monarch butterfly populations: a risk assessment. *Proc. Natl. Acad. Sci. USA*, **98**, 11937–11942.

Siegel, J. P., Maddox, J.V. and Ruesink, W.G. (1987) Survivorship of the European corn borer, *Ostrinia nubilalis* (Hübner) (Lepidoptera: Pyralidae) in central Illinois. *Environ. Entomol.*, **16**, 1071–1075.

Sharma, T. (2009) Toxic effect of *Ocimum sanctum* plant extract against *Acrida exaltata* (Orthoptera: Acrididae) adults. *J. Environ. Res. Develop.*, **4**, 1008–1012.

Shehu, A., Obeng-Ofori, D. and Eziah, V.Y. (2010) Biological efficacy of Calneem™ oil against the tropical warehouse moth *Ephestia cautella* (Lepidoptera: Pyralidae) in stored maize. *Int. J. Trop. Insect Sci.*, **30**, 207–213.

Stanley-Horn, D.E., Dively, G.P., Hellmich, R.L., Mattila, H.R., Sears, M.K., Rose, R., Jesse, L. C.H., Losey, J.F., Obrycki, J.J. and Lewis, L. (2001) Assessing the impact of Cry1Ab-expressing corn pollen on monarch butterfly larvae in field studies. *Proc. Nat. Acad. Sci. USA*, **98**, 11931–11936.

Storer, N.P., Dively, G.P. and Herman, R.A. (2008) Landscape effects of insect-resistant genetically modified crops. In: J. Romeis, A.M. Shelton and G.G. Kennedy (eds.), *Integration of Insect-Resistant Genetically Modified Crops within IPM Programs*, Springer, New York, pp. 273–302.

Sun, X.L. and Peng, H. (2007) Recent advances in biological pest insects by using viruses in China. *Virol. Sin.*, **22**, 158–162.

Suthisut, D., Fields, P.G. and Chandrapatya, A. (2011) Fumigant toxicity of essential oils from three Thai plants (Zingiberaceae) and their major compounds against *Sitophilus zeamais*, *Tribolium castaneum* and two parasitoids. *J. Stored Prod. Res.*, **47**, 222–230.

Toepfer, S., Gueldenzoph, C., Ehlers, R. U. and Kuhlmann, U. (2005) Screening of entomo- pathogenic nematodes for virulence against the invasive western corn rootworm, *Diabrotica virgifera virgifera* (Coleoptera: Chrysomelidae) in Europe. *Bull. Entomol. Res.*, **95**, 473–482.

Todd, K.J., Goettel, M.S., Svircev, A.M. and Woo, S.G. (2010) *The Use and Regulation of Microbial Pesticides in Representative Jurisdiction Worldwide.* IOBC-Global.org.

Trjapitzin, V.A. (2008) A review of Encyrtid wasps (Hymenoptera, Chalcidoidea, Encyrtidae) of Macaronesia. *Entomol. Rev.*, **88**, 218–232.

Usha Rani, P. and Devanand, P. (2011) Efficiency of different plant foliar extracts on grain protection and seed germination in maize. *Res. J. Seed Sci.*, **4**, 1–14.

Valicente, F.H., Peixoto, M.J.V.V.D.E., Paiva and Kitajima, E. (1989) Identificação e purificação de um virus de poliedrose nuclear da lagarta do cartucho, *Spodoptera frugiperda. An. Soc. Entomol. Brasil.*, **18**, 71–82.

Viana, P. A. Ribeiro, P. E. de A. Green leaves extract of neem (*Azadirachta indica*) and application time on damage and development of *Spodoptera frugiperda* (J.E. Smith, 1797) (Lepidoptera : Noctuidae) in maize. *Rev. Brasil. Milho e Sorgo,* **9**, 27–37.

Vidal, S., Kuhlmann, U. and Edwards, R. (2005) *Western Corn Rootworm: Ecology and Management.* CABI Publishers, Wallingford, UK.

Vijay Kumar and Kanta, U. (2011) Effectiveness of Trichogramma Chilonis Ishii in the Suppression of Chilo Partellus (Swinhoe) in Summer Maize. *J. Biol. Control*, **25**, 92–97.

Wagner, B.L. and Lewis, L.C. (2000) Colonization of corn, *Zea mays*, by the entomopathogenic fungus *Beauveria bassiana. Appl. Environ. Microbiol.*, **66**, 3468–3473.

Yan, Z.G. and Wang. C.Z. (2006) Similar attractiveness of maize volatiles induced by *Helicoverpa armigera* and *Pseudaletia separata* to the generalist parasitoid *Campoletis chlorideae. Entomol. Exp. Appl.*, 118, 87–96.

Zhang, G.Y., Sun, X.L., Zhang, Z.X., Zhang, Z.F. and Wan, F.F. (1995) Production and effectiveness of the new formulation of *Helicoverpa* virus pesticide-emulsifiable suspension. *Virol. Sin.*, **10**, 242–247.

4

Current Status and Potential of Biopesticides in Pest Management in Pulses

Hem Saxena* and P. Duraimurugan

Division of Crop Protection, Indian Institute of Pulses Research, Kanpur - 208024, India
*E-mail: hem_saxena@yahoo.com

INTRODUCTION

With the globalization and liberalization of agriculture the biological management of the crop pests has attained immense importance. The biopesticides can be intensively used without any possibility of development of resistance. In view of their specificity and safety, there will be least environmental, ecological and health hazards (Pawar and Borikar, 2005). Hence, use of biopesticides is environmentally safe, economical and sound method of tackling pest problems. Bioagents are safe to mankind and animals, do not pollute the environment, do not kill beneficial parasitoids and predators and generally pests do not develop resistance to these products (Lingappa *et al.,* 2005). Due to these qualities, several ecofriendly approaches have been developed.

As the present chapter deals with the pulses, it is pertinent to mention that pulses in India have long been considered as the poor man's only source of protein. Pulses are grown on 22-23 million ha of area with an annual production of 13–15 million tonnes. India accounts for 33% of the world area and 22% of the world production of pulses. The major pulse crops grown in India are chickpea, pigeonpea, lentil, moong bean, urd bean and field pea. About 90% of the global pigeonpea, 65% of chickpea and 37% of lentil area falls in India, corresponding to 93, 68 and 32% of the global production, respectively (FAOSTAT, 2009). Due to stagnant production, the net availability of pulses has come down from 60 gm/day/person in 1951 to 31 gm/day/person (Indian Council of Medical Research recommends 65 gm/day/capita) in 2008 (Reddy, 2009).

Depending on the domestic shortfall in pulses production, India's net imports of pulses have ranged from 1–3 mt, while exports are one-tenth of the volume of imports. Imports of pulses increased from 0.58–3.1 mt between 1994-96 and 2007-09 and are projected to increase to 4 mt by 2012 (Table 4.1).

Table 4.1. Sources of supply of pulses in India

Year	Kharif			Rabi			Total			Imports (Million Tonnes)	Exports (Million Tonnes)	Total Consumption (Million Tonnes)
	Area (Million ha)	Prodn (Million Tonnes)	Yield (Kg/ha)	Area (Million ha)	Prodn (Million Tonnes)	Yield (Kg/ha)	Area (Million ha)	Prodn (Million Tonnes)	Yield (Kg/ha)			
2000-01	10.7	4.5	417	9.7	6.6	684	20.4	11.1	544	0.4	0.2	11.2
2001-02	10.7	4.8	453	10.9	8.5	762	21.7	13.4	609	2.2	0.2	15.4
2002-03	10.0	4.2	417	10.6	7.0	661	20.5	11.1	543	2.0	0.2	13.0
2003-04	11.7	6.2	528	11.8	8.7	745	23.4	14.9	637	1.7	0.2	16.5
2004-05	11.3	4.7	417	11.4	8.4	735	22.8	13.1	577	1.3	0.3	14.2
2005-06	10.6	4.7	439	11.8	8.5	716	22.4	13.1	585	1.6	0.4	14.3
2006-07	10.7	4.8	449	12.5	9.4	751	23.2	14.2	612	3.7	0.4	17.5
2007-08	11.5	6.4	557	12.1	8.4	709	23.6	14.8	688	2.8	0.2	17.4
2008-09	10.4	5.0	484	12.6	9.2	726	23.0	14.2	617	2.3	0.1	16.4

Source: Department of Agriculture and Cooperation (2009)

Pulses undergo umpteen abiotic and biotic stresses in India (Table 4.2). More than 250 insect species are reported to affect pulses in India, which inflict heavy crop losses. On an average 2-2.4 million tonnes of pulses with a monetary value of nearly ₹ 6000 crore are lost annually due to ravages of insect pest complex in India. Among them, pod borer (*Helicovera armigera)* causes the most harm, followed by pod fly, wilt and root rot (Reddy, 2009). The main control measure has been the use of conventional chemical pesticides that have hazardous impact on environment and ecology. In recent past, a variety of chemical, biological and cultural methods have been found to reduce pest and disease damage, but emphasis has been to use the ecofriendly approaches like biopesticides.

Table 4.2. Important biotic and abiotic stresses identified in major pulse crops in India

Crop	Season	Stress	
		Biotic	**Abiotic**
Chickpea	Timely sown	FW, root rot, chickpea stunt, BGM, pod borer	Low termperature
	Early sown	FW, root rot, AB, or chickpea stunt, pod borer	Terminal drought, salt stress
	Late sown	FW, pod borer	Terminal drought, cold
Pigeonpea	Kharif-early	FW, PB, pod-borer complex	Waterlogging
	Medium late	FW, SM, pod-borer complex	Cold, terminal drought, waterlogging
	Pre-rabi	FW, ALB, Pod fly	Cold, terminal drought
Moong	Kharif	MYMY, CLS, WB, sucking insect pests	Pre-harvest sprouting, terminal drought
	Zaid	MYMV root and stem rot, stem agromyza, sucking insect pests	Pre-harvest sprouting, temperature stress, drought
	Rabi	PM, Rust, CLS	Terminal drought
Urad	Kharif	MYMV, anthracnose, WB, LCV	Terminal drought
	Zaid	MYMV, root and stem rot, stem agromyza	Pre-harvest sprouting, temperature stress, drought
	Rabi/rice fallow	Spot	Terminal drought
Lentil		FW, root rot, rust	Moisture temperature

FW = Fusarium wilt, PB = Phytophthora blight, SM= Sterility mosaic, ALB = Alternaria leaf blight, MYMV - Moongbean yellow mosaic virus, BGM = Botrytis gray mould, AB - Ascochyta blight.

Source: Reddy (2006)

Under the umbrella of biopesticides today parasitoids, predators and microbial organisms are the main agents used. Microbials include viruses, bacteria, fungi, protozoa and nematodes, which produce variable toxins. They are formulated to be applied as conventional insecticidal sprays, dusts, liquid drenches, liquid concentrates, wettable powders, or granules (Weinzierl and Henn, 2005). Biopesticides are inherently less harmful than conventional pesticides and are designed to affect only one specific pest or, in some cases, a few target organisms. They are often effective in very small quantities and often decompose quickly, thereby resulting in lower exposures and largely avoiding the pollution problems caused by conventional pesticides. Their advantage over many chemical pesticides is that they are not toxic to birds, mammals and fish (Mishra and Mishra, 2004). Use of these pathogens may vary considerably between crops and locations depending upon climate, symptomatology and economic threshold of crop damage. When used as a component of Integrated Pest Management (IPM) programs, biopesticides can greatly decrease the use of conventional pesticides, while crop yields remain high (Chandel *et al.*, 2001; Rabindra *et al.*, 2005; Saxena *et al.*, 2010).

PARASITOIDS AND PREDATORS

A large number of natural enemies (Table 4.3) do occur in pulses which live on insect pests of pulses (Saxena, 2009). In case of chickpea larval parasitism is common. About half a dozen of larval parasitoids are known to occur on *H. armigera* on chickpea in India. *Campoletis chlorideae* Uchida is one of the most important and potential larval parasite, which parasitizes 20–76% larvae of *H. armigera* in the field. It prefers late 1st or early 2nd instar host larvae for parasitization. Parasitoids pupate outside as a cocoon within a week and it lasts for about 5–7 days. Studies made on its biology revealed that the mean larval period is 10.1 days (range 9–13 days) and pupal period is 4.7 days (4–5days). Female lived longer, 4.3 days than male (3.8 days). Female laid 29 eggs during its life span. Sex ratio was found in favour of male (1:0.4). Maximum parasitized host larvae died (50%) between 6 and 8 days of parasitization leading to low recovery of parasitoids. Under field condition, the parasitoid population attains two definite peaks, first during early growth stage of chickpea (pre-winter) that is 46–50 standard week and second coinciding with reproductive phase of the crop (post-winter) that is 8–15 standard week. The maximum activity of parasitoids was noted during 49 standard week with 60% parasitization and second peak during 10 standard week with 35% parasitization. The low level of parasitization during reproductive phase of the crop does not keep pace with the increasing pest population, which needs manipulation of parasitoid population through inundative/inoculative release.

Table 4.3. Natural enemies recorded on the insect pests of pulses

Crop	Pest	Natural enemy	Stage
Pigeonpea	*Helicoverpa armigera*	*Aspergillus niger*	Larval
		A. flavus	Larval
		Copidosoma floridanum	Larval
	Melanagromyza obtusa	*Euderus agromyzae*	Larval/pupal
		Erytoma spp.	Larval/pupal
		Euderus spp.	Larval/pupal
		Ormyrus orientalis	Larval/pupal
	Eucosma critica	*Apanteles taragamae*	Larval/pupal
	Clavigralla gibbosa	*Gryon* spp.	Egg
	Nezara viridula	*Gryon* spp.	Egg
	Maruca vitrata	*Bracon (Habrobracon) hebetor*	Larval
Chickpea	*Helicoverpa armigera*	*Campoletis chlorideae*	Larval
		Apanteles ruficrus	Larval
		Bracon (Habrobracon) hebetor	Larval
			Larval
	Autographa nigrisigna	*Carcelia illota*	Larval
		Nuclear Polyhedrosis virus	Larval
		Unidentified Tachinid Parasite	
Mungbean & Urdbean	*Diacrisia obliqua*	*Meteorus* spp.	Larval
	Nezara viridula	*Gryon* spp.	Egg
	Clavigralla gibbosa	*Gryon* spp.	Egg
	Spodoptera litura	*Unidentified Hymenopterous Parasite*	Larval
Cowpea	*Aphis craccivora*	Syrphid	Adult & Nymph
		Coccinellid	Adult & Nymph

(*Source*: Saxena, 2009)

In North India the pupa of gram pod borer, *H. armigera* undergoes facultative diapause in severe winter, resulting into a prolonged pupal period to the extent of more that 100 days. This result into poor availability of host for potent larval parasitoid, *Campoletis chlorideae* during December and January, resulting into significant decrease in parasitoid population in the absence of sufficient food. During post-winter months *H. armigera* moths emerge out from hibernating pupa and breed freely till the parasite is in position to put any significant check on its multiplication. It is the population of *H.*

armigera, which causes, heavy crop losses. During 2004, heavy parasitization of *Copidosoma floridanum*, a larval parasitoid of *H. armigera* was reported for the first time at IIPR, Kanpur (Saxena, 2007). Recently, the larval ectoparasitoid, *Bracon hebetor* was found on *H. armigera* larvae in chickpea during post-winter months of March-April coinciding with the pod formation stage of chickpea (Saxena and Duraimurugan, 2009). This parasitoid is seen attacking spotted pod borer, *Maruca vitrata* in pigeonpea as well (Mohapatra *et al.*, 2008). It has also been observed that intercropping of chickpea with nectar rich plants like corriander enhances the parasitoid population.

In pigeonpea up to 22% larvae were parasitized by Diptera and less than 5% by Hymenoptera. The larval-pupal parasitoids, *Goniophthalmus halli* Mesnil and *Carcelia illota* (Curran) are the main parasitoids of *H. armigera* in pigeonpea (Saxena and Duraimurugan, 2007). *Apanteles taragamae*, a larval parasitoid of pigeonpea leaf tier, *Eucosma critica* was found active from last week of October to December when 18.4–23.8% larvae were parasitised. Highest parasitisation level of 29.8% was recorded during 2nd week of November.

Parasitoid complex of podfly of pigeonpea consists of three larval/ pupal parasitoids, *Euderus* sp., *Erytoma* sp. and *Ormysus orientates*. The parasitoids occur from end of December that gradually increase to a maximum of 34% during first week of March. The pupae collected at the time of maturity of crop exhibited 16% parasitisation by three parasitoids; *Euderus agromyzae, Ormyrus orientatis* and *Eurytoma* sp. In late stages of pigeonpea crop, severe super parasitism of late larval instars (18.18–50.0%) by unidentified dipteran was observed in our studies, which was responsible for the mortality of the pest during middle of March. Hundreds of parasitoid maggots (milky white) were found just beneath individual larval skin, laden in hexagonal cavity. In general, dipteran parasitoids have been noted dominant on *Helicoverpa* in pigeonpea ecosystem.

More than 60 species of arthropods are listed as predators but chrysopids and coccinellids are considered to be major predators occurring in fields. A carabid predator of *Maruca vitrata* was recorded in mung bean under field condition (Duraimurugan and Srinivasan, 2009). A number of bird species such as cattle egret, *Bulbulus ibis* (Linn), rosy pastor, *Sturnus roseus* (Linn.), common myna, *Acridotheres tristis* (Linn.), racket tailed drongo, *Dicrurus paradiseus* (Linn.), bank myna, *A. giginianus* (Latham) and river stern, *Sterna aurantia* Gray have also been observed feeding on *H. armigera* larvae and, therefore, erection of perches in chickpea fields have been suggested to encourage activity of these insectivorous birds. Clever and Ambrose (2002) studied the functional response of the predator *Rhynocoris fuscipes* a reduviid bug against *H. armigera* in laboratory. They found that the prey density was negatively correlated with the handling time of the predator. Ants are major predator of pre-pupae and pupae of *H. armigera.* Physical factors impeding adult emergence may also account for 10–20% pupal mortality.

A survey was conducted in chickpea, pigeonpea and lentil crops in 22 districts of North India to record the occurance of gram pod borer, *Helicoverpa armigera*, Bihar hairy caterpillar, *Spilarctia obliqua*, plant hopper of *Pentatomorpha* group, termite *Odontotermes* spp, and cut worm, *Agrotis ipsilon* in chickpea; *Campoletis chlorlidae* was recorded as natural enemy feeding on *H. armigera* larvae. In pigeonpea crop, mainly *H. armigera*, leaf webber, *Grapholita* (*Cydia*) *critica*, *Myllocerus* spp, spotted pod borer, *Maruca testulalis,* plume moth, *Exelastis atomosa*, tur pod bug, *Clavigralla gibbosa*, jassid *Amrsca bigutulta*, termite *Odontotermes* spp, green bug *Nezara viridula*, tur podfly *Melanogromyza*, blue butterfly *Lampides boeticus*, cow bug *Tricentrus bicolor* and *Pentatomorpha* bug were recorded while on lentil crop, black aphid, was recorded as major pest and cut worm as minor pests. Twenty insect pests and a total number of 16 parasitoids and predators were observed in these crops during the period of study. The bio-agents recorded belonged to Order Dictyoptera, Neuroptera, Hemiptera, Hymenoptera, Diptera and Coleoptera (Agrawal *et al.*, 2010).

MICROBIAL BIOPESTICIDES

Many microbial pest control products for control of arthropod pests have been studied and developed, which offer one of the sustainable and ecologically acceptable means of crop protection for modern agriculture. Annual sales of microbial pesticides are reported to be $750 million globally, amounting to about 2.5% of the chemical market (Evans, 2008). Four groups of microbial pathogens are typically found in insects; viruses, bacteria, fungi and protozoans.

Viruses

Insect pathogenic viruses belong to the family Baculoviridae, they are occluded in many sided inclusion bodies to afford environmental stability to virions and are used as biocontrol agents against insect pests. Of the different types of viruses, the nuclear polyhedrosis virus (NPV) has the greatest potential since they are more virulent, killing the insects much faster than granulosis virus (GV) and cytoplasmic virus (CPV). The NPV infection in *Helicoverpa armigera* was first reported from Anand, Gujarat in India during 1968. In fact, it was the first virus that has been registered for use on agricultural crops. The virus kills the insect when these are ingested. It multiplies rapidly and spread in nucleus. Diseased larvae move to the periphery of the plant and die showing typical symptom by hanging from the plant upside down in the form of inverted 'V' usually referred as "Top Disease". The NPV of *H. armigera* is effectively controlling these pod borers in chickpea and pigeonpea. In these two crops, HaNPV @ 250 LE and 350 LE/ha has been determined as an optimum dose for field application in respective crop, and found at par with HaNPV @ 500 LE/ha and endosulfan 0.07% in reducing pod borer damage and increase in yield (Pal *et al.*, 1998; Saxena and Ahmad, 2005).

Table 4.4. Interactive effect of different adjuvents on 50% Larval mortality of *H. armigera* by *Ha*NPV after indicated days of treatment

Adjuvant	Jaggary (0.5%)	Sugarcane juice (0.5%)	Cotton seed (0.1%)	Saccharin (0.01%)
Tannic acid 0.01%	6	8	7	7
Charcoal 0.05%	5	14	8	7
Na_2CO_3 0.01%	7	6	8	6
Boric acid 0.01%	7	6	7	6
*Ha*NPV alone	8	8	8	10

(Source: Saxena *et al.*, 2010)

Different adjuvants like pH or gut composition changer (tannic acid, charcoal, sodium carbonate and boric acid) and phagostimulants (jaggary, sugarcane juice, cotton seed kernel extract, and saccharin) have been studied in increasing the efficacy of HaNPV. Addition of jaggary (0.5%) and charcoal with HaNPV was the best, killing 50% larvae 5 days after treatment (DAT) as against 8–10 days with HaNPV alone. It was followed by addition of sugar cane juice with sodium carbonate/boric acid or jaggary with tannic acid causing 50% mortality 6 DAT. Thus jaggary/sugarcane juices as phagostimulant and charcoal/tannic acid/boric acid/sodium carbonate as gut composition changer have proved efficient adjuvants in enhancing the efficacy of HaNPV (Table 4.4). Addition of gut composition changer tannic acid

Table 4.5. Effect of adjuvants on efficacy of HaNPV against *H. armigera* in chickpea

Treatment	Pod damage (%)	Yield (kg/ha)
NPV 250LE + Tannic acid 0.01%	6.5	850
NPV 500LE + Tannic acid 0.01%	5.4	927
NPV 250LE + Charcoal 0.05%	5.6	1002
NPV 500LE + Charcoal 0.05%	4.5	1013
NPV 250LE + Boric acid 0.01%	6.8	1023
NPV 500LE + Boric acid 0.01%	6.2	1056
Control	13.4	683

(*Source*: Saxena *et al.,* 2010)

(0.01%), boric acid (0.01%) or charcoal (0.05%) with HaNPV enhanced the efficacy of virus in controlling pod borer, manifesting 4.5–6.8% pod damage as against 13.4% in control (Saxena *et al.,* 2010) (Table 4.5).

Table 4.6. Field Demonstration on application of HaNPV in chickpea

Treatment	Pod damage (%)	Avoidable damage (%)	Yield (kg/ha)	Yield increase over control (%)	CBR
1999-2000					
*Ha*NPV	3.2	10.6	2028	41.7	1:16.73
Control	13.8	-	1431	-	
2000-01					
*Ha*NPV	12.4	8.3	1403	57.5	1:3.34
Control	10.7	-	891	-	
2001-02					
*Ha*NPV	5.7	8.8	1494	82.6	1:3.40
Control	14.5	-	818	-	
2002-03					
*Ha*NPV	5.7	5.7	1862	25.1	1:2.24
Control	11.4	-	1488	-	

(*Source* : Saxena *et al.*; 2010)

Addition of surfactant (teepol 0.05%) and UV retardant (tinopal 0.01%) with phagostimulant (jaggary 0.5%) further increase efficiency and stability of NPV. Three weekly sprays of HaNPV @ 250 LE/ha with phagostimulant (jaggary 0.5%), surfactant (teepol 0.05%) and UV retardant (tinopal 0.01%) are equivalent to two sprays of endosulfan (0.07%) in their efficacy. This followed by HaNPV plus half dose of endosulfan or sequential application of HaNPV followed by endosulfan and HaNPV reduces pod damage and increases yield in chickpea. The NPV stored at 4°C remains effective for 18 months without losing its virulence as against only 4–6 months at room temperature. Thus, the efficacy of HaNPV could be increased by adding phagostimulants and UV protectants. Addition of boric acid (0.25–0.5%) does increase the efficacy of NPV (Saxena *et al.*, 2010).

Demonstration trials on HaNPV against *H. armigera* in chickpea conducted in Kanpur, Kanpur Dehat, Unnao and Hamirpur areas has revealed that application of *Ha*NPV prevented pod damage by more than 10% and increased the yield between 25–82% over control. The CBR ranged from 1:2.24–1:16.73 (Table 4.6).

Table 4.7. On farm evaluation of pest control modules

Particulars in IPM practices	2001-02	2002-03	2003-04	Mean
Pod damage				
BIPM	5.85	50.31	7.3	6.3
CIPM	5.2	46.93	6.8	6.1
FP	14.6	-	12.3	12.8
Yield				
BIPM	1561	1862	1729	1717
CIPM	1494	1821	1334	1550
FP	832	1488	1163	1161
% Increase in yield FP				
BIPM	87.62	25.11	48.66	53.8
CIPM	79.56	22.4	14.70	38.9
CBR				
BIPM	1:4.0	1:2.24	1:4.02	1:3.42
CIPM	1:4.5	1:2.32	1:4.49	1:3.77
Avoidable damage				
BIPM	60.3	50.3	40.6	50.4
CIPM	64.4	44.7	44.7	51.2

(*Source*: Ahmad and Rai, 2005).

Field demonstrations of a biointensive integrated pest management (BIPM) strategy using seed treatment with *T. harzianum* (Kanpur isolate (4 g) + Vitavax (1 g) with 1 kg seed; instillation of bird perches @ 40/ha; HaNPV 250 LE with adjuvant/ha as first spray; NSKE 5% 14 days later as second spray and chemointensive strategy (CIPM) using seed treatment with thiram (2g) + carbendazim 1 g/kg of seed and endosulfan 0.07% two spray modules were very effective in chickpea. Analysis on pod damage due to *H. armigera* revealed that lower damage was observed in BIPM strategy as compared to farmers practice (FP) and it was more or less similar to CIPM approach. Control operation resulted in avoiding pod damage by 50% or more. Average yield in BIPM was slightly higher (1717 kg/ha) than CIPM (1550 kg/ha) giving 53.8 and 38.9% yield increase over farmers practice (1161 kg/ha). The cost benefit ratio was almost similar in both the demonstrated modules (1:3.42 and 1:3.77, respectively) (Table 4.7). (Ahmad and Rai, 2005).

Bacteria

Most bacteria belong to spore forming species that provide potential biocontrol and belong to the genus *Bacillus*. *Bacillus thuringiensis* (Bt) is one of the earliest bacterial biopesticide produced commercially. The insecticidal activities of crystalliferous spore forming bacterium, *Bacillus thuringiensis* is primarily caused by parasporal crystal (delta endotoxin) produced during sporulation and has been found to be effective against several species of lepidopteran pests. *Bt* is the only bacterium that has been developed successfully as a commercial insecticide against lepidopterans including *Helicoverpa armigera*. In India, this biopesticide has been used in the management of gram pod borer *H. armigera* in chickpea and pigeonpea (Saxena and Ahmad, 1995; 1998). Laboratory evaluation of Delfin, a commercial *Bt* formulation at three different dosages (1, 1.5 and 2 kg/ha) against all the instars of *H. armigera* has shown that early instars (up to 3rd stadium) are highly susceptible to the treatment, while the late instars (3rd to 5th) are least affected even at higher dosage. Among seven different commercial preparations of Bt tested each in two dosages (1 and 1.5 kg/ha) against 3rd instars, BTK1 and BTK 2 @ 1 kg/ha kills 80% larvae in comparison to Dipel and Delfin (66.7%) and BTT and endosulfan (60%). A field study of Bt (0.5 kg/ha) against first instars of *Diacrisia obliqua* has also revealed maximum efficacy against 1st instars (100%) within 48 h rather than against 4-day-old larvae (Saxena *et al.*, 1992), which is down to 40–60% against these older larvae. Similarly, the toxicity of Bt shows similar trend against *Autographa nigrisigna* where 1.0 kg/ha application brought about 100% mortality of 1st and 2nd instar and 85% in 3rd instars (Saxena and Ahmad, 1997, 2000).

Some field studies also show that combination of Bt formulations with *Ha*NPV and endosulfan against *H. armigera* was more effective in chickpea. The higher yields varying from 1927–1863 kg/ha was also recorded. Commercial Bt formulations evaluated against *H. armigera* infesting chickpea, BTK 1 and BTK 2 @ 1.5 kg/ha have been found highly effective with lower pod damage (9.87–11.32%) comparable to endosulfan (13.65%). Next in order of effectiveness were Delfin and Dipel with 17.56 and 18.67%, respectively. The highest yield was recorded with BTK 1 (1666 kg/ha) closely comparable to that of BTK 2 (1633 kg/ha) and endosulfan (1550 kg/ha) (Saxena and Ahmad, 1998).

Bt transgenics are very well known globally now. However, in case of pulse crops some studies were conducted at ICRISAT with *Bt*-transgenic chickpea and pigeonpea plants. These studies showed that *H. armigera* suffered no increase in mortality on the transgenic plants and only up to 40% reduction in larval weight as compared to control plants. These effects are insignificant when compared to commercialized Bt cotton plants and due to low Cry protein expression levels. Hence, these plants have not made any significant impact.

Fungi

Entomopathogenic fungi have four major classes, Phycomycetes, Ascomycetes, basidiomycetes and deuteromycetes. These classes include organism causing mycosis in insects. Fungal pathogen particularly *Beauveria bassiana, Metarhizium anisopliae* and *Nomuraea riley* have been found promising in the control of pests of pulses. *Nomuraea riley* occur in epizootic form on *H. armigera, Spodoptera litura* and *Spodoptera exigua* in field conditions (Vimala Devi, 2001). Recently, *Aspergillus flavus* and *Aspergillus niger* were reported from the larvae of *H. armigera* in pulse ecosystem at IIPR, Kanpur. Both the fungus are very effective against *H. armigera* (Saxena and Chaudhary, 2002; Saxena, *et al.,* 2005; Pandey *et al.,* 2007). *Beauveria bassiana,* a white muscardine fungus is an important entomopathogenic fungus that causes disease in a variety of insects of diverse group. Laboratory evaluation of four concentrations of *B. bassiana* (2.68×10^7, 2.68×10^6, 2.68×10^5 and 2.68×10^4 spores/ml) against 1st to 5th instar larvae were effective against all the instars of *H. armigera* (Saxena *et al.,* 1989). Larval mortality increased with increasing spore concentrations. Younger larvae were highly susceptible with comparatively less incubation period required for fungus development on the host larvae and vice versa. Absolute mortality of 1st instars was noted in all treatments and the corresponding values for incubation period was lower (3–7 days). In case of 5th instar, larval mortality ranged from 28 to 4% with corresponding incubation period being higher (7–15 days). *B. bassiana* at 2.68 × 10^7 spores/ml was most effective. This study reveals that higher susceptibility of younger instars to the fungal infection is advantageous as the pest control in early stages is less likely to cause economic injury to the crop. Overall, *Metarhizium anisopliae, Beauveria bassiana, and Nomuraea rileyi* seem to control *H. armigera* in pulses effectively under pot controlled conditions. In these studies, the mass production of cell wall was carried out by two stage fermentation for the production of various enzyme and its inhibitors. The *M. anisopliae* and *B. bassiana* isolates from fields showed greater than 70% mortality of *H. arimgera* in the bioassay. The percentage mortality declined when there was a decrease in the enzyme activities of *B. bassiana. N. rileyi* isolates did not exhibit detectable chitinase levels even up to 120 h (Revathi *et al.*, 2011).

Laboratory evaluation of *B. bassiana* against pod-sucking bugs of pigeonpea was conducted by allowing the adult and nymphs of *Clavigralla gibbosa* and *Nezara viridula* to crawl on the spores of *B. bassiana* for 1–6 h. It was found that 100% adults and nymphs were killed between 3–5 h (Saxena and Ahmad, 2002). Dispel, a commercial formulation of *B. bassiana* was found to be highly effective against 2nd instar larvae of *Diacrisia obliqua* in laboratory.

Table 4.8. Field evaluation of *Beauveria bassiana* (Balsamo) Vaillemin against *H. armigera* infesting chickpea

Concentration (spores/ml)	Pod damage (%)	Yield (kg/ha)
2.68 × 10^7	6.8	2377
2.68 × 10^6	8.7	2192
2.68 × 10^5	10.5	2135
2.68 × 10^4	11.9	2025
Control	16.3	1844

(*Source*: Saxena and Ahmad, 1997)

Field trials on the effectiveness of the fungus at different concentrations against *H. armigera* revealed that *B. bassiana* at a concentration of 2.68 × 10^7 spores/ml was highly effective in increasing yield (2377 kg/ha) with significantly lowest pod damage of 6.8% (Table 4.8) (Saxena and Ahmad, 1997).

Nematodes

Entomopathogenic nematodes (EPN) belong to the families Steinernematidae and Heterorhabditidae and are being successfully used as biological control agents for *H. armigera* (Hussaini, 2001; Ali *et al.,* 2005). Recently two heat tolerant new species *Steinernema masoodi* and *Steinernema seemai* (Rhabditida: Steinernematidae) have been identified and described, which are highly virulent, causing up to 67% mortality of *H. armigera* larvae within two days and multiply well on *Helicoverpa armigera* larvae under laboratory condition The mortality of *Helicoverpa armigera* larvae by *Steinernema masoodi, S. seemai, S. carpocapsae, S. glaseri* and *S. thermophilum* occurred within 28, 32, 36, 38 and 56 h, respectively. A single larval cadaver of *H. armigera* yielded 1.24, 0.80, 1.16, 0.91 and 0.45 lakhs infective juveniles of EPN, respectively. Maas production technique on semisynthetic diet based on locally available material has been developed which is quite cheap, easy to handle and efficient than diet tested by others. The EPN has been found highly infective to pre-pupal and pupal stages. The infection of EPN to the pupal stage of this insect has so far not been reported in India (Ali *et al.,* 2005). *S. masoodi* and *S. seemae* have been evaluated against 10 insect species, *Callosobruchus chinensis* (gram dhora) *Callosobruchus* sp. (Bruchid), *Diacrisia obliqua* (Bihar hairy caterpillar), *Lampides boeticus* (blue butterfly), *Maruca vitrata* (spotted borer), *Mylabris pustulata* (blister beetle), *Myllocerus* sp. (grey weevil), *Poecilocerus pictus* (aak grasshopper), *Rhizopertha dominica* (lesser grain borer) and *Tribolium casteneum* (rust red flour beetle) (Ali *et al.,* 2005).

The viability of EPN formulations varies depending upon the base material. For example, in sponge cubes, soil, cadavers and dust the viability was 12–15, 15–18, 4–6 and 8 weeks, respectively. A new EPN based formulation of water dispersible granules (WDG) was prepared for *S. masoodi* and sprayed over the surface of chickpea foliage, which was effective for 3 h. A dose of 6×10^9 IJs of *S. masoodi*/pot was sufficient to kill the physiologically mature *H. armigera* larvae, which drop down to soil for pupation (Ali *et al.,* 2005). However, *Heterorhabdities indica* was found inefective against *Mylabris pustulata* and *Helicoverpa armigera* in pigeonpea (Sathiah *et al.,* 2007).

Protozoans

Protozoan pathogens naturally infect a wide range of insect hosts. Although these pathogens can kill their insect hosts, many are more important for their chronic, debilitating effects. One important and common consequence of protozoan infection is a reduction in the number of offspring produced by infected insects. *Nosema* sp. and *Vairiomorpha* sp. are the two protozoans reported on *H. armigera.* The role of protozoa as microbial control is limited because of the difficulties in their mass production for field release.

BOTANICALS

Among plant products, neem products have been extensively used in crop protection. They have been tested in pulses as well against *H. armigera* in laboratory as well as in field, Four commercial neem formulations, Repelin 1.0%, Neem Guard 0.4%, Neem EC 0.15% and Neem WDP 0.15% were evaluated under laboratory for their antifeedant properties against *H. armigera*. Neem 25 EC was the best antifeedant with maximum protection of leaf area (89.54%) from larval feeding followed by Neem 25 WDP and Repelin (83.46%) (Grzywack *et al.,* 2005; Saxena *et al.,* 2010). Neem has been advantageous in view of their high degree of antifeedant property and safety to environment. Among different oils (pine, cedar wood, castor, mineral, turpentine, linseed, arachis, eucalyptus; each at 0.5%), garlic extract (2%) and multineem (0.5%) have been evaluated against 2nd instars of *H. armigera*. Multineem has been the best among these to deter feeding and mineral oil was least effective. Three commercially available neem formulations, Repelin, Nimbecidine and Achook under field conditions have also prevented pod damage and were as effective as endosulfan (Table 4.9) (Saxena *et al.,* 2010).

Bruchid or pulse beetle, *Callosobruchus* spp. (Coleoptera : Bruchidae) is a serious stored grain pest of pulses. The attack of these beetles starts in the field and the infestation is carried over to the storage. Two species of bruchids, *Callosobruchus chinensis* (L) and *C. maculatus* (F) have been found most serious pest of pigeonpea, chickpea, mung bean and lentil. Mixing of grains with plant products, edible/nonedible oils and intert materials showed promise in preventing bruchid damage under storage. It reduces oviposition, adult emergence, grain weight loss and prolonged the developmental period of bruchids. Among the plant products seed treatment (20 g/kg of seed) with dhatura leaf powder, custard apple (*Annona* spp.) seed powder, sadabahar (*vinca rosea*) leaf powder, Kaner (*Neerium* spp.) leaf powder, neem (*Azadirachta indica*)leaf powder, *Eucalyptus* spp. Leaf powder, asgan leaf powder and orange peel powder are effective in providing protection against bruchids in pigeonpea and mungbean. Among the oils mustard oil, neem oil and mahua oil (3–12 ml/kg of seed) act as surface protectants against *C. chinensis* . Inert materials showed promise in preventing bruchid damage. Lime, pond clay, morang (fine sand), fly ash, paddy straw ash and Ipomea ash (20 gm/kg of seed) were effective in reducing *C. chinensis* population and its damage in pigeonpea and mungbean. Seed treatment with 2% neem seed kernel (NSK) powder was effective for six months against *C. chinensis* (Yadav, 1984). Seed treatment with 1% neem, castor or Karanj oil has shown ovipostional deterrence up to 100 days and suppressed hatching of *C. chinensis*.

Table 4.9. Field efficacy of different neem formulations against *H. armigera* in chickpea

Treatments	Pod damage (%)
Repelin 0.5%	11.7
Repelin 1.0%	8.6
Nimbecidine 0.2%	14.7
Nimbecidine 0.4%	10.3
Achook 1.0%	7.7
Endosulfan 0.07%	8.1
Control	26.1

(*Source*: Saxena *et al.*, 2010)

BIORATIONALS

Biopesticides like NPV and Bt formulations as well as insect growth regulators like Jadoo (Diflubenzuron + Alphamethrin) and Rimon (Novaluron) were tested against early instars of *H. armigera*. Rimon 0.01% was most effective causing 85% mortality followed by Bt (80%) and NPV (60%). Insect growth regulator (Jadoo 0.01%) though caused only 30% mortality, it induced heavy deformity in the emerged adults. Rimon and *Bt* showed effectiveness as Endosulfan (80% mortality). Jadoo with heavy deformities in adults can be advantageous to reduce next generation population build up (Saxena, 2009).

While studying the performance of newer biorationals, lowest pod damage of 33.8% in diflubenzuson at 178 g a.i./ha with highest yield of 742 kg/ha in chickpea was recorded. HaNPV (250 LE/ha) was next best in these studies (pod damage 41.1% and yield 694 kg/ha) followed by "Jadoo" at 1.0 kg/ha (pod-damage of 44.2% and yield 662 kg/ha) (Saxena, 2009).

CONCLUSIONS

Biopesticides in pest management is an environmentally safe, economical and sound method of tackling pest problems. Biopesticides provide ecofriendly alternatives to chemical insecticides but they face a number of constraints in their development, manufacture and utilization.

The Baculoviruses particularly NPVs have tremendous scope for development as biopesticides for the management of *H. armigera* and *S. litura*. Multinational companies who have the production capabilities are not interested in view of the limited market potential. A few indigenous small producers who ventured into commercial production of NPV of *H. armigera* and *S. litura* did not succeed in producing quality virus. Most of the samples tested had extremely low virus content even no virus at all. Many had unacceptable levels of spores of microsporidians. NPV production is done *in vivo* in respective host larvae and hence producers should employ properly trained manpower for production of host insects as well as the virus. Most of the Bt products now being sold in India are very expensive since they are imported. One or two products developed indigenously in India are not as effective as the imported ones. Few institutions in India including the TNAU and the BARC have isolated indigenous Bt strains which are as potent as the standard HD 1. There is ample scope for development of techniques of indigenous production and formulation of native Bt. Species of entomopathogenic fungi like *Metarhizium, Beauveria, Verticillium* and *Nomuraea* isolated from different agroecosystems in India have shown promise in pest management. We should genetically improve these strains and develop techniques of mass production and stable formulations. India with the rich biodiversity of organisms has tremendous potential for development of native strains of microbials well adapted to the Indian subcontinent. There has been success in the use of pathogens such as *Bacillus thuringiensis, N. rileyi* and *Helicoverpa armigera* nuclear polyhedrosis virus (*Ha*NPV). However, the relatively high cost and rapid inactivation by ultraviolet light often lead to poor performance under field conditions. In the case of *Ha*NPV, the difficulty in obtaining consistently high levels of purity and virulence necessary to achieve satisfactory control has limited its usefulness. The increasing prevalence of resistance to insecticides and awareness of environmental concerns has given a new impetus to the development of suitable biopesticides for use in IPM of *H. armigera*.

Educational programmes in institutions should give special emphasis on development of biopesticides for pulse crop pest control. There is a need to develop inter-institutional collaborations in order to pool all the resources available in different institutions and develop microbial pest management technologies which would ultimately reduce the pressure on chemical pesticide use. With the globalization and liberalization of agriculture the microbial management of the insect-pests using biopesticides has attained immense importance and has very bright future. For the promotion of biopesticides, their intensive commercial production and formulation with strict enforcement of quality control measures are necessary.

REFERENCES

Agrawal, N., Srivastava, M., Tripathi, A. and Singh, A. (2010) Survey and monitoring of pests, parasites and predators of pulse crops in central and eastern Uttar Pradesh. *J. Plant Prot. Sci.*, **2**, 45–52.

Ahmad, R. and Rai, A.B. (2005) *25 Years of Research on Helicoverpa at IIPR.* Indian Institute of Pulses Research, Kanpur. 54 pp.

Ali, S.S., Ahmad, R., Hussain, M.A. and Pervez, R. (2005). *Pest Management in Pulses through Entomopathogenic Nematodes.* Indian Institute of Pulses Research, Kanpur.

Chandel, S.F., Ahmad, R. and Saxena, Hem. (2001). Evaluation of biopesticides under farmer's field condition against *H. armigera* infesting chickpea. Abstract, *Proceedings of Symposium on Bio-control Based Pest Management for Quality Crop Protection in the Current Millennium,* PAU, Ludhiana, July 18-19, 2001.

Claver, M.A. and Ambrose, D.P. (2002) Functional response of the predator , *Rynocoris fuscipes* (Heterodera: Reduviidae) to three pest of pigeonpea. (*Cajanus cajan*). *Shashpa*, **9**, 47–51.

Duraimurugan P. and Srinivasan, T. (2009) Carabid predator of *Maruca vitrata* in mungbean- A new record. *Pulses Newslett.*, **20**, 5.

Grzywacz, D., Richards, A., Rabindra, R.J., Saxena, Hem and Rupela, O.P. (2005) Efficacy of biopesticides and natural plant products for *Heliothis/ Helicoverpa* control. In: H.C. Sharma (ed.), *Heliothis/Helicoverpa Management-Emerging Trends and Strategies for Future Research,* Oxford and IBH Publishing Co. Pvt. Ltd., New Delhi, pp. 371–389.

Hussaini, S.S. (2001) Scope of entomopathogenic nematodes against crop pests In: R.J. Rabindra, J.S. Kennedy, N.Sathiah, B.Rajasekaran and M.R. Srinivasan (eds.), *Microbial Control of Crop Pests*, Graphic Skill Publisher, Coimbatore, pp. 180–221.

Lingappa, S., Saxena, Hem and Vimla Devi, P.S. (2005) Role of biocontrol agents in management of *Helicoverpa armigera* (Hubner). In: Hem Saxena , A.B. Rai, R. Ahmad and Sanjeev Gupta (eds.), *Recent Advances in Helicoverpa Management*, Indian Society of Pulses Research and Development , IIPR, Kanpur, pp.159–184.

Mishra, U and Mishra, G.K. (2004) Biopesticides for ecofriendly pest management. *Agric. Ext. Rev.,* 20–26.

Mohapatra, S.D., Duraimurugan, P. and Saxena, Hem (2008) Natural parasitization of *Maruca vitrata* (Geyer) by *Bracon hebetor*. Pulses Newslett., **19**, 1.

Pal, V.S., Krishna, Kant, Lal, S.S. and Saxena, Hem (1998) Control of chickpea pod borer *Helicoverpa armigera* with *Helicoverpa* Nuclear Polyhedrosis virus under farmers field conditions. *Proceedings of National Symposium on Entomology in 21st Century* held on April 30–May 2, 1998, at Rajasthan College of Agriculture, Udaipur.

Pandey, Kirti, Saxena, Hem and Kumar, Prabhat (2007) A preliminary study on the effectiveness of *Aspergillus flavus* against *Helicoverpa armigera* (Hubner) on chickpea. *J. Ecofriendly Agric.*, **2**, 75–76.

Pawar, V.M. and Borikar, P.S. (2005) Microbial options for the management of *Helicoverpa armigera* (Hubner) In: Hem Saxena, A.B. Rai, R. Ahmad and Sanjeev Gupta (eds.), *Recent Advances in Helicoverpa Management*, Indian Society of Pulses Research and Development, IIPR, Kanpur, pp. 193–231.

Rabindra, R.J., Singh, America, Saxena, Hem and Ballal, R.C. (2005), Biological control of insect pests and diseases in food legumes. In Abstracts – *4th International Food Legumes Research Conference* October 18–22, New-Delhi.

Reddy, A.M. (2009) Pulses production technology: Status and way forward. *Econ. Polit. Weekly*, **64**, 73–80.

Revathi, N., Ravikumar, G., Kalaiselvi, M., Gomathi, D., Uma, C. (2011) Pathogenicity of three entomopathogenic fungi against *Helicoverpa armigera*. *J. Plant Pathol. Microbiol.,* **2**, 114. doi:10.4172/2157-7471.1000114.

Sathiah, N, Balagurunathan, R, Santham,G, Kennedy, S.S, Hussaini S.S, Nagesh, M and Rabindra, R.J. (2007) Evaluation of *Heterorhabdities indica* Poinar against *Mylabris pustulata* and *Helicoverpa armigera* (Hubner) infesting pigeonpea in Tamil Nadu. In: S. Ignacimuthu and S. Jayaraj (eds.), *Biotechnology and Insect Pest Management*, Elite Publishing House, New Delhi, pp. 202–204.

Saxena, Hem (2007) Record of new natural enemies of *Helicoverpa armigera* in pulses. Abstracts: *National Symposium on Legumes for Ecological Sustainablity : Emerging Challenges and Opportunities,* November 3–5, 2007, IIPR Kanpur.

Saxena, Hem (2009) Biological control. In: *25 Years of Pulses Research at IIPR*, Indian Institute of Pulses Research, Kanpur. pp. 131–142.

Saxena, Hem and Ahmad, R. (1997) Evaluation of *Bacillus thuringensis* formulations against gram semilooper *Autographa nigrisigna* Walker (Noctuideae : Lepidoptera). *J. Biol. Control*, **11**, 89–91.

Saxena, Hem and Ahmad, R. (1997) Field evaluation of *Beauveria bassiana* (Bals.) Vuillemin against *Helicoverpa armigera* Hubner infesting chickpea. *J. Biol. Control,* **11**, 93–96.

Saxena, Hem and Ahmad, R. (1998) Comparative performance of *Bacillus thuringiensis* (Bt) formulations against *Helicoverpa armigera* infesting chickpea. Proceedings of National Symposium on Management of Biotic and Abiotic Stresses in Pulse Crops held at IIPR, Kanpur on 26–28th June, 1998.

Saxena, Hem and Ahmad, R. (2000) Toxicity and persistence of *Bacillus thuringiensis* Berliner against gram semilopper, *Autographa nigrisigna* Walker. *Indian J. Pulses Res.*, **13**, 64–65.

Saxena, Hem and Ahmad, R. (2002) Studies on exposure duration of *Beauveria bassiana* to the mortality pod sucking bugs of pigeonpea. *Indian J. Pulses Res.*, **15**, 93–94.

Saxena, Hem and Chaudhary, R.G. (2002) Record of new entomogenous fungi from the larvae of *Helicoverpa armigera. IIPR News Lett.*, **13**, 3.

Saxena, Hem and Ahmad, R. (2003) Comparative performance of biopesticide for the management of *Helicoverpa armigera* infesting chickpea. Proceedings of International Chickpea Conference held at Raipur, Chattisgarh, India on Jan. 20–23, 2003.

Saxena, Hem and Ahmad, R. (2005) NPV production, formulation and quality control. In: Hem Saxena, A.B. Rai, R, Ahmad and Sanjeev Gupta (eds.), *Recent Advances in Helicoverpa Management*, Indian Society of Pulses Research and Development, Indian Institute of Pulses Research, Kanpur, pp. 185–192.

Saxena, Hem and Duraimurugan, P. (2007) Natural enemies of *Helicoverpa armigera* in short duration pigeonpea. *Pulses Newslett.*, **18**, 6.

Saxena, Hem and Duraimurugan, P. (2009) Record of *Bracon hebetor* Say (Hymenoptera: Braconidae) on *Helicoverpa armigera* (Hübner) in chickpea. *J. Biol. Control,* **23**, 191–192.

Saxena, Hem, Ahmad, R. and Sachan, J.N. (1989) Studies on *Beauveria bassiana* against insect pests of pulses. *Proceedings of National Symposium on New Frontiers in Pulses Research and Development* held at Directorate of Pulses Research, Kanpur from 10–12 November, 1989.

Saxena, Hem, Ahmad, R. and Sachan, J.N. (1992) Persistence and toxicity of *Bacillus thuringiensis* Berliner against Bihar hairy caterpillar *Diacrisia obliqua* Walker. Abstract *Proceedings of National Symposium on Recent Advances in Integrated Pest Management* held at PAU, Ludhiana, Oct. 12–15, 1992.

Saxena, Hem, Ahmad, R. and Sachan, J.N. (1995) Comparative efficacy of different *Bacillus thuringiensis* (Bt.) formulations against *Helicoverpa = Heliothis) armigera* Hubner infesting pigeonpea. Abstract *Proceedings of the 82nd session of the Indian Science Congress.*

Saxena, Hem, Chaudhary, R.G. and Dar, M.H. (2005) Bioefficiency of two entomogenous fungi against *Helicoverpa armigera.* Abstract *4th International Food Legumes Research Conference.* October 18–22, 2005, New Delhi.

Saxena, Hem, Singh, S.K., Duraimurugan, P., Mohapatra, S.D. and Lakshmi, S. P. (2010) *Pulse Entomology in India: Retrospective and Perspective.* Indian Institute of Pulses Research, Kanpur.

Vimala Devi, P.S. (2001) Prospects of using *Nomuraea rileyi* for the management of crop pests. In: R.J. Rabindra, J.S. Kennedy, N. Sathiah, B. Rajasekaran and M.R. Srinivasan (eds.), *Microbial Control of Crop Pests*, Graphic Skill Publisher, Coimbatore, pp. 80–94.

Weinzierl, R and Henn, T. (2005) *Microbial Insecticides.* Florida Cooperative Extension Service, Institute of Food and Agricultural Sciences, University of Florida. Fl.

Yadav, T.D. (1984) Efficacy of neem *Azadivachta indica* kernel powder as seed protectant against pulse beetle *Neem Newslett.*, **1**, 5–6.

5

Current Status and Potential of Biopesticides in Pest Management in Oilseeds

H. BASAPPA[1*] AND R. K. PATIL[2]

[1]Directorate of Oilseeds Research (ICAR), Rajendranagar, Hyderabad-500030, India
[2]Department of Entomology, University of Agricultural Sciences, Dharwad-580005, India
*E-mail: hbasappa@gmail.com; hbasappa@yahoo.com

INTRODUCTION

During 2012/13 global production of oilseeds has shown uptrend due to large supplies of soybeans and rapeseed. According to Foreign Agriculture Service/USDA (2013) World production of major oilseeds has been 469.43 million metric tonnes, that is about 6.0% larger than last season (Table 5.1). Major oils produced have been rapeseed and soybean (Table 5.2).

In terms of country wise scenario, United states has recorded decline in oilseed production from 2010/11 (100.38 million metric tonnes) to 2012/13 (92.63 million metric tonnes), which is a reduction of about 8%. Brazil, China and India, however, has shown an upward trend in the production (Table 5.3).

India has the fifth largest vegetable oil economy in the world yet it is also the largest importer of vegetable oils. The country is bestowed with all favorable agro- climatic conditions to grow all the nine annual major oilseed crops. Most of these crops are grown under rain fed conditions and thus have low productivity. Apart from these annual oilseed crops, a wide range of other minor oilseeds of horticultural and forest origin, including coconut and oil palm are grown in the country, besides by-products of non-oilseed crops. India as an important producer, importer and exporter plays an important role in the international trade related to vegetable oils and their value added products. The net domestic availability of vegetable oils during 2009-10 was only 6.5 million tonnes as against the actual consumption of 15.6 million tonnes and the gap of 8.8 million tonnes was met through imports which is causing a big drain on our foreign exchange besides affecting the domestic oilseed production and market. Most of the oilseeds and their by-products have export potential due to demand in the international market.

Table 5.1. World production of major oilseeds (Million Metric Tonnes)

	2008/09	2009/10	2010/11	2011/12	2012/13
Oilseed, Copra	5.88	5.88	6.02	5.66	5.96
Oilseed, Cottonseed	41.08	38.91	43.55	46.41	45.30
Oilseed, Palm Kernel	11.75	12.28	12.73	13.51	14.53
Oilseed, Peanut	34.89	33.60	36.14	35.13	37.05
Oilseed, Rapeseed	57.82	60.98	60.57	61.12	61.14
Oilseed, Soyabean	211.60	260.40	263.92	239.46	269.11
Oilseed, Sunflowerseed	33.50	32.21	33.63	40.64	36.36
Total	396.52	444.25	456.56	441.93	469.43

Source: USDA (2013)

Table 5.2. World production of major oils (Million Metric Tonnes)

	2008/09	2009/10	2010/11	2011/12	2012/13
Oil, Copra	3.54	3.63	3.81	3.50	3.75
Oil, Cottonseed	4.75	4.60	4.97	5.27	5.28
Oil, Olive	2.78	3.08	3.25	3.24	2.67
Oil, Palm	44.05	45.88	48.68	51.86	55.29
Oil, Palm Kernel	5.17	5.51	5.63	5.97	6.35
Oil, Peanut	5.05	4.72	5.08	5.06	5.25
Oil, Rapeseed	20.59	22.56	23.52	24.30	24.05
Oil, Soyabean	35.88	38.78	41.27	42.35	42.92
Oil, Sunflowerseed	12.08	12.28	12.42	15.26	14.08
Total	133.90	141.04	148.62	156.80	159.64

Source: USDA (2013)

Table 5.3. Country view of major production of oilseeds (Million Metric Tonnes)

	2008/09	2009/10	2010/11	2011/12	2012/13
United States	89.20	98.90	100.38	92.35	92.63
Brazil	60.31	71.44	79.04	70.02	86.01
Argentina	35.51	57.94	54.22	44.78	55.62
China	58.12	57.84	58.10	59.24	58.72
India	33.40	32.37	34.95	35.37	36.05
Other	119.99	125.75	129.86	140.19	140.40
Total	396.52	444.25	456.56	441.93	469.43

Source: USDA (2013)

Vegetable oil scenario is very complex in India and is greatly influenced by market forces, vagaries of weather, technology, poor resource base of the farmers, biotic and abiotic stresses. Insect pests are the major biotic constraints in oilseed production. Wide range of insect pests are associated with oilseed ecosystem but key pests include leaf miner, thrips, and tobacco caterpillar in groundnut; head borer and thrips in sunflower; aphids in safflower; leaf webber and capsule borer in sesame and aphids in rapeseed-mustard; semilooper and capsule borer in castor and bud fly in linseed.

Polyphagous pests like red hairy caterpillar, Bihar hairy caterpillar, root grubs, termites, cutworms and serpentine leaf miner also attack these crops. In oilseed ecosystem, several beneficial insect species like parasitoids, predators and pollinators play important role in the sustainable production of oilseed crops. Potential natural enemies have been recorded on economically important insect pests of different oilseed crops which suppress the insect pest population by interfering in the perpetuation of insect pests at different growth stages.

There is a need for reducing toxic stress due to insecticides in the environment by avoiding indiscriminate use of insecticides and by adopting biointensive integrated pest management (BIPM) approach in the management of insect pests of crops. BIPM approach represents a change in the basic philosophy of pest control from pest eradication to economic accommodation and management involving combinations of biologically based methods tailored into farming system. BIPM modules involving eco-friendly, economically viable and socially acceptable biologically based pest management methods like use of biopesticides, resistant varieties, cultural practices , suitable barrier and trap crops, inter-cropping, use of pheromone traps, conserving natural enemies, keeping bird perches to attract insectivorous predatory birds and maintaining field bunds and crop free from weeds as they favours the pest build up on the main crop are some of the adoptable strategies against insect pests of all the oilseed crops. The rationale behind replacing conventional pesticides with biopesticides is that the latter are more likely to be selective to attack specific pest targets, more efficient than chemical pesticides and biodegradable. Eco-friendly bio pesticides like botanicals, Bt derivatives, entomopathegenic fungi, nuclear polyhedrosis viruses (NPVs), entomopathegenic nematodes and insect growth regulators (IGRs) with insecticidal properties are now used in many countries around the world for the management of insect pests in various crops (Dhaliwal and Koul, 2007).

Botanical biopesticides are naturally occurring secondary metabolites synthesised by plants species, which act on the insect growth and survival thus do not affect natural enemies and pollinators. Natural bio pesticidal products are available as an alternative to synthetic chemical formulations. More than 2121 plant species can be utilized as biorationals which contain toxic principles out of which, 1005 insecticidal, 384 antifeedants, 297 repellents, 27 attractants and 31 are growth inhibitors (Grainge *et al.*, 1985). Biopesticides of microbial origin are products containing natural organisms, or their genes or metabolites that are used to protect crops against damaging pests. In nature more than 3000 microorganisms have been reported to cause diseases in insects and play important role in regulating pest population and many of them have been commercialized and several formulations are commercially available (Basappa *et al.*, 2005).

Biological based insect pest management practices involving biopesticides were working on sustainable basis before introduction of synthetic pesticides in the pest management. There is increasing demand for pesticide residue free food products in the international market, hence, biopesticides are important in pest management which needs serious consideration (Basappa, 2009a). Biopesticides will play important role in the production of insecticide residue free oilseed crops since all the oilseeds and their value added products have high export potential due to confectionary, medicinal and industrial importance. During 2008-09 India exported highest soybean defatted meal, cake, seed and flour worth ₹8958.08 crores followed by castor oil and cake (₹2253 crores), sesame seed (₹1510.36 crores), groundnut seeds, meal, butter (₹1387.21 crores), rapeseed and mustard products (₹ 1303.79 crores), niger seed (₹64.50 crores), sunflower ssed (₹31.92

crores), linseed (₹15.44 crores) and safflower products (₹0.42 crores) (Damodaram and Hegde, 2010). Biopesticides of plant and microbial origin are very important components of BIPM for sustainable oilseed production. Awareness has to be created among farmers about importance of these biopesticides in BIPM of oilseed crops through strong farmers' participatory extension programme (Basappa, 2008a). Current status and potential of biopesticide utilization in the ecofriendly management of major insect pests of nine annual oilseed crops like groundnut (*Arachis hypogaea* L.), soybean (*Glycine max* L.), rapeseed-mustard (*Brassica* spp.), sunflower (*Helianthus annuus* L.), sesame (*Sesamum indicum* L.), castor (*Ricinus communis* L.), linseed (*Linum usitatissimum* L.), safflower (*Carthamus tinctorius* L.) and niger (*Guizotia abyssinica* (L.f.) Cass) is discussed herein.

GROUNDNUT

India is one of the major producers of groundnut, which ranks first with global share of 26.1% in area and second rank in production with global share of 10.4%. respectively at the global level. More than 80% area of the total groundnut cultivation in the country is under rainfed with wide fluctuations in productivity depending upon rainfall and its distribution. On groundnut more than 360 species of insect pests are known to feed out of which a dozen of them are economically important in India (Wightman and Amin, 1988, Ranga Rao et al.,1995). Groundnut leaf miner (*Aproaererma modicella*), tobacco caterpillar (*Spodoptera litura*), gram pod borer (*Helicoverpa armigera*), red hairy caterpillars (*Amsacta albistriga, Amsacta moorei*), sucking insect pests, white grubs and termites are major insect pests causing severe loss to the crop. In the northern states of India major insect pests of groundnut include white grubs and termites while in the southern states leaf miner, red headed hairy caterpillar, tobacco caterpillar and of late, gram caterpillars are of greater importance. Aphids, thrips and leafhoppers are of minor concern as direct pests but they take a heavy toll as vectors of virus diseases. Biological based pest management approaches involving biopesticides in groundnut ecosystem include release of *Trichogramma chilonis* or *T. brasiliensis* immediately after red hairy caterpillar adult moths emergence from soil @50000 adult parasitoids/ ha, foliar spray of *Amsacta albistriga* NPV (AaNPV) @ 4.3 × 10^{10} POB/ml on early stage larvae or *Bacillus thuringiensis* var. *kurstaki* @ 2 g/l and spray NSKE (5%) against grown up larvae in addition to vegetative trapping along with twigs of *Jatropha, Ipomea* and *Calotropis*, use of pheromone traps @10 traps/ha for monitoring of *S. litura* and *H. armigera,* spray biopesticides like NSKE(5%), SlNPV and HaNPV @250LE/ha against *S. litura* and *H. armigera,* respectively, apply entomopathogenic fungus, *Nomuraea rileyi* against *S. litura* @ 2×10^{11} conidia/ha twice at 10 days interval coinciding with incidence of larvae in the field (Lingappa and Patil, 2002, Navi, 2002), spot application of *B. popillae* @ 5×10^{13} spores/ha and *Metarhizium anisopliae* and *Beauveria brongniartii* @10^{15} conidia/ha against root grub (Vyas *et al.*, 1986).

Neem seed extract was as effective as phosphamidon against *A. modicella* and *A. craccivora.* Pearl millet plant sap has prevented oviposition by leaf miner up to 74%. Among 43 plant species from 27 families having growth-regulating hormone against *S. litura, Azadirachta indica, Pongamia glabra* and *Parthenium hysterophorus* hold promise. Extracts of *Acorus calamus*, *Dhatura metal,* garlic, tulsi, parthenium, cassia and onion have shown good larvicidal properties against *S. litura* (Lingappa *et al.*, 1993). The neem based insecticides like, Margoside OK and CK suppressed *S. litura* damage and enhanced yield. Aqueous extracts of NSKE (5%), *Vitex negundo* (5%) and *A. squamosa* (5%) caused more than 52% neonate larval mortality. Application of NSKE (5%), *Vitex negundo* (5%), *Argemone mexicana* and *Calotropis gigantia* in combination with Sl NPV (250 LE/ha) recorded reduced defoliation by *S. litura* compared to their individual treatments. All the combinations and *V. negundo* and *A. mexicana* individually recorded at par yield with monocrotophos, but only the cost effectiveness of *V. negundo* and *A. mexicana* was comparable to monocrotophos (Patil, 2000).

There exist a positive interaction between the selected botanicals with SlNPV and *N. rileyi* for enhanced bioefficacy against *S. litura* in groundnut and this phenomenon can suitably be exploited in IPM. Spaying leaf extracts of pongamia (10%) + aloe (5%) with NSKE (10%) + cow urine (30%) has considerably reduced the larval population of *S. litura* among other indigenous methods. The botanicals combine well with NPV, *N. rileyi* and other control tactics and give increased effectiveness against larvae of *S. litura.* The neem oil and monocrotophos combination was found to be very effective in reducing the defoliators with increased yield (2011 kg/ha) compared to monocrotophos (1878 kg/ha) and control categories (Sahayaraj and Amalraj, 2005). Among biorationals, average defoliation was lowest in spinosad 45 SC followed by *N. rileyi* and SlNPV with benefit cost ratio of 4.0, 4.0 and 4.1, respectively. Combined application of reduced dose of indoxacarb 0.00725% + *Btk* 0.1% reduced 58.3% of *S. litura* larval population over control. The next treatment that closely followed was indoxacarb 0.00725% + azadirachtin 0.15% with 55.0% reduction of larval population. SINPV 125 LE/ ha + *Btk* 0.1%, *Btk* 0.2%, SINPV 250 LE/ha' and *N. rileyi* 1.2 × 10^8 spores/ml showed a reduction of larval population by 40.35, 35.7, 33.6 and 32.2%, respectively (Hegde *et el.*, 2009; Satyanarayana *et al.*, 2010). Spray application of *Verticillium lecanii* was found effective against sucking insect pests like leafhopper and thrips with highest yield of 1713 kg/ha compared to 1205 kg/ha in control (DGR, 2010). Drenching of *B. bassiana* @ 4 kg/ ha was found to be effective against white grub in groundnut (Kapadia *et al.*, 2010).

In addition to application of biopesticides as one of the components of BIPM modules involve cultivars like Girnar-1 against sucking pests (aphids, leafhoppers and thrips), BG-2 against defoliators (tobacco caterpillar, gram pod borer and leaf miner), soybean or cowpea as intercrops after 3-4 rows of groundnut to trap leaf miner, thrips and leaf hoppers, grow pearl millet or sorghum or maize as border crop to check thrips as vectors of PSND, keeping field and bunds free from parthenium, as well as inter culturing and manual weeding before peg formation will help to keep entire insect pest complex of groundnut below economic threshold level (ETL) for sustainable production of groundnut (Basappa, 2008a). Groundnut crop canopy will provide suitable microhabitat for increasing the efficacy of biopesticides and self sustainability of entomopathegenic microbial agents. Since groundnut is edible oilseed crop with export potential hence there is tremendous scope for successful utilization of biopesticides for the effective management of insect pests in groundnut ecosystem.

SOYBEAN

There has been overall increase in soybean production globally and all major countries like Unites States, Brazil, Argentina, China, India, Paraguy and Canada have shown increased production in 2012/13 (Table 5.4). Soybean is newly introduced crop in India and making rapid strides in oilseed scenario of the country with fifth rank in area and production in the world with global share of 8.3% area and 2.8% production. Area under cultivation is increasing in many parts of the country. About a dozen major insect-pets and diseases attack the crop. The major insect pests are stem fly, defoliators and pod borer. Yield losses due to these pests range from 25 to 100%. Defoliators can be effectively managed by spraying the crop with Bt or *Beauveria bassiana* based microbial insecticides @ 1 kg/ha. In case of simultaneous infestation of defoliators and stem borer/sap feeders, spraying tank mix formulation of *Bt* + monocrotophos @ 0.8 l/ha is recommended. Entomopathogenic fungus, *Nomuraea rileyi* @ 2 × 10^8 conidia/ha is also found effective against *S. litura, H. armigera* and *Thysanoplusia orichalcea.* The net return for every rupee investment from *N. rileyi* application was nearly 4 times more than in insecticidal treatment (Lingappa and Patil, 2002). *H. armigera* can be managed by spraying the crop with *Ha*NPV @250LE/ha or Bt based biopesticides @1 kg/ha at the early instar stage of larvae (Basappa *et al.*, 2005).

Table 5.4. Major countries producing Soybean (Million Metric Tonnes)

	2008/09	2009/10	2010/11	2011/12	2012/13
United States	80,749	91,417	90,605	84,192	82,055
Brazil	57,800	69,000	75,300	66,500	83,500
Argentina	32,000	54,500	49,000	40,100	51,000
China	15,540	14,900	15,100	14,480	12,600
India	9,100	9,700	9,800	11,000	11,500
Paraguay	3,647	6,462	7,128	4,350	8,350
Canada	3,336	3,501	4,445	4,298	4,930
Other	9,430	10,763	12,546	14,539	15,171
Total	211,602	260,403	263,924	239,459	269,106

Source: USDA (2013)

Diflubenzuron at 600 g/ha was the most effective treatment in suppressing the defoliator population followed by lufenuron 400 and 800 ml /ha (Kuldeep *et al.*, 2005). Combination of endosulfan (0.05%) with four bacterial formulations (Btk (Dipel 8L), Btk1 (Bioasp), Btt(Biotox) and Bta(Centari)) proved to be effective against *S. obliqua* (Agnihotri and Khan, 2006). When additives like teepol (30 kg/ha) and sandovit (30 kg/ha) were added to NSKE (5%), it was found more effective against *S. obliqua* (Pande *et al.*, 2008). Entomopathegenic nematodes *Heterorhabditis indicus* and *Steinernema carpocapsae* were found effective against semiloopers (*Chrysodeixis acuta* and *Gessonia gemma*) and *S. litura* with higher yield levels ranging from 2952 to 3108 kg/ha compared to 2201 kg/ha in control (DSR, 2010). Neem seed based products proved to be effective against the stem borers next to insecticide, triazophos application at 30, 50 and 70 days after sowing

(DAS) which was superior with maximum grain yield. The cost benefit ratio was highest (3.03:1) in one spray of neem seed water extract (NSWE) applied at 70 DAS which was economically viable than the other treatments (Arvind Kumar *et al.*, 2010). Soybean crop will also provide favourable microclimate for self sustainability of entomopathogenic microbial agents due to closed and ground level canopy hence there is tremendous scope for successful utilization of biopesticides for the effective management of insect pests in soybean ecosystem for sustainable production to meet the growing demand for pesticide residue free soybean products for domestic consumption and export purpose.

RAPESEED-MUSTARD

India is one of the important countries in the world which ranks first in rapeseed-mustard area with a global share of 25.6% and second in production with global share of 18.8%. Among insect pests attacking the crop, aphid (*Lipaphis erysimi*), musturd sawfly (*Athalia proxima*), painted bug (*Bagrada hilaris*) pea leaf miner, (*Chromatomyia horticola*) and Bihar hairy caterpillar (*Spilosoma obliqua*) are of major concern. Early sown crops (before 20th October) suffered less damage from aphid than crop sown later under varied agroecological conditions in Northern India. In rapeseed-mustard predators such as *Coccinella septumpunctata, C. transversalis*, syrphid fly, *Chrysoperla carnea* and parasitoid like *Diaeretiella rapae* had a regulatory impact on mustard aphid population in nature from February to April. Nicotine sulphate, rotenone and pyrethrins have shown variable toxicity to aphids. Extracts of *Brassica campestris and B. juncea* leaves containing glucosinolates had adverse effect on the nymphs, adult longevity and fecundity of mustard aphid. Bitter gourd seed emulsion (2%) caused 100% mortality of mustard sawfly larvae, whereas, ether extract of *Acorus calamus* gave 95-100% mortality after 72 h of spraying. *Eupatorium adenophorum* leaf extract affects different systems of mustard aphid through various mechanisms (Sudip Dey *et al.*, 2005). NSKE (5%) was effective in reducing 75.54% mustard aphid population than neem oil (2%) with 46.87% population reduction (Akhauri and Singh,2009). Entomopathegenic fungus *V. lecanii* @10^{8} CS/ml was found effective against mustard aphid with 1958 kg/ha seed yield (DRMR, 2010). Since aphid is key pest in rapeseed-mustard there is wide potential for biopesticide utilization for sustainable production as well as conservation of beneficial organisms in the rapeseed-mustard ecosystem.

SUNFLOWER

Global sunflower trade has been higher in European Union, Ukraine and Russia. In European Union, sunflower seed product supply was much higher (6885 thousand metric tonnes) compared to China (2440 metric tonnes) and India (700 thousand metric tonnes) in 2012/13. Sunflower is one of the introduced crops in to India which witnessed fastest growth in terms of area and production. Sunflower is also making rapid strides in the oilseed scenario of the Indian subcontinent due to its wide adaptability to different agro-ecological niches and cropping systems, day neutral nature and high responsiveness to better management practices.

In India, 129 species of arthropods and 13 species of vertebrates were found to be associated with sunflower crop which include 67 species of insects, 5 species of vertebrates and 1 mite species as herbivores, 48 species of natural enemies (7 parasitoids, 24 insect predators, 9 spiders and 8 predatory birds) and 8 species of pollinators (Basappa, 2008b). Though several insect pests are recorded in sunflower crop capitulum borer (*H. armigera*), tobacco caterpillar (*S. litura*), Bihar hairy caterpillar (*Spilosoma obliqua*), semilooper (*Thysanoplusia orichalcea*), white fly (*Bemisia tabaci*) and green leaf hoppers (*Amrasca biguttula biguttula*), have been reported as the major insect pests of economic importance in India (Basappa and Prasad, 2005).

Wide array of natural enemies and pollinators are associated with sunflower ecosystem hence biorational strategies involving both botanical and microbial origin biopesticides, biocontrol agents, use of pheromones, cultural practices as well as novel insecticides are essential for keeping insect pest load below ETL. The conservation of activity of natural enemies and pollinators associated with sunflower is possible by using botanicals for insect pest and disease management (Basappa and Chattopadhyay, 1999). Several species of coccinellids, *viz. Cheilomenes sexmaculata*, *Menochilus sexmaculata*, *Brumus suturalis*, *Chilocorus nigritus*, *Coccinella septumpunctata*, *Scymnus nubillus* keep a good check on sucking pests like aphids and leaf hopper nymphs, whitefly and thrips. In addition, predatory lygaeid *Geocoris tricolor* and *Anthocoris* sp., several mantids such as *Eumantissa giglio*, *Ceriagrion coromandeliarum*, *Ichneura* sp., *Dolichopus* sp. and *Therevia* sp. play vital role in suppression of sucking pests (Basappa, 2007a).

In nature, several egg, larval and pupal parasitoids have been observed to interfere in the succession of *S. litura* at different growth stages. Among them, the egg parasitoid, *Trichogramma evanescens minutum*; larval parasitoids, *Apanteles prodeniae, Cotesia* sp., *Peribaea* sp.; predatory stink bug, *Cantheconidia furcellata* and pupal parasitoids, *Tetrastichus ayyari, Metopius* sp. and *Trichospilus pupivora* are important. Several general predators like spiders, *Chrysoperla,* predatory bugs, insectivorous birds, etc. also play important role in the suppression of the pest. Though more than 75 parasitoids and 30 predators have been reported on *H. armigera* in different crops, very few can be multiplied and used in the biological control programme. Predators like *Chrysoperla carnea,* Coccinellids, reduviid bugs, spiders and parasitoids such as *Bracon* spp., *Campoletis* spp. and *Trichogramma* spp., have been found to be predominant regulatory biocontrol agents against *H. armigera* (Basappa *et al.*, 2005).

Spraying biopesticides like NSKE (5%) or other botanicals against sucking pests, defoliators and head borer is found to be effective in keeping pest population below ETL. Several botanicals of neem origin have been evaluated against *H. armigera* in sunflower. Application of NSKE (5%) and neem based pesticides not only reduce damage due to *H. armigera* but also help in conserving the activity of natural enemies as well as honey bees. *Clerodendrum inerme* dust (25%) and plant extracts (10%) not only proved to be effective against *S. litura* and *H. armigera* larvae but also produced abnormal pupae and adults. Among biorationals, seed extracts of neem and melia (5%), oils of *Jatropha curcas* L. (2%), and karanja (*Pongamia pinnata* (L.) oils were very effective against leafhopper and whitefly. Castor (*Ricinus communis* L.), cotton (*Gossypium hirsutum* L.) and kapok (*Ceiba pentandra* L. Gaertn.) oils (2%), leaf extracts of neem and melia (10%) and *Clerodendrum inerme* Linn. (5%) were also moderately effective against sucking pests in sunflower (Basappa *et al.*, 2009a).

The microbial agents like *S. litura* NPV (SlNPV) 250 LE/ha, entomopathogenic fungi, *Nomuraea rileyi*, bacteria, *Bacilus thuringiensis* and *Bacillus cereus* can also be used for the management of this pest. Among microbial agents, which are effective against *H. armigera,* nuclear polyhedrosis virus of *H. armigera* (HaNPV), *Bacillus thuringiensis* (Bt) are promising. A significant reduction in pest density is achieved with the spray of HaNPV @ 250 LE/ha. HaNPV @ 250 LE/ha with UV protectants gave better protection than without UV protectants. Three sprays of HaNPV @ 500 LE/ha brought down pest population completely and recorded the highest yield. It is most important to mix UV-protectants with HaNPV formulation and to apply it in the evening hours. Two sprays of HaNPV - first at button stage of the crop followed by the second after 10 days would reduce the pest population. Bt liquid formulation @ 2 l/ha reduced head borer damage with a concomitant increase in the seed yield. Coverage of capitulum and upper plant parts thoroughly with NPV + Bt might give better control of this pest. Only 6.66% ear head damage was noticed when sunflower crop was sprayed with HaNPV @ 250 LE/ha + endosulfan 35 EC @ 1 ml/l water twice with an interval of 15 days beginning with flowering. Two sprays of NPV at 250 LE in combination with 0.5 kg Bt var. *kurstaki* formulation per ha also gave effective control of the sunflower head borer. Sequential application of endosulfan (0.035%) followed by *Sl*NPV (250 LE/ha) is effective compared to endosulfan and *Sl*NPV alone. Similarly, sequential application of NSKE (5%) followed by *Sl*NPV is more effective than NSKE and *Sl*NPV alone. *Sl*NPV @ 250 LE/ha with 0.2 % jaggery is more effective than alone in reducing larval populations of the pest (Basappa, 2004 and Basappa *et al.*, 2005). Biopesticides, *viz. B. thuringenisis kurstaki (Btk)* (250 and 500 ppm), NPV (1×10^7 and1×10^8 POB/ml) and azadirachtin (500 and 1000 ppm) affected *H. armigera* food intake, faeces and weight gain (Ravi, 2006). Safer biopesticides to natural enemies need to be identified. Neem seed kernal extract (NSKE 5%), custard apple seed extract (CASE 5%), pongamia seed extract (PSE 5%), neem oil (2%), pongamia oil (2%) and *B. thuringenisis* (DIPEL) were found to be safe to *Chilomenus sexmaculatus* and *Trichogramma chilonis* (Basappa, 2007c). Among biopesticides evaluated against *H. armigera*, HaNPV (DOR isolate) @ 2×10^8 POBs/ml was found to be superior with higher seed yield followed by NSKE (5%) and pongamia seed extract (PSE 5%) (AICRP, 2011b).

Biopesticides can be important component of cost effective and ecofriendly biorational approaches apart from summer ploughing which exposes resting stages of insects to predatory birds as well as to hot sun. Other ecofriendly biorational components include cultural practices like growing sunflower on slopes of ridges of 6–8 cm height to reduce damage due to cutworms, seed treatment with imidacloprid @ 5g/kg seed keeps sucking pests below ETL without affecting natural enemies, hand collection and destruction of egg masses, early and late instar of *S. litura* and *S. obliqua* as well as late instar larvae of other defoliators and *H. armigera,* intercropping of sunflower with cotton will result in lower thrips and leafhopper infestation on sunflower whereas *H. armigera and S. litura* will be low on sunflower in the sunflower + pigeonpea intercropping system (Basappa and Prasad,

2005). Since crop can be grown in different seasons in different regions hence it is possible to grow this crop by adopting biointensive management tactics which protect the crop as well as the environment from pesticide toxic stress (Singh and Ballal, 1999). Sunflower oil is one of the preferred oils by the consumer and confectionary value of sunflower seed is high in western countries hence there is untapped high export potential hence biopesticides are the strongest pest management option for sustainable production of sunflower apart from conserving the activity of wide array of natural enemies and pollinators associated with crop.

SAFFLOWER

Safflower is one of the oldest crops known to human beings in the world and grown for centuries in India. India ranks first in the area and production with a global share of 51.1% area and 39.4% production. It is a multi-purpose crop with unexploited potential and world adaptability. Major producers of safflower are India, United States, Mexico, Kazakhstan, Argentina, China, Ethiopia and Australia. Traditionally, the crop was grown for its seeds and flower petals were used for colouring and flavouring foods. For the last fifty years, safflower crop has been cultivated mainly for the vegetable oil extracted from its seeds. All the parts of the plant find useful applications in herbal medicine specifically in preparations to treat physical disorders. Safflower produces oil rich in polyunsaturated fatty acids which play an important role in reducing blood cholesterol level and is considered as a healthy cooking medium. A total of 101 insect pests have been recorded on safflower throughout the world. Though several insect pests have been reported on safflower in India, safflower aphid, *Uroleucon carthami* is the most destructive pest. The other economically important pests are safflower caterpillar (*Perigaea capensis*), gram pod borer (*H. armigera*) and capsule fly (*Acanthiophilus helianthi*). A number of coccinellids such as *Brumus suturalis, Coccinella septempunctata* and *Menochilus sexmaculata* have been reported predating on the safflower aphids. In safflower ecosystem (Singh *et al.*, 1999), lady bird beetles (coccinellids), *Chrysoperla* sp. and syrphids are the main predators of safflower aphid hence exploitation of natural enemies by selective restraining from use of insecticides need to be aimed. Bioratioanl strategies in safflower include sowing of the crop during October middle to reduce aphid damage, spraying of NSKE (5%) at 45 days after sowing followed by dimethoate (0.05%) spray at 15 days later against aphid, seed treatment with carbosulfan 25 DS @ 20 g/kg of seed to check the initial build up of aphid and the later population by peripheral application (1.8 m all around the field) with NSKE (5%) or dimethoate (0.05%) (DOR, 2007). Among biorationals, chill garlic extract was found to be effective against leaf eating caterpillars in safflower with seed yield of 1009 kg/ha followed by NSKE (5%), azadirachtin (1500 ppm), neem fruit extract and Dipel 8L with seed yield of 989, 986, 919 and 816 kg/ha, respectively (AICRP, 2010). There is potential for biopesticide use in safflower for the pest management as all the parts of the plant find useful applications in herbal medicine, colouring and flavouring foods and healthy cooking medium.

SESAME

Sesame is an important minor oilseed crop, usually grown as rainfed crop on marginal lands in India. India ranks first in area with a global share of 25.2% and second in production with a global share of 18.8%. Sesame is attacked by a number of pests at various growth stages. Of these, sesame shoot webber and pod borer, *Antigastra catalaunalis* Duponchel; sesame gall fly, *Asphondylia sesami* Felt; sphingid moth, *Acherontia styx* Westwood; and hairy caterpillars, *Spilosoma obliqua* Walker. and *Amsacta moorei* Butler are most serious throughout India.

Wide array of natural enemies are associated with sesame shoot webber. *Bracon hebator, B. brevicornis* and *Phanerotoma handecasisella* were reported on shoot webber larvae. Predators like *Cantheconidea furcellata* and *Cicindella* sp. were reported on larvae. Parasitoids, *Trathala flavo-orbitalis*, *Campoplex* sp., *Erioborus* sp., *Temelucha biguttula*, *Apanteles* spp. and *Cremastus flavo-orbitalis* were found parasitizing the sesame shoot webber and pod borer. Larval endoparasite, *C. flavo-orbitalis* was found to keep the population of *A. catalaunalis* under check almost throughout the year except during January and February when both the host and parasite are uncommon in field. The other minor parasitoids recorded were *B. gelechiae*, *Elasmus brevicornis* and *Meteorus* sp. *Diadegma* sp. was found parasitizing up to 50% *A. catalaunalis* larvae.

In sesame biopesticide like Dipel @ 400 g/acre exerted a sustained control over the incidence of *A. catalaunalis* and recorded higher yield compared to insecticides. Neem preparations like Margocide C.K. (5 ml/l), Neem gold (3 ml/l), neem oil (10ml/l) or NSKE (20 ml/l) were found to be effective on *A. catalaunalis* and neem gold and nimbicidine (5 ml/l) on gall fly. More than 16 species of parasitoids and predators have been found to attack the webber and 20–50% of the larvae were found parasitized and greater activity of predators can also be part of BIPM in sesame ecosystem. The botanical pesticides were found to be effective against *Spilarctia obiqua* and superior to control in seed yield. Pest management approaches consisting of sequential application of Bt 1 kg/ha, NSKE (5%), endosulfan (0.07%) and NSKE (5%) were most effective and economical against *A. catalaunalis* (Sundria, 2009). NSKE @ 5%, NLE. @ 2%, nimbecidine @ 0.15, neem oil @ 2%, mahuwa oil @ 2% and KSKE. @ 5% reduced the larval population of Bihar hairy caterpillar *Spilarctia obliqua* to the tune of 23.0, 17.6, 16.4, 15.3, 14.3 and 11.5%, respectively (Kumar and Ali, 2010). NSKE (5%), neem oil (3%) and azadirachtin (0.004%) were found to be effective against *A. catalaunalis* with higher seed yield than control (AICRP, 2011a). The crop has good export potential, hence biologically based pest management is highly essential to reduce residue problem in seeds.

NIGER

Niger is grown as sole crop in the neglected and undulated hilly slopes as well as intercrop or mixed crop along with row crops in India. Niger caterpillar , *Condica conducta* and *S. litura* defoliate the crop and cause considerable yield loss. In niger spraying of NSKE (5%) will keep defoliators and sucking pests below ETL as well as conserve the activity of larval parasitoids, *Apanteles ruficrus* and *Chilonis* spp. on *Condica conducta* and *S. litura* which had 82.2 and 33% parasitisation, respectively (Basappa and Singh, 1990). Early sowing and proper weeding of the fields reduces hiding places of pests, using rice bran poison bait to attract and kill the cutworms, hand picking and destruction of egg masses and gregarious early instar larvae of *S. litura* will not disturb natural enemies (Basappa *et al.*, 2000). Niger seed is exported as feed to western countries hence biopesticide utilisation will help to produce residue free seed.

CASTOR

Castor is one of the industrially important oilseed crops cultivated in India. Castor productivity in India is more than the world average. Among the major castor producing countries in the world, India ranks first in both area and production with a global share of 59.3 and 64%, respectively. In the castor ecosystem, 74 species of arthropods and 8 species of vertebrates were found to be associated which include 27 species of insects and 1 mite species as herbivores, 51 species of natural enemies (26 parasitoids, 14 insect predators, 3 spiders and 8 predatory birds) and 2 species of pollinators (Basappa *et al.*, 2009b). In castor ecosystem, biopesticides of plant origin like azadirachtin (0.3%), *Clerodendrum inerme* dusts 100, 15 and 10% were as effective as cartap hydrochloride against *Achae janata* with cost:benefit ratio of 1:2.61, 1:5.75, 1:4.02 and 1:4.14, respectively compared to 1:1.96 in cartap hydrochloride (Basappa and Lingappa, 2005). All the neem based products proved their superiority to *B. glabra* cold alcohol extract in reducing larval population of *A. janata*. The benefit from every rupee spent was two times in all cases of neem products (Basappa and Lingappa, 2004a).

Dust and cold alcohol extract of *C. inerme*, cold alcohol extract of *B. glabra*, neem seed kernel extract, neem leaf extract and commercial neem products were tested and all of them were found safe to egg parasitoid *T. chilonis* as well as larval parasitoid, *M. maculipennis* (Basappa and Lingappa, 2002a). Application of reduced dose of fenvalerate (0.005%) insecticide in combination with microbial biopesticide, *Bacillus cereus* (3.2×10^8 cells/ml) was more effective with cost:benefit ratio of 1:3.40 and it was found relatively safer to natural enemies than reversing the sequence and insecticide and pathogen alone (Basappa and Lingappa, 2003). In castor the BIPM package involving release of *T. chilonis* @ 2 lakh/ha at 30 days after sowing (DAS) followed by fenvelerate (0.01%) spray at 45 DAS and second release of *T. chilonis* @ 3 lakh/ha at 65 DAS proved highly cost effective with cost:benefit ratio of 1:5 and also encouraged activity of larval paprasitoid, *M. maculipennis* (Basappa and Lingappa, 2002).

BIPM package involving sequential application of fenvelerate (0.01%) at 35 DAS followed by neem based biopesticide (0.3%) at 60 DAS was effective against castor semilooper and with highest cost:benefit ratio of 1:4 when compared to methyl parathion (Basappa and Lingappa, 2004b). BIPM modules involving, summer ploughing, use of NSKE (5%), keeping bird perches @ 15/ha to attract predatory birds, wilt resistant variety, hand picking and destruction of egg masses and early stage larvae of *S. litura*, *S. obliqua* and other hairy caterpillars and spraying of carbendazim (0.05%) against botrytis disease was superior over farmers practice (non-IPM module) in keeping insect pests and diseases below ETL and it was relatively safer to the parasitoids and predators than IPM modules with chemical insecticides and non-IPM module (Basappa, 2003; Basappa, 2006). The yield in the BIPM module was 667 kg/ha with cost:benefit ratio of 1:3.12 compared to 548 kg yield with a cost:benefit ratio of

1:2.25 in case of non-IPM module (Basappa, 2007b). Castor semilooper and capsule borer incidence was low in the intercropping system compared to sole crop due to higher activity of natural enemies. BIPM module proved to be relatively safer to natural enemies with 16.13 and 66.10% average field parasitisation of *Achaea janata* eggs and larvae by *Trichograma chilonis* and *Microplitis maculipennis* compared to 6.92 and 21.2, 8.80 and 24.0, 8.92 and 22.35% in PIPM module, non-IPM modules 1 and 2, respectively. *M. maculipennis* cocoon number per plant increased with increase in duration after treatment in the BIPM module whereas it was decreased in CIPM and non-IPM modules. Similarly, average parasitisation by *Charops obtusus* (3.54%), *Rhogas* spp.(4.60%), *Apanteles hyposidrae* (13.85%), *Euplectrus maternus* (7.92%) on *A. janata* larvae and *Cotesia flavipes* (8.96%) on *S. litura* was high in the BIPM module compared to CIPM module (1.65, 1.59, 3.50, 2.37 and 2.49%) and non-IPM modules 1(1.16, 0.80, 2.70, 1.20 and 2.68%) and 2 (1.50, 1.19, 2.80, 3.82 and 1.99%), respectively. BIPM module had higher population of predators like *Chrysoperla carnea, Cheilomenes sexmaculata, Cantheoconidea furcellata, Rhynocoris kumarii* and spiders/10 plants (8.45, 3.41, 3.27 and 7.10) than CIPM module (1.75, 2.90, 1.38, 1.45 and 1.40) and non-IPM modules 1 (1.20, 1.82, 1.00, 1.00 and 1.80) and 2 (1.97, 2.41, 1.32, 1.22 and 2.90), respectively. Black drongo (*Dicrurus adsinilis*) and Indian myna (*Acridotheres tristis*) were the predominant predatory birds which used bird perches frequently while predating on *A. janata, S. litura* and *Conogathus punctiferalis* larvae in BIPM and CIPM modules compared to non-IPM modules (Basappa, 2009b). Biointensive integrated pest management (BIPM) module involving biopesticides, resistant varieties, cultural practices and mechanical method was found to be effective against major insect pests in castor seed production with highest cost:benefit ratio (1:4.40) followed by CIPM module (1:4.25), pesticide alone (1:3.34) and farmers' practice (1:1.92). BIPM module was found to be safer to natural enemies than other modules (Basappa, 2010)

LINSEED

Linseed bud fly, *Dasyneura lini* has been considered as a national key pest as losses caused are very alarming. In linseed, application of tobacco leaf extract (0.5%) after imidacloprid 200 SL and vice versa proved effective against budfly. Other BIPM components like early sowing, summer ploughing, crop rotation, intercropping with gram, lowered the bud fly damage. Both grubs and adults of *C.*

septempunctata and *Menochilus sexmaculatus* preyed on the maggots and reduced aphid and bud fly population up to 35 and 15%, respectively. Bird perches do attract insectivorous birds like pond heron (*Ardeola grayii*) which predates on *A. ipsilon* and black drongo (*Dicrurus adsimilis*) on the larvae of *H. armigera* and reduce capsule damage. Commercial neem formulation (5%) and NSKE (5%) were found effective and economically viable against budfly (Malik *et al.*, 2009). Plant origin biopesticides like NSKE (5%), neem seed extract (5%) and nimbicidine (0.5%) were found effective against linseed budfly with seed yield ranging from 1135 to 1214.3 kg/ha compared to 658.6 kg/ha in control (AICRP, 2009).

CONCLUSIONS

In recent years, newer approaches to deal with crop losses due to pests include biological control, integrated pest management and biotechnology. The approach stresses on a more ecological or 'system approach' based on the study of population biology at the local farm level. The approach will thus involve a true integration of renewable technologies such as host-plant resistance and natural biological control, which can be made available to even the most resource-poor farmers. The Biointensive integrated pest management (BIPM) approach involving biopesticides and other ecofriendly pest management options is essential for the management of insect pests and increase oilseed production. In future major emphasis will be laid on biopesticides as a biotechnological approach which is only a part of the total system approach.

Though biopesticides are ecofriendly pest management options which reduce the pesticide toxic stress in the oilseed ecosystem, yet there are several limitations. The effectiveness of all botanicals decreases with increase in exposure period thus botanical biopesticides have short persistence which lead to repeated application. In none of the neem based and other botanicals the persistence was beyond 3 days to cause 75% mortality in castor semilooper (Basappa and Lingappa, 2002b). There is ample scope for the utilization of microbial biopesticides in the management of pest complex of oilseed crops in India but there is lack of systematic approach in the utilization of bio pesticides of microbial origin in the pest management of oilseed crops (Basappa and Masood Khan, 2003). Systematic research and development efforts are needed in the development of ecofriendly, economically viable and socially acceptable biopesticides for pest management. For development of new and improved biorational approaches for the management of insect pests in the oilseed ecosystem enhancing our understanding of natural biological control agents and development of improved methods for conserving beneficial organisms within oilseed production systems is very essential. In depth knowledge is essential on biorationals like microbial pesticides, pesticides derived from plants and various new types of pesticides that have low non-target impacts and degrade into non-toxic components. Application of nanotechnology in developing nanocides for insect pest management in oilseeds can be a future line of research. The multidisciplinary research programme is required with a goal to develop bio pesticides that are cost effective and environmentally compatible.

Regular training programmes like Hands on Training on Mass production and Quality testing methods of biopesticides and biointensive integrated pest management (BIPM) of insect pests of oilseed crops are organized in India at the Directorate of Oilseeds Research, Hyderabad for personals working on biopesticides in the biocontrol laboratories of state agriculture departments, KVKs, state agricultural Universities and private pesticide industries. This provides sound knowledge on theoretical and practical aspects of biopesticide mass production, quality testing methods and utilisation for the management of insect pests (Basappa *et al.*, 2009c). Farmers' participatory training programme on proper mass production and utilization of biopesticides for the management of insect pests of oilseed crops is essential for sustainable production of oilseed crops.

REFERENCES

Agnihotri, M. and Khan, M.A. (2006) Bioefficacy of combined effect of biopesticides and insecticides against *Spilarctia obliqua,* on soybean. *Ann. Pl. Prot. Sci.*, 14, 45–48.

AICRP (2009) *Linseed, Annual Report.* PC Unit Linseed, Kanpur, 272 pp.

AICRP (2010) *Safflower, Annual Report.* Directorate of Oilseeds Research, Hyderabad,160 pp.

AICRP (2011a) *Sesame and Niger, Annual Report.* PC Unit Sesame and Niger, Jabjalpur, 296 pp.

AICRP (2011b) *Sunflower, Annual Report.* Directorate of Oilseeds Research, Hyderabad, 251 pp.

Akhauri, R.K and Singh, N.K. (2009) Bioefficacy of some chemical insecticides and bioproducts against mustard aphid, *Lipaphis erysimi* Kalt in yellow mustard. *J. Oilseeds Res.*, **26** (Special issue), 456–457.

Basappa, H. (2003) *Integrated Pest Management in Castor.* Directorate of Oilseeds Research, Hyderabad, India, 52 pp.

Basappa, H. (2004) Integrated pest management in sunflower: An Indian scenario. *Proceedings, 16th International Sunflower Conference*, Fargo, North Dakota, **2**, 853–859.

Basappa, H. (2006) Evaluation of IPM modules in the castor growing areas of Mahaboobnagar district of Andhra Pradesh. *Indian J. Plant Prot.*, **34**, 10–14.

Basappa, H. (2007a) Population of predators associated with leafhopper in sunflower. *J. Biol. Control*, **21**, 85–87.

Basappa, H. (2007b) Validation of integrated pest management modules for castor (*Ricinus communis*) in Andhra Pradesh. *Indian J. Agric. Sci.*, **77**, 25–30.

Basappa, H. (2007c) Toxicity of biopesticides and synthetic insecticides to egg parasitoid, *Trichogramma chilonis* Ishii and coccinellid predator, *Cheilomenes sexmaculata* (Fabricius). *J. Biol. Control*, **21**, 31–36.

Basappa. H. (2008a) Role of biopesticides in IPM of oilseed crops. In: O. Koul, G.S. Dhaliwal and V.K. Kaul (eds.), *Sustainable Crop Protection – Biopesticide Strategies*, Kalyani Publishers, New Delhi, pp. 261–283.

Basappa, H. (2008b) Biodiversity of arthropods and vertebrates in sunflower ecosystem, *Leslie Coleman Memorial National Symposium on Plant Protection*, University of Agricultural Sciences, Bangalore, Dec., 4–6, 2008.

Basappa, H. (2009a) Biorational approaches for the management of insect pests of oilseed crops. *National Symposium on Biotechnological and Biorational Approaches for Pest Management in Agriculture* University of Agricultural Sciences, Dharwad, Nov. 6–7, 2009.

Basappa, H. (2009b) Impact of integrated pest management modules on the activity of natural enemies in the castor (*Ricinus communis* L.) ecosystem. *J. Biol. Control,* **23**, 221–228.

Basappa, H. (2010) Evaluation of IPM modules in the farmers' participatory castor seed production involving tribal farmers in the Mahaboobnagar District of Andhra Pradesh. *J. Oilseeds Res.*, **27**, 278–281.

Basappa, H. and Chattopadhyay, C. (1999) Role of botanicals in the integrated pest management in sunflower. In: H. Basappa, Harvir Singh and C. Chattopadhyay (eds.), *Integrated Pest Management in Sunflower*, Directorate of Oilseeds Research, Hyderabad, pp. 93-103.

Basappa, H. and Linappa, S. (2002a) Management strategies for castor semilooper, *Achaea janata* Linn. (Lepidoptera: Noctuidae) in castor. *Indian J. Plant Prot.*, **30**, 51–54.

Basappa, H. and Lingappa, S. (2002b) Persistent toxity of botanical insecticides against castor semilooper, *Achaea janata* Linn. (Lepidoptera: Noctuidae). *Indian J. Plant Prot.*, **30**, 187–190.

Basappa, H and Lingappa, S. (2003) Evaluation of *Bacillus cereus* alone and in combination with certain insecticides against castor semilooper, *Achea janata* Linnaeus and its parasitoid. In: O. Koul, G.S. Dhaliwal, S.S. Marwaha and J.K. Arora (eds.), *Biopesticides and Pest Management*, Vol. 2, Campus Books International, New Delhi, pp. 147–155.

Basappa, H and Lingappa, S. (2004a) Evaluation of various neem preparations against castor semilooper, *Achaea janata* Linn. (Lepidoptera : Noctuidae) under field conditions. *J. Res. ANGRAU*, **32**, 7–11.

Basappa, H and Lingappa, S. (2004b) IPM modules for the management of castor semilooper, *Achaea janata* Linn. and their impact on larval parasitoid, *Microplitis maculipennis* Zepligate. *Indian J. Plant Prot.*, **32**, 131–132.

Basappa, H and Lingappa, S. (2005) Field evaluation of *Clerodendrum inerme* against castor semilooper, *Achaea janata* and its safety to larval parasitoid, *Microplitis maculipennis Indian J. Plant Prot.*, **33**, 223–225.

Basappa, H and Masood Khan, M.A. (2003) Role of microbial biopesticides in the management of key pests of certain major oilseed crops. In: O. Koul, G.S. Dhaliwal, S.S. Marwaha and J.K. Arora (eds.), *Biopesticides and Pest Management*, Vol. 2, Campus Books International, New Delhi, pp. 27–39.

Basappa, H and Santha Lakshmi Prasad, M. (2005) *Insect Pests and Diseases of Sunflower and their Management*, Directorate of Oilseeds Research, Hyderabad, India.

Basappa, H and Singh, V. (1990) Record of insect pests on Niger (*Guizotia abyssinica*). *J. Oilseeds Res.*, **7**, 86–87.

Basappa, H, Singh, H. and Patil, B.R. (2009a) Non-conventional oilseeds as source of biopesticides for the management of sucking pests in sunflower. *J. Oilseeds Res.*, **26**, 431–433.

Basappa, H, Lakshminarayana, M., Singh, V and Singh, H. (2000) Pest complex of niger (*Guizotia abbyssinica* Gass) and assessment of yield losses. *National Seminar on Oilseeds and Oils Research and Development needs in the Millenium*, February 2–4, 2000, ISOR, Hyderabad.

Basappa, H, Patil, B.R., Kotagi, K. and Reddy, R.G.Y. (2009b) Biodiversity of arthropod pests and their natural enemies in castor ecosystem. *J. Oilseeds Res.*, **26**, 707–709.

Basappa, H., Vimla Devi, P.S., Prasad, R.D. and Raoof, M.A. (2009c) *Training Manual: Hands on Training on Mass production and Quality Testing Methods of Biopesticides*. Directorate of Oilseeds Research, Hyderabad, India.

Basappa, H., Vimla Devi, P.S, Prasad, R.D., Singh, H., Basu, M., Duhoon, S.S.S, Srivastava, R.L, Singh, V., Raoof, M.A, Lakshminarayana, M,, Santha Lakshmi, Prasad, M., Sharma, A.N and Malik, Y.P. (2005)

Biocontrol in oilseed crops. In: D.M. Hegde (ed.), Directorate of Oilseeds Research, Hyderabad, pp. 24–34.

Damodaram, T. and Hegde, D.M, (2010) *Oilseeds Situation : A Statistical Compendium*. Directorate of Oilseeds Research, Hyderabad.

Dey, S., Sinha, B. and Kalita, J. (2005) Effect of *Eupatorium adenophorum* Spreng leaf extracts on the mustard aphid, *Lipaphis erysimi* Kalt: a scanning electron microscope study. *Micros. Res. Techn.*, **66**, 31–36.

DGR (2010) *Annual Report*. Directorate of Groundnut Research, Junagadh, 152 pp.

Dhaliwal, G.S. and Koul, O. (2007) Biopesticides and Pest management: Conventional and Biotechnological Approaches. Kalyani Publishers, New Delhi.

DRMR (2010) *Annual Report*. Directorate of Rapeseed-Mustard Research, Bharatpur, 65 pp.

DSR (2010) *Annual Report*. Directorate of Soybean Research, Indore, 113 pp.

DOR (2007) *Safflower: Package of Practice for Increasing Production*. Directorate of Oilseeds Research, Hyderabad.

Grainge, M., Ahmed, S., Metchekl, W.C, and Hylin, J.W. (1985) *Plant Species Possessing Pest Control Properties*. EWS/UH Data Base Resource Systems, Institute East West Centre, Honolulu.

Hegde, M., Patil, R.K, Parmeshwarappa, K.G and Awaknavar, J.S. (2009) Mnagement of *Spodoptera litura* on groundnut crop, *Arachis hypgaea* L. in northern transitional zone of Karnataka. *J. Oilseeds Res.*, **26**, 456–457.

Kapadia, M.N., Butani, P.G, Jetbva, D.M., Virani, V.R. and. Beria, N.N. (2010) Management of white grub, *Apogonia rauca* (F.) in groundnut. *Ann. Plant Prot. Sci.*, **8**, 86–89

Kuldeep, Azizur-Rahman, S.M., Joshi, N.K and Shri-Ram (2005) Efficacy of biorational insecticides against tobacco caterpillar, *Spodoptera litura* infesting soybean crops. *J. Appl. Biosci.*, **31**, 21–24.

Kumar, R. and Ali, S. (2010) Efficacy of botanical pesticides against *Spilarctia obliqua* in *Sesamum indicum*. *Ann. Plant Prot. Sci.*, **18**, 223–282.

Lingappa, S., Basappa, H. and Holihosur, H.N. (1993) Use of botanicals in IPM of oilseeds. *Group Discussion on IPM Strategies in Oilseeds in India*, 23–24 December, 1993. Punjab Agricultural University, Ludhiana.

Lingappa, S and Patil, R.K. (2002) *Nomuraea rileyi-A Potential Mycoinsecticide*, University of Agricultural Sciences, Dharwad.

Malik, Y.P., Chandra, R., Srivastava, R.L. and Saxena, M.C. (2009) Evaluation of neem based biorational insecticides against budfly, *Dasynuera lini* infestation in linseed, *Linum usitatissimum* L. *J. Oilseeds Res.*, **26**, 481–482.

Navi, S.S. (2002) Ecosystem manipulation to enhance the efficiency of *Nomuraea rileyi* (Farlow) Samson. M.Sc.(Agri) Thesis, University of Agricultural Sciences, Dharwad.

Pande, R., Firake, D.M. and Yadav, S. (2008) Effect of NSKE with additives, synthetic and biopesticides against *Spilarctia oblique,* defoliators of soybean. *Ann. Plant Prot. Sci.*, **16**, 488–490.

Ranga Rao, G.V., Wightman, J.A., Arjuna Rao, P., Mahesh Babu, V., Sitharama Rao, G. and Rameshwara Rao, V. (1995) Integrated pest management of groundnut pests, Paper presented at *National Symposium on Plant protection and Environment.* February, 2–4, 1995, Madras, Tamilnadu.

Ravi. M, Dhandapani. N, Sathaiah, N. and Murugan, M. (2006) Consumption, digestion and utilization of biopesticides toxicated sunflower leaves by *Helicoverpa armigera. Ann. Plant Prot. Sci.*, **14**, 294–296.

Sahayaraj, K. and Amalraj, A. (2005) Impact of monocrotophos and neem oil mixture on defoliator management in groundnut. *J. Food Agric. Environ.*, **3**, 313–315.

Satyanarayana, N.V.V, Ramatbandra Rao, G and Arjuna Rao, P. (2010) Incidence and management of *Spodoptera litura* (Fab.) on post rainy season groundnut, *Ann. Plant Prot. Sci.*, **18**, 22–25.

Singh, S.P. and Ballal, C.R. (1999) Role of biocontrol in IPM. In: H. Basappa, Harvir Singh and C. Chattopadyay (eds.), *Integrated Pest Management in Sunflower*, Directorate of Oilseeds Research, Hyderabad, India, pp. 57–65.

Singh, V., Basappa, H. and Hegde, D.M. (1999) Safflower pests and their integrated management in India. *Seasame and Safflower Newsletter*, **14**, 117–124.

Sundria, M.M. (2009) Efficacy of botanical pesticides and endosulfan against sesame capsule borer, *Antigastra catalaunalis* under field conditions. *J. Oilseeds Res.*, **26**, 465–466.

USDA (2013) *Oilseeds: World market and Trade.* Foreign Agricultural Service, Circular Series, FOP045-13.

Vyas, H.G., Joshi, D.P., Patel, K.C., Yadav, D.N. and Dodia, J.F. (1986) Natural incidence of milky disease of white grubs in Gujarat. *Indian J. Agric. Sci.*, **56**, 213–214.

Wightman, J.A. and Amin, P.W. (1988) Groundnut pests and their control in semi-arid tropics. *Trop. Pest Manage.* **34**, 218–226.

6

Biopesticides in Insect Pest Management in Cotton

VIJAY KUMAR*[1], P. S. SHERA[2] AND A. K. DHAWAN[1]

[1]Department of Entomology, Punjab Agricultural University, Ludhiana-141004, India

[2]Department of Plant Breeding, Punjab Agricultural University, Ludhiana-141004, India

*E-mail: vijay_ento@pau.edu

INTRODUCTION

Cotton (*Gossypium* spp.), is an important commercial crop grown for its oil and fiber in India and plays a vital role in social and monetary affairs of the country. India is the only country in the world where all the four cultivated species of cotton, viz. *Gossypium arboreum* L., *G. hirsutum* L., *G. herbaceum* L. and *G. barbadense* L. along with intra-and inter-specific hybrids are cultivated along the diverse agroclimatic conditions, varying from 8–32°N latitude and 70–80°E longitude on commercial scale spanning around 12.20 million ha. Among the cotton growing countries of the world, India has a place of prominence with respect to area under this crop contributing about 34% of the world cotton area, but with regard to production, it is ranked next to China, the top producer. The production increased from a meagre 2.8 million bales (170 kg lint/bale) in 1947–48 to a high of 17.6 million bales in 1996–97 and an all time record of 34.5 million bales during 2011–12. Among cotton growing states, Gujarat leads in production with 11.7 million bales followed by Maharashtra (7.5 million bales) and Andhra Pradesh (5.5 million bales) in 2011–12.

However, the productivity of cotton in India (482 kg/ha) is still far less than other major cotton growing countries of the world, viz. Australia (1802 kg/ha), Brazil (1446 kg/ha), China (1326 kg/ha), USA (865 kg/ha), Pakistan (721 kg/ha) and Uzbekistan (682 kg/ha) (AICCIP, 2012).

Besides other causes, major bottleneck in cotton cultivation is biotic stresses due to attack of insect pests, diseases and competition from weeds which plays a significant role in achieving optimum yield potential. As many as 1326 species of insects have been listed on the cotton crop worldwide (Hargreaves, 1948) and 162 species have been reported on the cotton crop in India, of which 24 species have attained pest status which also includes those, of little or no economic importance, sporadic, casual or accidental visitors so far as cotton is concerned (Sundramurthy and Chitra, 1992). More than 130 species of arthropods including 54 pestiferous species were found to be associated with cotton crop in Punjab. Among these, the bollworm complex comprising of American bollworm *Helicoverpa armigera* (Hübner); spotted bollworm *Earias vittella* (Fabricius); spiny bollworm *E. insulana* (Boisduval) and pink bollworm *Pectinophora gossypiella* (Saunders) remained as key key pests in the state before the introduction of Bt cotton. The monetary value of yield losses due to insect pests has been estimated to be ₹ 2,87,600/- million annually (Dhaliwal *et al.,* 2004). The extent of losses caused by sucking pests, bollworms and both sucking pests as well as bollworms have been recorded up to 12, 44 and 52%, respectively on *hirsutum* cotton and 47% due to bollworms alone on *arboreum* cotton (Dhawan *et al.,* 1988).

Due to non availability of resistant cultivars and effective bioagents along with constraints in adoption of cultural practices, insecticides are the mainstay of integrated pest management (IPM) in cotton. However, insecticide resistance has rendered insecticides ineffective, thus increasing need for repeated applications, wastage of resources and consequent environmental pollution. The total consumption of pesticides in Punjab has increased from 3200 MT technical grade in 1980–81 to 7200 MT in 2001-02, cotton being the maximum shareholder. However, there has been reduction in pesticide load ever since transgenic cotton has been commercialized in Punjab. The total consumption of pesticides was 4500 MT during 2005-06 in Punjab and got reduced to 2600 MT in 2010 (Anonymous, 2010). Besides increasing cost of production and environmental problems, the excessive and indiscriminate use of insecticides for the control of these pests has resulted in development of insecticidal resistance particularly in *H.*

amigera (Kapoor *et al.*, 2002; Gill and Dhawan, 2006), decline in natural enemies' population and resurgence of the pests like whitefly, *Bemisia tabaci* (Gennadiaus) (Dhawan and Simwat, 1997; Butter and Kular 1999; Virk *et al.*, 2004) and jassid, *Amrasca biguttula biguttula* (Ishida) (Dhawan and Simwat, 1997).

Therefore, for a country that consumed 50% of insecticide alone on cotton crop despite growing on 5% of cropped area, any technology that reduces dependence upon chemicals was an immediate priority which could not only check the menace of this pest but also help in scaling down the pesticide load for cotton sustainability. Prominently, Bt cotton has now been playing a major role in effectively protecting the crop from bollworms, especially the *H. armigera*, thus preventing yield losses. Indian cotton farmers, who were displeased with the intractable problem of bollworms in cotton, responded positively to Bt cotton and strongly supported the technology, to the extent that in the history of agriculture, there has never been ever before, an example of such a rapid technology adoption, especially with reference to the spread in area under Bt cotton in India. The introduction of genetically engineered Bt cotton hybrids, viz. Bollgard (Cry1Ac) and Bollgard-II (Cry1*Ac* + Cry2*Ab*) conferring in-built resistance to bollworms and both bollworms as well as tobacco caterpillar, respectively has given a new impetus to IPM programme with significant reduction in number of sprays against these insect pests. The integration of Bt cotton as a component of IPM module augmented with suitable cultural, mechanical control strategies, sowing of *refugia* around the main field and use of biopesticides will not only reduce pesticide load and make cotton cultivation economical but will also get opportunity to access global market.

Keeping in view these problems, the need of the hour is to give up strategy of over reliance on chemicals in this era of World Trade Organization (WTO) by switching over from traditional method of chemical control to ecological and economical sound integrated pest management (IPM) strategies involving use of biopesticides, thereby, conserving natural biodiversity. Biopesticides include many aspects of pest control such as microbial, encouragement of natural enemies (parasitoids, predators), plant-derived pesticides, secondary metabolites from microorganisms, pheromones, use of biorational compounds (semiochemicals, insect growth regulators, etc.) and transgenic crop varieties in pest management. The recent developments in the management of insect pests of cotton based on use of biopesticides, an important component of IPM, are discussed below.

PARASITOIDS AND PREDATORS

Biological control agents have played a pivotal role in regulating pest populations. A rich fauna of naturally occurring bioagents have been reported on insect pests of cotton crop. Out of several hundred species of natural enemies recorded in cotton ecosystem, only a few species are economically important for reducing the population of insect pests (Singh, 1990, 1992). The predators, *viz.* coccinellids, chrysopids and syrphids feed upon soft bodied insects like aphids, whiteflies, thrips, mites, eggs and early instar larvae of bollworms (Jayaswal and Sundaramurthy, 1992). However, not much field work has been done on their utilization for pest control on cotton. Parasitoids, on the other hand, have been utilized on large scale for bollworm control. To conserve natural enemies, an approach of no spray during first three months after sowing to conserve natural enemies and if needed, use of relatively safe insecticide and avoiding use of broad spectrum organophosphates especially as early season sprays have been advocated as a part of insecticide resistance management strategy in Punjab (PAU, 2008).

Parasitoids

The work on mass rearing of useful parasitoids under controlled conditions and releasing them on large scale in the field for the control of insect pests especially bollworms have also been exploited in cotton agroecosystem. Brar *et al.* (2003, 2004) observed that efficacy of *Trichogramma chilonis* alone to be less effective under Punjab conditions, but can be integrated with insecticides. According to Singh *et al.* (2004), field releases of bioagents, i.e. *T. chilonis* and *Chrysoperla carnea* failed to provide desired kill of bollworm complex. Hence, strenuous efforts have been made by cotton entomologists to evaluate different biocontrol based IPM modules for the control of bollworm complex of cotton. Eight releases of *T. chilonis* at 150,000/ha at weekly interval during July-August to October + one release of *C. carnea* at 10,000/ha + need based application of insecticides for the control of sucking pests (3 sprays) and for bollworms (3 sprays) during September-October was found effective and on par with spray schedule (3 sprays for sucking pests and 6 sprays for bollworms) in reducing bollworm damage and increasing the yield (Brar *et al.*, 2001). Eleven to thirteen releases of multiple insecticide tolerant/Bathinda strain of *T. chilonis* at the rate of 1,50,000/ha at weekly interval during July to October along with 6–9 insecticidal sprays proved significantly better than spray schedule (6 to 9 sprays) for the control

Table 6.1. Efficacy of some parasitoids against insect pest of cotton

Sr. No	Parasitoid	Host species*	Parasitism (%)	Reference(s)
1	*Apanteles angaleti* Mues.	PBW	1.0–14.3	Sekhon, 1980
		PBW	20.0–25.0	Varma and Shenhmar, 1988
2	*Aphelinus* sp.	SBW	1.0–10.0	Varma and Shenhmar, 1988
3	*Agathis fabiae* (Nixon)	SBW	2.0	Varma and Shenhmar, 1988
4	*Bracon* sp.	ABW	5.0–10.0	Bilapate, 1985
		SBW	7.55	
5	*B. greeni* Ashmead	PBW	1.0–6.6	Sekhon, 1980
		SBW	2.0–6.3	Varma and Shenhmar, 1988
		PBW	7.0–10.0	
6	*Chelonus* sp.	PBW	0.5–3.6	Sekhon, 1980
		PBW	3.0	Varma and Shenhmar, 1988
		ABW	7.6	Bilapate, 1985
7	*Campoletis chlorideae* Uchida	ABW	-	Bilapate, 1989
8	*Carcelia illota* Curran	ABW	6.1–12.0	Natarajan and Sundaramurthy, 1990b
9	*Chalcis* sp.	ABW	3.2	Bilapate, 1985
10	*Camptothlipsis* sp.	PBW	0.4–7.1	Sekhon, 1980
11	*C. gossypiella* Gupta and Bhut	PBW	7.0	Varma and Shenhmar, 1988
		PBW	5.0	
12	*Eritmocerus serius* Silv.	Whitefly nymphs	40.2	Rao *et al.*, 1989
13	*Elasmus johnstoni* Ferri	PBW	0.3–0.7	Sekhon, 1980
		PBW	10.0–12.0	Sekhon and Varma, 1983
14	*Erythmelus*	SBW	5.0	Varma and Shenhmar, 1988
	Empoasca Subha Rao	SBW	2.0	
15	*Goniozus* sp.	PBW	0.3–16.7	Sekhon, 1980
16	*Pyemotes ventricosus* (Newport)	PBW	29.0–39.0	Dhawan *et al.*, 1981
17	*Rogas* sp.	SBW	2.0–12.5	Sekhon, 1980
18	*R. aligharensi* Quadri	SBW	59.07	Bilapate, 1985
		PBW	0.4–1.0	
19	*Rogas aligarhensis* Quadri	BW	33.0	Varma and Shenhmar, 1988
20	*Scambus lineipes* Morley	ABW	0.3–16.7	Sekhon, 1980
21	*T. chilonis* Ishii	SBW	30.0–35.0	Varma and Shenhmar, 1988
		ABW	35.0–83.0	Singh, 1990

Sr. No	Parasitoid	Host species*	Parasitism (%)	Reference(s)
22	*Trichoramma achaeae* Nagaraja and Nagarkatti	SBW	5.9–27.2	Sekhon, 1980
		SBW	4.0–60.8	
23	*T. chilotraeae* Nagaraja and Nagarkatti	SBW	30.0–35.0	Sekhon and Varma, 1983
		SBW	2.7–15.0	Sekhon, 1980
24	*T. brasiliensis* Ashmed	SBW	5.0	Varma and Shenhmar, 1988
25	*Trichogrammatoidea* Sp. Near guamensis Nagaraja	PBW	21.4–25.0	Sekhon , 1980
		SBW	4.9–41.9	Sekhon and Varma, 1983
		SBW	35.0–45.0	

*PBW= Pink bollworm; SBW= Spotted bollworm;
ABW= American bollworm, BW=Bollworms.

of bollworms and increasing egg parasitism and seed cotton yield (Brar *et al.,* 2007). Combined releases of *T. chilonis* (1.5 lakh parasitized eggs/ha) with different rates of *C. carnea* were comparable with endosulfan (0.07%) and reduced the infestation in squares, flowers and green bolls in Maharashtra, India (Panchbhai *et al.*, 2005). Two releases of *T. chilonis* with two sprays of Btk and *Beauveria bassiana* resulted in the reduction of larval population (0.33/plant), less damage on intact squares (9.12%), bolls (14.20%) and loculi (10.68%) and recorded higher yield (782 kg/ha). Similarly, lowest larval population, minimum damage of *H. armigera* and higher seed cotton yield was obtained in 2 releases of *C. carnea* and 2 sprays of *Bacillus thuringiensis* var. *kurstaki* (Btk) treated plots followed by *C. carnea* in conjunction with HaNPV and Btk alone treated plots in Tamil Nadu, India (Balakrishnan *et al.,* 2004).

The parasitoid *Encarsia* spp. was found to be the most important mortality factor in the pupal stage of *B. tabaci* particularly in hot, dry conditions (de Albergaria *et al.,* 2003). However, *B. tabaci* (B biotype) can not be controlled solely by the use of naturally occurring parasitoids but can be integrated with chemical control to reduce crop damage. Several species of egg, larval and pupal parasitoids of bollworms, which cause parasitization of host to a variable degree in the field, are given in Table 6.1. Among these species, *Trichogramma achaeae* Nagaraja and Nagarkatti, *Trichogramma chilonis* Ishii, *Bracon greeni* Ashmead, *Bracon hebetor* Say, *B. brevicornis* Wesm., *Apanteles angaleti* Mues, *Chelonus* sp., *Chelonus narayani* Subha Rao, *Campoletis chlorideae* Uchida and *Rogas aligarhensis* Quadri are some important indigenous parasitoides, while *Trichogramma brasiliensis* Ashmead, *Bracon kirkpatricki* (Wilk.) and *Chelonus blackburni* Cameron are exotic species, which have given good control of boll-worms in the field (Jayaswal and Sundaramurthy, 1992). The use of broad spectrum insecticides has reduced the population of parasitoids in cotton ecosystem. Therefore, conservation and augmentation of natural enemies by inoculative and inundative releases is necessary for achieving effective control of cotton bollworms.

Different experiments have been conducted on the use of natural enemies on cotton. *B. greeni* gave satisfactory control of *Earias* sp. at Karnal in Haryana (Khan and Roa, 1960). The egg parasitoid, *T. chilonis* reduced the infestation of pink bollworm by 25% in colonized fields as compared to control plots (Cherian and Margabandhu, 1943). *T. brasiliensis* parasitized up to 89% eggs of spotted bollworms (Sangwan *et al.,* 1974). Varma *et al.* (1976) reported the effectiveness of this egg parasitiod against spotted bollworms in Punjab. Mass releases of *T. chilonis* and *T. achaeae* at 40,000 to 50,000/ha/release and larval parasitoids, *C. blackburni* and *B. kirkpatricki* at 15,000 to 20,000/ha/release during initial stage of formation of fruiting bodies reduced the incidence of all 3 species of bollworms (Agarwal and Gupta, 1986). Inundative releases of *T. brasiliensis, B. kirkpatricki* and *C. blackburni* at 7–10 days interval parasitized 37–89 and 64.7–72.0% population of pink bollworm and spotted bollworms in Haryana and Punjab, respectively (Singh, 1985). These parasitoids were released and also recovered from bollworms in southern states. Seven releases of *T. brasiliensis, A. angaleti* and *C. blackburni* at 10-day interval from 45 days age of crop reduced the infestation of bollworms by 55.5% in Maharashtra (Dhumal *et al.*, 1982).

T. brasiliensis parasitized 64 and 70% eggs of *E. vittella* and *H. armigera,* respectively in the laboratory, but field results were not encouraging. *R. aligarhensis* parasitized 9.3–35.5% larvae of *E. vittella* in Gujarat (Singh, 1990). *T. chilonis* has been reported to be effective against bollworms in the early phase of crop growth in Gujarat particularly on semidwarf varieties (AKH 81, LRK 516) and dwarf hybrids (DHB 1 and CD HB 2) and may, therefore, enhance the effectiveness of the parasitoid (Sundaramurthy and Chitra, 1992). The inundative releases of *T. chilonis* at 2 lakh/ha or *Chrysopa scelestes* Banks at 1 lakh/ha at weekly interval during peak oviposition period of pest have been suggested for control of *H. armigera* (Singh, 1990). In Punjab, release of *T. chilonis* at the rate of 1.5 lakh/ha/release at 10-day interval during July-October reduced bollworm damage by 20–45% (Varma and Gill, 1992). In large field trials, field releases during effective boll formation period on *hirsutum* cotton were compared with 6–8 sprays of chemicals and untreated check for control of pink bollworm and spotted bollworms during 1987, 1989 and 1990. It was observed that chemicals and biocontrol agents were significantly better than control for management of bollworms and increase in the seed cotton yield. Parasitoids were, however, less effective than insecticides. These experiments revealed that parasitoids with mass releases are effective at initial stage of bollworm infestation, which may thus delay the application of insecticide spray and reduce insecticide pressure on cotton. Extreme environmental conditions, insecticide toxicity and initial high cost of biological control agents are some of other limitations in their success.

Predators

Naturally occuring native predators viz., *Chilomenes sexmaculatus* and *Chrysoperla carnea* offer significant control of the early season sucking insect pests. A predatory prey ratio of 1:5 in respect of jassids and 0:1 for aphids was found optimal for natural control in presence of coccinellids and chrysopids. As the use of broad spectrum insecticides like organophosphorus components for sucking pest control eliminates these natural enemies, strategy of using sucking pest tolerant genotypes in conjuction with natural enemy exploitation is advocated. Hymenopterous and tachinid parasitoids, *Campoletis chlorideae*, *Microchelonus* spp, *Palexorista laxa*, *Carcelia illota* and *Goniophthalmus halli* are common on *H. armigera* larvae with parasitisation ranging from 9–12% while *Rogas aligarhensis* parasitisation on *Earias vittella* larvae varies between 4 and 18%. Pink bollworm control by *Apanteles angaleti* and *Bracon greeni* is 2 and 8%, respectively. Spiders and birds also execute a fair amount of natural control of cotton insect pests, however, their potential remains unestimated so far. Mesbah *et al.* (2008) found six insect predators, *Coccinella undecimpunctata*, *Scymnus interruptus*, *Paederus alfierii*, *Syrphus* spp., *Orius* spp. and *Chrysoperla carnea*. *C. undecimpunctata* showed significant correlation with *Spodoptera littoralis* eggs. No parasitism was recorded on *P. gossypiella* and *E. insulana* larvae. Similarly, Joshi *et al.* (2010) recorded various effective biological control agents *Aenasius bambawalei*, *Anagyrus kamali*, *Cryptolaemus montrouzieri*, *Chrysoperla carnea* against *Phenacoccus solenopsis*.

MICROBIAL AGENTS

The exploitation of disease causing organisms, i.e. microbial pathogens to reduce the population of insect pests below the damaging levels is referred as microbial control. So far over 3000 microorganisms have been reported to cause diseases in insects. Nevertheless, scientists familiar with specific pathogen groups agree that a very large number of insect pathogens remain undiscovered or unidentified. Various disease causing organisms perform an important role in the dynamics and natural regulation of insect populations. Among different microbial agents, the utilization of nuclear polyhedrosis virus (NPV) and bacterium *Bacillus thuringenesis* Berliner (Bt) formulations have been tested extensively for the control of cotton pests in USA, Canada, Egypt, Brazil, etc. The research on the utilization of these microbes for pest control on cotton in India is scanty. The world's largest programme for the use of an entomopath-ogen to control a pest on a single crop involved the management of soybean caterpillar, *Anticarsia gemmatalis* (Hubner) by its nucleopoly-drovirus (AgNPV). The use of AgNPV in Brazil has generated substantial economic, ecological and social benefits (Moscardi, 2007).

Nuclear Polyhedrosis Virus

Viruses have been isolated from more than 1000 species of insects from at least 13 different insect orders. World over, about 525 insect species in 52 families and 8 orders are known to be infected by nuclear polyhedrosis virus (NPV) and of this large number of species, 455 belong to Lepidoptera. Among the lepidopterans, noctuids comprise of the largest family with 107 species (Dhaliwal and Koul, 2010). Baculoviruses (NPVs) are invertebrate-specific pathogens of which over 500 isolates/species have been documented. Several wild type NPVs have been registered and used commercially with limited success (Gard, 1997).

Tobacco caterpillar. A locally produced wettable powder formulation of a nuclear polyhedrosis virus (NPV) controls the *Spodoptera littoralis* on cotton (Jones *et al.,* 1994). The highest mortality (91.7%) of fourth instar larvae of *Spodoptera litura* of nuclear polyhedrosis virus (NPV) treated food plants and a semi-synthetic diet was recorded (Baskaran *et al.*, 1996). Mixture of GV and NPV infecting 3rd to early 4th instar larvae of *S. littoralis* caused 100% mortality. High volume of NPV at 250–500 LE/ha containing 0.05% Triton-X-100 as emulsifier after 4 PM reduced 80.6–87.7% population of 2nd to 3rd instar larvae of *S. litura* on cotton. NPV at 250 LE/ha was as effective as chlorpyriphos 20 EC (2.5 l/ha) and caused 58.5 and 51.4% mortality, respectively of 4th and 5th instar larvae of this pest. The spray of NPV at 125 LE/ha in combination with methamidophos at 1.25 l/ha was as effective as normal dose (2.5 l/ha) of insecticide killing 67.4–72.0% larvae of tobacco caterpillar (Jayaral *et al.*, 1981). Moroccan isolate of NPV isolated from *Autographa californica* (Speyer) was effective against larvae of the *S. littoralis* in cotton (Abul Nasr *et al.*, 1980). Feeding of *S. litura* on the cotton leaves treated with NPVat the rate of 250-300 LE/ha on 45 days old cotton crop resulted in 69% mortality (Kulkarni and Hugar, 1999). The nuclear polyhedrosis viruses of *S. litura* caused only 0.94–1.21% mortality of *Chrysoperla carnea*, indicating that these viruses can be used safely in conjunction with *C. carnea* for integrated control of cotton pests (Balasubramani and Swamiappan, 1998).

American bollworm. The NPV spray at 250–500 LE/ha containing 0.5% jaggery and 0.1% Ranipal at weekly interval during peak oviposition period of pest has been recommended for control of *H. armigera* on cotton (Reddy and Krishnamurthy, 1989). Similarly, alternate sprays of NPV at 180 LE/ha and fenvalerate effectively control this pest. However, NPV alone was significantly less effective than insecticide for control of *H. armigera* (Jayaswal and Sundaramurthy, 1992). Sunlight and crop environment significantly reduced the activity of NPV *of H. armigera* after 2 days of application. The addition of many adjuvants like crude sugar at 15% + larval extract of *H. armigera* at 4% to virus spray has been recommended to increase the persistence of virus under natural sunlight and shade conditions (Dhandapani *et al.*, 1992). The nuclear polyhedrosis viruses of *H. armigera* caused only 0.94–1.21% mortality of *C. carnea*, indicating that these viruses can be used safely in conjunction with *C. carnea* for integrated control of cotton pests (Balasubramani and Swamiappan, 1998). PAU isolates (NPV of *H. armigera*) with an LC_{50} value of 0.3593 LE/litre was most effective against *H. armigera* (Arora *et al.*, 1997). The crude cane sugar (15–20%) when added to NPV and applied as an ultra low volume (ULV) spray against *H. armigera* infesting cotton attracted the formicids, *Camponotus sericeus*, *Solenopsis geminata*, *Prenolepis longicornis* and *Monomorium indicum*. All formicids, except *C. sericeus*, preferred to prey upon virus-killed cadavers. *C. sericeus* preferred healthy larvae of *H. armigera* whereas *P. longicornis* preferred the dead/diseased larvae (Dhandapani *et al.*, 1994).

Srinivasa *et al.* (2008) revealed that HaNPV of Pest Control India, Pvt Limited, Bangalore, Margo Bio Pesticides, Bangalore and Project Directorate of Biological Control, Hebbal, Bangalore were on par with each other in controlling *H. armigera.* Kambrekar *et al.* (2008) observed an additive effect of Bt and NPV, which resulted in higher larval mortality of *H. armigera* on Bt plants. NPV sprays applied alone resulted in significantly less seed cotton yield and higher larval population and extent of damage as compared to NPV in combination with insecticides or NPV alternated with insecticides which was on par with insecticidal sprays (Dhawan *et al.*, 2001). Early season application of NPV led to increase *Heliothis* the susceltilility of *H. armigera* (Meade *et al.*, 1995). One application of nuclear polyhedrosis virus, a wettable powder and liquid suspension at the rate of virus at 10×10^{10} PIB/mu (1 mu = 0.067 ha) gave effective control of *H. armigera* on cotton in China. The effectiveness of a nuclear polyhedrosis virus of *H. armigera* on cotton was greater and lasted longer than that of a bacterial insecticide (Zhang *et al.*, 1988). *H. zea* NPV (Elcar) affects the reproductive capacity of adult females of *Helicoverpa* spp surviving treatment in the larval stage in USA. Combinations of *H. zea* (Boddie) nuclear polyhedrosis virus (NPV) and spray adjuvants significantly increased the mortality of larvae of *Helicoverpa* spp. on cotton. (Luttrell *et al.*, 1982). Application of *P. fluorescens* as a seed treatment and foliar spray either alone or in combination with HaNPV significantly reduced larval and pupal weights of *H. armigera* (Jeyarani *et al.*, 2005).

The utilization of viruses in pest control has many limitations. Being obligate parasites, they need to be multiplied on living hosts. They are to be stored at low temperature for maintaining their effectiveness. They are stomach poisons and are mostly effective against early instar larvae of pests (Rabindra and Jayaraj, 1986), which necessitates their timely application near egg hatching to get good results. They are slow to act. The incubation period of disease lasts for 4–8 days. Death of host takes place in 5–10 days. Moreover, viruses are inactivated by sunlight and alkaline surface of cotton leaves. Addition of proper adjuvants to virus spray is necessary to increase the intake of virus by insects, to prepare good emulsion and to protect it from solar radiation and alkaline leaf surface of cotton.

Bacteria

More than 100 bacteria have been identified as arthropod pathogens, among which *Bacillus thuringiensis* (Bt) has received maximum attention as microbial control agent. The crystal proteins produced by subspecies of *kurstaki* and *aizawai* are effective against most species of Lepidoptera. Since the discovery of the dipteran specific subspecies *israelensis*, the recorded host rage of Bt has expanded considerably with the addition of more than 100 susceptible species of mosquitoes and black flies. The more recently discovered subspecies *tenebrionis* and *sandiego* produce toxins specific in their activity against coleopterans. The bacteria, *B. thuringiensis*, like NPV, are stomach poisons and specific in action. These bacteria produce toxic crystal at the time of sporulation in the gut of caterpillars which kill them. Out of toxins produced by the bacteria, *delta* endotoxin is most effective against caterpillars. Many bacterial formulations, i.e. Thuricide, Dipel, Dendrobacillin, Endobacterium Bactospeine, Tarmik 3, Endobacterium 3 and Delfin have been tested in USA, Canada, Brazil, Egypt, etc. In India, few studies have been carried out with respect to use of *B. thuringiensis*.

American bollworm. The spray of Thuricide/Biotrol at 1.25 1/ha effectively controlled *H. armigera* (Reddy and Krishnamurthy, 1989). But the application of *B. thuringiensis* sub sp. *kenyae* (6.5 x 10^8 spores/g) at 1 g/1 containing molasses (2 g/1) was not effective against bollworms (Satpute *et al.*, 1989). However, Biolep, Dipel and Bioasp formulations of Bt at the rate of 1.5 kg/ha were as effective as insecticide (quinalphos) against young larvae of *H. armigera* but, against grown up larvae, they were significantly inferior (Dhawan and Simwat, 1998).

Pink bollworm. *B. thuringiensis* has been found effective against pink bollworm, and reduced the boll damage and larval population of the pest by 42–53 and 56.1%, respectively. The bacteria with addition of small quantity of fenvalerate caused 70–95% mortality of larvae of pest and increased the lint yield by 21–35% (Luo *et al.,* 1986). Bioinsecticides Agerin (*Bacillus thuringiensis* var. *kurstaki*) in combination with low doses of insect growth regulators have additive effect in controlling larvae of *Agrotis ipsilon* in cotton and can be used as a component of IPM to protect beneficial and natural enemies and to minimize the environmental pollution (El-Khawas, 2001). Bacterial treatments have also certain limitations with respect to their use under field conditions. *B. thuringiensis* is inactivated by sunlight. It is slow to act and takes 3–7 days to cause death of pest. It persists for short period. It is more effective against early instar larvae. Addition of feeding stimulants like sugar, cotton leaf extract, etc. is necessary to increase the efficacy of bacteria.

Fungi

There are over 72,000 described species of fungi and the total number of species may be as high as 1.5 million. Until now, over 800 species of entomopathogenic fungi have been identified.

Sucking pests. The fungi *Verticillium lecanii* (Zimmermann) and *Beauveria bassiana* (Vuillemin) are the effective biological control agents against *Phenacoccus solenopsis* Tinsley (Joshi *et al.,* 2010). Similarly, nymphs and adults of *Paracoccus marginatus* gave highest mortality of 80% at 7 days after treatment (DAT) in case of *Lecanicillium lecanii* whereas *Beauveria bassiana* and *Metarhizium anisopliae* induced 75 and 70% mortality, respectively, at 9 DAT. A maximum of 100% mortality of nymphs were recorded at 9 DAT by all the 3 fungi (Banu *et al.,* 2010). A combination of biopesticides, viz., *Verticillium lecanii* 1.25 kg/ha and *Beauveria bassiana* 1 kg/ha with thiamethoxam (0.008%) was effective against whitefly, *Bemisia tabaci,* infesting hybrid cotton in India (Boricha *et al.,* 2010). The patented *Beauveria bassiana* JW-1 strain of fungus was effective against *Bemisia tabaci, Tetranychus urticae* and *Helicoverpa* spp on cotton (Hinz and Wright, 1997). The supernatant of *Beauveria bassiana* SFB-205 reduced the population of *Aphis gossypii* Glover (Kim *et al.,* 2010). Similarly, highest virulence of an isolate of *L. attenuatum* CS625 against *Aphis gossypii* was observed in Korea.

Bollworms. The entomopathogenic fungus *M. anisopliae* was most virulent to the first and second instar larvae of *H. armigera* (Gundannavar *et al.,* 2007). The fungus *Nomuraea rileyi* performed better at its higher concentration (108 conidia/ml) against *H. armigera* (Gundannavar *et al.,* 2008). The fungus, *B. bassiana*, proved virulent against eggs and larvae of *Pectinophora gossypiella.* The eggs appeared more susceptible to infection than larvae and about 90% of the hatched larvae from treated eggs failed to develop to adults and died (Rashad and Aly, 1997). Natural mortality of *H. armigera* due to *Nomuraea rileyi* could be up to 8% during cooler months and years of epizootics.

Other insect pests. The fungi *Metarhizium anisopliae* strain M.a.-4 and *Beauveria bassiana* showed potential as potent biological control agent of workers of the subterranean termite, *Odontotermes obesus* on cotton (Kakde *et al.,* 2005). *Beauveria bassiana* reduce diseases caused by soilborne plant pathogens, such as *Pythium, Rhizoctonia,* and *Fusarium* in cotton. Pure sprays of *Metarhizium anisoplae* (Ma456) and *Beauveria bassiana* (Bb734) gave good control (88.0 and 85.7%, respectively) of cotton spider mites, mainly *Tetranychus truncatus* and *T. turkestani* in China (Shi-WeiBing *et al.,* 2008). Similarly, the strains of *Beauveria bassiana* were effective against western tarnished plant bug, *Lygus hesperus* in California (Spurgeon, 2010). *B. bassiana* 302 and *Trichoderma harzianum* T24 were most effective isolates against larvae of *Spodoptera littoralis* in Egypt (El Katatny, 2010). An infection index of *Beauveria bassiana* (87.83%) against *Anthonomus grandis* on cotton in Brazil was also recorded.

Nematodes

The two major groups of entomopathogenic nematodes are *Steinernema* (55 species) and *Heterorhabditis* (12 species). More than 1000 species of protozoa pathogenic to insects have been described and many more remain to be discovered (Dhaliwal and Koul, 2010). Cotton crop treated with *Steinernema riobravis* at the rate of 2.5 and 3.2 billion nematodes/ha provided 100% mortality of *P. gossypiella* after 17 days of nematode application in Arizona and Texas, USA. The larval mortality of 53.3 was recorded with *S. carpocapsae* in Texa USA (Gouge *et al.*, 1998).

BOTANICAL PEST CONTROL

It has been estimated that there are at least 250,000 species of plants in the world, of which only 10% have been examined chemically. So far, more than 6500 plant species have been screened for anti-insect properties and of these, nearly 2500 species belonging to 235 plant families exhibit measurable to considerable pest control activity. Various isolated compounds from these plants indicate that approximately 360 compounds are insecticidal and more than 900 isolates are feeding deterrents alone (Koul, 2005). Botanicals are incorporated in integrated pest management programs as products in crop protectant rotations. In view of the environmental hazards of broad spectrum synthetic insecticides, studies on the exploitation of plant origin insecticides like neem and its formulations have also been carried out for the control of cotton insect pests as they are comparatively safe to beneficial arthropods without any environmental impact. The most important components identified in neem are the tetranortriterpenoids, the azadirachtins. These occur at concentrations of 0.1–0.9% in seed core and it has been established that 30–60 g azadirachtin per ha is sufficient to combat and repel key pests of various crops. Inspite of high selectivity, neem derivatives are known to affect more than 500 species of insects belonging to different orders (Koul and Wahab, 2004).

Products derived from leaves and kernels of neem (*Azadirachta indica* A. Juss.) are becoming popular in plant protection programmes for cotton, mainly because synthetic pesticides have several undesirable effects. Neem derived pesticides have been found to be effective and economical in controlling major cotton pests. Neem products act both as systemic and as contact poisons, and their effects are antifeedant, toxicological, repellent, sterility inducing or insect growth inhibiting. Furthermore, neem products appear to be environmentally safe and IPM compatible, and have the potential to be adopted on a broad scale, together with other measures, to provide a low cost management strategy (Gahukar, 2000). All botanicals serve similar purposes of biocontrol agents towards conservation of native as well as augmented bioagents and reduction in insecticide use vis-à-vis their selection pressure on pest population. Their high photo instability, suspected quality and inconsistent pest control efficiency are serious problems requiring research-cum-demonstration before an effective component of IPM. Botanicals are helpful in case of the documented resistance of the diamondback moth to *Bacillus thuringiensis* and spinosad due to overuse. Recently, the use of microencapsulation has allowed slower release rates of oils to be achieved, thus prolonging protection time.

Whitefly

Neem oil, Karanj oil and several neem based formulations have been developed and tested against cotton pests. Neem products can be effectively used for pest management alone or in combination/ alternation with recommended insecticides as best alternative to reduce the pesticide usage. Neem seed kernel extract at the rate of 5%, neem formulations at the rate of 2 1/ha and neem or karanj oil at the rate of 1%, having antifeedent/deterrent properties are recommended against sucking pests as well as bollworms. Neemark (0.4%) was as effective as endosulfan in controlling both pests, while Neemark (0.5%) was more effective than fenvalerate in controlling *B. tabaci.* Similarly, foliar and systemic treatments with 2% neem extracts (Margosan-O) were sufficient to prevent most individuals developing into adults. Settling and feeding of *B. tabaci* were inhibited (Meisner *et al.*, 1992). Flint and Parks (1989) also observed significant reduction (60%) in the numbers of immature whiteflies on cotton leaves after applications of aqueous sprays containing 160 ppm of Margosan-O (containing azadirachtin) at the rate of 138.7 ml a.i./ha at Arizona. Nimbalkar *et al.* (1993) observed that neem oil at 0.5% and neem seed extract at 5% reduced population of *B. tabaci* to the extent of 56.36 and 43.70%, respectively. Neem oil (0.5%) also gave 75.20% reduction in population. At 7th day neem seed extract (5%) was equally effective with the fenvalerate (0.015%), monocrotophos (0.1%), phorate (1kg a.i./ha at 10 cm depth), phosalone (0.1%) in reducing adult population of whitefly. Dimetry *et al.* (1998) observed that neem seed extracts were effective against *B. tabaci* infesting tomato. Similarly, neem seed extract and neem oil gave considerable reductions of adult and nymphal populations of *B. tabaci* on cotton in Maharashtra, India (Nimbalkar *et al.*, 1993). NeemAzal (1%) and RakshakGold (1%) were effective against *B. tabaci* during flowering phase. In addition, the alternate application of neem products and conventional insecticides can reduce 50% of synthetic insecticide load for its control (Mann *et al.,* 2001). Neemark at 0.4% was as effective as endosulfan in controlling *B. tabaci* and *Helicoverpa* on cotton in India (Phadke *et al.*, 1988). Neem products (Neemguard and Neemark) at 3 ml/litre used in conjunction with monocrotophos (36 WSC, 1 ml/litre) proved effective against *B. tabaci* than monocrotophos alone (Nandihalli *et al.*, 1990).

Neem oil (0.5%) and NSKE 5% were found effective against whitefly (Verma *et al.*, 1989; Uthamasamy and Parameswaran, 1989; Natarajan and Sundaramurthy, 1990; Raju *et al.*, 1992; Uthamsamy and Gajenderan, 1992; Vidya *et al.*, 1993; Jayaraj, 1993). However, Reddy and Rao (1993) reported that neem oil 0.5% was not effective against whitefly. Similarly, Patil *et al.* (1993) reported ineffectiveness of neem oil 0.5% and NSKE 5% for the control of this pest. Neem oil 0.5% and NSKE 5% were equally effective as endosulfan 35EC (0.75 l/ha) and monocrotophos 36 SL (0.5 l/ha) (Vidya *et al.*, 1993). Both these neem treatments were also comparable with dimethoate 0.1%, monocrotophos 0.1%, endosulfan 0.1%, methyl demeton 0.1%, fenvalerate at 0.015% and phorate at 1.0 kg a.i./ha for control of nymphs and adults of whitefly (Nimbalkar *et al.*, 1993). Similarly, Verma *et al.* (1989) reported that neem oil 0.5% + teepol 0.1% was as effective as amitraj and quinalphos for control of whitefly. Neem oil 2% (Rao *et al.*, 1990), neem oil 3% (Uthamasamy and Gajendran, 1992), fish oil resin soap 2.5% and nicotine sulphate 0.25% (Patil *et al.*, 1993) have also been found effective for control of whitefly. The spray of Neemark (1.25 and 1.875 l/ha), Indne 20 EC (0.625 and 1.25 l/ha), and RD 9 Repelin 100 EC (2.5 and 5.0 l/ha) were significantly at par with dimethoate (0.625 l/ha) at 7 days after spraying at higher doses for control of whitefly (Dhawan and Simwat, 1992a). Neemark at 0.4 and 0.5% was on par with endosulfan 0.1%. Neemark at 0.5% was rather more effective than fenvalerate (0.0125%) for control of this pest (Phadke *et al.*, 1988). Margocide OK-80 gave effective control of whitefly (Muthukrishnan *et al.*, 1992).

Butler *et al.* (1991) observed that cotton seed oil formulations at the rate of 2 and 5% and castor oil formulation at the rate of 5% reduced the number of immature forms of *B. tabaci* by 92, 95 and 84%, respectively after 5 days of application after treatment in Gujarat, India. They also reduced the number of *Aphis gossypii* by an average of 87–95%. The oil from seeds of sugar apple (*Annona squamosa*), was effective in controlling the silver leaf whitefly, *Bemisia argentifolii* Bellows in greenhouse conditions like recommended insecticide. It was also very effective in controlling *A. gossypii.* (Lin ChienYih *et al.*, 2009). The efficacy of plant and animal products like fish oil resin soap was the most effective in reducing immature and adult populations of *B. tabaci* and aphids, and gave the maximum seed cotton yield in Karnataka, India (Patil *et al.*, 1993).

Cotton Jassid

The spray of Neemark (1.25 and 1.875 l/ha), Indne 20 EC (0.625 and 1.25 l/ha) and RD 9 Repelin 100 E (2.5 and 5.0 l/ha) were significantly at par with dimethoate (0.625 l/ha) on 7th and 10th day after spraying for control of jassid (Dhawan *et al.*, 1992a). Neem oil (0.5%) containing 0.1% Teepol as surfactant and NSKE (neem seed kernel extract) 5% have been found to be effective for control of cotton jassid (Verma *et al.*, 1989; Uthamasamy and Parameswaran, 1989; Natarajan and Sundarmurthy 1990a; Raju *et al.*, 1992; Uthamasamy and Gajendran, 1992). Vidya *et al.* (1993) have also reported equal efficacy of NSKE 5%, neem oil 0.5%, endosulfan 35 EC (0.75 l/ha) and monocrotophos 36SL (0.5 l/ha) for control of this pest. The spray of mixture of neem oil and Karanj oil (10:1) at 3% effectively controlled cotton jassid (Uthamasamy and Gajendran, 1992). Spraying of Navneem 95EC (*Azadirachta indica*) at the rate of 500 ml/ha reduced *Amrasca biguttula biguttula* populations during the vegetative phase (Dhawan *et al.*, 1992b).

Aphids

It has been observed that various neem treatments like 3% neem leaf extract, 3% neem kernel extract, 3% neem oil, neem oil with 3% sesame oil in a 5:1 ratio, neem oil with 3% sesame oil in a 10:1 ratio and 0.07% endosulfan were equally effective in reducing *A. gossypii* on cotton in India (Venkatetan *et al.*, 1987). Neem oil (0.5%) and NSKE 5% provide effective control of aphids on cotton (Verma *et al.*, 1989; Uthamasamy and Parameswaran, 1989; Natarajan and Sundaramurthy, 1990; Raju *et al.*, 1992; Uthamsamy and Gajenderan, 1992;

Vidya *et al.*, 1993). Neem oil 0.5% and NSKE 5% were equally effective as endosulfan 35EC (0.75 l/ha) and monocrotophos 36SL (0.5 l/ha) for control of aphids (Vidya *et al.*, 1993). Neem leaf extract 3% containing 0.025% Teepol or Triton X 100 as emulsifiers (Venkatesan *et al.*, 1987), nicotine sulphate 0.2% (Reddy and Rao, 1993) and neem oil 3% alone or in combination with Karanj oil (10:1) 3% (Uthamsamy and Gajenderan, 1992) were also effective against cotton aphid.

Pink Bollworm and Spotted Bollworm Complex

Neem seed kernel extract (NSKE) was as effective as deltamethrin against *H. armigera*. Among various oils evaluated, neem oil (2%) has been shown as most effective in reducing the borer, *Earias vittella* damage (Sardara and Kumar, 1989). High doses of azadirachtin up to 10 µg/larva resulted in 100% larval mortality of *Spodoptera exempta*, but this effect was delayed and prolonged. The efficacy of neem seed extracts from 1 to 5%, neem oil (0.5%), karanj seed extract (1–5%), Indiana (1%) along with endosulfan (0.06%) has been established against cotton bollworms. Endosulfan and 4% neem seed extract (4%) did show least bad kapas production (Nimbalkar *et al.*, 1990). Neemrich 20 EC at 1.5 l/ha, Navneem 95EC at 0.75 l/ha and RD 9-Repelin at 2.5 l/ha were equally effective as quinalphos 25EC (2 l/ha) for control of pink bollworm and spotted bollworms, and increased the yield of seed cotton over control. Neemark 1.25 l/ha and lower dosages of above botanicals were comparatively poor treatments (Dhawan *et al.*, 1992b.; Dhawan and Simwat, 1992b; Dhawan and Simwat, 1993a). The spraying of combination of Neemark (1.25 l/ha), Indne 20 EC (0.75 l/ha) and RD 9 Repelin (2.5 l/ha) with quinalphos 25 EC (2 l/ha) and cypermethrin 25 EC (0.2 l/ha) was significantly more effective than above insecticides for reduction in bollworm incidence and increase in yield of seed cotton (Dhawan and Simwat, 1993a). Neem oil 0.5% was effective against bollworms (Reddy and Rao, 1993). According to Nimbalkar *et al.* (1993), neem oil 0.5%, NSKE 5% and Karanj seed extract 5% were not effective against bollworms of cotton. Fish oil resin soap 2.5% and nicotine sulphate 0.25% were effective treatments against bollworms (Patil *et al.*, 1993). However, number of commercial neem formulations, i.e. Neemrich, Repelin, Neemark, Rakshak Nimbecidine, Neemgold, Neemazal, Achook, Neemax, Neem powder were comparatively less effective than conventional insecticides and did not give desired level of kill against *H. armigera* (Simwat and Dhawan 1992; Dhawan and Simwat 1998; Anonymous, 2002).

The spray of mixture of neem oil and Karanj oil (10:1) at 3% was found to be effective for control of *Earias* spp. (Uthamsamy and Gajenderan, 1992) and has also been found to be effective for control of *H. armigera*. Neemark at 0.4% was as effective as endosulfan 0.1%, whereas Neemark at 0.5% was more effective than fenvalerate 0.0125% for control of this pest (Phadke *et al.*, 1988). Neemark (0.5 and 0.75%), RD 9 Repelin (1.0 and 2.0%), quinalphos (0.2%) and cypermethrin (0.02%) caused 42.1, 48.5, 77.5 and 72.1% mortality of third instar larvae of *H. armigera* 3 days after spraying, respectively indicating comparative less effectiveness of botanicals than insecticides (Simwat and Dhawan, 1992a). In another trial also, RD 9 Repelin (1.0 and 2.0%), Neemark (0.5 and 0.75%) and Neemrich 20EC (0.1 and 0.15%) caused 53.3-70.0% mortality of younger larvae of *H. armigera* against 100% mortality in case of quinalphos (0.2%) (Dhawan and Simwat, 1993a). The combined use of botanicals at 0.5% with quinalphos (0.2%) has been reported to be effective against *H. armigera*. These botanical treatments also exhibited highest repellency to ovipositing moths (Rosaiah and Reddy, 1993). The spray of RD 9 Repelin (1.0%) alone or in mixture with half dose of insecticide has been suggested for control of *H. armigera* on cotton (Subramaniam *et al.*, 1993).

Neem oil (0.4 or 0.5%) or neem oil (0.5%) with 0.09% deltamethrin alone or in combination were effective against bollworms in Tamil Nadu, India (Samuthiravelu and David, 1990). Seed extracts of neem and *Pongamia glabra*, 1% indiara, 0.5% neem oil and 0.06% endosulfan were evaluated against bollworms on cotton in Maharashtra, India but none of the plant extracts gave better control than endosulfan (Nimbalkar *et al.*, 1993). Dhawan *et al.*, (1998) also evaluated different biopesticides against *H. armigera*. Neem based insecticides were significantly inferior to standards, quinalphos and chlorpyriphos against young and mature larvae except NeemAzal which was as effective as quinalphos against young larvae.

Other Insect Pests. Significant reduction (56.60 to 62.26%) in the population of grey weevil, *Myllocerus maculosus* was observed with 5000 ml Servo Agrospray/ha; 2500 ml of 1500 ppm Ozoneem/ha and 1500 ml of 10000 ppm Ozoneem/ha (Shanthipriya *et al.*, 2007). The final instar nymphs of aged 0–24 h were more susceptible to neem oil than 72 to 96-h-old and 144 to 168-h-old nymphs of *Dysdercus koenigii* (Gujar and Mehrotra, 1993). Addition of a food spray product to neem increased numbers of beneficial insects in cotton fields and improved cotton yields in Africa (Vodouhe *et al.*, 2009).

The flower extract of *Chrysanthemum coronarium* and its fractions have shown insecticidal effect against *Spodoptera littoralis* (Shonouda *et al.*, 2008). The insecticidal activity of crude extracts and fractions of *Argemone mexicana* (Papaveraceae) was investigated against *Spodoptera litura*. Pupation was nil in the chloroform and acetone extracts. The adult life span was least in the acetone solvent, followed by hexane and petroleum ether extracts (Malarvannan *et al.*, 2008).

These studies have revealed that botanical insecticides are effective against cotton insect pests, but to a lesser extent than insecticides. But their combined use with insecticides was as effective as insecticides for control of bollworms. Botanical pesticides, however, have delayed action. Scarce raw stocks, problematic collection and processing of neem seeds, lack of standard processing technology, high costs, lack of multispectrum base and inadequate research are some of the other limitations which need to be looked into before realizing the potential of botanical insecticides in IPM on cotton.

BIORATIONAL PESTICIDES

Biorational pesticides or reduced risk pesticides include all chemicals that affect the growth, development, biology and ecology of pest species differently from that of beneficial species. Two groups of biorationals, behavior modifying chemicals or semiochemicals and insect growth regulators have gained prominence recently. Most of the biorational pesticides are preferable to conventional pesticides because of their specificity to target pests, their effectiveness at low rates and their nonpersistent characteristics in the environment. In view of their specificity, the market potential of these chemicals is limited and the industry is not keen to invest in the development and use of new chemicals, which requires huge investment in terms of resources and time. However, the biorational pesticides are becoming more reliable and look set for playing a vital role in ecofriendly management of crop pests in future (Dhaliwal and Koul, 2010).

Several important classes of biorationals have been produced mostly from soil microorganisms by fermentation procedures. Avermectins are a group of macrocyclic lactones isolated from fermentation of the soil actinomycete microorganisms, *Streptomyces avermitilis*. Avermectins are among the most potent antihelmintic, insecticidal and acaricidal compounds known. These compounds exhibit a broad spectrum of activity against agricultural pests such as aphids, mites, caterpillars, and other pests that feed on crops and ornamental plants. The spinosyns, a new class of insecticidal macrocyclic lactones, derived by fermentation of the soil actinomycetes, *Saccharopolyspora spinosa*, are active against many pests of crops, ornamentals, forestry, greenhouse, garden and households. The most effective product of this group is spinosad, which has shown contact and stomach activity against Coleoptera, Diptera, Hymenoptera, Lepidoptera, Siphonaptera and Thysanoptera. It is particularly effective against lepidopterans such as heliothine complex and is one of the most powerful chemicals for controlling thrips and leaf miners (Horowitz and Ishaaya, 2004).

The biorational approaches in pest management like pheromones, new molecules with novel mode of action and insect growth regulators fit idyllically in IPM programme. The use of pheromones in pest control can be explored as a component of IPM by monitoring of pest population with pheromone baited traps; mass trapping using large number of traps to reduce population levels and mating disruption technique. The monitoring of bollworms and tobacco caterpillar with sex pheromones traps have been recommended as a monitoring tool in IPM module as it helps in decision making for their effective management (PAU, 2008). The disruption of mating using PB-Rope L dispensers containing sex pheromone (Z,Z)-7, 11 hexadecadienyl acetate and (Z,E)-7, 11 hexadecadienyl acetate proved to be an effective, safe, easy to apply and eco-friendly method for the management of pink bollworm on cotton (Sohi *et al.,* 1999). The efficacy of the new molecules with novel mode of action is being tested on regular basis in the cotton belt of Punjab. Few promising new molecules, floniacmid 50WG for cotton jassid, flubendiamide for *P. gossypiella, Earias* sp. and early instars of *H. armigera*, naturalyte (spinosad) as well as rynaxypur (coragen) have been recommended for the control of *H. armigera* and tobacco caterpillar in Punjab (PAU, 2008). Efficacy of several insect growth regulators, viz. diflubenzuron, teflubenzuron, lufenuron, flufenouron, RH-2485, buprofezin have been tested against different insect pests under laboratory and field conditions in Punjab (Dhawan and Simwat, 1998; Butter and Kular, 1998; Singh *et al.,* 2005; Aulakh and Butter, 2006). Among the most recent novel insecticides with selective properties, novaluron, a chitin synthesis inhibitor has been found very effective and is recommended for the control of *S. litura* in Punjab (PAU, 2008) which can fit, ideally in IPM strategies as a potential component thereby reducing dependence on conventional insecticides.

Sex Pheromones

There is great scope for use of sex pheromones in integration or in alternation with insecticides against bollworms in IPM on cotton. However, there are some limitations also in the field use of sex pheromones. Only, gossyplure, the sex pheromone of pink bollworm is commercially available in several formulations in foreign countries, whereas no such progress has been made for other species of bollworms, while all the 3 species of bollworms attack cotton crop in India. Moreover, high costs, extreme climatic conditions and cultivation of large number of varieties with different maturity periods, and staggered sowing may limit the prospects of use of sex pheromones for bollworm control on cotton in India.

Bollworms. Sex pheromone of pink bollworm, i.e. gossyplure is commercially available in several formulations, viz. Nomate PBW (7.6%), microencapsulated formulation, rubber capsules, laminated chips/flakes, stirrup PBW (0.632%), checkmate PBW (80.6%), polyethylene tube dispensers, rope dispensers and twistie dispensers. Extensive and large scale trials on use of gossyplure in mating disruption technique/air permeation method for control of pink bollworm have been conducted in USA, Australia, Egypt, U.K., Brazil, Mexico, China, USSR, etc. Some experiments on field use of gossyplure in the form of Nomate PBW (7.6%) and stirrup PBW (0.632%) have also been conducted in India for control of pink bollworm. Aerial application of Nomate PBW at 50–75 g/ha, eight times starting form pin head square stage of crop at 14–30 days interval reduced the moth catch, fecundity and increased the yield of seed cotton by 33.7%.

One application of pheromone (gossyplure) provided 100% mating disruption and control of *Pectinophora gossypiella* on cotton in Pakistan. Overall, a 50% reduction in insecticide application was obtained (Ahmad and Attique, 1993). Mitsubishi PBW Rope was used for an area wide control of *P. gossypiella* on cotton in Arizona. The amount of rope used was decreased progressively to zero after 3 years (Antilla *et al.*, 1996). However, Busoli (1993) showed that 2 pheromones hexalure and virelure can be used together successfully to capture *P. gossypiella* and *Heliothis virescens*. Maximum catches of *P. gossypiella* were recorded with Modified Sharma traps baited with Pherocon PBW caps as dispensers, and Stickem as a trapping medium in Punjab, India (Dhawan and Sidhu, 1993). Season long control of the *P. gossypiella*, *E. insulana* and *E. vittella* was obtained with controlled release pheromone formulation with a saving of 5 insecticide applications on cotton in Pakistan (Cork and Hall, 1998). Mafra Neto and Habib (1996) used oil traps containing lures with a high dose of pheromone for mass trapping of pink bollworms in Brazil. Two applications of the Shin Etsu rope formulation (1000 ropes/ha), containing 50 mg pheromone/rope, effectively disrupted mating for 90 days, reduced damage to cotton bolls, and minimized the number of chemical treatments needed for *E. insulana* in a cotton field in Israel (Nakache *et al.*, 1992). Cotton plots treated with PB-ROPE (polyethylene tubes at 55 g a.i./ha) had the least infestation of flowers (0.9%) and green bolls (1.67%) by *P. gossypiella* on cotton in Pakistan (Qureshi *et al.*, 1988; Teague and Tugwell (1997)). Combination of HP rope containing 175 mg of *H. armigera* and 65 g of *P. gossypiella* pheromone could completely suppress mating over 75 days in Israel (Kehat *et al.*, 1998). Similarly another combinations of 2 sex pheromone dispensers a commercial laminated plastic dispenser containing 10 mg of 7 pheromone components and a polyvinyl chloride (PVC) dispenser containing 6.25 mg of 2 components and a wire cone trap consistently captured significantly greater numbers of males of *H. virescencs* in Texas (Lopez *et al.*, 1988).

Dhawan and Simwat (1993) found that the use of sex pheromone gossyplure alongwith cypermethrin gave good control of *Pectinophora gossypiella* on cotton in the Punjab, India. Race (1995) revealed that application of the bisexual pheromone with Larvin reduced the numbers of *Heliothis zea* larvae by 50–75% within 7 days on cotton in Texas. Similarly, spraying of Stirrup PBW at 10 g a.i./ha with knapsack sprayer in combination with insecticides, quinalphos, fenvalerate and cypermethrin at the recommended dosages (Regupathy and Mahadevan, 1992; Srinivasan and Venugopal, 1992; Dhawan and Simwat, 1993b) was effective against pink bollworm. The combined spray of sex pheromone and insecticides has been reported to give better control of bollworm and increase the yield of seed cotton with reduced use of insecticides as compared to insecticide treatments alone.

Tobacco caterpillar. The rope dispensers containing 160 mg of total pheromone can be used to protect cotton from a strong flying insect like *Spodoptera exigua*, even in small fields (Mitchell *et al.*, 1997). Dunkelblum *et al.*, (1987) observed effective mating disruption of *Spodoptera littoralis* with the use of polymeric aerosol formulation containing the pheromone at 75 g a.i./ha placed at 500 release points/ha and giving a daily release rate of 3 g/ha.

Other insect pests. A mixture of sex pheromone (grandlure 0.8%) and cypermethrin (6.4%) controlled *Anthonomus grandis* up to 80 days on cotton in Brazil (Santos and Hofer, 1996). The use of the Boll Weevil Attract and Control Tube (a pheromone trap) eliminated *Anthonomus grandis* on cotton farms in Nicaragua (Daxl *et al.*, 1995). Pheromone trapping helped in reduction in overwintered boll weevils by 99.9% (Fuchs *et al.*, 1999). An attract and kill device, containing feeding stimulants, a slow release insecticide and a 40 mg grandlure pheromone dispenser was effective against *A. grandis* on cotton (Langston, 1996). The potential natural enemies comprising of parasitoids, i.e. *Cotesia* sp. and *Palexorista laxa*; predators, i.e. *Cantheconidia furcellata* and *Oxyopes* sp.; and pathogens like *Beauveria bassiana* and *Nomuraea rileyi* were used against green semilooper, *Anomis flava* in Orissa, India (Mohapatra and Sahu, 2004) and found very effective.

Insect Growth Regulators

Several insect growth regulators, viz. diflubenzuron, triflumeron, methoprene, hydropene, lufenuron, novaluron, etc. have been evaluated against cotton insect pests in USA, Egypt, etc. Some experiments have also been conducted in India.

Bollworms. Diflubenzuron (0.4 kg a.i./ha) was effective against pink bollworm and American bollworm. Six alternate sprays of diflubenzuron and fenvalerate (0.075 kg a.i./ha) at 15 days interval initiated at ETL of 10% bollworm incidence, effectively controlled these pests (Balasubramanian *et al.*, 1987). Diflubenzuron (Dimilin 25 WP) at 0.125–0.315 kg a.i/ha was significantly better for controlling the incidence of pink bollworm and spotted bollworms over control, but was significantly less effective in comparison with endosulfan (0.875 kg a.i./ha) and quinalphos (0.5 kg a.i./ha). Similarly, diflubenzuron at 0.075 and 0.100 kg a.i./ha was tested in combination with phenthoate (0.5 kg a.i./ha), endosulfan (0.875 kg a.i./ha), quinalphos (0.5 kg a.i./ha), monocrotophos (0.5 kg a.i./ha) against pink bollworm and spotted bollworm complex. The yield of seed cotton increased significantly in combination of diflubenzuron with endosulfan and monocrotophos on LH 900 as compared to insecticides alone (Dhawan *et al.*, 1992b; Simwat and Dhawan, 1992b). The spraying of triflumeron (1.3 kg a.i./ha) in combination with cyflutrhin (0.125 kg a.i./ha) effectively controlled all the 3 species of bollworms and registered 73.5% boll opening (Kulkarni *et al.*, 1986). Insect growth regulator, diflubenzuron alone is less effective and not economical as compared to insecticides for control of bollworms. Mixing of diflubenzuron has added effect in combination with only some insecticides. A new chitin inhibitor teflubenzuron (Nomoult 15 SC) at the rate of 0.1 kg a.i./ha was effective against bollworm complex (Butter and Kular, 1998). The oviposition period of females developed from third instar larvae of *H. armigera* was shortened significantly on cotton treated with insect growth regulator, lufenuron (Butter and Singh, 2004). An insect growth regulator novaluron (10 EC) at 50 and 37.5 g ai/ha rendered effective control of *S. litura* on Bt cotton in Punjab, India (Dhawan *et al.*, 2008). Murthy and Ram (2002) found novaluron effective against against American bollworm on cotton. The natural bioproduct, spinosad, (Spintor 24 SC) at the rate of 50 ml/fed resulted in 100% mortality of fresh egg masses of *Spodoptera littoralis* in Egypt (Temerak, 2005). The insect growth regulator triprene exhibited a high level of efficiency followed by methoprene and diflubenzuron against *P. gossypiella* (Abdel *et al.,* 1988). Eggs of the *Heliothis virescens* deposited on cotton plants treated with fenoxycarb did not produce viable larvae (Masner *et. al.,* 1987).

Sucking pests. Aerial application of the growth regulator buprofezin against *Bemisia tabaci* in cotton was effective in controlling nymphs, but required the addition of amitraz to control adults successfully in Ethiopia (Ledson *et. al.,* 1994). The juvenoid insect growth regulators kinoprene and fenoxycarb and the insecticide synergist piperonyl butoxide (PBO) slowed population growth, delayed adult emergence and significantly altered population structures of *Aphis gossypii* (Satoh *et. al.,* 1995). The early stage nymphs of *Bemisia tabaci* were highly susceptible to buprofezin (Beevi and Balasubramanian, 1995). The insect growth regulator buprofezin at the rate of 0.00625% was effective against both glasshouse aleyrodids *Trialeurodes vaporariorum* and *Bemisia tabaci* (Stenseth and Singh, 1990). However, Horowitz and Ishaaya (1992) advocate restricted use of buprofezin against *B. tabaci* on cotton in Israel.

Other insect pests. An insect growth regulator methoprene, formulated as emulsifiable concentrate was more effective against *Spodoptera littoralis* (Kassem *et. al.,* 1987). Topical application of precocene I to eggs of *Dysdercus koenigii* resulted in a reduction in the size of emerging nymphs (Gawade *et al.,* 1993). Diflubenzuron had a latent effect, while chlorfluazuron acted faster on fourth-instar larvae of *Agrotis ipsilon* on cotton leaves in Egypt (Mostafa *et. al.,* 1991). Newly emerged females of *Myllocerus undecimpustulatus* Desbr fed with diflubenzuron treated leaves laid significantly fewer eggs (mean

10.7/female) than did newly emerged ones fed with untreated leaves (436 eggs/female) (Thangavelu, 1982). Among the 4 chitin synthesis inhibitors diflubenzuron, teflubenzuron, triflumuron and penfluron, the latter possess high contact activity against *D. koenigii* in cotton (Gujar and Mehrotra 1989). A formulation of feeding substrate containing cotton products, grandlure (boll weevil pheromone), a sticker and ultraviolet protectant (Nufilm 17) and *Beauveria bassiana*, an entomopathogenic fungus was effective against *Anthonomus grandis* (Wright and Chandler, 1992). Shternshis (2002) recorded various biological control agents (microbial insecticides and fungicides) based on use of microbial metabolites (such as those produced by *Bacillus thuringiensis*, *Metarhizium anisopliae*, *Beauveria bassiana*, *Streptomyces avermitilis* and *Verticillium lecanii*) against cotton ball weevil, *A. grandis* in Russia.

BIOCHEMICAL CONSTITUENTS

The allelochemical compounds known to exert adverse effects on pests in cotton include gossypol, gossypurin, helicocides and related terpenoids, *p*-hemigossypolone, tannins, anthocyanins, flavonoids and phenolics. Though the gossypol is a main inhibitor in the growth and development of cotton bollworm but on other hand a low level of gossypol may stimulate the growth and development of bollworms. Vilkova *et al.* (1988) concluded that high gossypol cotton cultivars have detrimental effects on insect development, viz. increasing incubation periods, causing greater mortality among young larvae and lowering larval weight compared with low gossypol. The reduction of growth and weight in spotted bollworm larvae on G-27 variety of *G. arboretum* (high gossypol) was found responsible for conferring resistance to spotted bollworm. However, Reddy *et al.*, (1989) was the first to report variability in the quantity of gossypol, hemigossupolone, helicocide-1 and helicocide-2 as important biochemical compounds imparting resistance to spotted bollworms in *G. arboreum*. Rajeshwari *et al.* (1999) observed low bollworm incidence in genotypes with high phenols, tannins and gossypol.

Gossypol not only influenced the abundance of cotton bollworms but also had significant effect on the population density of cotton jassid. Free gossypol in cotton leaves was found to be significant negatively correlated with the number of eggs of cotton jassid. The grey weevil affected 92.7–100% plants and 37.6–78.1% leaves in gossypol-free lines and 8.8–13.8% plants and 1.7–5.1% leaves in glanded lines. The high gossypol content of leaves is responsible for imparting considerable resistance against grey weevil in cotton (Agarwal and Gupta, 1986). Cotton cytotypes with lower jassid grades possessed high total phenol and flavanol contents in leaves, stem and squares. Tannins are other most important components to impart resistance against several cotton pests. Tannin content of the bolls was significantly negatively correlated with adult emergence in spotted bollworm (Sharma *et al.,* 1989). Tannin content was also found negatively correlated with population build up of whitefly (Butter *et al.,* 1992). The studies conducted at Andhra Pradesh Agricultural University, Hyderabad also highlighted the role of tannins in cotton for imparting resistance against whitefly. Higher tannin content was related with low incidence of serpentine leaf miner and where low tannin content was recorded, and incidence and maggot population was also low (Sushila and Patil, 2001). Rajeshwari *et al.* (1999) observed negative and significant correlation in tannin content in leaves and square damage by *H. armigera.* Silica plays an important role in imparting resistance against cotton jassid. Negative correlation between the silica content of leaves and jassid population was also recorded.

Sharma and Agarwal (1983) observed that jassid population is negatively correlated with total sugar and free amino acid content. Sugar content in the leaves is positively correlated with the susceptibility of varieties to whitefly in cotton (Jayaswal and Pundarikshudu, 1989). Leaf nutritional constituents, viz. moisture, proteins and total sugars showed positive role in leafhopper survival and oviposition. Phenolic compounds have been reported to have significant effect on the abundance of several insect species, e.g. total phenols show negative correlation with whitefly population (Butter *et al.,* 1992). Rajeshwari *et al.* (1999) observed negative and significant correlation in phenol content in square and damage by *H. armigera.* Ananthakrishnan *et al.* (1990) also observed characteristic antibiotic effects against the bollworms with resorcinol and phloroglucinol (phenolics) which included reduction in consumption index and approximate digestibility, extended larval duration and high mortality. Total sugars, tannins, flavonols, phenols and gossypol were negatively correlated to population density of whitefly (Butter *et al.,* 1992). Levels of non-reducing sugars, tannins, silica and free gossypol in the leaf were negatively correlated with ovipositional preference of jassids.

A greater understanding of the role of these biochemcicals interaction in insect behaviour and physiology will undoubtedly open new vistas in the development of resistant varieties against cotton insect pests. The biochemistry of insect-plant interactions will presents opportunities for future research and its application in sustainable pest management programme in cotton.

TRANSGENIC COTTON

Advances in molecular biology have made it possible to introduce desirable traits from diverse sources into crop plants, thereby developing genetically engineered plants or 'transgenic plants' conferring resistance to biotic and abiotic stresses. Most commercially available insect-resistant transgenic plants express genes coding for soil inhabiting spore forming bacterium, *Bacillus thuringiensis* Berliner (Bt) toxins, that negatively influence the survival and development of a target herbivore. Gene from *B. thuringiensis* var. *kurstaki* that produces Cry lAc protein which is toxic specifically to lepidopteran species was inserted into cotton plant. The first Bt cotton plants were developed by the Chinese Academy of Agricultural Sciences (CAAS) in China and by Monsanto Company in the USA in the early 1990s. However, regulatory approval and large scale commercial cultivation of Bt crops which included Bt cotton along with Bt corn and Bt potato developed by Monsanto, took place in the USA in 1996. Thus, 1996 marked the beginning of commercialization of transgenic crops (Manjunath, 2011). Amongst the genes that have been deployed in insect resistant transgenic cotton, so far *cry1A*c provides acceptable control of *H. armigera* and other lepidopteran insect pests that include the two bollworms, *Earias* spp. and *P. gossypiella.* It was however, not effective on the defoliating tobacco caterpillar, *Spodoptera litura* (Fabricius). Gene *cry2Ab* present in Bollgard II along with *cry1Ac* is moderately toxic to *H. armigera* and *S. litura.* However, the expression levels of Cry2Ab are very high as compared to the Cry1Ac and, therefore, Bollgard II is highly toxic to *H. armigera* and toxic to *S. litura.* The seasonal expression of Cry2Ab in BG II ranged between 14.0 and 83.0 μg/g fresh weight in leaves and 5.5–62.0 μg/g fresh weight in various fruiting parts. Thus, compared with Cry1Ac levels, which ranged from 0.01–8.0 μg/g in various plant parts, the Cry2Ab level are atleast 10-fold higher than Cry1Ac at any given point in any of the plant parts during the season (Kranthi *et al.,* 2009).

In India, Monsanto provided the Bt-gene, *cry1Ac* (Bollgard), incorporated in the seeds of Coker 312 to its Indian licensee, Maharashtra Hybrid Seed Company Ltd. (Mahyco), along with the technology. Through routine back-crossing with the parental lines of

Mahyco's proprietary hybrids, the gene was transferred from Coker 312 into the natural genotypes. Thus, the native parental lines were converted into Bt plants. Converted parental lines were strongly stabilized through multiple backcrossing before producing the Bt hybrids in the usual manner as cotton hybrids are produced (Manjunath, 2011). It was first approved for commercial cultivation by the Genetic Engineering Approval Committee (GEAC), Ministry of Environment and Forests, Government of India in Central and South zones in 2002 and in North zone in 2005 (Karihaloo and Kumar, 2009) and the country has ten years of impressive benefits from this crop.

Since its commercialization during 2002, the number of events, Bt cotton hybrids and companies marketing approved hybrids have all increased significantly in India. So far, six transgene events have been approved by GEAC in India (Table 6.2). The events, MON 531 having single gene (*cry1Ac*) and MON 15985 having dual toxin genes (*cry1Ac* and *cry2Ab*) have been developed by Mahyco / Monsanto and were approved in 2002 and 2006, respectively. These two events have been licensed to other seed companies. The third event, Event 1 (modified *cry1Ac*) has been developed by Indian Institute of Technology, Kharagpur and is being utilized by JK Agri-Genetics Ltd. The event was approved for sale for the first time in 2006. The fourth event, GFM event (fusion *cry1Ab/cry1Ac*) has been sourced from China and is being utilized by Nath Bio-Genes (I) Ltd. It was also approved for sale for the first time in 2006. The fifth event, BNLA-601 (*cry1Ac*) developed by Central Institute of Cotton Research (CICR), Nagpur and University of Agricultural Sciences (UAS), Dharwad (Karnataka) was approved for commercial cultivation in an indigenous public sector *G. hirsutum* variety *Bikaneri Narma* Bt in contrast to the above four events, which were all incorporated in cotton hybrids. It was approved for commercial release in all three zones of country in 2008. In 2009, a public sector developed Bt cotton hybrid NHH 44 Bt based on same event was also approved by GEAC. The sixth event, MLS-9124 (*cry1C* gene) has been developed indigenously by Metahelix Life Sciences Ltd., Bangaluru and was approved for commercial cultivation in Central and South zone in 2009 (Choudhary and Gaur, 2011). Since 2002, cotton scenario has been dominated by single gene Bt cotton hybrids predominantly from Monsanto's 531 event and to some extent from JK Seeds's Event 1 and Nath Seeds's GFM event. But now, there is increasing tendency among farmers to adopt Bt hybrids expressing dual toxins (*cry1Ac* and *cry2Ab* genes) belonging to event MON 15985. Bollgard II is being considered as an improved pest management technology for its enhanced efficacy and also as an efficient resistance management strategy.

Table 6.2. Commercial release of different Bt cotton events by GEAC in India

Event	Insect resistant gene	Year of approval	Developer
MON531	*cry1Ac*	2002	Mahyco/Monsanto
MON15985	*cry1Ac* and *cry2Ab*	2006	Mahyco/Monsanto
Event-1	modified *cry1Ac*	2006	JK Agri-Genetics/ IIT Kharagpur
GFM Event	*cry1Ab/cry1Ac*	2006	Nath Bio-genes (I)/CAAS China
BNLA-601	*cry1Ac*	2008	CICR, Nagpur and UAS Dharwad
MLS-9124	*cry1C*	2009	Metahelix Life Sciences, Bangaluru

Source: www.isaaa.org

The commercial exploitation of Bt cotton in India has significantly established its positive impact in all cotton growing areas of the world. The pesticide load on the cotton crop has substantially declined owing to effective control of bollworm complex which in turn has led to reduction in cost of cultivation and environmental pollution. Besides this, social and economical status of farmers has been uplifted due to yield advantage and higher economic returns. Farmers' earnings and profitability from Bt cultivation have been significantly higher than those from cultivation of non-Bt cotton.

Results from the extensive field trials in India conducted from 2002–03 to 2008–09 in AICCIP under the aegis of ICAR clearly indicated that Bt cotton hybrids were highly effective in controlling bollworms thereby resulting in higher retention of first formed bolls due to low fruiting point and boll damage (Anonymous, 2005–2009). In a field study with natural bollworm incidence (Benedict *et al.,* 1996), significantly less larvae of *Heliothis virescens* were recorded in Bt cotton lines carrying either *cry1Ac* or *cry1Ab* (< 0.55/60 plants) compared to non-Bt cotton lines (> 198 larvae/60 plants). Similarly, significantly lower flower bud and square injury was recorded in Bt cotton lines carrying either *cry1Ac* or *cry1Ab* gene (< 4.07 and < 1.04%, respectively) compared to non-Bt cotton lines (> 20.60 and 11.77%, respectively). Wu *et al.* (2000) evaluated the transgenic cotton cultivars developed by CAAS, China against *H. armigera* in Hebei and Henan provinces of China. The control efficacy of GK-1 and GK-12 to the second, third and fourth generation cotton bollworms was up to

88.71–95.45%, 92.75–97.65%, and 93.33%, respectively. Jackson *et al.* (2003) observed that the reduction in percentage of squares infested with larvae of *H. zea* was approximately 10 and 19-fold for BG I and BG II lines, respectively, while the percentage of squares sustaining damage was reduced 6-fold by BG I and 16-fold by BG II genotypes compared with the conventional cultivar. In another study, Jackson *et al.* (2004) reported that *H. zea* larval population was statistically on par between Bollgard and conventional cotton varieties under untreated conditions. However, BG II genotype showed reduced larval population by 91 and 94% compared to Bollgard and conventional varieties, respectively. Mann *et al.* (2010) also observed that transgenic Bt cultivars showed lower square, boll and loculi damage than conventional cultivars.

Since its commercialization, increase in area from 50,000 ha in 2002 to 10.6 million ha in 2011 represents an unprecedented 212-fold increase in ten years (James, 2011). Bt cotton has literally revolutionized cotton production in India. Thus, at a national level, Bt cotton is a major factor contributing to higher cotton production which increased from 15.8 million bales in 2001-02, to 24.4 million bales in 2005-06, 28 million bales in 2006-07, and 34.5 million bales in 2011-12. Similarly, the productivity of cotton in India has also rallied to a higher level. The productivity of cotton increased from 308 kg per hectare in 2001–2002 to 554 kg per hectare in 2007–08 and 507 kg per hectare in 2010–2011, major share being attributed to Bt cotton (AICCIP, 2012). The global study of benefits generated by biotech crops conducted by Brookes and Barfoot (2012), estimates that India enhanced farm income from Bt cotton by US$ 9.4 billion in the period 2002 to 2010 and US$ 2.5 billion in 2010 alone. Besides, it has halved insecticide consumption, contributed to the doubling of yield and transformed India from a cotton importer to a major exporter.

For sustaining the production and productivity of cotton crop, an imperative necessity is to adopt management practices according to their modification in relative pest status. Bt cotton has literally revived cotton cultivation in India. In transgenic cotton, the need to spray against bollworms especially *Helicoverpa* spp. is drastically reduced thereby allowing other pests, normally controlled inadvertently by sprays targeting *Helicoverpa* spp., to emerge as pests. This includes aphids, jassids, whiteflies, mealybug, mirids and other foliage feeders and lint stainers. Use of insecticides against these surviving and emerging pests may disrupt natural enemies, creating outbreaks of mites, whitefly, thrips or aphids. Hence, in both conventional (non-transgenic) and transgenic cotton, there remains a strong incentive to use IPM to help reduce reliance on insecticides. Furthermore, an IPM approach helps to avoid creating situations that increase the risk of losses due to these pests. The integration of Bt cotton as a component of IPM module augmented with suitable cultural, mechanical control strategies, sowing of *refugia* around the main field and need based application of pesticides will not only reduce pesticide load and make cotton cultivation economical but will also help in management of resistance in insects against Bt cotton. Considerable effort has been directed toward such alternatives, and new technology has been implemented. In this context, studies should be conducted in different parts of the country to validate and promote IPM technology in Bt cotton through farmer's participatory approach to sustain productivity of cotton. Though transgenic cotton crop has enormous potential and such innovations have revolutionized crop protection technology, yet several issues concerning ecological and ethical issues, development of resistance and secondary pest problems need to be addressed.

INTEGRATED PEST MANAGEMENT

The use of insecticides has been the main tool of pest control on cotton in India upto mid 1970s, which created many side problems (Rajak, 1992; Mehrotra, 1993). The research, thereafter, was oriented towards developing and integrating nonchemical methods with minimum chemical use to reduce the gravity of side problems. The choice of some or several methods in the IPM strategy depends upon the locality, insect species complex and efficacy and economics of pest control measures. There is a great potential for management of different insect pests of cotton with biopesticides based IPM technology.

Bollworms

Application of neem products or *B. thuringiensis* in combination with at least one spray of a synthetic pyrethroid in a 4 spray schedule gave effective control of bollworm complex and an increase in cotton yield (Gupta and Kirti, 1996). Two releases of *Trichogramma chilonis* with two sprays of Btk and *Beauveria bassiana* resulted in the reduction of larval population (0.33/plant), less damage on shed squares (42.78%), intact squares (9.12%), bolls (14.20%) and loculi (10.68%) and of course provided higher yield (782 kg/ha) (Balakrishnan *et al.*, 2004). IPM strategies on Bt cotton in 112 and 120 villages of Ferozepur, Bathinda and Mansa districts covering an area of 2591 and 5720 ha during 2005 and 2006, respectively resulted in additional profit of ₹ 4725 and 4267/ha to the farmers of IPM villages as against non-IPM villages (Singh *et al.*, 2008). Use of trap crops (okra), neem seed kernel extract (NSKE) and *Bacillus thuringiensis* provide effective management of *Helicoverpa armigera*.in Tamil Nadu, India (Duraimurugan and Regupathy, 2005). IPM module comprising of application of neem (*Azadirachta indica*) seed extract (NSE) at 5%, endosulfan 35 EC at 0.06%, novaluron (Rimon 10 EC) at 0.015, release of *Trichogramma chilonis* at 1.5 lac/ha, *H. armigera* nuclear polyhedrosis virus (HaNPV) at 500 LE/ha and beta cyfluthrin 2.5 EC at 0.0025% decreased the number of eggs and larvae of *H. armigera*, *E. vittella* larvae, *P. gossypiella* larvae, damage to green fruiting bodies, open bolls and loculi in Maharashtra, India (Lande *et al.*, 2008). Econeem + NPV and econeem + Bt reduced the damage to fruiting bodies by 70.36 and 71.99% and increased crop yield to 15.64 and 15.79 q/ha, respectively. The use of pheromones in conjunction with inundative releases of *Trichogramma chilonis* and PB/SB-ROPE suppressed *P. gossypiella* and *Earias* sp. infestations to a subeconomic level in cotton fields in Pakistan (Nazir *et al.*, 1996). Neem seed kernel extract 6% with HNPV 500 LE/ha was the most effective treatment, followed by a half concentration of NSKE (3%) + HNPV 500 LE/ha, NSKE 6% + a half concn of HNPV (250 LE/ha) and NSKE 6% for the control of *P. gossypiella* (Sarode *et al.*, 1995).

Sucking Pests

IPM module comprising of various cultural practices, installation of pheromone traps at the rate of 5 traps/ha, two sprays of neem seed kernel extract at 45 and 55 days; spray of HaNPV at 250 LE/ha at 65 DAG; two releases of *Trichogramma chilonis* at 1.5 lac/ha at 75 and 85 DAG and need based spray of conventional insecticides helped in the reduction of sucking insect pests, namely *Aphis gossypii*, *Amrasca bigutulla bigutulla*, *Bemisia tabaci*, *Thrips tabaci* and bollworms in Rajasthan, India. The IPM module also encouraged the population of coccinellids, *Chrysoperla carnea* and spiders (Ameta *et al.*, 2006). Simlarly, weekly release of *T. chilonis*, weekly spray of neem and cultural practices in Haryana also resulted in 50% reduction in the number of insecticidal sprays and incidence of insect pests (Bhupender and Saini, 2008). Cotton crop treated with NSKE 5% and trap crop (pigeonpea) treated with NPV (250 LE/ha) resulted in minimum eggs (5.22/plant), larva (0.86/plant), damage in fruiting bodies (0.90%) and locule damage (6.60%) in Maharashtra, India (Jadhav *et al.*, 2008).

Application of Bt and neem on cotton crop is effective against *H. armigera* and *H. punctigera* as well as major predators like coccinellids, chrysopids, etc. at Queensland, Australia (Ma *et al.*, 2000). The use of biopesticides *Bacillus thuringiensis* (Bt), nuclear polyhedrosis virus (NPV)) reduced the use of synthetic insecticides by 50% without sacrificing cotton yield. The use of transgenic cotton affected a 25% saving on synthetic insecticide usage (Mensah, 2002). Alternate application of neem formulation, containing 1500 ppm azadirachtin (1 litre/ha) and bacterial formulations significantly reduced the incidence of sucking pests (Parthiban and Ananthan, 1998).

The integrated use of *T. chilonis, C. carnea*, NPV (at the rate of 250 LE/ha), nimbecidine (2%), dipel and synthetic chemicals imposed at different intervals on the basis of pheromone trap threshold level in Karnataka, India was most effective IPM strategy against *H. armigera* (Reddy and Manjunatha, 2000). Similarly, combinations of nimbecidine 2% + NPV at 250 larval equivalents (LE)/ha and dipel 81 + NPV at the rate of 250 LE/ha were the most effective treatments against *H. armigera* (Reddy and Manjunatha, 2000). The impact of IPM module consisted mainly of balanced fertilizer use, regular scouting and pheromone monitoring, need based application of neem products alternating with chemical insecticides and release of *Beauveria bassiana* and *Trichogramma chilonis* and conservation of natural enemies in irrigated cotton. The overall reduction in pesticide use over 3 seasons was reflected in the high population of natural enemies, i.e. 0.35 ova + larvae of *Chrysoperla carnea* and 0.16 adult spiders per plant in IPM compared to 0.14 ova + larvae of *C. carnea* and 0.07 adult spiders per plant in farmer practice. Better parasitization of *H. armigera* ova by *T. chilonis* was in the IPM block (22.4%) compared to natural parasitization (2.2%) in farmer practice (Tanwar *et al.,* 2007).

The desirability of controlling pests by the use of biopesticides is no longer questioned. A variety of techniques have been developed and refined for controlling different insect pests. The value of pesticides in protection of crops can be questioned when it is noted that estimated percentage losses in food production attributed to pests are as great today as they were 50 years ago when organic pesticides began to be widely used. The future prospect and challenges of IPM is that the research must continue on genetic approaches to pest resistance in plants. Good biopesticide management practices are essential to reduce the losses and to minimize the negative effect of these toxic chemicals on health, environment and socio-economic status of the farmers. Continued improvements are needed in bopesticide application technology and in methods to manage or prevent resistance in pests. Biological control is an emerging technology to control pests, diseases and weeds. Genetic engineering will play a vital role in production of transgenic biocontrol agents having biocontrol potential and ecological acceptability. Hopefully, all the stake holders including scientist, industry, government and farming community will work hand-in-hand to develop and implement innovative IPM strategies targeted towards a sustainable crop protection technology in the coming years and will share the responsibility to protect public health and environment.

CONCLUSIONS

A complex of naturally occurring biopesticides, viz. predators, parasitoids, or pathogens (entomopathogenic viruses, bacteria, fungi and nematodes), sex pheromones, botanicals and Bt cotton have been found to be effective in suppression of many insect pests of cotton and considered as potential, economically viable and ecologically safe tools of insect pest management. However, widespread pesticide use, difficulties involved in mass production, resitricted host range, slow action, poor quality, limited persistence in the field and inability to withstand adverse environmental conditions have been experienced as the partical difficulties in commercial deployment of some of these biopesticides for the control insect pests. Biocontrol agents that are genetically approved for various traits such as environmental hardiness, easy mass production, broad host range and host finding ability and adaptation to exteremes of climatic conditions and capable of tolerating sublethal doses of pesticides can play an important role in cotton pest management. Recent advances in the genetically improved bioagents such as pesticide resistant *Trichogramma* spp., improvement of Bt for increasing in host range, baculoviruses with more toxic proteins and incorporation of benomyl resistance into fungus, *M. anisopliae* which kill the insect in fast rate.

Several species of parasitoids and other predatory fauna in cotton ecosystem play a major role in regulation of pest population. During the last 2–3 years, with the adoption of Bt cotton in India, there has been significant increase in its area. Therefore, studies on impact of transgenic cotton on arthropod biodiversity in cotton ecosystem needs immediate attention. Genetically modified cotton having genetic resistance provides innate control of bollworms and tobacco caterpillar but sucking pests are still posing a great threat. Therefore, special attention is required for the validation and promotion of IPM technologies in transgenic cotton keeping in view the upliftment of new pests, i.e. mealybugs and secondary pest problems like thrips etc. Botanical biopesticides have a good potential for insect pest management. Among botanicals, neem products can be effectively used for pest management alone or in combination/alternation with recommended insecticides as best alternative to reduce the pesticide usage. Neem seed kernel extract which is very effective against insect pests can be prepared at village level with simple technology.The efficiency of neem products can further be increased by using additives and synergists. More systematic research efforts are needed to determine the frequency of application, spray interval and compatibility with biocontrol agents.The in depth study is required to identify more active principles as well as various factors responsible for inconsistency in the bioefficacy of neem formulations.There is need to educate the farmers about side effects of toxic chemicals and benefits of adopting neem as component of integrated pest management in a scientific manner. Many of the recent biorationals have exhibited broad spectrum of activity, and suitable standards for quality and performance have been laid out for many biorational compounds. It is emphasized that pest problems should be tackled with IPM approach based on sound ecological principles involving integration of all the available control strategies with minimum use of insecticides so that pest population can be managed below damaging level for achieving sustained production in the cotton ecosystem.

REFERENCES

Abdel Sattar, M.M. and El Guindy, M. (1988) Joint action of insect growth regulator pyrethroid mixtures on the biological activity of the pink bollworm, *Pectinophora gossypiella* Saund. *Agric. Res. Rev.*, **66**, 33–38.

Abul Nasr, S., Elnagar, S. and Nasr, S. Abul (1980) The susceptibility of the cotton leafworm, *Spodoptera littoralis* (Boisd.), to different isolates of nuclear polyhedrosis virus. *Z. Angew. Entomol.*, **90**, 289–292.

Agarwal, R.A. and Gupta, G.P. (1986) Recent advances in cotton pest management. *Pl. Prot. Bull.*, **38**, 375–378.

Ahmad, Z. and Attique, M.R. (1993) Control of pink bollworm with gossyplure in Punjab, Pakistan. *Bull. OILB/SROP,* **16**, 141–148.

AICCIP (2012) *Annual Report 2011-12.* All India Coordinated Cotton Improvement Project. Central Institute for Cotton Research, Regional Station, Coimbatore.

Ameta, O.P., Sharma, K.C., Rana, B.S. and Bambawale, O.M. (2006) Validation and popularization of IPM technology in cotton through farmers participatory approach in tribal area of Southern Rajasthan. *Ann. Agric. Res.,* **27**, 162–166.

Ananthakrishnan, T.N., Senrayan, R., Annadurai, R.S. and Murugesan, S. (1990) Antibiotic effects of resorcinol, gallic acid and phloroglucinal on *Helothis armigera* Hubner (Insecta: Noctuidae). *Proc. Indian Acad. Sci. (Anim. Sci.),* **99**, 39–52.

Anonymous (2002) *Report on Cotton Research in Punjab.* Punjab Agricultural University, Ludhiana.

Anonymous (2009) *Evaluation of* Bt *cotton Hybrids-North Zone (2004-05 to 2008-09).* All India Coordinated Cotton Improvement Project. Central Institute for Cotton Research, Regional Station, Coimbatore.

Anonymous (2010) *Agriculture at a glance 2010-11.* Department of Agriculture, Punjab, Chandigarh.

Antilla, L., Whitlow, M., Staten, R.T., El Lissy, O. and Myers, F. (1996) An integrated approach to areawide pink bollworm management in Arizona. *Proc.Beltwide Cotton Conferences*, Nashville, TN, USA, January 9-12, **2,** 1083–1085.

Arora, R., Battu, G.S. and Bath, D.S. (1997) Comparative evaluation of some native isolates of a nuclear polyhedrosis virus against *Heliothis armigera* Hubner. *J. Entomol. Res.,* **21**, 183–186.

Aulakh, S.S. and Butter, N.S. (2006) Toxicity of insect growth regulator, Lufenuron to *Helicoverpa armigera* (Hubner) on Cotton. *J. Insect Sci.,* **19**, 178-80.

Balakrishnan, N., Baskaran, R.K.M. and Mahadevan, N.R. (2004) Efficacy of *Trichogramma chilonis* Ishii in combination with biopesticides against *Helicoverpa armigera* (Hubner) in rainfed cotton ecosystem. *J. Biol. Control,* **18**, 121–127.

Balakrishnan, N., Baskaran, R.K.M. and Mahadevan, N.R. (2004) Field efficacy of *Chrysoperla carnea* (Stephens) in combination with biopesticides against *Helicoverpa armigera* (Hubner) on cotton under rainfed condition. *J. Biol. Control,* **18**, 147–153.

Balasubramani, V. and Swamiappan, M. (1998) Safety of Ha NPV and SI NPV to green lacewing *Chrysoperla carnea* Stephens (Chrysopidae: Neuroptera). *Insect Environ.*, **4**, 70–71.

Balasubramanian, G., Balasubramaniana, M., Venkatesan, S. and Gopalan, M. (1987) Control of bollworms of cotton with diflubenzuron. *Madras Agric. J.,* **74**, 490–492.

Banu, J.G., Suruliveru, T., Amutha, M. and Gapalakrishnan, N. (2010) Susceptibility of cotton mealy bug, *Paracoccus marginatus* to entomopathogenic fungi. *Ann. Plant Prot. Sci.*, **18**, 247–248.

Baskaran, R.K.M., Venugopal, M.S. and Mahadevan, N.R. (1996) Effect of host plants on the infectivity and yield of nuclear polyhedrosis virus on *Spodoptera litura* (Fabricius). *J. Biol. Control,* **10,** 73–78.

Beevi, S.P. and Balasubramanian, M. (1995) Effect of buprofezin, an novel insect growth regulator, against cotton whitefly *Bemisia tabaci* Genn. *Entomon.*, **20**, 11–14.

Benedict, J.H., Sachs, E.S., Altman, D.W., Deaton, W.R., Kohel, R.J., Ring, D.R. and Berberich, S.A. (1996) Field performance of cotton expressing transgenic Cry1A insecticidal proteins for resistance to *Heliothis virescens* and *Helicoverpa zea* (Lepidoptera: Noctuidae). *J. Econ. Entomol.,* **89**, 230–238.

Bhupender Singh and Saini, R.K. (2008) Effect of different management practices on bollworms incidence and yield of seed cotton. *J. Insect Sci.,* **21**, 17–23.

Bilapate, G.C. (1985) Studies on overlapping generations of *Earias vittella* (Fab.) in pesticides free cotton field. *J. Maharashtra agric. Univ.*, **10**, 147–150.

Bilapate, G.C. (1989) Investigation on *Heliothis armigera* (Hubner) in Marathwada. XXIV. Key mortality factors on cotton and pigeon pea. *J. Maharashtra Agric. Univ.*, **14**, 258–262.

Boricha, H.V., Raghvani, K.L., Joshi, M.D., Makadia, R.R. and Varmora, J.M. (2010) Bio efficacy of bio pesticides against whitefly, *Bemisia tabaci* infesting cotton. *Int. J. Plant Prot.,* **3**, 8–10.

Brar, K.S., Sekhon, B.S., Singh, J., Shenhmar, M. and Bakhetia, D.R.C. (2001) Evaluation of biocontrol based IPM modules for the control of bollworm complex on cotton. *J. Insect Sci.,* **14**, 19–22.

Brar, K.S., Singh, J., Shenhmar, M., Singh. J. and Singh, J. (2003) Management of cotton bollworms with egg parasitoids, *Trichogramma chilonis* Ishii in Punjab In: S Ignacimuthu and S Jayaraj (eds.), *Biological Control of Insect Pests.* Phoenix Publishing House Pvt Ltd. New Delhi., pp. 22–24.

Brar, K.S., Virk, J.S., Jalali, S.K., Sohi, A.S. and Shenhmar, M. (2007) Evaluation of multiple insecticide tolerant strain of *Trichogramma chilonis* Ishii for the management of cotton bollworms. *J. Insect Sci.*, **20**, 208-211.

Brar, K.S., Virk, J.S., Shenhmar, M., Singh, J. and Sohi, A.S. (2004) Integration of *Trichogramma chilonis* and insecticides for the management of bollworms. Fourth International Congress of Entomological Sciences. University of Agriculture, Faislabad, Pakistan September 22-23, 2004. pp. 24-25.

Brookes, G. and Barfoot, P. (2012) *GM Crops: Global Socio-economic and Environmental Impacts 1996-2010.* PG Economics Ltd, Dorchester, UK. Available at: *http:// www*.pgeconomics.co.uk/pdf/2012globalimpact studyfinal.pdf.

Busoli, A.C. (1993) Control of *Pectinophora gossypiella* (Saunders) and *Heliothis* spp. in cotton crop, using pheromones and by the mating disruption method. *Anais da Soci. Entomol. Brasil*, **22**, 139–148.

Butler, G.D., Puri, S.N. and Henneberry, T.J. (1991) Plant derived oil and detergent solutions as control agents for *Bemisia tabaci* and *Aphis gossypii* on cotton. *Southwest. Entomol.,* **16**, 331–337.

Butter, N.S., Vir, B.K., Kaur, G., Singh, T.H. and Raheja, R.K. (1992) Biochemical basis of resistance to whitefly, *Bemisia tabaci* (Genn.) (Aleyrodidae: Hemiptera) in cotton. *Trop. Agric.*, **69**, 119–122.

Butter, N.S. and Gurmeet Singh (2004) Toxicity of lufenuron to *Helicoverpa armigera* (Hubner) on cotton. *Indian J Entomol.,* **66**, 168–171.

Butter, N.S. and Kular, J.S. (1998) Field evaluation of chitin inhibitor, Teflubezuron against bollworm complex on cotton. *J. Insect Sci.*, **11**, 152–153.

Butter, N.S. and Kular, J.S. (1999) Resurgence of whitefly in cotton and its management *Indian J. Entomol.,* **61**, 85–90.

Cherian, M.C. and Margabandhu, V. (1943) Preliminary trials with *Trichogramma* parasites for the control of cotton bollworms. *Madras Agric. J.,* **31**, 107–111.

Choudhary, B. and Gaur, K. (2011) Bt *Cotton Events & Hybrids in India, 2002 to 2010.* ISAAA Biotech Information Centre, International Service for the Acquisition of Agri-Biotech Applications (ISAAA), New Delhi, India.

Cork, A. and Hall, D.R. (1998) Application of pheromones for crop pest management in the Indian sub continent. *J. Asia Pac. Entomol.*, **1**, 35–49.

Daxl, R., Centeno, B.R., Caceres, J.B. (1995) Performance of the boll weevil attract & control tube (BWACT) in a 3 year area wide Nicaraguan boll weevil control program. *Proc. Beltwide Cotton Conferences*, San Antonio, TX, USA, January 4-7, 1995: **2**, 933–937.

De Oliveira, M.R.V., Amancio, E., Laumann, R.A., and Gomes, L. de O. (2003) Natural enemies of *Bemisia tabaci* (Gennadius) B biotype and *Trialeurodes vaporariorum* (Westwood) (Hemiptera: Aleyrodidae) in Brasilia, Brazil. *Neotrop. Entomol.,* **32**, 151–154.

Dhaliwal, G.S. and Koul, O. (2010) *Quest for Pest Management: From Green Revolution to Gene Revolution*. Kalyani Publishers, New Delhi.

Dhaliwal, G.S., Arora, R. and Dhawan, A.K. (2004) Crop losses due to insect pests in Indian agriculture. *Indian J. Ecol.,* **31**, 1–7.

Dhandapani, N., Jayaraj, S. and Rabindra, R.J. (1992) Influence of sunlight and crop environment on the efficacy of nuclear polyhedrosis virus against *Helicoverpa armigera* (Hbn.) on cotton. In: *Proc. National Seminar on Changing Scenario in Pests and Pest Management in India.* January 31- February 1, 1992, Hyderabad, p. 33.

Dhandapani, N., Jayaraj, S. and Rabindra, R.J. (1994) Activity of ants on cotton plants sprayed with nuclear polyhedrosis virus and adjuvants against *Heliothis armigera* (Hubner). *J. Entomol. Res.,* **18**, 65–68.

Dhawan, A.K. and Simwat, G.S. (1992a) Field evaluation of Neemrich 20EC for management of insect pests on cotton during reproductive phase.

Proc. First National Symposium Allelopathy in Agroecosystems Agriculture and Forestry, February 12-14, 1992. CCS Haryana Agricultural University, Hisar, India, pp. 152–153.

Dhawan, A.K. and Simwat, G.S. (1992b) Management of sucking pests of cotton with natural plant products. In: *Proc. First National Symposium Allelopathy in Agroecosystems Agriculture and Forestry,* February 12-14, 1992 CCS Haryana Agricultural University, Hisar, India, pp. 154-155.

Dhawan, A.K. and Sidhu, A.S. (1993a) Catches of pink bollworm, *Pectinophora gossypiella* (Saunders) males with synthetic gossyplure. *J. Insect Sci.*. **6**, 37–40.

Dhawan, A.K. and Simwat, G.S. (1993b) Management of pink bollworm (*Pectinophora gossypiella*) through a sprayable formulation of gossyplure. *Indian J. Agric. Sci.,* **63**, 193-194.

Dhawan, A.K. and Simwat, G.S. (1997) Insecticide induce resurgence of sucking pests in upland cotton in Punjab, India. In: *Third International Congress of Entomological Sciences,* March 18 20, 1997 (Abstract). National Agricultural Centre, Islamabad.

Dhawan, A.K. and Simwat, G.S. (1998) Evaluation of different biopesticides against cotton bollworms, *Helicoverpa armigera* (Hubner). In: G.S. Dhaliwal, N.S. Randhawa, R. Arora and A.K. Dhawan (eds.), *Ecological Agriculture and Sustainable Development*, IES & CRID, Chandigarh, pp. 274–280.

Dhawan, A.K., Ravinder Singh and Shera, P.S. (2008) Field efficacy of novaluron against tobacco caterpillar, *Spodoptera litura* (Fab.) on Bt cotton. *Pestici. Res. J.,* **20**, 75–78.

Dhawan, A.K., Sidhu, A.S. and Simwat, G.S. (1988) Assesment of avoidable loss in cotton due to sucking pests and bollworms. *Indian J. Agric. Sci.,* **58**, 290–292.

Dhawan, A.K., Simwat, G.S. and Dhaliwal, G.S. (1998) Evaluation of different biopesticides against cotton bollworm, *Helicoverpa armigera* Hubner. In: G.S. Dhaliwal, R. Arora, N.S. Randhawa and A.K. Dhawan (eds.), *Ecological Agriculture and Sustainable Development,* IES & CRID, Chandigarh. pp. 274–280.

Dhawan, A.K., Simwat, G.S., Tauro, P. and Narwal, S.S. (1992a) Management of sucking pests of cotton with natural plant products. In: *Proceedings First Natural Sym Allelopathy in Agroecosystems (Agriculture and Forestry)* held at CCS Haryana Agricultural University, Hisar, India, pp. 154-155.

Dhawan, A.K., Srimannarayana, G.S., Simwat, G.S. and Nagaiah, K. (1992b) Management of cotton pests on upland cotton *Gossypium hirsutum* with Navneem 95EC. In: *Proc. First National Symposium Allelopathy in Agroecosystems Agriculture and Forestry,* February 12 14, 1992 held at CCS Haryana Agricultural University, Hisar 125 004, India. pp. 156-157.

Dhawan, A.K., Sithanantham, Baumgartner, J. and Rao Ranga, G.V. (2001) Evaluation of nuclear polyhedrosis virus against *Helicoverpa armigera* (Hubner) on upland cotton, *Gossypium hirsutum* Linnaeus. (Abstract). *Conference on Biopesticides: Emerging trends* February 7-9, 2001, Chandigarh, p. 119.

Dhumal, V.S., Jadhaw, D.B., Khurhade, P.M., Khandge, S.V., Naik, L.M. and Ajri, D.S. (1982) Field studies on the comparative efficacy of biological control agents and recommended insecticides against bollworms on cotton. In: *Proc. Workshop on Methodology of Biological Control in Cotton, Sugarcane and Groundnut.* October 4-5, 1982.

Dimetry, N.Z., Gomaa, A.A., Salem, A.A. and Moniem, A.S.H. (1998) Bioactivity of some formulations of neem seed extracts against the whitefly, *Bemisia tabaci* Genn., *Neem Newsl.,* **15,** 38.

Dunkelblum, E., Kehat, M. and Gothilf, S. (1987) Mating disruption of the cotton leafworm, *Spodoptera littoralis,* monitored by direct night observations. *Bull. SROP.,* **10**, 27–29.

Duraimurugan, P. and Regupathy, A. (2005) Joint action of *Beauveria bassiana* (Balsamo) Vuillemin and synthetic pyrethroids against *Helicoverpa armigera* (Hubner) (Noctuidae: Lepidoptera). *J. Food Agric. Environ.,* **3,** 65–67.

El Katatny, M.H. (2010) Virulence potential of some fungal isolates and their control promise against the Egyptian cotton leaf worm, *Spodoptera littoralis. Arch. Phytopathol. Plant Prot.,* **43**, 332–356.

El Khawas, M.A.M. (2001) Joint action due to combination of bacterial insecticide, agerin with insect growth regulator or chemical insecticide on *Agrotis ipsilon* (Hufn.) larvae (Lepidoptera: Noctuidae). *Ann. Agric. Sci.,* **39**, 1723–1739.

Flint, H.M. and Parks, N.J. (1989) Effect of azadirachtin from the neem *Azadirachta indica* tree on immature sweetpotato whitefly, *Bemisia tabaci* Homoptera: Aleyrodidae and other selected pest species on cotton. *J. Agric. Entomol.,* **6,** 211–215.

Fuchs, T.W., Minzenmayer, R. and Sansone, C. (1999) Use of pheromone trapping data to evaluate Boll Weevil Eradication in the Southern Rolling Plains of Texas. *Proc. Beltwide Cotton Conferences*, Orlando, Florida, USA, 3-7 January, 1999, **2**, 838–840.

Gahukar, R.T. (2000) Use of neem products/pesticides in cotton pest management. *Int. J. Pest Manage.*, **46**, 149–160.

Gard, I.E. (1997) Field testing a genetically modified baculovirus. Microbial insecticides: novelty or necessity? Proc. Symposium held at the University of Warwick, Coventry, UK, 16-18 April 1997, pp. 101–114.

Gawade, M.B., Naik, D.G. and Kapadi, A.H. (1993) Antijuvenile hormone effect of precocene I on red cotton bug *Dysdercus koenigii. Botanical Pesticides in Integrated Pest Management*, pp. 318–320.

Gill, H.K. and Dhawan, A.K. (2006) Global status of insecticide resistance in *Helicoverpa armigera* on cotton. *J. Cotton Res. Dev.*, **20**, 226–231.

Gouge, D.H., Smith, K.A., Lee, L.L. and Henneberry, T.J. (1998) Control of pink bollworm *Pectinophora gossypiella* (Saunders) (Lepidoptera: Gelechiidae) with entomopathogenic nematodes (Rhabditida: Steinernematidae) and gossyplure in cotton. *Recent Res. Develop. Entomol.*, **2**, 55–85.

Gujar, G.T. and Mehrotra, K.N. (1989) Effects of four chitin synthesis inhibitors on the red cotton bug, *Dysdercus koenigii* Fabricius. *Entomon.*, **14**, 71–73.

Gujar, G.T. and Mehrotra, K.N. (1993) Toxicity and morphogenetic effects of neem oil on the red cotton bug *Dysdercus koenigii* F. *Botanical Pesticides in Integrated Pest Management,* pp. 217-224.

Gundannavar, K.P., Lingappa, S. and Giraddi, R.S. (2007) Study of dose mortality response between instars of *Helicoverpa armigera* (Hubner) and *Metarhizium anisopliae* (Metschinkoff) sorokin. *Karnataka J. Agric. Sci.,* **20**, 140–141.

Gundannavar, K.P., Lingappa, S., Giraddi, R.S. and Kulkarni, K.A. (2008) Susceptibility of *Helicoverpa armigera* (Hubner) to *Nomuraea rileyi* (Farlow) Samson. *J. Entomol. Res.,* **32**, 11–13.

Gupta, G.P. and Kirti Sharma (1996) Utilization of biopesticides in managing the cotton pest complex in India. *Proc. Beltwide Cotton Conferences,* Nashville, TN, USA, January 9-12, 1996, **2**, 1135–1140.

Hargreaves, H. (1948) *List of Recorded Cotton Insects of the World.* Commonwealth Institute of Entomology, London.

Hinz, S.E. and Wright, J.E. (1997) Naturalis L: a biological product (Beauveria bassiana JW 1) for the control of cotton pests. *Proce. Beltwide Cotton Conferences,* New Orleans, LA, USA, January 6 10, **2**. 1997, 1300–1302.

Horowitz, A.R. and Ishaaya, I. (1992) Susceptibility of the sweetpotato whitefly (Homoptera: Aleyrodidae) to buprofezin during the cotton season. *J. Econ. Entomol.*, **85**, 318–324.

Horowitz, A.R. and Ishaaya, I. (2004) Biorational insecticides-Mechanisms, selectivity and importance in pest management. In: A.R. Horowitz and I. Ishaaya (eds.), *Insect Pest Management: Field and Protected Crops,* Springer Verlag, Berlin, Germany, pp. 1–28.

Jackson, R.E., Bradley, J.R. and Van, D.W.J. (2003) Field performance of transgenic cotton expressing one or two *Bacillus thuringiensis* endotoxins against bollworm, *Helicoverpa zea* (Boddie). *J. Cotton Sci.,* **7**, 57–64.

Jackson, R.E., Bradley, J.R., Van, D.J.W. and Gould, F. (2004) Comparative production of *Helicoverpa zea* (Lepidoptera: Noctuidae) from transgenic cotton expressing either one or two *Bacillus thuringiensis* proteins with and without insecticide over sprays. *J. Econ. Entomol.,* **97**, 1719–1725.

Jadhav, S.B., Sadawarte, A.K. and Bhalkare, S.K. (2008) Evaluation of push pull technique for the management of *Helicoverpa armigera* (Hubner) in cotton. *Indian J. Entomol.,* **70**, 360–364.

James, C. (2011) *Global Status of Commercialized Biotech/GM crops*: 2011. ISAAA Briefs No. 43. International Service for the Acquisition of Agri-Biotech Applications (ISAAA), Ithaca, NY.

Jayaraj, S. (1993) Neem in pest control: Progress and prospectives. *In: World Neem Conf. Souvenir.* February 24-28, 1993, Bangalore, pp. 37–43.

Jayaraj, S., Santharam, G., Narayanan, K., Sundarajan, K. and Balagurunathan, R. (1981) Effectiveness of polyhedrosis virus against field population of tobacco caterpillar, *Spodoptera litura* on cotton. *Andhra Agric. J.,* **27**, 26–29.

Jayaswal, A.P. and Pundarikakshudu, R. (1989) *Evaluation of Some Cotton Germplasm for their Resistance to Whitefly, Bemisia tabaci* (Genn.) Proceedings Biennial Workshop, AICCIP, Hyderabad.

Jayaswal, A.P. and Sundaramurthy, V.T. (1992) *Achievements in Insect Management in Cotton.* All India Coordinated Cotton Improvement Project Silver Jubilee (1967-1992), September 17-19, 1992, Nagpur, pp. 117–151.

Jeyarani, S., Sathiah, N. and Karuppuchamy, P. (2005) Influence of *Pseudomonas fluorescens* and a nucleopolyhedrovirus on cotton bollworm *Helicoverpa armigera* (Hubner). *Trop. Agric. Res.,* **17**, 221–230.

Jones, K.A., Irving, N.S., Grzywacz, D., Moawad, G.M., Hussein, A.H. and Fargahly, A. (1994) Application rate trials with a nuclear polyhedrosis virus to control *Spodoptera littoralis* (Boisd.) on cotton in Egypt. *Crop Prot.,* **13**, 337–340.

Joshi, M.D., Butani, P.G., Patel, V.N. and Jeyakumar, P. (2010) Cotton mealy bug, *Phenacoccus solenopsis* Tinsley a review. *Agric. Rev.,* **31**, 113–119.

Kakde, T.D., Wadaskar, R.M., Gawande, S.P., Gaikwad, S.J. and Vidyapeeth, P.D.K. (2005) Susceptibility evaluation of termites, Odontotermes obesus (Rambur) against the fungal entomopathogens. *Resist. Pest Manage. Newsl.,* **14**, 26–29.

Kambrekar, D.N., Kulkarni, K.A. and Giraddi, R.S. (2008) Influence of Bt cotton on HaNPV isolates. *Ann. Plant Prot. Sci.,* **16**, 95–101.

Kapoor, S.K., Singh, J., Russel, D., Singh, B. and Kalra, R.L. (2002) Susceptibility change of *Helicoverpa armigera* (Hubner) to different insecticides in Punjab. *Pestic. Res. J.,* **14**, 177–180.

Karihaloo, J.L. and Kumar, P.A. (2009) Bt *Cotton in India-A Status Report* (Second Edition). Asia-Pacific Consortium on Agricultural Biotechnology (APCoAB), New Delhi, India.

Kassem, S.M.I., Aly, M.I., Bakry, N.S. and Zeid, M. (1987) The efficiency of methoprene formulated as microencapsulated or emulsifiable concentrate and their mixtures with certain insecticides. *Alexandria J. Agric. Res.,* **32**, 309–321.

Kehat, M., Anshelevich, L., Gordon, D., Harel, M. and Dunkelblum, E. (1998) Evaluation of Shin Etsu twist tie rope dispensers by the mating table technique for disrupting mating of the cotton bollworm, *Helicoverpa armigera* (Lepidoptera: Noctuidae), and the pink bollworm, *Pectinophora gossypiella* (Lepidoptera: Gelechiidae). *Bull. Entomol. Res.*, **88**, 141–148.

Khan, Q. and Rao, V.P. (1960) Insect and mite pests. In: R.H. Dastur, R.D. Asana, K. Sawhney, S.M. Sikka, R.S. Vasudeva, Q. Khan, V.P. Rao and B.L. Sethi, *Cotton in India A Monograph*. Indian Central Cotton Committee, Bombay, pp. 217–301.

Kim, J.S., Roh JongYul, Choi JaeYoung, Wang Yong, Shim HeeJin, Je YeonHo (2010) Correlation of the aphicidal activity of *Beauveria bassiana* SFB 205 supernatant with enzymes. *Fungal Biol.*, **114**, 120–128.

Koul, O. (2005) *Insect Antifeedants*. CRC Press, Boca Raton, Florida, USA.

Koul, O. and Wahab, S. (2004) *Neem: Today and in the New Millennium*. Kulwer Academic Publishers, Dordrecht, The Netherlands.

Kranthi, S., Dhawad, C.S., Naidu, S., Bharose, A.A., Chaudhary, A., Sangode, V., Nehare, S.K., Bajaj, S.R. and Kranthi, K.R. (2009) Susceptibility of the cotton bollworm, *Helicoverpa armigera* (Hubner) (Lepidoptera: Noctuidae) to the *Bacillus thuringiensis* toxin Cry2Ab before and after the introduction of Bollgard-II. *Crop Prot.*, **28**, 371–375.

Kulkarni, G.G. and Hugar, P.S. (1999) Effect of host plants on POBs yield and mortality of *Spodoptera litura* larvae infected with NPV. *Karnataka J. Agric. Sci.*, **12**, 60–63.

Kulkarni, K.A., Thippeswamy, C., Thirumalraju, G.T., Kumar, N.G., Thimmaiah, G. and Abraham, E.S. (1986) Effect of synthetic pyrethroid insecticides and chitin synthesis inhibitor on cotton bollworms and fibre properties. *Mysore J. Agric. Sci.*, **20**, 28–37.

Lande, G.K., Katole, S.R., Bhalkare, S.K. and Bisane, K.D. (2008) Evaluation of biorational IPM modules against bollworm complex on cotton. *J. Cotton Res. Develop.*, **22**, 91–96.

Langston, W.C. (1996) Boll weevil control with boll weevil attract and control tubes 1995. *Proce. Beltwide Cotton Conferences*, Nashville, TN, USA, January 9–12, 1996: **2**, 805.

Ledson, T.M. and Thornback, N. (1994) Control of *Bemisia tabaci* Gennadius (Sternorrhyncha: Aleyrodidae) in Ethiopian cotton by the insect growth regulator buprofezin. *Int. Pest Control*, **36**, 42–43.

Lin, ChienYih, Wu, DerChung, Yu, JihZu, Chen, BingHuei, Wang, ChinLing, Ko, WenHsiung (2009) Control of silverleaf whitefly, cotton aphid and Kanzawa spider mite with oil and extracts from seeds of sugar apple. *Neotropical Entomol.*, **38**, 531–536.

Lopez, J.D. Jr, Goodenough, J.L. and Hendricks, D.E. (1988) Evaluation of two pheromone dispensers in combination with plastic and wire cone traps for tobacco budworm (Lepidoptera: Noctuidae). *J Econ. Entomol.*, **81**, 1750–1753.

Luo, S.B., Van, J.P., Chal, C.J., Liang, S.P., Zhang, V.M., Zhang, V. and Le, G.K. (1986) Control of pink bollworm *Pectinophora gossypiella* with *Bacillus thuringiensis* in cotton fields. *Chinese J. Boil. Control,* **2**, 167–169.

Luttrell, R.G., Yearian, W.C. and Young, S.Y. (1982) Effects of Elcar (*Heliothis zea* nuclear polyhedrosis virus) treatments on *Heliothis* spp. *J. Georgia Entomol. Soc.,* **17**, 211–221.

Luttrell, R.G., Yearian, W.C. and Young, S.Y. (1982) Mortality of *Heliothis* spp. larvae treated with *Heliothis zea* nuclear polyhedrosis virus spray adjuvant combinations on cotton and soybean. *J. Georgia Entomol. Soc.,* **17**, 447–453.

Ma, D.L., Gordh, G. and Zalucki, M.P. (2000) Toxicity of biorational insecticides to *Helicoverpa* spp. (Lepidoptera: Noctuidae) and predators in cotton field. *Int. J. Pest Manage.*, **46**, 237–240.

Mafra Neto, A. and Habib, M. (1996) Evidence that mass trapping suppresses pink bollworm populations in cotton fields. *Entomol. Exp. Appl.,* **81**, 315-323.

Malarvannan, S., Giridharan, R., Sekar, S., Prabavathy, V.R. and Sudha Nair (2008) Bioefficacy of crude and fractions of Argemone mexicana against tobacco caterpillar, *Spodoptera Litura* Fab. (Noctuidae: Lepidoptera). *J. Biopestic.,* **1**, 55–62.

Manjunath, T.M. (2011) *Q & A on* Bt *cotton in India: Answers to more than 85 questions. Second Edition.* Association of Biotechnology Led Enterprises, Agricultural Group, Bengaluru.

Mann, G.S., Dhaliwal, G.S. and Dhawan, A.K. (2001) Effect of alternate application of neem products and insecticides on population of *Bemisia tabaci* Gennadius and its impact on bollworm damage in upland cotton. *Ann. Plant. Prot. Sci.,* **9**, 22–25.

Mann, R.S., Gill, R.S., Dhawan, A.K. and Shera, P.S. (2010) Relative abundance and damage by target and non target insects on Bollgard and Bollgard II cotton cultivars. *Crop Prot.,* **29,** 793–801.

Masner, P., Angst, M. and Dorn, S. (1987) Fenoxycarb, an insect growth regulator with juvenile hormone activity: a candidate for *Heliothis virescens* (F.) control on cotton. *Pestic. Sci.,* **18**, 89–94.

Meade, T., Felton, G.W. and Young, S.Y. (1995) *Host Plant Modification of Susceptibility to the Heliothis Nuclear Polyhedrosis Virus.* Special Report, Agricultural Experiment Station, Division of Agriculture, University of Arkansas, Arkansas.

Mehrotra, K.N. (1993) Status of insecticide resistance in insect pests. In: G.S. Dhaliwal and B. Singh (eds.), *Pesticides: Their Ecological Impact in Developing Countries.* Commonwealth Publishers, New Delhi, pp. 30–50.

Meisner, J., Klein, M. and Ben Moshe, E. (1992) Effect of Margosan O on the development of the leafhopper Asymmetrasca decedens. *Phytoparasitica,* **20**, 15– 23.

Mensah, R.K. (2002) Development of an integrated pest management programme for cotton. Part 2: Integration of a lucerne/cotton interplant system, food supplement sprays with biological and synthetic insecticides. *Inte. J. Pest Manage.*, **48**, 95–105.

Mesbah, A.H., Marie, S.S. and El Husseini, M.M. (2008) Biocontrol agents of three lepidopterous insect pests on some crops during summer season at Khafr Elshikh Governorate. *Egyptian J. Biol. Pest Control,* **18**, 171–176.

Mitchell, E.R., Kehat, M., Tingle, F.C. and McLaughlin, J.R. (1997) Suppression of mating by beet armyworm (Noctuidae: Lepidoptera) in cotton with pheromone. *J. Agric. Entomol.*, **14**, 17–28.

Mohapatra, L.N. and Sahu, B.B. (2004) Natural enemies of green semilooper, *Anomis flava* (Fab.) in cotton ecosystem of Orissa. *Insect Environ.*, **10**, 36–37.

Mostafa, S.A., Ali, F.A.F., Abbass, M.G. and Azeem, M.A. (1991) Effect of some insect growth regulators on the greasy cutworm, *Agrotis ipsilon* (Hufn). *Egyptian J. Agric. Res.*, **69**, 1–10.

Murthy, K.S.R.K. and Ram, G.M. (2002) Studies on the efficacy of a new chitin synthesis inhibitor Rimon (novaluron 10 EC) on American bollworm *Helicoverpa armigera* Hubn. attacking cotton. (Abstract), *Symposium Resources Management in Plant Protection during Twenty first Century*, Hyderabad, India.

Muthukrishnan, N. and Balasubramanian, M. (1992) Relationship between the pheromone trap catches of *Pectinophora gossypiella* and *Spodoptera litura* their field infestation and larval population. *Madras Agric. J.*, **79**, 321–326.

Nakache, J., Dunkelblum, E., Kehat, M., Anshelevich, L. and Harel, M. (1992) Mating disruption of the spiny bollworm, *Earias insulana* (Lepidoptera: Noctuidae), with Shin Etsu twist tie ropes in Israel. *Bull. Entomol. Res.*, **82**, 369–373.

Nandihalli, B.S., Hugar, P. and Patil, B.V. (1990) Evaluation of neem and neem products against cotton whitefly, *Bemisia tabaci* (Gennadius). *Karnataka J. Agric. Sci.*, **3**, 58–61.

Natarajan, K. and Sundaramurthy, V.T. (1990a) Effect of neem oil on cotton whitefly, *Bemisia tabaci. Indian J. Agric. Sci.*, **60**, 290–291.

Nazir Ahmad, Mohammad Ashraf, Talib Hussain, Bilquis Fatima, Nasrullah (1996) Significance of pheromones and parasites for the control of cotton bollworms. *Pakistan J. Zool.*, **28**, 355-357.

Nimbalkar, S.A., Khodke, S.M., Taley, Y.M. and Patil, K.J. (1990) Biofficacy of neem and karanj seed extracts along with other commercial plant products against bollworms on cotton, In: *Proc Symposium Botanical Pesticides in IPM*, Rajamundry, pp. 252–255.

Panchbhai, P.R., Sharnagat, B.K., Dhawad, C.S., Wadaskar, R.M. and Bagade, I.B. (2005) Effect of augmentation of *Trichogramma chilonis* Ishii and *Chrysoperla carnea* (Stephens), two natural enemies of

Helicoverpa armigera (Hubner) on larval infestation and yield in cotton. *Pest Manage. Eco. Zool.,* **13**, 77–81.

Parthiban, M. and Ananthan, G. (1998) Alternate application of neem and biopesticide in cotton pest management. *Insect Environ.,* **3**, 97–98.

Patil, B.V., Nandihalli, B.S., Hugar, P., Thimmanagouda, B.R. and Lingappa, S. (1993) Efficacy of some plant and animal products against cotton pest complex. *Botanical Pesticides in Integrated Pest Management.* 1993, 183–189.

PAU (2008) *Package of Practices for Crops of Punjab (Kharif 2008).* Punjab Agricultural University, Ludhiana.

Phadke, A.D., Khandal, V.S. and Rahalkar, S.R. (1988) Use of a neem *Azadirachta indica* product in insecticide resistance management IRM in cotton. *Pesticides,* **22,** 36–37.

Qureshi, Z.A., Arif, M.D., Ahmed, N. and Najeebullah (1988) Control of pink bollworm, *Pectinophora gossypiella* (Saunders) by mating disruption technique. *Pakistan J. Sci. Ind. Res.,* **31**, 711–713.

Rabindra, R.J. and Jayaraj, S. (1986) Multiplication and use of nuclear polyhedrosis viruses for the control of important crop pests. In. S. Jayaraj, (ed.), *Pest and Disease Management: Oilseeds, Pulses, Millets and Cotton.* Tamil Nadu Agic. Univ., Coimbatore, pp. 52–61.

Race, J.L. (1995) Cotton moth monitoring and control. *Proc. Beltwide Cotton Conferences,* San Antonio, TX, USA, January 4-7, 1995, **2**, 798–799.

Rajak, R.L. (1992) Plant protection-Future research strategies, Keynote Address at *National Seminar on Changing Scenario in Pests and Pest Management in India.* January 31- February 1, 1992, Hyderabad,

Rajeshwari, V.R., Subha Rao, I.V., Naidu, N.V., Gopal Singh, B. and Ranganadhacharyulu, N. (1999) Study on biochemical factors on incidence of American bollworm (*Helicoverpa armigera* (Hubner) in cotton. *J. Indian Soc. Cotton Improv.,* **24**, 36–42.

Raju, N., Saivaraj, K. and Kareem, A.A. (1992) Effect of neem products and insecticides against jassid, *Amrasca biguttula biguttula* (Ishida) on MCU 7 cotton. In*: All India Coordinated Cotton Improvement Project Silver Jubilee Souvenir,* September 17-19, 1992, Nagpur, p. 129.

Rao, N.V., Reddy, A.S. and Rao, K.T. (1989) Natura enemies of cotton whitefly, *Bemisia tabaci* Genn. In relation to pest population and wheather factos. *J. Biol. Control,* **3**, 10–12.

Rao, N.V., Reddy, A.S. and Reddy, P.S. (1990) Relative efficacy of some new insecticides on insect pests of cotton. *Indian J. Plant Prot.,* **18**, 53–58.

Rashad, A.M. and Aly, S.H. (1997) Susceptibility of eggs and larvae of the *Pectinophora gossypiella* (Lepidoptera Gelechidae) to the *Beauveria bassiana. Anal. Insti. Cerc Cereale Prote Plant.,* **28**, 193–201.

Ravi, M., Dhandapani, N., Sathiah, N. and Murugan, M. (2008) Effect of some selected biopesticides on growth and development of *Spodoptera litura* fab. (Lepidoptera: Noctuidae) larvae. *Entomon.,* **33**, 71–74.

Reddy, A.S. and Rao, N.V. (1993) Efficacy and selectivity of insecticides of natural origin on pests of cotton, *Gossypium hirsutum*. In: M.S. Chari and G. Ramaprasad, (eds.), *Botanical Pesticides in Integrated Pest Management*, Rajahmundry, pp. 170–174.

Reddy, G.P.V. and Krishnamurthy, M.M. (1989) Insect pest management in cotton. *Pesticides,* **23**, 18–19.

Reddy, G.V.P. and Manjunatha, M. (2000) Laboratory and field studies on the integrated pest management of *Helicoverpa armigera* (Hubner) in cotton, based on pheromone trap catch threshold level. *J. Appl. Entomol.,* **124**, 213–221.

Regupathy, A. and Mahadevan, N.R. 1992. Evaluation of gossyplure against pink bollworm (*Pectinophora gossypiella*) of cotton. *In: Proc. National Seminar on Changing Scenario in Pests and Pest Management in India.* January 31- February 1, 1992, Hyderabad, p.32.

Rosaiah, B. and Reddy, A.S. (1993) Utility of botanicals and biocides as additives to insecticides in *Helicoverpa armigera* (Hubner) management on cottn. *In: World Neem Conf. Souvenir.* February 24-28, 1993, Bangalore, pp. 16-17.

Samuthiravelu and David, B.V. (1990) Bio efficacy of neem oil and deltamethrin against spotted boll worm *Earias vittella* (Fab.) (Noctuidae: Lepidoptera) on cotton (MCU 5). *Madras Agric. J.,* **77**, 294–298.

Sangwan, H.S., Varma, S.N. and Sharma, V.K. (1974) Possibility of integration of exotic parasite, *Trichogramma brasiliensis* Ashmead for the control of cotton bollworm. *Indian J. Entomol.,* **34**, 69–70.

Santos, W.J. and Hofer, D. (1996) Study of the effectiveness of boll weevil control applying drops of a mixture of sex pheromone (Grandlure) and cypermethrin through a pistol. *Proc. Beltwide Cotton Conferences,* Nashville, TN, USA, January 9-12, **2,** 712–715.

Sarode, S.V., Patil, P.P. and Borkar, S.L. (1995) Evaluation of neem seed kernel extract in combinations with Heliothis nuclear polyhedrosis virus against cotton bollworms. *J. Entomol. Res.,* **19**, 219–222.

Satoh, G.T., Plapp, F.W., Jr Slosser, J.E. (1995) Potential of juvenoid insect growth regulators for managing cotton aphids (Homoptera: Aphididae). *J. Econ. Entomol.,* **88**, 254–258.

Satpute, U.S., Sarnaik, D.N. and Bhalerao, P.D. (1989) New insecticides for management of cotton bollworm. *Pesticides,* **23**, 26–27.

Sekhon, B.S. (1980) Role of parasitoids and predators in the management of cotton bollworms in the Punjab. Ph.D. Thesis, Punjab Agricultural University, Ludhiana.

Shanthipriya, B. and Misra, H.P. (2007) Management of cotton grey weevil, *Myllocerus maculosus* Desb. using bio pesticides. *J. Plant Prot. Environ.,* **4**, 159–160.

Sharma, H.C. and Agarwal, R.A. (1983) Oviposition behaviour of spotted bollworm, *Earias vittella* Fab. on some cotton genotypes. *Insect Sci. Applic.*, **4**, 373–376.

Sharma, P.K., Singh Mukand and Dhawan, A.K. (1989) Management of bollworms in *Gossypium arboreum* L. through nutrient and irrigation in Punjab. *J. Res. Punjab Agric. Univ.*, **26**, 204–205.

Shi WeiBing, Zhang LiLi and Feng MingGuang (2008) Field trials of four formulations of *Beauveria bassiana* and *Metarhizium anisoplae* for control of cotton spider mites (Acari: Tetranychidae) in the Tarim Basin of China. *Biol. Control,* **45**, 48–55.

Shonouda, M.L., Osman, S., Salama, O. and Ayoub, A. (2008) Insecticidal effect of *Chrysanthemum coronarium* L. flowers on the pest *Spodoptera littoralis* Boisd and its parasitoid *Microplitis rufiventris* Kok. with identifying the chemical composition. *J. Appl. Sci.*, **8**, 1859–1866.

Shternshis, M.V. (2002) Biological preparations based on microbial metabolites. *Zash Karantin Rast.*, **9**, 18–19.

Simwat, G.S. and Dhawan, A.K. (1992a) Use of plant insecticides against *Helicoverpa armigera* (Hubner) infesting cotton and chickpea. In: *Proc First National Symposium on Allelopathy in Agroecosytems (Agriculture & Forestry)*, February12 14, 1992, Hisar. pp 158–159.

Simwat, G.S. and Gill, B.S. (1992b) Population build up of cotton jassid and whitefly on *Gossypium arboreum* varieties and its correlation with weather factors. *J. Res. Punjab Agric. Univ.*, **29**, 70–76.

Singh, D, Singh, K. and Singh, G. (2005) Survival of cotton whitefly, *Bemisia tabaci* (Gennadius) following application of buprofezin *J. Insect Sci.*, **18**, 40–43.

Singh, J., Sohi, A.S. and Russel, D. (2004) Magnitude of *Helicoverpa armigera* (Hubner) problem on cotton in Asia. In: *Proc Seminar on Recent Advances in Spray Technology in Irrigated Cotton in Asia*, Ludhiana, Punjab, pp. 34-44.

Singh, K., Dhawan, A.K., Kumar, T., Arora, P.K. and Singh, R. and (2008) IPRM on Bt cotton: its adoption and impact on production and economics (Abstract), *Proc 2nd Congress on Insect Science.* February 21-22, 2008. Punjab Agricultural University, Ludhiana

Singh, R. and Agarwal, R.A. 1988. Role of biochemical components of resistant and susceptible cotton and okra in ovipositional preference of cotton leafhopper. *Proc. Indian Acad. Sci (Anim. Sci.),* **97,** 545–550.

Singh, S.P. (1985) Biological control of insect pests of national importance. In: S. Jayaraj (ed.), *Integrated Pest and Disease Management,* TNAU, Coimbatore, pp. 309–318.

Singh, S.P. (1990) *Heliothis armigera* (Hubner), Prospects of its biological suppression. *In:* S. Jayaraj, S. Uthamasamy, M. Gopalan and R.J. Rabindra (eds.), *Proc. National Workshop on Heliothis Management*, TNAU, Coimbatore, pp. 223–229.

Singh, S.P. (1992) Potential of biological control in integrated pest management in India. *In: Proc. Natn. Symp. Recent Advances in Integrated Pest Management.* October 12-15, 1992, Ludhiana, pp. 59-60.

Skehon, B.S. and Varma, G.C. (1983) Parasitoids of *Pectinophora gossypiella* (Lepidoptera : Gelechiidae) and *Earias* spp. (Lepidoptera, Noctuidae) in the Punjab. *Entomophaga,* **28,** 45–53.

Sohi, A.S., Singh, J., Brar, D.S. and Russell, D. (1999) Further studies on mating disruption in pink bollworm, *Pectinophora gossypiella* (Saunders) using sex pheromone as a component of IPM programme in irrigated cotton fields in Punjab. *Pest Manage. Econ. Zool.,* **7**, 31–38.

Spurgeon, D.W. (2010) Efficacy of *Beauveria bassiana* against *Lygus hesperus* (Hemiptera: Miridae) at low temperatures. *J. Entomol. Sci.,* **45**, 211–219.

Srinivasa, M., Babu, C.S.J., Anitha, C.N. and Girish, G. (2008) Laboratory evaluation of available commercial formulations of HaNPV against *Helicoverpa armigera* (Hub.). *J. Biopestic.,* **1**, 138–139.

Srinivasan, M.R. and Venugpal, M.S. (1992) Enhancing pink bollworm and control with combination of insecticides with stirrup PBW. *In: Proc. Natn. Symp. Recent Advances in Integrated Pest Management.* October 12-15, 1992, Ludhiana, p. 108.

Stenseth, C. and Singh, H.M. (1990) Buprofezin against the glasshouse whitefly and the cotton whitefly. *Gartneryrke,* **80**, 18–19.

Subramaniam, T.S., Sivaprasad, B. and Prasad, V.V.L.N. (1993) Neemoil 93EC (RD 9 Repelin)-an effective botanical for pest management. *In: World Neem Conf. Souvenir.* February 24-28, 1993, Bangalore, p. 17.

Sundramurthy and Chitra, (1992) Integrated Pest management in cotton. *Indian J. Plant Prot.,* **20**, 1–17.

Sushila, N. and Patil, B.V. (2001) Studies on host plant resistance in cotton to serpentine leaf miner, *Liriomyza trifoli* (Bugress). *J. Cotton Res. Dev.,* **15**, 186–190.

Tanwar, R.K., Bambawale, O.M., Jeyakumar, P., Dhandapani, A. Vikas-Kanwar; Sharma, O.P. and Monga, D. (2007) Impact of IPM on natural enemies in irrigated cotton of North India. *Entomon.,* **32**, 25–32.

Teague, T.G. and Tugwell, N.P. (1997) Establishing trap crops to aggregate and suppress spring populations of boll weevil an alternative tactic for Midsouth eradication programs. *Proc. Beltwide Cotton Conferences*, New Orleans, LA, USA, January 6 10, 1997: **2**, 1205–1207.

Temerak, S.A. (2005) Ovicidal activity of the natural bio-product spinosad through field observation of tagged egg masses of Spodoptera littoralis on cotton, in five Governorates of Egypt. *Assiut J. Agric. Sci.,* **36**, 85–95.

Thangavelu, K. (1982) Ovicidal effect of diflubenzuron on ash weevil. *Curr. Sci.,* **51**, 439–440.

Uthamasamy, S. and Gajendran, G. (1992) Efficacy of neem products in the management of pests of cotton. In: *Proc. National Seminar on Changing*

Scenario in Pests and Pest Management in India. January 31- February 1, 1992, Hyderabad, p. 45.

Uthamasamy, S. and Parameswaran, S. (1989) *Cotton Insect Pest Management* (In Tamil). C.P.P.S., TNAU, Coimbatore.

Varma, G.C. and Shenhmar, M. (1988) Parasitoids of *Pectinophora gossypiella* (Saunders) (Lep.: Gelechiidae) and *Earias* spp. (Lep.: Noctuidae) in Punjab. *J. Res. (PAU),* **25,** 592-594.

Varma, G.C. and Gill, J.S. (1992) Manipulation of natural enemies in the management of insect pests of sugarcane and cotton in Punjab. In: *Proc. Natn. Symp. Recent Advances in Integrated Pest Management.* October 12-15, 1992, Ludhiana, pp. 61–62.

Varma, G.C., Bindra, O.S. and Singh, S. (1976) A preliminary study on the control of cotton bollworms with *Trichogramma brasiliensis. Cotton Dev.*, **6**, 17–18.

Venkatetan, S., Balasubramanian, G., Jayaraj, S. and Gopalan, M. (1987) Studies on the efficacy of neem products against the aphid *Aphis gossypii* Glover on cotton. *Madras Agric. J.*, **74**, 255–257.

Verma, R.S., Das, S.B., Shaw, S.S., Mandloi, K.C. and Badaya, A.K. (1989) Chemical control of whitefly (*Bemisia tabaci*) on cotton (Khandwa, M.P,). *J. Cotton Res. Dev.*, **3**, 49–51.

Vidya, F., Sethuraman, K., Mariappan, V., Karunakaran, K.R. and Vidhyseakaran, P. (1993) Management of pests and diseases of cotton and soybean using neem products. In: *World Neem Conf. Souvenir.* February 24-28, Bangalore, p. 24.

Vilkova, N.A., Kunzetsova, T.L., Ismalilov, A.L. and Islambekov, S.Y. (1988) Effect of cotton cultivars with high content of gossypol on development of the cotton bollworm, *Helicoverpa armigera* (Hbn.) (Lepidoptera: Noctuidae). *Entomol. Obazr.*, **4**, 689–698.

Virk, J.S., Brar, K.S. and Sohi, A.S. (2004) Impact of insecticides on the resurgence of jassid and whitefly in cotton. *Indian J. Entomol.*, **66**, 319–322.

Vodouhe, D.S., Mensah, R., Sanfilippo, D. and Assogba, G. (2009) A new tool for improving organic cotton yields in Africa. *Pestic. News.*, **84**, 6–9.

Wright, J.E. and Chandler, L.D. (1992) Laboratory evaluation of the entomopathogenic fungus, *Beauveria bassiana* against the boll weevil (Curculionidae: Coleoptera). *J. Invertebr. Pathol.*, **58**, 448–449

Wu, K. M., Guo, Y. Y. and Wang, W. G. (2000) Field resistance evaluations of Bt transgenic cotton GK series to cotton bollworm. *Acta Phytophy. Sin.*, **27**, 317–321.

Zhang, G.Y., Wang, X.R., Fu, C.J., Li, Z.X., Yang, B.Y. and Ma, X.H. (1988) Field trials of using NPV to control *Heliothis armigera* (Lepidoptera: Noctuidae) on cotton in Xinjiang. *Chinese J. Biol. Control,* **4**, 165–167.

7

Current Status and Potential of Biopesticides in Management of Sugarcane Insect Pests

SAROJ JAIPAL

Chaudhary Charan Singh Haryana Agricultural University, Regional Research Station, Uchani, Karnal-132001 (Haryana) India
E-mail: jaipalsaroj@gmail.com

INTRODUCTION

Sugarcane (*Saccharum officinarum*) is a tropical grass native to Asia where it has been grown for over 4,000 years. By 400 BC, methods for manufacturing sugar from sugarcane had been developed in India. Europeans were introduced to sugar during the Crusades. By the 11th century AD, sugar was being imported throughout Europe. Christopher Columbus likely brought the plant to the West Indies. Today, over 75% of the world's sugar comes from sugarcane. Sugarcane is grown primarily in the tropics and subtropics. In the United States, sugarcane is grown commercially in Florida, Louisiana, Texas and Hawaii. Cane for sugar was 26.7 million tons in 2011. In 2010, cane for sugar was valued at nearly $1.1 billion. World sugar production for 2010/2012 is estimated at 168,247 metric tons raw value (Fig. 7.1), up from last year's production. Brazil is, by far, the world's largest producer of sugarcane, accounting for one third of world production. Asian production, which includes India, China and Thailand, accounts for another one-third of the world production. India is the world's second-largest sugar producer and biggest consumer. It is estimated that India can produce around 24.5 million tonnes in 2012-13, compared with 25.8 million tonnes in 2011–2012 and is expected to reach > 35 million tonnes in 2020. India needs around 21.5 million to 22 million tonnes for annual consumption (Fig. 7.2).

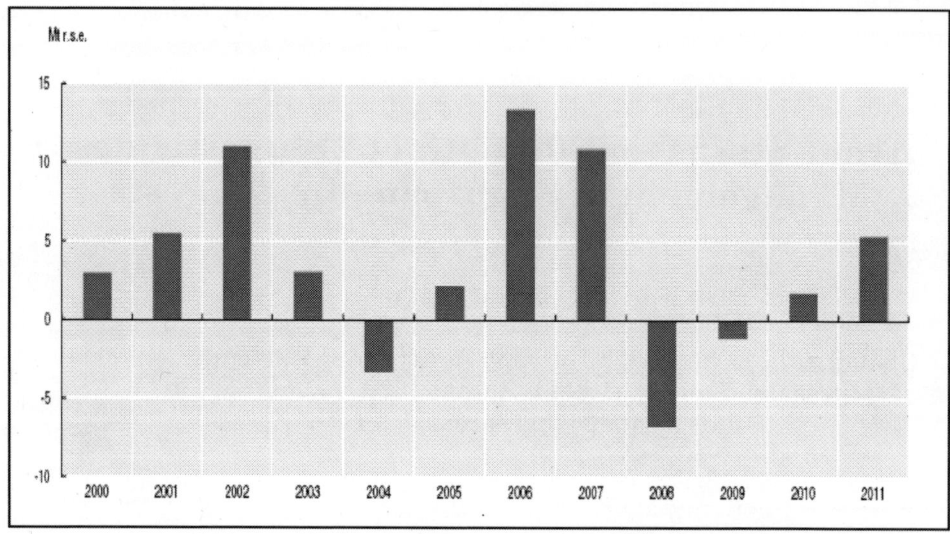

Fig. 7.1. World sugar production (*Source*: OECD and FAO, 2011)

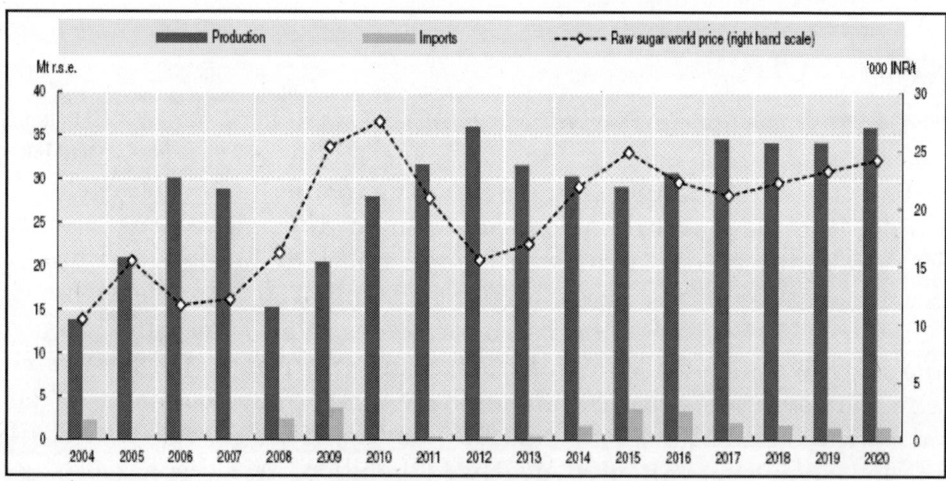

Fig. 7.2. Evolution of India's sugar production, consumption and imports to 2020 (*Source*: OECD and FAO, 2011)

Sugarcane crop in country like India, owing to its climatic conditions and its particular environment, suffers severe damage due to pests leading to radical yield and sugar declines. Sugarcane is grown in many countries worldwide and is known to host more than 1500 insects and 80 diseases, but the vast majority has restricted geographic distributions (Goebell and Sallam, 2011). More than 200 species of insects have been reported as economically important on this crop (Long and Hensley, 1972), of which about two dozen species are considered of economic importance in India (Table 7.1). Some of these pests have assumed considerable significance in subtropics because of availability of a conducive environment for their build up. The industrially important cultivars, have also invariably been found more susceptible to a host of insects particularly the tissue borers which are not even effectively controlled with pesticides. The overgrown cane canopy during grand growth and maturity phases of crop also restricts the application of pesticides. These limitations, on the other hand, have proved blessings in disguise to encourage the use of alternative non-chemical methods.

Worldwide moth borers are the key pests of sugarcane crop. In India, 17 insect pests have been recognized to cause reduction in sugar cane yields (Table 7.2). On the basis of symptoms of damage, these are conveniently grouped as stem and top and shoot borers. In India, the major stem borers like stalk borer (*Chilo auricilius*), internode borer (*Sacchariphagus indicus*) and Gurdaspur borer (*Acigona steniellus*) have distinct geographical distribution, while the top borer (*Scirpophaga excerptalis*) and shoot borer (*Chilo infucscatellus*) are of wide occurrence. The minor borers namely, the green borer (*Raphimetopus ablutella*), pink borer (*Sesamia inferens*) and plassey borer (*Chilo tumidicostalis*) are also present but restrictive in distributed. The root borer (*Polyocha depressella*), a minor pest earlier, has assumed serious proportions in recent years. There are two aspects to tissue borer problems in sugarcane according to the stage of crop growth. At formative phase, very young shoots whether virgin or ratoon are subject to attack by most species, almost in conjunction. Irrespective of the species involved, the larvae destroy the apical meristem resulting into death of shoots. The leaf spindles first turn brown to die and subsequently develop into characteristic 'dead hearts'. This does not necessarily affect crop yield because more tillers emerge which under normal physiological conditions are able to compensate to a considerable extent for early loss of shoots. However, heavy loss of shoots in young crop may occur in fields due to physiological stresses or shortening of tillering phase. Beyond formative phase, when stems have already formed, significant losses are caused by the larvae as they eat their way along the spindles and stems cutting holes and galleries, impairing growth, destroying meristematic, transport and storage tissues, causing breakage of canes and thereby reducing both cane yield and quality. Also wound injury caused by the borers provides an opportunity for the entry of microorganisms into the plant tissues.

Microbes not only deteriorate juice quality but also augment sugar losses.

Table 7.1. Status of major sugarcane insect-pests in India

Sr. No.	Insect pest / Species	Pest status	Distribution
	Borers		
1	Root borer, *Polyocha (Emalocera) depressella*	Major	Punjab, Haryana, Bihar, Madhya Pradesh, UP, Gujarat, Maharashtra, Karnataka, Andhra, TN
		Minor	Assam, Orissa
2	Early shoot borer, *Chilo infuscatellus*	Major	All sugarcane areas
3	Top borer, *Scirpophaga excerptalis*	Major	Punjab, Haryana, UP, WB, Assam, MP, Gujrat, Maharashtra
		Minor	TN, Karnataka, AP, Kerala
4	Stalk /stem/tarai borer, *Chilo auricilius*	Major	Haryana, Punjab, Uttar Pradesh, Bihar
		Minor	Nagaland, HP, WB, Orissa
5	Internode borer, *Chilo sacchariphagus indicus*	Major	Tamil Nadu, Kerala, Karnataka, Andhra Pradesh, Maharashtra, Uttar Pradesh
		Minor	Bihar, West Bengal
6	Gurdaspur borer, *Acigona steniellus*	Major	Punjab, Haryana West Uttar Pradesh
		Minor	Rajasthan, Himachal Pradesh, Maharashtra
7	Pink borer, *Sesamia inferens*	Major	UP, Bihar, Rajasthan, Maharashtra, Karnataka
		Minor	Remaining sugarcane states
8	Plassey borer, *C. tumidicostalis*	Major	Assam, West Bengal, Bihar
		Minor	Nagaland, Orissa
9	Green borer, *Rhaphimetopus ablutella*	Minor	Northern India
	Sucking pests		
10	Black bug, *Cavelerius sweeti*	Major	Haryana, Punjab, Uttar Pradesh, West Bengal
		Minor	Bihar, MP, Rajasthan, HP. Assam
	Black bug, *Dimorphopterus gibbus*	Major	Maharashtra, Uttar Pradesh
		Minor	Punjab, Bihar, Rajasthan, MP
11	Thrips, *Baliothrips serratus*	Minor	Haryana, Punjab, Uttar Pradesh, Bihar
	Thrips, *Stenchaetothrips indicus*	Sporadic	Haryana, Panjab, Uttar Pradesh
	Thrips, *Hoplothrips tolerabilis*	Minor	Haryana, Panjab, Uttar Pradesh
	Whitefly, *Aleurolobus barodensis*	Major	Haryana, Punjab, Uttar Pradesh, Bihar, Gujrat
		Minor	Karnataka, Tamil Nadu, Andhra Pradesh

Sr. No.	Insect pest / Species	Pest status	Distribution
	Whitefly, *Neomuskellia bergii*	Major Minor	Bihar, TN, Karnataka, Maharashtra, MP Haryana, Punjab, UP
13	Scale insect, *Melanaspis glomerata* Scale insect, *Aulacaspis tengalensis*	Minor	Tamil Nadu, Haryana (in pockets), Maharashtra, MP, Gujarat, Andhra, Karnataka, Bihar, West Bengal and Assam
14	Mealy bug, *Sacch-aricocus sacchari*	Minor	All sugarcane areas
	Mealy bug, *Kiritschenkella sacchari*	Spor-adic	Bengal, Bihar, TN, Karnataka, Maharashtra, UP, Orissa
15	Leaf hopper, *Pyrilla perpusilla*	Major	All sugarcane areas
	White wooly aphid, *Ceratovacuna lanigera*	Major Minor	Maharashtra, Karnataka Western UP, Uttaranchal
	Subterranean pests		
	Termite, *Microtermes obesi*	Major Minor	Haryana, Punjab, TN, AP, Karnataka, UP MP
	Termite, *Odontotermes obesus*	Major	All sugarcane areas
	White grub, *Holotrichia consanguinea* *H. serrata*	Major	Haryana, Panjab, UP, AP, Maharashtra, TN, Karnataka, Rajasthan, Bihar
	Defoliators		
	Grass hopper, *Heiroglyphus banian*	Minor	All sugarcane areas
	Grass hopper, *Poeki-lecerus heiroglyphicus*	Spor-adic	All sugarcane areas
	Army worm, *Mythimna separata*	Spor-adic	All sugarcane areas
	Non insect pests		
	Red mite, *Oligonichus indicus* Yellow mite, *Tetranychus indicus*	Spor-adic	Haryana, Punjab, Uttar Pradesh

Black bug, *Cavelerius sweeti* and leaf hopper, *Pyrilla perpusilla* are the main sap sucking pests of sugarcane. The nymphs and adults of black bug hide in the central leaf whorl, under the sheathing basis of leaves and in crop residue continuously feeding on the leaves which in turn become pale yellow with brown rust irregular spots. Severe infestation causes drastic reduction in crop growth. *Pyrilla* has been reported to occur sporadically in a severe form at intervals of 5–8 years. The infested leaves besides turning pale yellow and then brown to dry also develop sooty mould leading to arrest of photosynthetic activity and hence of sugar and yield. Scale insect juveniles desap the parenchymatous cells reducing their size and content. The severely infested canes show pithiness and are found to contain less juice. Aphids, whiteflies and mealy bugs also attack sporadically the cane crop, the former occurring particularly under unusual weather conditions and recurrent infestations have started showing during dry monsoon.

Table 7.2. Extent of losses due to different insect and pests in India

S. No.	Name of Pest	Reduction in cane yield (%)	Reduction in sugar recovery (%)
1.	Early shoot borer	22–33	2 CCS
2.	Internode borer	34.88	1.7–3.07
3.	Top shoot borer	21–37	0.2–4.1
4.	Stalk borer	upto 33	1.7–3.07
5.	Gurdaspur borer	5–15	0.1–0.8
6.	Root borer	35.00	0.3–2.90
7.	Scale insect	32.60	1.5–2.5
8.	Mealy bug	poor germination up to 35	brix loss 16.20
9.	Black bug		0.1–2.8
10.	Pyrilla	31.60	2.0–3.0
11.	Arboridia sp.	14.70	1.0–1.5
12.	White Fly	86.00	1.4–1.8
13.	White grub(H)	80	5.0–6.0
14.	Whiter grub(L)	100	complete drying
15.	Termite	33	4.5
16.	Rodents	22.27	-
17.	Sugarcane woolly aphid	7 to 39	1.2–3.43

Sourcs: http://dsd.dacnet.nic.in/PestManage.htm

Though armyworms, weevils, and grasshoppers sporadically defoliate the crop, the subterranean termites and white grubs are recognized as the major cane pests. These pests have been reported to inflict varying degree of losses in yield and sugar (Table 7.2) depending chiefly on factors like the variety under cultivation, stage of crop attacked and the environmental conditions. Generally speaking, the losses are in terms of reduced cane tonnage and reduced available cane sugar per unit weight of millable canes.

Biopesticide use in sugarcane

In order to circumvent the insect pest damage to sugarcane crop, systematic approaches based on interactions between the insect pests, natural enemies, host plant and climate have been attempted. The strategies are oriented to develop towards a simple integrated system, making use of ecological, cultural, mechanical and biological tactics in conjunction with mass trapping and need based application of insecticides. Biological control agents, together with biorational products present the sugarcane farmers with an opportunity to use environmentally benign, target specific pest control strategies at a comparatively low cost. These strategies have become very crucial for profitable cane production as also for favourable environmental consequences.

The hardy nature of sugarcane crop enables it to withstand damages by sizable population densities of different pests. The economic injury levels for different pests, set at comparatively higher levels, provide an allowance for the sustained action of natural mortality factors in the crop. Therefore, the use of biological materials has become the core component of sugarcane IPM. Biological control agents alone have shown immense utility. A large number of natural enemies' biodiversity has been identified to be associated with the crop (Table 7.3). Mass production techniques have been evolved for some of the important natural agents so that they are available in sufficient number for field use. Biological control of some of the important insect pests of sugarcane such as *Pyrilla*, moth borers and soil pests has been taken up extensively (Table 7.4). The control of leafhopper, *Pyrilla perpusilla* is a notable example in sugarcane which has virtually eliminated the use of pesticides for its control in subtropical India. Trichogrammatids have been recorded as a natural mortality factor of a number of tissue borers in sugarcane. Innundative liberations of this parasitoid have been attempted against major moth borer species. The protocol for its mass production has been quite simplified and made economical. A cottage production system is now available with our laboratory for entrepreneurs and consumers. Several sugar factories have started producing the parasitoid for their own use.

Sugarcane crop has also shown possibilities for the use of three kinds of biopesticides- biochemical, microbial and botanicals. Biochemical pesticides are similar to naturally occurring chemicals and are non-toxic, such as insect pheromones used to locate mates, while microbials like granulosis virus, *Bacillus* sp., *Beauveria*, *Trichoderma*, *Metarhizium Steinernema*, *Heterorhabditis*, etc. come from pathogenic viruses, bacteria, fungi and nematodes that occur naturally in the crop

Table 7.3. Natural enemies of insect pests of sugarcane and extent of their parasitism

Name of the pest	Number recorded	Key natural enemies and their extent of parasitism
Chilo infuscatellus	50	*Trichogramma* sp. 4-27% *Telenomus* sp. 15-22% *Cotesia flavipes* 4-14% *Sturmiopsis inferens* 6.5–26.2% *Granulosis virus* 6–14% *Beauveria bassiana* 5–15%
C. sacchariphagus indicus	39	*Trichogramma* sp. 2–78 % *Cotesia flavipes* *Xanthopimpla* sp. *Granulosis* virus up to 31%
C. tumidicostalis	09	*Trichogramma* sp. 30–71% *Cotesia flavipes* 16-35% ha
Scirpophaga excerptallis	61	*Trichogramma* sp. 4–15 % *Telenomus beneficiens* 5–90% *Cotesia flavipes* 6–15% *Isotima javensis* 5–40% *Granulosis virus* 3–12% *Beauveria bassiana* 3.7–19%
C. auricilius	33	*Trichogramma sp.* 2–15% *Cotesia flavipes* 8–19% *Campyloneurus mutator* 60–78% *Sturmiopsis inferens* 4–50% *Beauveria bassiana* 2.8–21.3%
Acigona steniellus	18	*Cotesia flavipes* 2–3.5% *Stenobaracon nicevillei* 2–33% *Isotima javensis* 1–25% *Tetrastichus ayyari* 8% *T. israeli* 8%
Pyrilla perpusilla	31	*Cheiloneurus pyrillae* up to 60% *Ooencyrtus papillionis* 71.5–79% *Tetrastichus pyrillae* 3–41% *Epiricania melanoleuca* 15–100% *Metarrhizium anisopliae* up to 90%
Melanaspis glomerāta	23	*Adelencyrtus mayurai, Botryoideclava bharatiya, Chilocorus nigritus, Pharoscymnus horni* and *Fusarium monilliforme*
Holotrichia serrata	23	*Scolia* sp., *Campsomeriella collaris, Bacillus popilliae, Metarhizium anisopliae, Heterorhabditis* sp. and *Steinernema* sp.

ecosystems. These pathogens along with parasitoids and predators are a factor in regulation of populations of a number of sugarcane pests especially the tissue borers and soil pests. All these are being used to a greatly varying extent experimentally and a few commercially for insect control in sugarcane. The potential for production and utilization of these naturally occurring pathogens in country like India are enormous. The Recent advances and past research done on some potent biopesticides for management of various categories of pest insects of sugarcane crop are discussed as under:

Subterranean Pests

Subterranean termites are the major problem which affect sugarcane crop from its germination through shoot emergence to maturity. At germination stage, the termite losses up to 90–100% have been recorded in sugarcane (Ahmed, 2005). *Microtermes obesi* and *Odontotermes obesus* are the frequently recorded species from sugar cane crop in India. For the control of termites, many methods have been adopted, among which chemicals are successful and dominated means of the control since the emergence of epoch-making chemical group (organochlorines) which opened new horizons in termite control.

From biopesticidal point of view, many farmers in Asia and Africa use extracts from plants such as neem, wild tobacco, dried chillies, *Calotropis procera* and wood ash, etc. for controlling and repelling termites. Some botanicals and entomopathogenic fungi are recently, being scientifically investigated as environmentally benign and easily biodegradable alternative for termite management in sugarcane. The reduction in infestation of termites in sugarcane with neem and *Calotropis* extracts has been established (Parihar, 1994; Deka and Singh, 2001; Singh *et al.*, 2002). Soil application of neem based biopesticides is now advocated in Haryana for moderate termite infestation in sugarcane crop. Several monoterpenes, sesquiterpenes and ethers of aromatic oximes have shown promise in controlling the pest (Pawar *et al.*, 1997).

Research on possibilities of using entomopathogenic fungi against termites in sugarcane crop gained momentum during the last two decades. *Beauveria bassiana* Vuill., *Metarhizium anisopliae* var. *anisopliae*, *M. flavoviridae* var. *minus*, *Paecilomyces lilacinus* and *P. fumosoroseus* have been shown as pathogenic to *O. obesus* (Khan *et al.*, 1993). Recent field tests with the white and green muscardines recorded up to 44% control (Anonymous, 2008; Jaipal, 2009). These fungi are usually safe for humans and other non-target organisms.

Table 7.4. Field use and release frequency of some promising bioagents

Bioagents	Insect pest	Time and dosage
(a) Parasitoids		
Trichogramma chilonis	Early shoot borer	4-6 releases at the rate of 50,000/ha at 10 days interval starting from 45th days after planting
	Stalk borer/ Gurdaspur borer/internode borer	8-10 releases at the rate of 50,000/ha at 10 days interval starting from July to October
T. japonicum	Top borer	4-6 releases at the rate of 50,000/ha at 10 days interval usually from July to October
Isotima javensis	Top borer	125 adults/ha thrice between July and August at weekly interval during exist hole disc formation period
Sturmiopsis inferens	Shoot borer	125 gravid females/ha at 10–15 days interval from 35–45 days after planting in South India
	Stalk borer/ Gurdaspur bor-er/ pink borer	125 gravid females/ha at 10-15 days interval from July to October in North India
(b) Entomopathogens		
Granulosis virus	Shoot borer	750 virosed larvae + 0.05% sandovit mixed in 200 l water is sprayed on 30th days of crop growth and then at fortnightly interval
Bacillus popillae	White grub	0.5 kg/ha for proper distri-bution of spores, 2 g of spore dust (containing 200 million spores) is deposited with a spacing of 10 feet
Metarhizium anisopliae	*Pyrilla*	Fungus conidia at the rate of 10^8/ml are applied in water suspension or infected adults are released in field during humid period
	White grubs	The required quantity of spores is mixed with 0.05% sandovit
	White grubs	Infective juveniles at the rate of 1 billion/ha
(c) Predators		
Chilocorus nigritus *Pharoscymnus horni*	Scale insect	1500 adult beetles/ha at the first appearance of the pest

Based on required toxicity tests no risks to humans are expected from the use of products containing *M. anisopliae* or *B. bassiana.* As a precaution, farmers while making field applications are required to wear protective clothing etc. These fungi are also not harmful to earthworms or to other such beneficial insects as ladybird beetles, green lacewings, parasitic wasps, honey bees, etc. Termites are highly social insects and engage in a variety of activities that necessitate frequent direct physical contact with other colony members. Trophallaxis (the exchange of regurgitated food), proctodeal trophallaxis (consumption of anal excretions) and grooming are regular necessary colony functions. It is through these grooming behaviours that the infective propagules of entomopathogenic fungi can also be transferred from one individual to another. For creating epizootics of the fungus, a portion of the colony is removed from the site of infestation, dusted with conidia and released back in the colony. As these conidia-carrying termites encounter other colony members, they are groomed and the conidia are transferred and, therefore, through this technique hundreds of termites are eliminated with a very minute quantity of the spore suspension of these fungi. Both, *B. bassiana* and *M. anisoplae* fungi can survive 35-40°C and hence seem quite suitable candidate for sugarcane microenvironments. If humidity is the problem in arid areas or deserts, special type of chemicals can be used in combination with fungi to provide a good control. Termite baits utilizing attractant and toxicant mixtures are sufficient to kill termites without creating bait shyness. Therefore, for field applications, there is a need for enhancing the effectiveness of entomopathogenic fungal products. *Metarhizium* and *Beauveria* are ubiquitous species but strain selection is vital, since a high level of variation exists among isolates in relation to pathogenicity, optimal temperature and viability.

In India, *Holotrichia consanguinea* and *H. serrata* are the main white grub species on sugarcane. These have caused even up to hundred percent crop losses to cane crops in states like Maharashtra and Gujrat. Recently, there are few new introductions. Young plant canes in some areas in Uttar Pradesh and Assam are now facing the problem of devastations due to these new species. There has been much speculation on the role played by antagonistic microorganisms and pathogens acting on the larvae in the soil. Pathogens are in general recorded more effective than entomophagous insects in the biological control of white grubs. An exotic strain of milky disease bacterium, *Bacillus popilliae* was exploited against the white grub, *Holotrichia serrata* (David, *et al.*, 1973) long back. A spore suspension of the bacterium at the rate of 1 million spores/ml/grub resulted in 68.4–84.4% mortality of second and third instar grubs. The entomopathogenic fungus, *Metarhizium anisopliae* has found widespread use as a biopesticide for control of sugarcane white grubs. In India, the indigenous strains isolated and identified from main sugarcane grub species are being investigated. Bioassays in the laboratory indicated effectiveness of both white and green muscardine fungi in controlling white grub and root borer (CCSHAU, 2004, 2005). Limited field tests showed up to 30 percent infectivity of grubs with *B. brongniarti* @ 1 × 10^{14} spores/kg of compost (PDBC, 2006). Talc based formulation of these are currently being investigated for field use in southern states (Thamarai Selvi *et al.*, 2010).

The last decade of the past century also witnessed research on the use of entomophillic nematodes. The species *Steinernema carpocapsae* and *Heterorhabditis indicus*, with their proven ability to seek their hosts in soil environments were isolated from field collections of sugarcane grubs. These lethal endoparasites entering the body of grubs through the spiracles, mouth and anus penetrated in to the haemoceol, releasing mutualistic bacteria that killed the grubs in 1–5 days depending on the grub stage. In the laboratory tests, the *Heterorhabditis* sp. was found to inflict 58–97% mortality to the eggs of *H. serrata. S. carpocapse*, *S. glaseri* and *H. indicus* could successfully be multiplied on both natural and artificial diets (Karunakar *et al.*, 1992) Restricted field applications in 2004 with *H. indica* at the rate of 2 million IJs/ha were found to reduce white grub population to half of that achieved with pesticide application and thus showed promise in white grub control (Karunakar *et al.*, 2000; Anonymous, 2005, 2006; Jaipal, 2009). It is expected that, these pathogens would supplement existing control measures or help to reduce dependence on chemical pesticides and, therefore, lower cost of production.

Tissue Borers

Insect pathogens, viz. viruses, bacteria, fungi and nematodes reportedly naturally affecting several borer species of sugarcane, have also been exploited. Granulosis virus spray at 10^8 and 10^9 IBS/ml significantly decreased the incidence of shoot borer in fields (Chaudhary and Singh, 1998). High volume sprays of the crude preparation of shoot borer GV were as effective as purified preparation in bringing about significant reduction in shoot borer infestation (Easwaramoorthy and Jayaraj, 1991). Young instar larvae were more susceptible to the virus, however, the loss in body weight of the larvae was more pronounced as the larvae matured indicating a positive correlation between the larval weight and the virus recovery. Attempts to formulate GV in wettable powder form and its storage over a period of time (12 months) could successfully be achieved without loss of activity of the virus (Easwaramoorthy and Santhalakshmi, 1999). Other granulosis viruses were isolated from the internode borer, *Chilo sacchariphagus indicus* and top borer, *Scirpophaga excerptalis*. This virus is seen to be active throughout the year (Easwaranmoorthy *et al.*, 1996b). The larvae of internode borer have also reported the concurrent natural occurrence of the fungus, *Hirsutella nodulosa* during most part of the cropping season save April through June (summer months) in southern India.

A comparison of the shoot borer granulosis virus with that isolated from the internode borer in subsequent years revealed host specificity. Restriction enzyme analysis produced quite distinct profiles with very few co-migration fragments. The per cent relative hybridisation between the two DNAs was in the range of 30-40 (Easwaramoorthy and Cory, 1990). Strains of these viruses collected from different locations in Tamil Nadu were further compared for DNA homology and their biological activity. Biological activity of the six strains slightly differed indicating greater homology and highly conserved nature of the viruses (Easwaramoorthy *et al.*, 1999). Immuno-diffusion, ELISA and SDS-PAGE studies also did not show significant differences in enveloped nucleocapsids of the two viruses. The electrophoretic profiles of enveloped nucleocapsids, however, showed differences in their polypeptide composition (Easwaramoorthy *et al.*, 2000). A haemocyte tissue culture from the *Chilo* sp. has been developed for multiplication of pure line GV of shoot borer for its genetic manipulation to increase its virulence and for use as a carrier for foreign toxic gene. Both the viruses did not produce any adverse effects on the albino rats and also did not cross infect other insects. The virus is also safe to seven parasitoids and two predators naturally occurring in sugarcane ecosystem and honey bees and silk worm. Commercial production of the virus has bben taken up by a few sugar factories e.g. Tamil Nadu Sugar Federation, Thiru Arooran Sugar Mills, Vadapathimangalam (Thanjuvar District) and Shakti Sugar Ltd. Shaktinagar in Tamil Nadu, India.

Beauveria spp. have been found pathogenic to several insect pests of sugarcane (Boas *et al.*, 1984; Varma *et al.*, 1988; Sivasankaran *et al.*, 1990; Easwaramoorthy and Santhalakshmi, 1993; Varma and Tandon, 1996). Its role as a potential bio-agent in management of sugarcane tissue borers has recently been reaffirmed (Siwach and Jaipal, 2001a, b, 2002; Jaipal, 2004, 2006, 2009). The success triggered interest in increasing the infection artificially. This required a small scale 'at site' cottage level production system using cheap, locally available materials and equipment. Resultantly, we developed an economical and simple farmers/sugar industry level production system utilizing sugar industry wastes for the mass production of this fungus. Such a system could form part of an integrated pest management package for many other pest problems as well.

Bacillus thuringiensis (Bt) infections were identified and the bacteria isolated in the midgut of field populations of Gurdaspur larvae quite long time back. As an effort to increase the infection artificially, innundative foliar applications of the various commercial preparation of the bacterium against Gurdaspur borer, in Haryana and Panjab were made. These resulted in 26–55% reduction in Gurdaspur borer infestation (Atwal and Sohi, 1969, CCSHAU, 1991).

Apart from pathgens, some botanicals were also tested for checking borer populations in sugarcane. The kernel extracts of *Thevitia* and oil formulations of *H. laurifolia* and *P. globra* caused 70–80% mortality of shoot borer larvae (Mukunthan, 1997). Leaf extract of *Eucalyptus rostrata* at 3% level was shown to influence the population of the top borer (Pandey and Suchita, 1998).

Recently, a major addition to sugarcane IPM strategy is the introduction of male trapping technique (behavioral tactic) through the use of synthetic sex pheromone-baited traps to capture male moths in flight to reduce oviposition and also to use information on moth numbers to schedule release of bio-agents or chemical application. The concept of using pheromone lures is based on the fact that during the mating stage, adult female insects emit specific chemical odor called pheromone to attract males. The pheromones are now synthesized *in vitro* and are used for insect pest management in crops such as monitoring, mass trapping and mating disruption. A range of pheromone traps and lures pheromones are currently available for sugarcane moth borers, viz. the early shoot borer, the stalk borer, the internode borer, the top borer and the pink borer. The uncommon phenology of sugarcane crop and the concealed and overlapping broods of the borer species damaging cane crop do not appear to be barriers for use of pheromones as with other strategies. Their eco-friendliness, high compatibility with other IPM strategies and possibility of use even at low pest density levels has prompted their application in sugarcane. The sex behaviour of four species of these borers, viz.

internode borer, stalk borer, shoot borer and top borer have been studied and among these the internode borer and stalk borer have been found to have potent sex attractants that can be used for mass trapping in the management of internode borer while disruption technique is likely to be useful in the case of stalk borer (David *et al.*, 1985). However, with more than 25 years into research, large scale application of pheromone technology in sugarcane has not gained momentum. The major constraints in their use are non-availability of quality products, trap types and design and unregulated release of pheromones which reduce the efficiency of the lures.

Foliage Pests

The role of entomopathogenic fungus, *M. anisopliae* in regulating the populations of leafhopper, *Pyrilla perpusilla* was documented in mid nineties (Kamar *et al.*, 1952). In repeated field tests from June to December in 1959, spray of artificially cultured fungus resulted in 7-80 and 28-90 per cent mortality of nymphs and adults which started appearing after a week of spraying. The moderate temperatures and high humidity during monsoon and post monsoon period increased the rate of infection. Moderate epizootics of the fungus were observed in Madhya Pradesh in post monsoon period of 1961. Foliar application of the fungus at the rate of 1.5×10^7 spores/ml of spray solution at Indian Institute of Sugarcane Research, Lucknow in Uttar Pradesh during 1985 witnessed cent per cent mortality of Pyrilla. The fungus proved safe to the ectoparasite, *Epiricania melanoleuca* and also was not found antagonistic to other naturally occurring fungi like *Hirsutella citriformis* and *Entomophthora* sp. (Varma *et al.*, 1992). Redistribution of Pyrilla nymphs and adults seeded with spores of the fungus in infested fields in Maharashtra harnessed equally effective suppression (Waghulde *et al.*, 1991).

Inoculative spraying of *B. bassiana* at the rate of 1.5×10^7 spores/ml during November resulted in 62% infection of the residual over wintering population of black bug (Varma and Agnihotri, 1993). In Haryana, foliar application of the fungus at the rate of 10^6 spores/ml caused 27.3% infection in natural black bug population after 115 days of treatment (Varma, 1998) as against 32.6% achieved with liberation of 5000/ha fungus seeded adults of the bug. Release of the lepidopteran ectoparasitoid *Epiricania melanoleuca* at the rate of 4000 –5000 cocoons or at the rate of 4–6 lakh eggs/ha checks multiplication of *Pyrilla perpusilla* (Paul, 2007).

Whitefly, *Aleurolobus barodensis* in recent years has assumed serious proportions on sugarcane in Bihar, Gujarat, Haryana, Karnataka, Maharashtra, Punjab, Tamil Nadu and Uttar Pradesh in India and in a mild form in Andhra Pradesh, where necotinoids have been tried for their control. Sugarcane wooly aphid, *Ceratovacuna lanigera* was observed in 2002 as severe pest in sugarcane crop in Maharashtra. Later this pest was found distributed in Karnataka, Andhra Pradesh, Tamil Nadu, Kerala, Uttraranchal and Bihar. Biological control of this pest was seen effective like release of the natural enemies *Dipha aphidivora* or *Micromus igorotus* in shade net cages (5 ×5 m) at the rate of 50/cage and allow them to develop and disperse by removing the cages. Release of *D. aphidivora* larvae at the rate of 1000/ha and *M. igorotus* larvae at the rate of 2500/ha was quite effective in the controlling this pest (Paul, 2007).

Conclusion

Despite the fact that the role of biopesticides has increasingly been recognized in sugarcane crop in recent past, their use is yet to gain the grounds. The potential pathogens have been studied partially and their effective use in the field, thus, remains elusive. With further research and more scientific experiments, use of biopesticides/ biological agents could potentially play a huge role in the future of sugarcane pest prevention. Many obstacles will need to be overcome before biopesticides use can reach its full potential. India has extremely rich biodiversity resources. Many of the vivid plant resources could be harnessed as botanical pesticides which are good alternatives to synthetic chemicals. In entomopathogens, a huge area remains undiscovered. For example their ecological role in agroecosystems is not yet fully understood which limits the ability to use them successfully for pest management. The methodologies for mass multiplication or application are largely underdeveloped. There is immense scope for increasing the inoculum potential. So far, the biotechnological interventions in the area of biopesticides research are almost completely unexplored. A large number of pathogens apart from being entomopathogenic, play additional roles as endophytes, antagonists of plant pathogens, associates with the rhizosphere and possibly even plant growth promoting agents in nature. This requires new research strategies and focus. Fungi are also potential candidates for future conservation biological control in sugarcane. However, in order to manipulate the environment for the benefit of populations of the entomopathogens, knowledge of the fundamental aspects of their ecology is essential.

This knowledge is scarce despite the large volume of recent research into their innundative applications. Baculoviruses especially the granulosis virus has the potential for use as biopesticide against the *C. infuscatellus*, *C. sacchariphagus indicus* and *S. excerptalis* in sugarcane. But their use is limited because of their slow killing action and the technical difficulties for their *in vitro* commercial production. Two approaches for the wider application of granulosis virus can be explored in future. In country like India, where use of genetically modified organisms is still restricted, the improvements will be initially mainly at the level of *in vitro* production and formulation. Baculoviruses improved by genetic modifications then gradually be introduced in due course.

The outlook for biopesticide use in sugarcane fields is complicated. Whatever the current scenario, unfortunately, except for a few parasitoids, their importance in regulating pest populations in cane crop is not yet properly understood by the farming community. The extension related problems are also manifold. The concept of using biopesticides as one of the IPM tactics is still very new to many extension agents and the growers. Traditionally, educational and extension efforts have not provided sugarcane growers with the necessary information on the multifarious acts of biopesticides and need immediate reprisal since farmers empowerment is the key to development of green products.

References

Ahmed, S. (2005) Comparative bioefficacy of botanicals and insecticides on termite Control. *Pak J. Entomol.*, **27**, 1–5.

CCSHAU (1991) *Annual Report Sugarcane Entomology*. CCS Haryana Agricultural University, Regional Research Station, Uchani, Karnal.

CCSHAU (2004) *Annual Report Sugarcane Entomology*. CCS Haryana Agricultural University, Regional Research Station, Uchani, Karnal.

CCSHAU (2005) *Annual Report Sugarcane Entomology*. CCS Haryana Agricultural University, Regional Research Station, Uchani, Karnal.

PDBC (2006) *Annual Report on Biological Control of Insects, Diseases and Weeds*. Project Directorate of Biological Control, Bangalore.

CCSHAU (2008) *Annual Report Sugarcane Entomology*. CCS Haryana Agricultural University, Regional Research Station, Uchani, Karnal.

Atwal, A.S. and Sohi, B.S. (1969) Effectiveness of *Bacillus thuringiensis* Berliner against *Bissetia steniellus* Hampson (Lep : Pyralidae). *J. Res. PAU*, **6**, 174–176.

Boas, A.M., Margues, E.J., Rebero, S.M.A. and Lima, R.O.R. (1984) The pathogenicity of *Beauveria bassiana* (Bals.) Vuilli. on larvae of *Castinia licas,* the giant moth borer of sugarcane. *Entomol. Newslett.*, **16**, 25–26.

Choudhary, A.K. and Singh, S.N. (1998). Efficiency of granulosis virus against the infestation of early shoot borer, *Chilo infuscatellus* Snellen. and yield of sugarcane. *Ann. Plant Prot. Sci.*, **6**, 211–213.

David, H., Alexander, K.C. and Ananthanarayana, K. (1973) Milky disease bacterium of the sugarcane white grub. *Curr. Sci.*, **42**, 695.

David, H., Nesbitt, B.F., Easwaramoorty, S. and Nandagopal, V. (1985) Application of sex pheromones in sugarcane pest management. *Proc. Anim. Sci.*, **94**, 333–339.

Deka, M.K. and Singh, S.N. (2001) Neem formulation is the management of sugarcane insects and pests. Proceedings 63rd Annu. Conv. Sugar Technol. Association, 28-27 August, Jailpur, India, pp. 33–38.

Easwaramoorthy, S. and Cory, J.S. (1990) Characterization of the DNA of granulosis viruses isolated from two closely related moths, *Chilo infuscatellus* and *C. Sacchariphagus indicus. Arch. Virol.*, **110**, 113–119.

Easwaramoorthy, S. and Jayaraj, S. (1991). Influence of spray fluid volume and purity on the effectiveness of a granulosis virus infecting sugarcane shoot borer *Chilo infuscatellus* Snellen. *Trop. Pest Manage.*, **37**, 134–137.

Easwaramoorthy, S. and Santhalakshmi, G. (1993) Occurrence of *Beauveria bassiana* on sugarcane root borer *Emmalocera depressella* Swinhoe. *J. Biol. Control*, **7**, 47–48.

Easwaramoorthy, S. and Santhalakshmi, G. (1999). Laboratory evaluation of a wettable formulation of granulosis virus of *Chilo infuscatellus* Snell. *Insect Environ.*, **5**, 3–5.

Easwaramoorthy, S., Srikanth, J. and Santhalakshmi, G. (1996) Seasonal occurance of the fungus *Hirsutella nodulosa* Petch. and granulosis virus of sugarcane internode borer, *Chilo sacchariphagus indicus* (Kapur). *Entomon,* **21**, 205–209.

Easwaramoorthy, S.; Suseela Gomathi, K.; Rabindra, R.J. and Sreenivasan, T.V. (1999) Homology and comparative activity of six isolates of granulosis viruses from sugarcane shoot borer and internode borer. *National Symposium on Emerging Trends in Biotechnological Applications for Integrated Pest Management*. Loyola College, Chennai. Feb. 25–26.

Easwaramoorthy, S.; Suseela Gomathi, K. and Sreenivasan, T.V. (2000) A biochemical comparison of the granulosis viruses from *Chilo infuscatellus* Snellen and *Chilo sacchariphagus indicus* (Kapur). *Proc. Entomocongress* 2000: *Perspectives* for the *New Millenium*, Trivandrum, Nov. 5-8.

Goebel. F. and Sallam, N. (2011) New pest threats for sugarcane in the new bioeconomy and how to manage them. *Curr. Opinions Environ. Sustain.*, **3**, 81–89.

Jagdeesh, S.G. and Srimannarayana, G. (1998) Neem as environmentally safe pesticide: An overview. *Int. Conf. on Pest and Pesticide Management for Sustainable Agric., Abstract ICPPMSA* 11-13 Dec., (1998), CSAUA & T, Kanpur (India).

Jaipal, S. (2004) Biological control of sugarcane insect pests. In: *Improving Sugarcane Productivity*. U.P. Council of Sugarcane Research, Shahjahanpur. pp 121–135.

Jaipal, S. (2006) An economical and simple method for the mass production of cane pest killing fungus, *Beauveria bassiana* and its utilization under IPM in sugarcane. *Proc. Sugar Technol. Assoc.*, **67**, 108–115.

Jaipal, S. (2009) Utilization of entomopathogenic fungi in management of pest insects of sugarcane crop. In: S.C. Bhardwaj, O.P. Ameta and Ashok Kumar (eds.), *Impact of Global Warming on Incidence and Management of Insect Pests in Agriculture*, pp. 10–11.

Jenkins, N.E; Heviefo, G; Langewald, J; Cherry, A.J. and Lomer, C.J. (1998) Development of mass production technology for aerial conidia for use as mycopesticides. *Biocontrol News Inf.*, **19**, 21–31.

Kamath, M.N.; Patel, M.K. and Dhande, G.W. (1952). Occurrence of green muscardine fungus on *Pyrilla sp. Curr. Sci.*, **21**, 317.

Karunakar, G.; David, H. and Easwaramoorthy, S. (1992). Influence of dosage of *Steinernema carpocapse* (Weiser), *S. glaseri* Steiner and *Heterorhabditis indicus* (Poinar, Karunakar and David) on mortality of the host and multiplication of infective juveniles in sugarcane internode borer, *Chilo sacchariphagus indicus* (Kapur). *J. Biol. Control*, **6**, 26–28.

Karunakar, G., Easwaramoorthy, S. and David, H. (2000) Host-parasite interaction between two species of white grubs infesting sugarcane and two species of entomopathogenic nematodes. *Sugar Technol.*, **2**, 12–16.

Khan, H.K, Jayaraj, S. and Gopalan, M. (1993). Muscardine fungi for the biocontrol of agroforestry termites- Odontotermes obesus Rambur. *Insect Sci. Applic.*, **14**, 529–533.

Long, W.H. and Hensley, S.D. (1972) Insect pests of sugar cane. *Annu. Rev. Entomol.*, **17**, 149–176.

Mukunthan, N. (1997) *Bioassay of Plant Derivatives*. SBI, Coimbatore Annual Report, Coimbatore.

Pandey, K.P. and Singh, S. (1998) Evaluation of natural plant product as an insecticide against sugarcane *Entomon*, **23**, 77–79.

Parihar, D.R. (1994) Termite management in arid zone of Rajasthan, India. *Pest Manage. Econ. Zool.*, **2**, 81–84.

Paul, A.V.N. (2007) Insect pests and their management. (http://nsdl.niscair.res.in/bitstream)

Pawar, P.V., Joseph, M. Sharma, R.N. and Parwardhan, S.A. (1997) Insecticidal and growth inhibiting activity of aromatic terpenoids, oxime ethers. *Pestic. Res. J.*, **9**, 1885-1889.

Singh, M. (1999) Incidence of termite infestation of sugarcane crops in district Shahjahanpur of U.P. and formulation of effective control measures using certain new synthetic chemicals and plant byproducts. Ph.D. Thesis Department of Animal Science M.J.P. Rohilkhand University Bareilly, UP, India.

Singh, M., Lal, K. and Singh, S.B. (2000) Effect of Madar (*Calotropis procera* L.) extract on the infestation of Termites (*Odantotermes* spp.) Unpublished.

Singh, M., Lal, K., Singh, S.B. and Singh, M. (2002) Effect of *Calotropis procera* extract on infestation of termite (*Odonototermes obesus*) in sugarcane hybrid. *Indian J. Agric. Sci.*, **72**, 439–441.

Singh, S.P. (1997) Studies on biological activities of some aromatic angiosperms against sugarcane pathogens. Ph.D. Thesis. Department of Botany, Gorakhpur University, Gorakhpur.

Singh, S.P., Murphy, S.T. and Ballal, C.R. (2001) *Augmentative Biocontrol.* Proceedings of the ICAR-CABI Workshop, Project Directorate of Biological Control, Bangalore.

Sivasankaran, P., Easwaramoorthy, S. and David, H. (1990). Pathogenicity and host range of *Beauveria nr. bassiana,* a gungal pathogen of *Chilo infuscatellus* Snellen. *J. Biol. Control*, **4**, 48–51.

Sivasankaran, P.; Easwaramoorthy, S. and David, H. (1998). Influence of temperature and relative humidity on the growth, sporulation and pathogenicity of *Beauveria nr. bassiana. J. Biol. Control*, **12**, 71–75.

Siwach , P. and Jaipal, S. (2001a). Possibility of using *Beauveria bassiana* for the control of sugarcane stalk borer, *Chilo auricilius* Ddgn. Abstract, Biocontrol based pest management for quality crop protection in current millennium' PAU, Ludhiana.

Siwach, P. and Jaipal, S. (2001b) Safety of *Beauveria bassiana* to larval parasitoid, *Sturmiopsis inferens* and egg parasitoid, *Trichogramma* spp. in sugarcane. Abstract, Biocontrol based pest management for quality crop protection in current millennium' PAU, Ludhiana.

Siwach, P. and Jaipal, S. (2002) Possibility of using entomogenous fungus *Beauveria Bassiana* for the control of cane insect pests. Proc. Natnl. Seminar on "Enhancing Cane productivity through Conventional and biotechnological approaches" held at Sugarcane Research Institute, UP council of Sugarcane Research, Shahjahanpur on March 21-22, 2002.

Thamarai Selvi, C., Rhichard Thilaga Raj, W. and Kandasamy, R. (2010) Laboratory culture and virulence of *Beauveria brongniarti* isolates on sugarcane white grub, *Holotrichia serrata* F. *J. Biopestic.,* **3**, 177–179.

Varma, A. (1988). Effect of entomogenous fungi on some pests of sugarcane and their possible value in controlling the pests. PhD. Thesis, Lucknow University, Lucknow.

Varma, A. (1998) An alternative method of propogating muscardine infection in field Population of susceptible sugarcane pests, *Cavelerius sweeti* Slater and Mugamoto and *Pyrilla perpusilla* Walker. *Indian J. Sugarcane Technol.*, **13**, 6–9.

Varma, A. and Tandon, B.K. (1996) Pathogenicity of three entomogenous fungi against insect pests of sugarcane. *J. Biol. Control*, **10**, 87–91.

Varma, A. Tandon, B.K. and Singh, K. (1988) First record of *Beauveria bassiana* (Bals.) Vuilli., an entomopathogenic fungus from the sugarcane defoliator, *Phytoscaphus* sp. (Coleoptera : Cuculionidae) from India. *Curr. Sci.*, **57**, 396.

Varma, A., Tandon, B.K. and Singh, K. (1992). Field evaluation of the entomogenous fungus, *Metarhizium anisopliae* (Deuteromycotina: Hyphomycete) of *Pyrilla perpusilla* (Homoptera : Lophopidae). *Indian Sugar*, **42**, 463–470.

Waghulde, J.K., Patil, A.S. and Hapase, D.G. (1991) Supression of *Pyrilla perpusilla* Wlk. in sugarcane with *Metarhizium anisopliae. Bharatiya Sugar*, 53–58.

8

Current Status and Potential of Biopesticides in Pest Management in Vegetable Crops

RAM SINGH* AND SUCHETA KHOKHAR

Department of Entomology, CCS Haryana Agricultural University, Hisar-125 004, India
*E-mail: ramjood@rediffmail.com; ramsjood@hau.ernet.in

INTRODUCTION

India is the second largest producer of vegetables after China with a production of 109.05 million tonnes. The existing area under vegetable cultivation in India is around 7.164 million ha. Vegetable crop contribute 58.93% of total production of horticultural crops. The per capita availability of vegetables increased from 76 g in 1951 to 256 g in 2001 as against 280 g/day recommended by the ICMR for a balanced diet. Increase in the production is mainly due to increase in the productivity, however, the productivity is still lower in India (15 t/ha) than the world average (15.7 t/ha) (Reddy, 2009). Among the many factors that hinder the production of the vegetables, the insect pests are considered as important. They cause annual global losses of 40%. Though modern pest control methods have been resorted to control the insect pests the losses are continued till due to development of insecticide resistance, resurgence and secondary pest outbreaks. Injudicious use of the synthetic insecticides also results in the crop residues having larger ecological impact.

The increased demand for the 'blemish free' products especially in the developed countries warrants for the management of the insect pests in ecologically sustainable manner without the harmful insecticide residues. This led to the identification and development of ecologically sustainable alternatives for management of insect pests on these important commodities. Biopesticides are viewed as the best alternatives for the management of the insect pests in these commodities as they have origin from the nature and have minimal impact on the ecosystem. The biopesticides include a wide array of the naturally occurring substances such as microbial insecticides derived from microorganisms, semiochemicals and botanicals derived from the natural flora, which are used for the control of insect pests of the vegetable crops in single or in combination with other strategies in a successful IPM programme (Dhaliwal and Koul, 2007).

SOLANACEOUS VEGETABLES

This is the major and economically important group of vegetables including tomato, *Lycopersicon esculentum*; potato, *Solanum tuberosum*; chillies, *Capsicum annum* and brinjal, *Solanum melongena*, extensively cultivated in the Indian sub-continent for their taste and nutritive value.

Tomato

Tomato, *L. esculentum* is a very popular vegetable in Indian diet. It has been ravaged by an array of pests; however, the major damage is caused by fruit borer, *Helicoverpa armigera*, serpentine leaf miner, *Liriomyza trifolii*, white flies, *Bemisia tabaci* and thrips, *Frankliniella occidentalis*. Several biopesticides have been tested under the laboratory and field conditions against these pests.

Bacillus thuriengeinsis. It has been observed that the tomato plots receiving the application of *Bacillus thuringiensis* subsp *kurstaki* @ 1 kg/ha showed the lowest fruit infestation (10.68%) compared to plots receiving the treatment comprising of HaNPV at the rate of 250 LE/ha (11.95%) and azadirachtin (14.68%) among the different biopesticides tested against *H. armigera* (Mehta *et al.*, 2000). Among the different biopesticides tested *B. t. k* at the rate of 1 kg/ha proved to be effective treatment in controlling the tomato fruit borer with fruit yield of 29.6 q/ha, and it is safer to natural enemies such as coccinellids, *Cocccinella septumpuncata*, which in turn plays very important role in the management of sucking pests (Dhaka *et al.*, 2010).

Viruses. The NPV used against tomato fruit borer resulted in the increased yields and also helped in reduction of the pesticide usage. Nuclear polyhedrosis virus applied with half the recommended dose of endosulfan (350 g a.i./ha) gave effective control of *H. armigera* on tomato than NPV tested alone (Satpathy *et al.*, 1999). HNPV + 0.07% endosulfan (15 days after spraying; DAS), HNPV + 0.035% endosulfan (7 DAS), HNPV + 0.07% endosulfan (7 DAS) and two sprays of 0.07% endosulfan at 15 days interval proved to be the best treatments as they recorded the lowest per cent fruit damage and the highest yields 465.78, 454.06, 435.03 and 432.43 q/ha, respectively (Chandrakar *et al.*, 1999). Similarly HaNPV (250 LE/ha) + monocrotophos (375 ml/ha) was found next best effective treatment in comparison with commercial insecticides such as endosulfan and monocrotophos at full recommended doses controlling *H. armigera* in tomato in Tamil Nadu. HaNPV (400LE/ha) showed a promising reduction in the number of larvae and regarded as the effective treatment in comparison with commercial insecticides like deltamethrin and endosulfan (Sharma and Bhardwaj, 2008). NPV mixed with chickpea flour (1%) + Jaggery (0.59%) significantly reduced the larval population of *H. armigera* (93.1%), lowest fruit damage (6.9%) and higher fruit yield compared to control (Shamshad *et al.*, 2011).

Entomopathogenic fungi. The field application of 5 rounds of *Nomuraea rileyi* @ 3.2 × 10^8 conidia /ml along with Triton X-100 (0.01%) at weekly intervals during the evening hours effectively controlled the fruit borer. The treatment reduced the fruit damage to 6.76% when compared to control (41.16%) and insecticide (8.02%) (Gopalakrishnan and Mohan, 1996). The highest larval mortality (100%) of 3rd instar larvae of *H.armigera* was observed in case of *N. rileyi* at a concentration of 1 × 10^8 spores/ml 5 days after treatment.

Many entomopathogenic fungi have been evaluated for their efficacy against many sucking pests such as whiteflies and mites infesting tomato under green house and field conditions. The laboratory evaluation of 40 isolates of entomopatogenic fungi from six genera against two spotted spider mite, *Tetranychus urticae* in green house tomato showed that only three isolates caused greater mortality over the control. These were *Metarhizium anisopilae, Hirsutella spp.* and *Verticilium lecanii* (Chandler *et al.*, 2005). Spraying of *Beauveria bassiana* 1 x 10^4 cfu/g two times with a week interval satisfactorily managed whitefly nymphs and adults. The biopesticides based on the entomopathogenic fungi *Paecilomyces fumosoroseus and Lecanicillium lecanii* proved effective against *B. tabaci* in tomato under field conditions. The variability in the infection of *B. bassiana* on *Spodoptera litura* was observed in case of tomato genotypes that showed the importance of plant factors in influencing growth and conidial germination of the fungus. This warrants the study of tritrophic interactions in selecting resistant genotypes and use of biopesticides.

Botanicals. The soil application of neem cake at 250 kg/ha at flowering reduced the fruit borer incidence to 13.2% as compared to 33.2% in control. Foliar sprays of neem and pongamia soap both at 1% starting from 25 DAP and repeated at 10 days interval proved to be effective in controlling the fruit borer incidence to less than 7% as compared to 33% in control (Krishnamoorthy and Krishna Kumar, 2002). Tobacco caterpillar *S. litura* is currently being managed with neem extract in South India (Joshi *et al.*, 1993). The extract was not only cheaper but also as effective as synthetic insecticides. Neem oil at 1% gave the lowest leaf miner infestation which proved better than commercial formulation spinosad and neem seed kernel extract at 5% against leaf miner *L. trifolii* (Wankhede *et al.*, 2007).

Biointensive IPM. Many of the biointensive integrated pest management strategies have been tried to control the insect pest complex on tomato.

Biointensive IPM	Pests controlled	Reference
4 sprays of *Bt* @ 1 kg/ha + *Trichogramma pretiosum*@ 50000/ha	*H. armigera*	Tyagi *et al.* (2010)
Bt @1 Kg/ha + HaNPV @250 LE/ha + Neem oil (3%)	*H. armigera, B. tabaci*	Praveen *et al.* (2001)
Need based application of dimethoate 35 EC @2ml/l	*H. armigera*	Amutha and Manisegaran (2006)
FYM @ 12.5t/ ha + *Azospirulum* 2 kg/ha + silica solubilizing bacteria @ 0.4 g a.i. / l as basal dose and mahua cake 300kg/ha in two splits @20 days interval		
Installation of pheromone traps @12 traps/ha at the rate of 20 DAT		
Six releases of *Trichogramma chilonis* at weekly intervals coincides with flowering time and based on the ETL of 4–6 moths/ trap		
Foliar application of HaNPV at the rate of 250 LE/ha		
Foliar application of NSKE 5% with the initiation of flowering		
B.t.k at the rate of 1 kg/ha + HaNPV at the rate of 250 LE/ha (2 sprays each alternatively)	*H. armigera*	Pathania *et al.* (2009)

Brinjal

Brinjal, *S. melongena* is commonly and most extensively cultivated vegetable crop throughout the country for its fleshy nutritional fruits. However, this crop has been attacked by a numerous pests among them the major pests include shoot and fruit borer, *Leucinodes orbonalis*, which is responsible for 70–92% loss of the crop. Hadda beetle, *Epilachna vigintioctopunctata*; aphids, *Aphis gossypii*); ash weevil, *Myllocerus subfasciatus*.

***Bt* formulations.** Many commercial formulations and natively isolated bacterium suspensions have been effective in controlling *L. orbonalis*. Yin (1993) and Patnaik and Singh (1997) reported that spraying of *Bt* emulsion against shoot and fruit borer in brinjal resulted in 78.8 to 100% control over untreated check. Five sprays of Delfin WG at the rate of 1 kg/ha given at 10 days interval starting from initiation of flowering were found to be most effective recording minimum fruit infestation (10.79%) and maximum marketable fruit yield of 120.4 q/ha. Five sprays of Dipel SL at the rate of 0.2% at 10 days interval resulted in minimum shoot (9.56%) as well as fruit infestation (11.78%) and maximum yield of marketable fruits (196.96 q/ha) and proved to be the most effective treatment. It was, however, at par with Delfin WG, Halt WP and Biolep WP at 0.2% concentration (Puranik *et al.*, 2002).

The *Bt* formulations used along with commercial insecticides have been effective in controlling the insect pests of brinjal. The combination of Dipel and carbaryl was found to be more effective against *L. orbonalis* compared to 0.04% carbaryl alone (Bhaskaran *et al.*, 1980). Minimum shoot borer infestation was recorded in 500 g *Bacillus thuringiensis* subsp. *kurstaki* (Btk)/ha + 126 g monocrotophos/ha treatments. Cartap hydrochloride at 250 g a.i./ha + 126 g monocrotophos/ha treatments had the least number of fruit borers (25.72%). Maximum fruit yield was recorded with application of 500 g Btk/ ha + 126 g monocrotophos/ha (169.10 q/ha) (Naitam and Mali, 2001).

Semiochemicals. The monitoring and mass trapping of the adult moths of the brinjal shoot and fruit borer using sex pheromone traps is a component of the widely used IPM strategies. The studies conducted by using synthetic sex pheromones showed effective monitoring and mass trapping of the adults of *L. orbonalis*. Leucinlure, produced using indigenously synthesized pheromone concentrates, trapped significantly more number of adults when used with PCI's portable water traps (87.83 adults/trap over 10 weeks), as compared to funnel traps (21.0) and delta traps (15.17). Leucinlure was found capable of attracting *L. orbonalis* adults for about 9 weeks under farmers' field conditions. The half-life of the lure was found to be 19 days, during release rate studies under laboratory conditions, with 11% of the initially loaded pheromone remaining in the dispenser after 63 days (Bhanu *et al.*, 2007).

Botanicals. Botanical biopesticides have been tested for their efficacy in controlling the insect pests and many of the commercial formulations derived from the neem are known to be efficient in controlling the insect pests of brinjal. Neem (*A. indica*) based products have comparable or better efficacy than endosulfan in controlling shoot borers. Fruit yields in nimbecidine (13.02 t/ha) and Neem Azal (12.80 t/ha) treated plots were higher than plots treated with endosulfan. However, the pest control after using monocrotophos and carbaryl does provide better fruit yields of 14.07 and 14.02 t/ha, respectively (Srinivasan *et al.*, 1998). The commercial formulations of neem, Nemmarin and Achook have been shown on par with the commercial insecticides in reducing the infestation of *L. orbonalis* (Gautam *et al.*, 2008). Field evaluation of botanical and microbial insecticides was carried out in comparison with the chemical insecticides and a satisfactory control of aphids, *A. gossypii* was obtained with the extracts of polygonum plants and pongamia leaves at 5% (Ghosh *et al.*, 2009). Results on antifeedant properties of two commercial neem formulations *viz.*, Neem Azal T/S and Neem Azal–F in comparison with standard insecticide endosulfan 35 EC against brinjal sphingid larvae, leaf beetle grubs and ash weevil adults revealed that Neem Azal-F 1 ml/l was the most effective (Kumar and Babu, 1999). Plant extracts of *Annona squamosa*, *Argemone americana* and *Calotropis gigantea* also reduce *Henosepilachna vigintioctopunctata* (Fabricius) populations in brinjal (Banerji *et al.*, 1985).

Commercial neem formulation, Neemazol at 0.5% has been reported as superior in reducing shoot and fruit damage which helps in obtaining maximum fruit yield. Neemazol 0.25% and NSKE 5% applications have also been shown as superior to monocrotophos treatments. The leaf extracts of *Anona*, *Calotropis* and *Jatropha* are not very effective against brinjal shoot and fruit borer (Rosaiah, 2001a). Patial *et al.* (2009) compared efficacy of biopesticides with commercial insecticides where neem seed kernel extract was on par with the commercial insecticides in reducing the fruit infestation by *L. orbonalis* on brinjal.

Biointensive IPM. The integrated pest management module consisting of different biopesticides have been tested against the important pests.

Biointensive IPM	Pests controlled	Reference
Soil application of neem cake + 3 sprays of NSKE at the rate of 5% + one spray of chemical insecticide + release of *T. chilonis* at the rate of 1 lakh/ha + installation of delta and sticky traps for trapping + collection and destruction of egg masses, grubs and adults of hadda beetle	*Eublema olivacea* *Epilachna dodecastigma* *Amrasca biguttula biguttula* *Aphis gossypii* *Leucinodes orbonalis*	Sardana *et al.* (2004)
spinosad + *Metarhizium anisoplae* + chelating agent +cartap hydrochloride	*L. orbonalis*	Suradkar *et al.* (2008)
Mass trapping of *L. orbonalis* moths with the help of plastic funnel traps at the rate of 1/100 m^2, baited with leucilure sex pheromone, + clipping of infested shoots at weekly interval starting at 20 days after transplanting (DAT) + spraying with NSKE 4% four times at an interval of 15 days starting at flowering + destruction of infested fruits after harvest	*L. orbonalis*	Pawar *et al.* (2009)
Erection of bird perches + pheromone traps at the rate of 12/acre for shoot and fruit borer + soil application of neem cake, need based sprays of 5% NSKE + releases of *Trichogramma* sp. @ 1.5 lakh/ha for shoot and fruit borer and clipping + collection and destruction of damaged shoots and fruits	*E. olivacea* *E. dodecastigma* *A. biguttula biguttula* *A.gossypii* *L. orbonalis*	Sardana *et al.* (2009)
NSKE 5% + *Bt* (delphin 85%) 0.05% + *T. chilonis* release at the rate of 50000/ha	*L. orbonalis*	Kadam *et al.* (2005)
Application of fresh neem cake in nursery at the rate of 3 kg/m^2 at land preparation + fresh neem cake at the rate of 1 kg/plant once in every 30 days after transplanting + foliar application of neem seed kernel extract at the rate of 7 ml a.i./l at an interval of 7 days beginning with 30 days after transplanting + root zone application of benzene at the rate of 1 ml a.i./plant once in every 30 days after transplanting + clipping and destruction of infested plant parts + single application of carbofuran at the rate of 5 g.a.i/plant on 30 days after transplanting.	*L. orbonalis*	Chakraborti (2001)

Potato

Potato, *Solanum tuberosum* is an important staple vegetable crop of north India known to be attacked by dozen serious pests, which sabotage the crop from germination to harvesting resulting in the severe losses in the tuber yields. Among them the important insect pests are potato tuber moth, *Phthorimaea operculella*; tabocco caterpillar, *Spodoptera litura* and cut worm, *Agrotis ipsilon*.

***Bt* formulations.** Laboratory evaluation of different *B. thuringiensis* subspecies revealed that *B. thuringiensis* subsp. *Kurstaki* (NCIM 2514) at 10^8 spores/ml concentration caused more than 85% mortality to the neonate larvae of the lepidopteran insects *S. litura* and *P. operculella* (Puntambekar *et al.*, 1997). *Bt* 10^9 spores/ml was found to be effective in controlling both the species of aphids, *Myzus persicae* and *A. gossypii* on potato under West Bengal conditions but it was inferior to chemical insectcides (Konar and Chettri, 2003).

Viruses. The granulosis virus has been reported infecting *P. opercullala (*PoGV). Laboratory bioassays at 25°C showed that both PoGV and a wettable powder (WP) formulation of *Btk* incorporated with carriers (water, talc, sand, diatomaceous earth, and kaolin clay), were effective against neonate larvae. Depending on the technique, 100% larval mortality was achieved at concentrations as low as 0.025 larval equivalents (LE) PoGV/kg tuber and 150 mg *Btk* WP/kg tuber (Arthurs *et al.*, 2008). The rate of application of PoGV solutions as biopesticides was also studied. For stored tubers, 1.5 × 1011 IB/t of tubers was used. The application rate of PoGV adopted under field conditions and found effective was 3.0×10^{12} IB/ha (Zeddam *et al.*, 2003).

Botanicals. Shobarani and Nandihalli (2004a) reported that nimbecidine at the rate of 5ml/l and NSKE 5% were significantly superior in reducing the shoot infestation caused by *L. orbonalis.* In the store, neem seed oil (500 ppm) was highly protective and was as effective as sevin.

Biointensive IPM. The biopesticides *Bt* (at 3.96×10^7 IU/ha), *B. bassiana* (at 4.5×10^{12} cfu/ha) and neem derivative Achook (at 3.38 g azadirachtin/ha) were statistically superior over the untreated control in suppressing *P. operculella* infestation but were less effective than synthetic insecticides. *B. bassiana* in combination with synthetic pesticides was equally effective in managing *P. operculella* infestation on potato foliage (Sharma *et al.*, 2006). Similar results were observed in case of *A. ipsilon* in potato (Badiyala and Sharma, 2007). Shobarani and Nandihalli (2004b) reported that bioagents Dipel 8L at the rate of

2.0 ml/l sprayed three times and inundative release of *T. chilonis* at the rate of 2.0 lakh/ha were found significantly superior in reducing the shoot infestation caused by *L. orbonalis* and recorded higher tuber yields of 33.05 and 29.72 q/ha, respectively

MALVACEOUS VEGETABLES

Okra, *Abelmoschus esculantus* is economically important vegetable grown throughout the country known for its nutritional value. The yield and quality of the crop have been hampered by various insect pests that attack the crop from early stage to harvesting. The important among them are leafhopper, *A. biguttula biguttula*; whiteflies, *B. tabaci*, which is known to be the vector of vein clearing disease and fruit borers, *Earias insulana* and *E. vittella* (Fabricius), which may in some cases result in 100% yield losses.

Bt Formulations

Commercial formulations of Bt var. *kurstaki* viz. Biolep followed by Bioasp proved to be most effective at all concentrations against *E. vittella* (Pandy *et al.*, 1999). The combination of commercial formulation of Bt (Dipel) with lower concentration of insecticides (endosulfan, fenvalerate, multineem, carbaryl, acephate), and Dipel alone were tested against okra shoot and fruit borer, *E. vittella.* The Dipel + endosulfan and dipel + fenvalerate were found very effective in reducing shoot and fruit infestation (Tomar, 1998). Biotox (Bt subsp. *thuringiensis* serotype) at 1.0 kg/ha alternated with malathion at 0.5 kg a.i./ha was effective against fruit and shoot borer of okra (Mishra and Mishra, 2002).

Viruses

Pokharkar and Chaudhary (1999) suggested that HaNPV (375 LE/ha) + Jaggery (1%) and HaNPV (250 LE/ ha) + Jaggery (1%) could be compared favourably for reduction in larval population, fruit damage and increase in total and marketable fruit yields. Gopalakrishnan (2001) recorded least damage by okra fruit borer in plot with 500 LE/ha HaNPV and obtained highest yield as compared to 250 LE/ha and organophosphate and untreated check.

Entomopathogenic Fungi

Spraying of *B. bassiana* 1×10^4 cfu/g two times with a week interval satisfactorily managed whitefly nymphs and adults (Dhandapani *et al.*, 2003). Oil based formulation of *M. anisopliae* reduced leafhopper populations that was superior in efficacy as compared to *B. bassiana* and *V. lecanii* oil based treatments. In aphid management, oil based formulation of *V. lecanii* was the best (7.75 aphids/3 leaves) followed by *B. bassiana* oil based (9.75 aphids/3 leaves) after second spray. Oil based formulation of *V. lecanii* was also significantly effective against whiteflies followed by oil based formulation of *B. bassiana* after second spray. However, in case of thrips oil based formulation of *B. bassiana* was more effective than *V. lecanii* oil based formulation treatments. *M. anisopliae* oil based formulation were apparently similar to *B. bassiana* treatemnts. However, all these formulations were next best to standard check (Naik and Shekharappa, 2009).

Botanical Insecticides

The botanical insecticides do not exert any adverse effect on the natural beneficial predatory fauna such as coccinellids, syrphids, and spiders and hence found suitable in okra IPM when they are tested extensively against these pests. Leaf extract of *Parthenium hysterophorus* (3%) was effective against leaf hopper. However, it was at par with rhizome extract of *Acorus calamus*, leaf extract of *Ocimum basilicum* and *Datura strumarium*, when used at 3% level (Pawar *et al.*, 2001). Nicotine sulphate also was effective in suppressing leafhopper population.

Okra seeds treated with neem oil 60 EC has been reported as effective in maintaining the population of the leafhopper and aphid below economic threshold level (Gandhi *et al.*, 2006). The vectors of okra yellow vein mosaic and leaf curl virus can be managed with neem based biopesticide Achook and barrier crop of bajra wasd effective against vectors of both the diseases but was inferior than insecticide imidacloprid with bajra as barrier crop (Kalita and Dhawan, 2010). The flyash based herbal pesticides also control the insect pests of okra. Turmeric powder (10%) + flyash followed by flyash + neem seed kernel extract were effective against *S. litura* in okra (Sankari and Narayanasamy, 2007). The use of neem formulations Neemax (neem seed kernel extract) at 1.0 kg/ha and Multineem (neem oil) at 2.5 l/ha and several other bioinsecticides like Biodart (B. *thuringiensis* subsp. *kurstaki*), Kalichakra (*M. anisopliae*), Ozoneem, spinosad, Daman (*B. bassiana*), cartap hydrochloride against insect pests of okra showed that biopesticides were safer to lady bird beetles and parasitoids in comparison with commercial formulations of spinosad and cartap hydrochloride (Mishra and Priya, 2007).

Biointensive IPM

Several biointensive IPM schedules have given promising results against mite pests of okra.

Biointensive IPM	Pests controlled	Reference
Installation of pheromone traps for shoot and fruit borer *(E. vittella)* at the rate of 5/acre about one month after sowing + Three sprays of NSKE at the rate of 5% at 35, 45 & 55 days interval + Five releases of *Trichogramma chilonis* at the rate of one lakh/ha at weekly interval starting at 40 DAS	*Earias vittella* *Amrasca biguttula biguttula* *Tetranychus urticae*	Sardana *et al.* (2005)
Seed treatment with imidacloprid (Tatamida 17.8 SL) at the rate of 10 ml/kg of seed + release of *T. chilonis* at the rate of 6 cc/ac + release of *Chrysoperla carnea* at the rate of 2000 eggs/acre + setting of pheromone traps Ervit lure at the rate of 5/ac + application of Bt (Spicturin) at the rate of 2ml/l	*Thrips tabaci* *A. biguttula biguttula* *Bemisia tabaci* *T. urticae* *E. vittella*	Preetha and Nandarajan (2007)
Bt at the rate of 1kg/ha + Neemarin at the rate of 2 l/ha + endosufan at the rate of 2l/ha + *Trichogramma* at the arte of 3 card/ha	*E. vittella*	Yadav *et al.* (2008)
Seed treated with thiamethoxam 70 WS 0.3% + release of *C. carnea* at the rate of 10,000 first instar larvae/ha + *T. chilonis* at the rate of 1.5 lakh/ha + mechanical collection + spraying with Bt at the rate of 1 kg/ha	*A. biguttula biguttula* *Aphis craccivora*	Pathan *et al.* (2010)
Release/application of the predator, *C. carnea* (25,000 larvae/ha/release) + Econeem 0.3% (0.5 l /ha) for three times at 15 days interval starting from 45 days after sowing	*H. armigera* *E. vittella* *A. biguttula biguttula* *B. tabaci* *A. craccivora*	Praveen and Dhandapani (2008)
Sorghum as a border crop + seed treatment with imidacloprid at the rate of 5g/kg of seed + spraying of NSKE 5% alternated with neem oil at the rate of 5%	*E. vittella*	Rao *et al.* (2008)
Intercropping okra with maize(6:1) + seed treatment with imidacloprid 70 WS at the rate of 10 g/kg of seed + removal of infested shoots along with *Earias* larvae + foliar application of neem oil at the rate of 0.5% + foliar spraying of Uplenchwar formulation at the rate of 1% + foliar application of endosulfan at the rate of 0.07%	*A. biguttula biguttula* *B. tabaci* *T. urticae* *E. vittella* *A. craccivora*	Patel *et al.* (2009)

CRUCIFEROUS VEGETABLES

The cruciferous crops include economically important crops like Cabbage, *Brassica oleracea* var. *capitata*; Cauliflower, *B. oleracea* var. *botrytis* and knol khol, *B. caulorapa*. These crops are ravaged by many pests like diamondback moth (DBM), *Plutella xylostella*; leaf webber, *Crocidolomia binotalis*; cabbage web worm, *Hellula undalis*; head borer, *S. litura* and aphid, *Brevicoryne brassicae*.

Bt Formulations

Commercial preparations of *Bacillus thuringiensis* (Bt) such as Thuricide HPSC and Dipel WP were more promising at 1 and 1.5 g/l of water than Bactospeine or Thuricide 90TS (Varma and Gill, 1977) against cabbage pests. Biotrol and Dipel at 0.5 kg/ha were also effective in controlling the pest on cabbage (Krishnaiah *et al.*, 1981). Dipel 8L containing *B. thuringiensis* subsp. *Kurstaki* (Btk) at the rate of 1 ml/l and Cen Tari (*B. thuringiensis* subsp. *aizawai*) at the rate of 1 g/l have been reported as effective among different commercial Bt formulations against DBM. Delfin has also been shown to control pest on cruciferous crops (Asokan *et al.*, 1996; Sannaveerappanavar and Viraktamath, 1997; Malathi *et al.*, 1999). However, indications of DBM resistnace in Northern India against Bt has been reported, specifically for Dipel serotype 3a, 3b and of its components Cry 1Ab. Combinations of Bt-based insecticides Dipel and Biobit (0.15%) with Ripcord (0.005%) and Ripcord alone (0.01%) are reported as most effective treatments against the cabbage aphid (Jasrotia and Suri, 2003). In case of other Bt biopesticides like Biolap and Elcar and neem-based pesticide Neemarin were found equally effective in controlling DBM. A combination of Btk at the rate of 0.02% + neem oil 5% has been shown to control *S. litura*, *C. binotalis* and *P. xylostella* in cauliflower with increase in yield to 4886.6 kg/ha and higher cost benefit ratio (1:4.6) (Babu *et al.*, 2008).

Viruses

A granulosis virus infecting *P. xylostella* (PxGV) was isolated and it has been capable of curtailing the development of resistance by interacting and synergizing with the insecticides. It was also observed that cross infection of *P. xylostella* by the multiple occluded nuclear polyhedrosis virus isolated from greater wax moth, *Galleria mellonella* (GmNPV) was found to initiate heterogonous infection of *P. xylostella* besides *C. binotalis* and *H. undalis*. Third instar larvae of *P. xylostella* were found highly susceptible to GmNPV with mortality of 78.85%

(Parthasarathy and Rabindra, 2002). A nuclear ployhedrosis virus obtained from *P. xylostella* (PxNPV) was also evaluated under field conditions and when applied at 1.7 × 10^8 POB/ml mixed with Indian ink as a sun light protectant gave the greatest reduction in insect population. Similarly, a granulosis virus strain infecting *Pieris brassicae* (PbGV) was isolated from the dry temperate region of north western Himalayas and tested for its efficacy against *P. brassicae* in combination with botanicals, showed that petroleum-ether extract of neem seed kernel (NSK) when combined with PbGV resulted in maximum reduction of LC_{50} value (4.39 × 10^2 OBs/ml) followed by methanolic extract (7.38 × 10^2 OBs/ml) and aqueous extract (9.36 × 10^3 OBs/ml) as compared with PbGV alone (1.85 × 10^4 OBs/ml) for 2nd instar larvae of *P. brassicae* (Bhandari *et al.*, 2009).

Entomopathogenic Fungi

Many of the fungal pathogens were tested for their efficacy against lepidopteran pests attacking cruciferous vegetables. The fungal pathogen *N. rileyi* was found to be as effective as commercial insecticide endosulfan in controlling *H. armigera, S. litura* and *P. xylostella* on cabbage when they occur in a complex. Entomopathogenic fungi such as *Beauveria bassiana, B. brogniarti, Paecilomyces fumosoroseus, V. lecanii* and *M. anisopliae* were also found highly pathogenic even to third instar larvae of *P. xylostella* (Gopalakrishnan, 1989) and the virulence of *B. bassiana, P. fumosoroseus* and *M. anisopliae* increased significantly after the *in vivo* passage in the larvae of *P. xylostella* (Sairbanu and Rabindra, 2002).

The fungus *P. farinosus* was alone studied in detail. The fungus could be exploited for the control of DBM under field conditions for reasons that unlike other fungi, *P. farinosus* causes death to larvae of DBM in just 48–72 h after infection and is pathogenic to different larval instars and pupae of DBM (Gopalakrishnan *et al.*, 1999a). Five rounds of this fungus at the rate of 7 × 10^8 spores/ml + Triton X–100 (0.01%) at weekly intervals significantly brought down the larval population to a mean of 1.1/ten plants as against 5.2/ten plants in unsprayed control plots. It also controlled other pests such as *C. binotalis, Trichoplusia ni, H. armigera* and *S. litura* (Gopalakrishnan *et al.*, 1999b).

Monitoring with Traps

The study on the DBM pheromone traps indicated that 8 and 12 males/trap/night in cabbage and cauliflower, respectively warranted an application of insecticide (Reddy and Guerrero, 2001).

Botanicals

Three applications of neem seed kernel extracts were adequate to reduce *P. xylostella* population of more than 40% to a negligible level (Moorthy *et al.,* 1998). Shankaramurthy *et al.* (2003) revealed that among the various commercial neem formulations, Soluneem (0.15 and 0.1%) and Econeem plus (0.25%) were highly detrimental to second instar larvae and caused 91.11, 84.45 and 84.45% mortality, respectively as against 100% in case of NSKE on fifth day after treatment and before pupation. Roopa *et al.* (2003) observed that, among the plant products, neem oil was superior to Ponagamia oil and malathion by recording 70.86% larval reduction at three days after first spray and recorded maximum yield of 31.86 tons/ha which was not only significantly superior over other plant products, viz. pongamia oil (27.60 t/ha), *A. calamus* (28.71 t/ha) but also with the standard check malathion (18.70 t/ha).Treatments with 1% *V. negundo* and *C. inerme* extracts significantly reduced *P. xylostella* larval density, and the percentage of infested plants, proving to be more effective than a standard insecticide Challenger 10EC (cypermethrin) (Yankanchi and Patil, 2009).

Biointensive IPM

The insect pests of cruciferous crops have been managed with a biointensive IPM program as reported by many authors.

Biointensive IPM	Pests controlled	Reference
Use of Indian mustard as a trap crop for every 25 rows of cabbage or cauliflower + spraying of NSKE 4% at primordial stage and two more sprays of NSKE 4% at 15 days interval	*Plutella xylostella* *Crocidolomia binotalis* *Hellula undalis* *Brevicoryne brassicae* *Bagrada hilaris*	Srinivasan and Krishnamoorthy (1991)
Growing of mustard as trap crop (25:1) + spraying of Bt products at 750 g a.i. /ha + SLNPV sprays at 250–750 LE/ha against *S. litura*	*P. xylostella* *C. binotalis* *H. undalis* *S. litura*	Arora *et al.* (2000)
Application of neem cake at the rate of 50g/m^2, as soil amendment + installation of pheromone traps for monitoring *S. litura* at the rate of 5/acre + hand collection and destruction of egg masses + spraying with NSKE 5% + application of SLNPV at the rate of 250 LE/ha	*P. xylostella* *C. binotalis* *H. undalis* *S. litura*	Ahuja *et al.* (2011)

Intercropping mustard with cabbage (1:10) + installation of bird perches at the rate of 50/ac + mechanical collection of egg masses and first instar larvae of *S. litura* and put them in cages + use of sex pheromone trap at the rate of 5/ha each of *S. litura* and *H. armigera* + 4 releases of *T. chilonis* at the rate of 1.5 lakh/ha/week + application of nimbecidine 0.03 EC at the rate of 5ml/l after four releases of parasitiod + application of Bt at the rate of 1 kg /ha	*H. armigera* *S. litura* *P. xylostella*	Thakor *et al.* (2009)
Two sparys of *Verticilium lecanii* at the rate of 1.5×10^{12} cfu/ha at 10 days interval + 5 releases of *T. bactrae* at the rate of 50,000 adults/ha/release at weekly interval + 2 intermittent sprays of Bt at the rate of 1 kg/ha at 10 days interval + a spray of HaNPV at the rate of 250 LE/ha (1.5 X 10^{12} POBs/ha)	*B. brassicae* *H. armigera* *P. xylostella*	Palande and Pokharkar (2009)

CUCURBITACEOUS CROPS

Cucurbits are an important group of vegetables, which includes pumpkin, *Cucurbita moschata*; ash gourd, *Benincasa hispida*; bitter gourd, *Momordica charantia*; snake gourd, *Trichosanthes cucumeria*; ridge gourd, *Luffa acutangula*; cucumber, *Cucumis sativus*, etc. popularly called as gourds. They are amenable for cultivation round the year. Many insect pests infest the crops and affect the appearance of the marketable yield. These pests include red pumpkin beetle *Aulacophora foveicollis*; fruity fly, *Bactrocera cucurbitae* and leaf miner, *Liriomyza trifolii.*

Fungal Pathogens

Many of the fungal pathogens have been reported to attack various insect pests that damage the cucurbit crops. The important fungal pathogens such as *Paecilomyces lilacinus, Aspergillus candidus and B. bassiana* were found to be effective against *B. cucurbitae.* (Jiji *et al.*, 2006). *B. bassiana* at 2×10^{6} cfu/g was effective against pumkin beetle (Anonymous, 2001). In fact, pumpkin beetles cause severe damage in gourds (70.2%) that has been controlled after treatment with *B. bassiana* at the arte of 3.0 g/l of water along with crop yield of 315.36 q/ha, as compared to untreated control (Vishwakarma *et al.*, 2011).

Mass Trapping

Installation of twenty yellow sticky traps smeared with castor oil adhesive has worked effectively against whitefly, *Aleurodicus dispersus* (Geetha *et al.*, 2002). Installation of traps containing methyl euganol (1 ml/l) + Vinegar/gur + malathion (1ml/l) has been effective in controlling *B. cucurbitae.*

Botanicals

Studies indicate that neem seed kernel extract at 5% concentration is promising in control of leaf miner and also gives highest yield of bittergourd. Karanj oil (29.68%) has also the potential as compared to monocrotophos (Rosaiah, 2001b). Spraying with *Ailanthus* and cashew leaf extracts has been recorded as effective in reducing fruit fly (*B. cucurbitae*) attack in bitter gourd (John *et al.*, 2007).

Biointensive IPM

A biointensive IPM schedule has been found quite effective in managing major insect pests of Cucurbitaceous crops.

Biointensive IPM	Pests controlled	Reference
Soil application of neem cake at the rate of 250 kg/ha at 15 DAS + installation of methyl eugenol traps to monitor fruit fly + 3 sprays of either 4% NSPE or 1% pongamia soap + collection and destruction of red pumpkin beetles + removal and destruction of damaged fruits + stirring the soil	*Bactrocera cucurbitae* *Liriomyza trifolii* *Aulacophora foevicolis*	Reddy (2009)

CONCLUSIONS

Successful cultivation of vegetables is hampered due to the incidence of several insect-pests. Diamondback moth (*Plutella xylostella*) in cabbage and cauliflower, fruit borer (*Helicoverpa armigera*) in tomato, shoot and fruit borer (*Leucinodes orbonalis*) in brinjal, thrips (*Thrips tabaci*) and red spider mites (*Tetranychus urticae*) in many crops have become difficult to control. There is need to re-evaluate the role of pesticides in vegetable production. Many of the IPM technologies using ecofriendly biopesticides (botanicals and bioagents) have been developed in some vegetables with less reliance on conventional pesticides. For effective control of *H. armigera* in tomato, the egg parasitoid, *Trichogramma pretiosum* and HaNPV could be integrated in IPM programme. Foliar sprays of 1% neem and pongamia soaps and 3.2 × 10^8 conidia/ml of *Nomuraea rileyi* were also effective in the management of *H. armigera.* Neem cake extract or neem seed powder extract, inundative releases of *Trichogramma chilonis* or *Bacillus thuringiensis* (Bt) formulation have been found effective in the management of shoot and fruit borer in brinjal. Neem cake, neem seed kernel extract (NSKE), neem and pongamia soaps and neem oil have reduced leafhopper and shoot and fruit borer damage in okra significantly. Releases of *T. chilonis* and application of *B. thuringiensis* formulation effectively controlled *Earias vittella* in okra. Field sprays of NSKE, Bt, neem oil, neem soap and pongamia soap were effective in checking the incidence of diamondback moth in cabbage and cauliflower. Similarly, there are several botanicals, parasitoids and microbials available for the management of pests of vegetable crops in India and worldover.

The application of modern biotechnological tools to overcome the disadvantages of these bio-pesticides will definitely increase their potential use in the vegetable pest management. The microbial insecticides such as NPV and GV and other transgenic crop plants expressing the genes of Bt are likely to play a vital role in increasing the productivity of the vegetable crops in future.

REFERENCES

Ahuja, D.B., Singh, S., Singh, R.V., Meenakshi, M., Usha Rani, A. and Bambawale, O.M. (2011) Farmers-led development and validation of IPM technology in cabbage. *Indian J. Hort.*, **68**, 578–580.

Amutha, M. and Manisegaran, S. (2006) Evaluation of IPM modules against *Helicoverpa armigera* (Hubner). *Ann. Plant Prot. Sci.*, **14**, 22–26.

Anonymous (2001) *Integrated Pest Management Package for Cucurbitaceous Crops*, Directorate of Extension. Ministry of Agriculture, Governament of India, Offset Press, New Delhi.

Arora, R., Battu, G.S. and Bath, D.S. (2000) Management of insect pests of cauliflower with biopesticides. *Indian J. Ecol.*, **27**, 156–162.

Arthurs, S.P., Lacey, L.A. and Rosa, F-de-la (2008) Evaluation of a granulovirus (PoGV) and *Bacillus thuringiensis* subsp. *kurstaki* for control of the potato tuberworm (Lepidoptera: Gelechiidae) in stored tubers. *J. Econ. Entomol.*, **101**, 1540–1546.

Asokan, R., Mohan, K.S. and Gopalakrishnan, C. (1996) Effect of commercial formulations of *Bacillus thuringiensis* Berliner on yield of cabbage. *Insect Environ.*, **2**, 58–59.

Babu, G.R., Krishnayya, P.V., Rao, P.A. and Rao, V.S. (2008) Field efficacy of *Bacillus thuringeinsis* var. *kurstaki* in combination with plant oils on pest complex of cauliflower. *Ann. Pl. Prot. Sci.*, **16**, 53–57.

Badiyala, A. and Sharma, D.C. (2007) Bioefficacy and persistent toxicity of some biopesticides and insecticides against cutworm, *Agrotis ipsilon* Hufner on spring potato. *Pestic. Res. J.*, **19**, 37–41.

Banerji, R., Misra, G. and Nigam, S.K. (1985) Role of indigenous plant material in pest management. *Pesticides,***19**, 32–37.

Bhandari, K., Sood, P., Mehta, P.K., Choudhary, A. and Prabhakar, C.S. (2009) Effect of botanical extracts on the biological activity of granulosis virus against *Pieris brassicae*. *Phytoparasitica,* **37**, 317–322.

Bhanu, K.R.M., Prabhakara, M.S. and Jayanth, K.P. (2007) Field evaluation of indigenously developed sex pheromone lures for mass trapping brinjal shoot and fruit borer *Leucinodes orbonalis* Guenee (Lepidoptera: Pyralidae). *Pest Manage. Hort. Ecol.,* **13,** 80–85.

Bhaskaran, B. N., Pawar, A. D. and Jadhav, D.R. (1980) Efficacy of different pesticides against brinjal shoot and fruit borer, *Leucinodes orbonalis*. *Indian J. Entomol.*, **42**, 31–34.

Chakraborti, S. (2001) A biorational approach for the management of brinjal fruit and shoot borer, *Leucinodes orbonalis* Guen. *J. Entomol. Res.*, **25**, 73–76.

Chandrakar, G., Ganguli, R.N., Kaushik, U.K. and Dubey, V.K. (1999) Management of tomato fruit borer, *Heliothis armigera* Hubner using biopesticides. *J. Appl. Zool. Res.,* **10,** 105–107.

Chandler, D., Davidson, G. and Jacobson, R.J. (2005) Laboratory and glass-house evaluation of entomopathogenic fungi against the two-spotted spider mite, *Tetranychus urticae* (Acari: Tetranychidae), on tomato, *Lycopersicon esculentum*. *Biocontrol Sci. Technol.,* **15**, 37–54.

Dhaka, S.S., Singh, G., Ali, N., Yadav, A. and Yadav, A. (2010) Field evaluation of insecticides and bio-pesticides against *Helicoverpa armigera* on tomato. *Ann. Plant Prot. Sci.*, **18,** 13–16.

Dhandapani, N., Shelkar , U.R. and Murugan, M. (2003) Bio-intensive pest management (BIPM) in major vegetable crops: an Indian perspective. *Fd. Agric. Environ..,* **1**, 33–339.

Gandhi, P.I., Gunasekaran, K. and TongMin, S. (2006) Neem oil as a potential seed dresser for managing Homopterous sucking pests of Okra (*Abelmoschus esculentus* (L.) Moench). *J. Pest Sci.*, **79**, 103–111.

Gautam, C.P.N., Verma, R.A., Gautam, R.D. and Khan, M.A. (2008) Comparative efficacy of insecticides, bio-pesticides and botanicals against *Leucinodes orbonalis* Guenee infesting brinjal. *Annl. Plant Prot. Sci.*, **16**, 309–311.

Geetha, B., Swamiappan, M. and Kempuchetty, N. (2002) Predators for the biological control of the spiraling whitefly, *Aleurodicus dispersus* on vegetables. In: *International Conference on Vegetables.* Nov.11-14. Bangalore, India.

Ghosh, S.K., Mahapatra, G.S. and Chakraborty, G. (2009) Field efficacy of plant extracts and microbial insecticides against *Aphis gossypii* Glover infesting okra [*Abelmoschus esculentus* (L.) Moench]. *Redia.*, **92**, 249–252.

Gopalakrishnan, C. (1989) Susceptibility of cabbage diamondback moth *Plutella xylostella* to the entomofungal pathogens *Verticillium lecanii* (Zimmerm.) Viegas. *Curr. Sci.*, **58**, 1256–1257.

Gopalakrishnan, C. (2001) Field efficacy of HaNPV against *Helicoverpa armigera* (Hubner) on okra. In: Proceeding of Second National Symposium on Integrated Pest Management in Horticultural Crops, New Molecules,Bio-pesticides and Environment, Indian Institute of Horticultural Research, Bangalore, October 17-19, 2001.

Gopalakrishnan, C. and Mohan, K.S. (1996). Field efficacy of entomopathogenic fungus *Nomuraea rileyi* against noctuid pests of tomato (Abstract). *Natl. Symp. Hort. Botechnol.* Bangalore.

Gopalakrishnan, C., Anusuya, D. and Narayanan, K. (1999a) Occurrence of entomopathogenic fungi, *Paecilomyces farinosus* (Holmskiold). Brown and Smith and *Zoophthora radicans* (Brefeld) Batko in the field population of *Plutella xylostella* L. on cabbage. *Entomon.*, **24,** 363–369.

Gopalakrishnan, C., Anusuya, D. and Narayanan, K. (1999b) *Paecilomyces farinosus* (Holmskiold). Brown and Smith, a potential pathogen for diamondback moth of cabbage, *Plutella xylostella.* Paper presented in International Seminar on IPM, Oct. 8-9, 1999. IICT, Hyderabad, pp.52.

Gopalakrishnan, C. and Mohan, S. (2002) Field efficacy of *Nomuraea rileyi* (Farlow) Samson against certain lepidopterous insect pests of cabbage. *Proceedings of the Symposium on Biological Control of Lepidopterous Pest*, July 17-18, 2002, Bangalore, pp. 247–250.

Jasrotia, P. and Suri, S.M. (2003) Preliminary screening of certain biopesticides against cabbage aphid, *Brevicoryne brassicae* (L.) attacking off season cabbage in Himachal Pradesh. *Insect Environ.,* **9**, 10–11.

Jiji, T., Praveen, R., Babu, K., Naseema, A. and Anitha, N. (2006) Three fungal strains pathogenic to fruit flies. Fruit Flies of Economic Importance: From Basic to Applied Knowledge, Proceedings of the 7th International Symposium on Fruit Flies of Economic Importance. 10-15 September 2006, Salvador, Brazil. pp. 175-177.

Joshi, B.G., Sriprakash, K., and Murthy, P.N.S. (1993) Efficacy of neem seed kernel suspension, neem seed bitter, pongamia cake water extract, *Bacillus thuringiensis* Berliner, nuclear polyhedrosis virus alone and in combination against tobacco caterpillar, *Spodoptera litura* F. in nurseries. In: M.S. Chari and G. Ramaprasad (eds.), *Botanical Pesticides in Integrated Pest Management*, Proc. Symposium on botanical pesticides in IPM, Rajamundry, India, pp. 154–158.

John, J., Leela, N.K. Sreekumar, K.M. Anesh, R.Y. and Hema, M. (2007) Phytotoxicity of leaf extracts of multipurpose trees against insect pests in bitter gourd (*Momordica charantia*) and brinjal (*Solanum melongena*). *Allelopathy J.,* **20**, 411–418.

Kadam, J.R. Bhosale, U.D. Chavan, A.P. and Mhaske, B.M. (2005) Bio-efficacy of insecticide sequences against pests of brinjal and their impact on natural enemies. *Ann. Plant Prot. Sci.*, **13**, 278–282.

Kalita, M.K. and Dhawan, P. (2010) Ecofriendly management of okra yellow vein mosaic and leaf curl virus diseases by combination of barrier crop and insecticidal sprays. *Adv. Plant Sci.,* **23**, 289–291.

Konar, A. and Chettri, M. (2003) Effectiveness of synthetic insecticides, biopesticides and azadirachtin against potato aphids. *J. Indian Potato Assoc.*, **30**, 331–334.

Krishna Kumar, N.K. and Krishnamoorthy, P.N. (2002) Advances in the use of botanicals for the IPM of major vegetable pests. In: *International Conference on* Vegetables, Nov. 11-14. Bangalore, India.

Krishnamoorthy, R.N. and Krishna Kumar, N.K. 2002. Advances in the use of botanicals in the IPM of major vegetable crops. *Int. Conf. Vegetables*, Bangalore.

Krishnaiah. K., Jaganmohan. N. and Prasad, V.G. (1981) Efficacy of *Bacillus thuringiensis* Berliner for the control of lepidopterous pests of vegetable crops. *Entomon,* **6**, 87– 93.

Kumar, S.P. and Babu. S. (1999) Evaluation of antifeedant properties of Neem Azal against certain leaf feeders of brinjal. *Pestology*, **22**, 62–64.

Malathi, S., Sriramulu. M. and Babu, T.R. (1999) Evaluation of certain eco-friendly insecticides against lepidopterous pests of cabbage. *Indian J. Entomol.*, **61**,127–133.

Mehta, P.K., Vaidya, D.N. and Kashyap, N.P. (2000) Management of tomato fruit borer, *Helicoverpa armigera* (Hubner) using insecticides and biopesticides. *Himachal J. Agric. Res.,* **26,** 50–53.

Mehetre, S.T. and Salunkhe, G.N. (1999) Comparative efficacy of *Trichogramma pretiosum* Riley and Ha NPV with endosulfan in controlling *Helicoverpa armigera* Hub. infesting tomato. *J. Maharastra. Agric. Univ.,* **23**, 175–176.

Mishra, N.C. and Mishra, S.N. (2002) Impact of biopesticides on insect pests and defenders of okra. *Indian J. Plant Prot.*, **30**, 99–101.

Misra, H.P. and Priya, B.S. (2007) Safety of biopesticides to semilooper infesting okra. *Insect Environ.* **13**, 110–111.

Moorthy, P.N.K., Kumar, N.K.K., Selvaraj, C. and Daniel, J.S. (1998) Neem seed kernel extract application for diamondback moth manage-ment : Transfer of technology for mechanical farming. *Pest Manage. Hort. Ecos.,* **4**, 128–130.

Naik, P.R.H. and Shekharappa (2009). Field evaluation of different entomopathogenic fungal formulations against sucking pests of okra. *Karnataka J. Agric. Sci.,* **22,** 575–578.

Naitam, N.R. and Mali, A.K. (2001) IPM of brinjal pests using insecticide mixtures and natural enemies. *Pest Manage. Hort. Ecos.,* **7**, 137–140.

Palande, P.R. and Pokharkar, D.S. (2009) Evaluation of BIPM module against cabbage pest complex. *Pestology,* **33**, 25–29.

Pandy, N.K., Ojha, P.K. and Upadhyay, K.D. (1999) Studies on the bio efficacy of *B.thuringiensis* formulations against *E.vittella* Fabr. under laboratory conditions. *RAU. J. Res.*, **9**, 57–61.

Parthasarathy, P. and Rabindra, R.J. (2002) Biological suppression of lepidopteran pest complex of cruciferous crops by heterologous infection of baculovirus. *National Symposium on Biological control of insect pests.* Entomological Research Institute, Chennai.

Pathan, N.M., Nalwandikar, P.K. and Shinde, S.T. (2010) Evaluation of components of integrated pest management against aphids and jassids infesting okra. *J. Entomol. Res.*, **34,** 317–325.

Pathania, S.S., Verma, A.K. and Gupta, P.R. (2009) Economics of IPM modules against *Helicoverpa armigera* (Hubner) on tomato. *Indian J. Entomol.,* **71**, 125–126.

Patial, A., Mehta, P.K. and Sharma, P.C. (2009) Field efficacy of some insecticides and biopesticides against *Leucinodes orbonalis* Guenee on brinjal. *Environ. Ecol.,* **27**, 320–325.

Patel, P.S., Patel, G.M. and Shukla, N. (2009) Evaluation of different modules for the management of pest complex of okra. *Pestology,* **33**, 31–37.

Patnaik, H.P. and Singh, K.M. (1997) Efficacy of *Bacillus thuringiensis* Berliner and conventional insecticides against brinjal shoot and fruit borer under different spraying schedules. *Orissa J. Hort.*, **25**, 18–21.

Pawar, D.B., Warade, S.D. and More, T.A. (2001) Efficacy of different plant extracts against leafhopper (*Amrasca devastans* Dist.) and fruit borer (*E. vittella*) of Okra. In: *Proceedings of the second National Symposium on Integrated Pest Management (IPM) in Horticultural Crops: New Molecules, Biopesticides and Environment.* October 17-19. Bangalore, India. 66 p.

Pawar, D.B., Bhalekar, M.N., Chandele, A.G., Ilhe, B.M. and Shinde, H.N. (2009) Sex pheromone based IPM technology for brinjal shoot and fruit borer *Leucinodes orbonalis* Guenee (Lepidoptera : Pyralidae). *Pest Manage. Hort. Ecos.,* **15**, 114–120.

Praveen, P.M. and Dhandapani, N. (2008) Eco-friendly management of major pests of Okra (*Abelmoschus esculentus* (L.) Moench). *J. Veg. Crop Prod.,* **7**, 3–15.

Praveen, P.M., Dhandapani N. and Kennedy, J.S. (2001) Efficacy of *Bacillus thuringiensis* (Berliner) formulations for the management of tomato fruit borer, *Helicoverpa armigera* (Hubner). *Pestology,* **25**, 58–61.

Preetha, G. and Nandarajan, L. (2007) Validation of IPM modules against sucking pests of okra in Karaikal. *Indian J. Entomol.*, **69,** 210–214.

Pokharkar, D.S. and Chaudhary, S.D. (1999) Application time of nuclear polyhedrosis virus against *Helicoverpa armigera* (Hubner) on tomato. *J. Mah. Agric. Univ.*, **24,** 147–151.

Puranik, T.R., Hadapad, A.B., Salunkhe, G.N. and Pokharkar, D.S. (2002) Management of shoot and fruit borer, *Leucinodes orbonalis* Guenee through *Bacillus thuringiensis* formulation on brinjal. *J. Entomol. Res.*, **26**, 229–232.

Puntambekar, U.S., Mukherjee S.N. and Ranjekar, P.K. (1997) Laboratory screening of different *Bacillus thuringiensis* strains against certain lepidopteran pests and subsequent field evaluation on the pod boring pest complex of pigeonpea (*Cajanus cajan*). *Antonie van Leeuwenhoek,* **71**, 319–323.

Rao, N.V., Rao, V. N. and Bhavani, V. (2008) Evaluation of integrated pest management modules against bhendi shoot and fruit borer, *Earias vittella* (Fabricius). *Pestology,* **32,** 16–18.

Reddy, P. P. (2009) Advances in Integrated Pest and Disease Management in Horticultural Crops. Vol. 2. Studium Press India Pvt. Ltd. India.

Reddy G.V.P. and Guerrero A. (2001) Optimum timing of insecticide applications against diamondback moth, *Plutella xylostella* in cole crops using threshold catches in sex pheromone traps. *Pest Manage. Sci.,* **57**, 90–94.

Rosaiah, B. (2001a) Evaluation of different botanicals against the pest complex of brinjal. *Pestology,* **25**, 14–16.

Rosaiah, B. (2001b) Effect of different plant products against serpentine leaf miner *Liriomyza trifolii* on ridge gourd *Luffa accutangula. Pestology,* **25,** 7–9.

Roopa, S. P., Basavana goud, K. and Kulkarni, K. A. (2003) Management of diamondback moth on cabbage using plant extracts and new insecticides. *Karnataka J. Agric. Sci.*, **16,** 307–310.

Sankari, S.A. and Narayanasamy, P. (2007) Bio-efficacy of flyash-based herbal pesticides against pests of rice and vegetables. *Curr Sci.*, **92**, 811-816.

Sannaveerappanavar, V.T. and Viraktamath, C.A. (1997) Management of insecticides resistant diamondback moth, *Plutella xylostella* L. (Lepidoptera: Yponomeutidae) *Mysore J. Agric. Sci.*, **31**, 230–235.

Sairbanu, B. and Rabindra, R.J. (2002) Effect of entomofungal pathogens after *in vivo* passage through the larvae of the diamondback moth, *Plutella xylostella* (Lepidoptera: Plutellidae). *National Symposium on Biological Control of Insect Pests.* Entomological Research Institute, Chennai. Feb. 7-8. 2002, pp. 141–148.

Sardana, H.R., Arora, S., Singh, D.K. and Kadu, L.N. (2004) Development and validation of adaptable IPM in egg plant, *S. melongena* L. in a farmers participatory approach. *Indian J. Plant Prot.*, **32**, 123–128.

Sardana, H.R., Bambawale, O.M., Kadu, L.N. and Singh, D.K. (2005) Development and validation of adaptable IPM in okra through farmer's participatory approach. *Ann. Plant Prot. Sci.*, **13**, 54–59.

Sardana, H.R., Jalali, K.S. and Bambawale, O.M. (2009) Validation of non-chemical IPM and INM technologies on brinjal through farmers' participatory approach. *Indian J. Plant Prot.*, **37**, 20–23.

Satpathy, S., Rai, S. and Chattopadhyay, M. (1999) Field evaluation of NPV and insecticides against tomato fruit borer, *Helicoverpa armigera* (Hubner). *Insect Environ.*, **5**, 117–118.

Shamshad, A., Kumar, R. and Satendra, K. (2011) Efficacy of biopesticides against *Helicoverpa armigera* (Hubner) in tomato. *Ann. Plant Prot. Sci.*, **19**, 29–32.

Sharma, D.C., Badiyala, A. and Choudhary, A. (2006) Bioefficacy and persistent toxicity of biopesticides and insecticides against potato tuber moth, *Phthorimaea operculella* Zell. on spring potato. *Pestic. Res. J.*, **18**, 43–46.

Sharma, K.C. and Bhardwaj, S.C. (2008) Management of *Helicoverpa armigera* (Hubner) in tomato. *Ann. Plant Prot. Sci.*, **16**, 33–39.

Shankaramurthy, M., Jagadeesh, P.S., Sannaveerappanavar, V.T. and Ashok Kumar, C.T. (2003) Bioefficacy of plant products and new insecticides against diamondback moth on cabbage. In : *Proceedings of the National Symposium on Fronterier Areas of Entomological Research*, B. Subrahmanyam, V.V. Ramamurthy and V.S. Singh (eds.) 5-7 November, 2003, pp. 364-365.

Shobharani, M. and Nandihalli, B.S. (2004a) Field evaluation of botanicals against potato shoot borer, *L. orbonalis* Guenee. *Andhra Agric. J.*, **50**, 506–507.

Shobharani, M. and Nandihalli, B.S. (2004b) Role of Bt and *Trichogramma chilonis* Ishii in the management of potato shoot borer, *L. orbonalis* Guenee. *Andhra Agric. J.*, **50**, 508–509.

Srinivasan, G., Babu, P.C.S., Reddy, P.P., Kumar, N.K.K. and Verghese, A. (1998) Management of brinjal shoot and fruit borer, *Leucinodes orbonalis* (Lepidoptera : Pyralidae) using neem products and insecticides. *Advances in IPM for Horticultural Crops, Proceedings of the First national Symposium on Pest Management in Horticultural Crops: Environmental Implications and Thrusts,* Bangalore, India, **13**, 87–93.

Srinivasan, K. and Krishna Moorthy, P.N. (1991) Indian mustard as a trap crop for management of major lepidopterous pests on cabbage. *Trop. Pest Manage.*, **37**, 26–32.

Suradkar, A., Naitam, N.R., Gotarkar, S.B. and Aherkar, S.K. (2008) Management of brinjal shoot and fruit borer, *Leucinodes orbonalis* Guenee using different IPM modules. *J. Entomol. Res.*, **32**, 125–126.

Thakor, S.B., Patel, I.S. and Jakhar, B.L. (2009) Evaluation of different modules for the management of major pests of cabbage. *Pestology,* **33**, 14–18.

Tomar, R.K.S. (1998) Efficacy and economics of biopesticides and insecticides combinations against okra shoot and fruit borer. *Indian J. Entomol.*, **60**, 25–28

Tyagi, A., Gaurav,S.S., Prasad,C.S., Mehraj-ud-din-Mir (2010) Efficacy and economics of biopesticides combinations and insecticide against tomato fruit borer. *Ann. Plant Prot. Sci.* **18**, 148–152.

Varma, G.C. and Gill, G.S. (1977) Laboratory studies in the comparative efficacy of the biotic insecticides for the control of *Plutella xylostella* (Linn) (Plutellidae : Lepidoptera). *J. Res. Punjab Agric. Univ.*, **14**, 304–308.

Vishwakarma, R., Chand, P. and Ghatak, S.S. (2011) Potential of plant extracts and entomopathogenic fungi against red pumpkin beetle, *Raphidopalpa foveicollis* (Lucas). *Ann. Plant Prot. Sci.*, **19**, 84–87.

Wankhede, S.M., Deotale, V.Y., Undirwade, D.B., Mane, P.N., Deotale, R.O. and Kahare, R.N. (2007) Performance of some insecticides and biopesticides against tomato leafminer, *Liriomyza trifolii* Burgess. *J. Soils Crops.* **17,** 136–138.

Yadav, J.B., Singh, R.S. and Tripathi, R.A. (2008) Economics of IPM modules against okra pests. *Ann. Plant Prot. Sci.*, **16**, 203–204.

Yankanchi, S.R. and Patil, S.R. (2009) Field efficacy of plant extracts on larval populations of *Plutella xylostella* L. and *Helicoverpa armigera* Hub. and their impact on cabbage infestation. *J. Biopestic.*, **2,** 32–36.

Yin, R.G. (1993) Bionomics of *Leucinodes orbonalis* Guenee and its control. *Entomol. Knowledge*, **30**, 91–92.

Zeddam, J.L., Vasquez S.R.M., Ramos, V.Z. and Lagnaoui, A. (2003) Quantification of viral production and rates of application of a granulosis virus used for the biological control of the potato moths, *Phthorimaea operculella* and *Tecia solanivora* (Lepidoptera: Gelechiidae). *Revista Chilena de Entomol.*, **29**, 29–36.

9

Biopesticides in Management of Fruit Pests: Progress and Potential

M. MANI AND D. R. SHARMA*

Division of Entomology and Nematology,
Indian Institute of Horticultural Research, Bangalore-560089, India
*Department of Horticulture, Punjab Agricultural University,
Ludhiana-141004, India
*E-mail:drajsharma@pau.edu

INTRODUCTION

Climatic conditions in India are very diverse and fruit crops of temperate, sub-tropical and tropical regions are frequently grown in different environmental regions. Thus, the fruit crops in such climatic diversity are often found to have various pest problems, as the pests have easy access to any fruit crop from other regions. It is estimated that there are about 100 major and minor fruit crops in India with a total fruit production of 715.16 lakh tonnes from 63.29 lakh ha during 2009-10 (Kumar *et al.*, 2010). The available information in the literature indicated that there are about 1000 insect and mite species on these fruit crops in India. This number is certainly affected with the collection and evaluation of germplasm and improvement in production technology coupled with the climatic changes as all these factors significantly influence the nefarious activities of the pest species. The wide geographical diversity and host range of perennial fruit crops also provide a stable environment, which offers a good opportunity for the biological control agents in pest management programmes (Bennet *et al.*, 1976).

In fruit crops, several predator and parasitoids of different insect pests have been identified and appeared to be ideal for the control of pests, but actually there are few instances where the biocontrol agents are being used to reduce the pest population below the economic injury level. The bioagents are generally used indirectly, except in few instances where they are used directly for the immediate control of pest species, e.g. predatory coccinellid, *Cryptolaemus montrouzieri* Mulsant has been used against mealybugs and scales on several fruit crops. Among different bioagents, parasitoids have received increasingly more attention than predatory insects. Hitherto many entomophagous spiders, mites, fungi, bacteria, protozoa, viruses, nematodes and other vertebrates have been found to play a significant role in balancing the pest population of the fruit crops. The biological pest suppression is not only an important tool for sustainable agriculture but also ecofriendly and safe for mankind. A complex of natural enemies affecting all life stages of insect pest would be desirable to promote population regulation below damaging level rather than reliance on a single parasitoid species to function on its own (Mani, 2001; Sharma, 2007). In this text, an attempt is made to collect the available information about the different biological agents being used, evaluated or experimented against various insect and mite pests of fruit crops.

CITRUS

Citrus Mealybug, *Planococcus citri* (Risso) (Hemiptera: Pseudococcidae)

The mealybug causes 38 to 65% damage on various citrus species in India. *P. citri* was known to be attacked by more than 20 natural enemies in citrus orchards. *Anagyrus* sp., *Blepyrus insularis* (Campere), *Diversinervus* sp., *Tetrastichus* sp., *Microterys* sp., *Cryptochaetum* sp., *Scymnus coccivora* Ayyar, *Pullus pallidicollis* Mulsant, *Nephus* sp., *Chrysopa* sp., *Micraspis cardoni* (Weise), *Pseudaspidimerus uttami* Kapur, *Cryptolaemus montrouzieri* Mulsant and *Spalgis epius* Westwood were recorded in Kodagu, Karnataka (Singh, 1980, 1987, 2003). In Assam, *C. montrouzieri* and *Entomophthora fumosa* Speare were observed on *P. citri* (Chowdhury and Majid, 1954). *Coccidoxenoides peregrinus* (Timberlake) (Krishnamoorthy and Mani, 1989), *Mallada boninensis* (Okamoto), *Plesiochrysa lacciperda* (Kimmins), *Anisochrysa basalis* Walker and *Chrysoperla carnea* (Stephens) (Krishnamoorthy and Mani, 1989a) were found attacking *P. citri* in citrus orchards in Karnataka.

The encyrtid parasitoid, *C. peregrinus* played a dominant role in the suppression of *P. citri* on acid lime and lemon (Mani, 1994b). The parasitoid was multiplied on 5-10 days old laboratory bred *P.citri.* Several plant products and deltamethrin were found safer to *C. perigrinus* (Mani *et al.*, 2002; Mani and Krishnamoorthy 2002). The exotic parasitoid *Leptomastix dactylopii* Howard was imported from West Indies into India during 1983 for trials against *P. citri* (Nagarkatti *et al.,* 1992). It was multiplied on 15-20 day old *P. citri* (Krishnamoorthy, 1989a) and *P. lilacinus* (Cockrell) (Mani, 1995a). The mummies of *L. dactylopii* could be stored for about 20 days at 15°C and 70-80% RH (Krishnamoorthy, 1989b). Releases of *L. dactylopii* gave excellent control of *P. citri* causing upto 100% parasitism in citrus orchards in Karnataka (Manjunath, 1985b; Krishnamoorthy and Singh, 1987; Nagarkatti *et al.*, 1992; Krishnamoorthy, 1990).

Leptomastix dactylopii released in April 1984 was recovered in large numbers from *P. citri* infesting acid lime and lemon in 1991-92 (Mani, 1994b). Later *L. dactylopii* was recovered from *P. citri* infesting several horticultural crops (Krishnamoorthy and Mani, 1989b). Dichlorvos, dicofol, several fungicides and plant products are safer to *L. dactylopii* (Mani *et al.*, 1993). But these selective pesticides were known to affect the longevity and progeny production to some extent in *L. dactylopii* (Mani *et al.*, 1995).

The Australian ladybird beetle *C. montrouzieri* was first introduced from Australia to India in 1898. It was widely used in controlling the citrus mealybug. A simple method involving the multiplication of the mealybug and *C. montrouzieri* on pumpkin fruits was standardized in India (Singh, 1978; Chacko *et al.*, 1978). Ten beetles per Coorg mandarin tree were recommended in Karnataka. By the fifth week of the release of the predator, mealybug population was reduced to a negligible level. Releases of the predator were to be repeated whenever the mealybug appeared (Singh, 1978; Mani and Krishnamoorthy, 2007a, 2008a,b). *Cryptolaemus montrouzieri* has a better detection of prey at higher density of prey (Verghese *et al.*, 2001). The threshold temperature for the activity of *C. montrouzieri* is about 20°C. The predator, *C. montrouzieri* was found to feed the parasitised mealybug in the early stage of *L. dactylopii* development but not the mummified mealybugs (Anonymous, 1986). *Scymnus coccivora* was found useful in the management of lime mealybug *P. citri* (Murali Baskaran *et al.*, 2007a).

Spherical Mealybug, *Nipaecoccus viridis* (Newstead) (Hemiptera: Pseudococcidae)

Over a dozen parasitoids were recorded on *N. viridis* in India. *Anagyrus dactylopii* Howard was found parasitising upto 90% in the field (Ali, 1957; Subba Rao *et al.*, 1965). A severe infestation of *N. viridis* on citrus was wiped out with the liberation of 10 beetles of *C. montrouzieri* per tree in Andhra Pradesh (Tirumala Rao and David, 1958). The population of *N. viridis* declined from 221.3 on 16th March to 1.40/plant on 10th June 1994 due to the action of *C. montrouzieri* and *Anagyrus* sp. in acid lime orchard in Karnataka (Mani and Krishnamoorthy, 1999g; 2002a).

Oriental Mealybug, *Planococcus lilacinus* (Cockerell) (Hemiptera: Pseudococcidae)

The encyrtid parasitoid *Tetracnemoidea indica* (Ayyar) was able to check the mealybug on acid lime in Karnataka (Mani, 1995b). Five day old mealybug nymphs were found suitable for the breeding of *T. indica* (Mani and Krishnamoorthy, 1995a). The parasitoid could be conserved by the application of fenvalerate (0.01%), neem seed kernel extract (5%) and the commonly used fungicides (Mani and Krishnamoorthy, 1996e).

Cottony Cushion Scale, *Icerya purchasi* Maskell (Hemiptera: Margarodidae)

In India, *I. purchasi* as a pest appeared in the Nilgiri hills, Tamil Nadu in 1927 on many crops including citrus. The vedalia beetle, *Rodolia cardinalis* Mulsant was introduced from California to India, and employed for controlling the scale between 1928 and 1931 in Nilgiri hills. The predator later spread to Mysore, Kerala and certain parts of Maharashtra. The scale was brought under complete control within a year of release. Again, a serious outbreak of the pest occurred in 1941 in the Kodaikanal hills of Tamil Nadu, and it was controlled by the same predator. Release of 8-10 beetles on each of the moderately infested plant was suggested. Besides *R. cardinalis*, the parasitoid *Cryptochaetum iceryae* (Williston) was also tried but proved unsuccessful in India (Rao and Kamath, 1966). *Rodolia amabilis* Kapur, *R. breviuscula* Weise, *R.netara* Kapur, *R. minima* Kapur, *R. fumida* Mulsant, *Cryptochaetum grandicorne* Rondani, *C. iceryae* and *Euzophera cocciphaga* Hampson were also recorded on *I. purchasi* (Rama-chandra Rao and Cherian, 1944; Subramanian, 1954, 1955). The fascinating story of control of cottony cushion scale had established biological control as a valid method of pest control (Bartlett, 1978). Recently, the pest appeared on several crops like citrus, mango, guava etc. without any natural enemy complex (Mani & Krishnamoorthy, 2001).

California Red Scale, *Aonidiella aurantii* (Maskell) (Hemiptera: Diaspididae)

About 15 natural enemies were known to attack *A. aurantii* in India. *Aphytis fisheri* DeBach and *Aphytis* sp. nr. *hispanicus* (Mercet) at Gauhati, Assam (DeBach, 1959), *Aphytis* spp., *Coccobius* sp., *Encarsia* sp. and *Pterotrix koebeli* (Howard) in Karnataka (CIBC, 1984), *Chilocorus nigritus* (Fabricius), *Scymnus quadrillum* Motschulsky (Pruthi and Mani, 1945), entomopathogenic fungus *Attractium indica* Chona and Munjal, *Chilocorus circumdatus* Schönherr and *C. hauseri* Weise (Rao, 1969) were reported on red scale. According to Krishnamoorthy (1993), *Aphytis melinus* DeBach and *A. linganensis* Compere, *A. fisheri*, *Cybocephalus* sp. and *Eryngiopus* sp. were also known to attack the red scale. *Aphytis melinus* when introduced into California from India had resulted in excellent control of the scale in citrus orchards in 1956-57. *Chilocorus nigritus* was the most voracious predator of *A. aurantii.* Release of adult beetles was found to be effective in the control of the red scale on citrus in Karnataka, Andhra Pradesh and Punjab (Krishnamoorthy, 1993; Batra *et al.*, 1987). Neem oil, pongamia oil, dicofol, sulfur, copper oxychloride and maneb were found safer to *A. melinus* and *C. nigrita* (Krishnamoorthy and Rajagopal, 1995; Mani, 1996).

Green Scale, *Coccus viridis* (Green) (Hemiptera: Coccidae)

It is a serious pest of citrus, sapota and coffee in India. *Microterys* sp., *M. flavus* (Howard), *Anicetus annulatus* Timberlake, *Gyranusa* sp., *Tetrastichus* sp., *Coccophagus* sp. (*lycimnia* group), *C. cowperii* Girault, *C. bogoriensis* Koningsberger, *Cheiloneuromyia javensis* Girault, *C. circumdatus, C. nigritus, Pharoscymnus horni* Weise, *C. montrouzieri*, *Chrysopa* sp., *Verticillium lecanii* (Zimmermann) Viegas, *Beauveria bassiana* (Balsamo) Vuillenim, *Entomophthora* sp., *Hypocrella olivacea* Petch, *Nectria* sp., *Podonectria* sp., *Fusarium* sp. and *Cladosporium* sp. were recorded from Kodagu, Karnataka (Singh, 1980, 1987). Among them, *Aneristus ceroplastae* Howard and *Encyrtus lecanorium* Mayr. caused upto 43% parasitism in the field (Tandon, 1985). The natural enemies were capable of suppressing *C. viridis* in undisturbed ecosystem. The predation by *C. circumdatus* and *C. nigritus* resulted in the effective suppression of scale population on acid lime in Karnataka (Mani and Krishnamoorthy, 1996b, 1998c). Releases of *C. nigritus* helped in bringing down the population of *C. viridis.*

At Kodagu, Karnataka, a combination of the fungal suspension *Verticillum lecanii* containing 10 x 10^6 spores/ml + 0.005% quinalphos + 0.05% Teepol sprayed just before the onset of rainy season gave effective control of *Coccus viridis* where the light interception was 900 -14100 lux units. Where the light interception was more than 4800-23900 lux, this treatment proved less effective (Singh, 1987, 1995).

Hemispherical Scale, *Saissetia coffeae* (Walker) (Hemiptera: Coccidae)

It is a minor pest of citrus in India and about 10 parasitoids and five predators were known to attack the scale. *Aneristus ceroplastae, Anysis saissetiae* (Ashmead), *Cardiogaster fusciventris* Motschulsky, *Coccophagus flavescens* Howard and *Encyrtus adustipennis* Motschulsky were reported parasitising this species (Pruthi and Mani, 1945).

Microterys sp., *Anysis* sp., *Coccophagus* sp., *C. cowperii, Chilocorus circumdatus, C. nigrita, Pharoscymnus horni* (Weise), *C. montrouzieri, Chrysopa* sp. and *V. lecanii* were recorded on *S. coffeae* in Kodagu, Karnataka (Singh, 1980, 1987). The encyrtid *Metaphycus helvolus* Campere has effected complete control of the scale in Peru (Beingolea, 1969).

Florida Red Scale, *Chrysomphalus ficus* Ashmead (=*C. aonidum* Linnaeus) (Hemiptera : Diaspididae)

It is of minor importance in India. About 10 natural enemies were known to attack *C. ficus* in India. *Microterys* sp., *Tetrastichus* sp., *Aphytis* sp., *A.mytilaspidis* (Le Baron), *Coelophora biplagiata* (Swartz), *Chilocorus circumdatus, C. nigritus, Pharoscymnus horni* and *Nectria* sp. were recorded on *C. ficus* in Kodagu, Karnataka (Singh, 1980, 1987, 1993a). *Chilocorus nigritus* gave substantial control of the scale in Seychelles (Bennet *et al.*, 1976).

Purple Scale, *Cornuaspis (=Lepidosaphes) beckii* (Newman) (Hemiptera: Diaspididae)

It is a minor pest in India but sometimes becomes serious in south India. *Microterys* sp., *Adelencyrtus femoralis* Compere and Annecke, *Aphytis* sp., *A. mytilaspidis, Tetrastichus* sp., *Coelophora biplagiata, Chilocorus circumdatus, C. nigritus, Pharoscymnus horni* and *Nectria* sp. were reported on *C. beckii* from Kodagu, Karnataka (Singh, 1980, 1987). In Wynaad (Kerala), it was found infected by entomopathogenic fungus, Sphaerostilbe auranticola Petch (Ramachandran, 1954).

Yellow Scale, *Aonidiella citrina* (Coquillet) (Hemiptera: Diaspididae)

Aphytis chrysomphali (Mercet), *Physcus* sp. nr. *flaviventris* (Howard), *Tetrastichus purpureus* Campere and *Comperiella bifasciata* (Howard) from Namkum (Bihar) (Glover, 1935) and *C. bifasciata* from *A. citrina* Coquillett at Bombay and Pune were reared from *Aonidiella citrina. Pharoscymnus horni* was considered to be the most promising predator in India (Compere, 1961).

Citrus Psylla, *Diaphorina citri* Kuwayana (Hemiptera : Psyllidae)

It is a major pest of citrus in Himachal Pradesh, Punjab and Maharashtra besides a vector of citrus 'greening disease'. Nymphs of the psyllid were predated upon by 12 species of coccinellids (Sharga, 1948). In Punjab, *Coccinella septumpunctata* Linnaeus, *Coccinella transversalis* Fabricius (=*Coccinella repanda* Thunberg), *Cheilomenes* (=*Menochilus*) *sexmaculata* (Fabricius), *Chilocorus nigritus*, *Brumus suturalis* Fabricius, a syrphid fly and *Chrysopa* sp. were found feeding on citrus psylla (Hussain and Nath, 1927; Bindra, 1970). The eulopid *Tamarixia* (=*Tetrastichus*) *radiata* (Waterson) was the most common parasitoid of citrus psylla causing upto 95% parasitism in Punjab (Atwal, 1962). The parasitoid was quite active from July to November (Hussain and Mathur, 1923; Bindra, 1969, 1970). The same parasitoid was introduced from India, and found established against *D. citri* in Reunion island (Etienne, 1978).

Citrus Blackfly, *Aleurocanthus woglumi* Ashby, spiny blackfly, *A. spiniferus* (Quaintance) and Hussain blackfly, *A. husaini* Corbett (Hemiptera: Aleyrodidae)

Aleurocanthus woglumi was reported serious on citrus in Kulu, Sikkim, Kashi hills, Pusa and Maharashtra. In Assam, *Encarsia divergence* Silvester and *Encarsia merceti* Silvester were dominant accounting for 90% parasitism. Parasitoids, *Encarsia* sp., *E. clypealis* Silvester, *Encarsia* sp.nr. *citrifolia, E.divergence* Silvester, *E. merceti* Silvester, *Amitus hesperidum* Silvester, and *Eretmocerus serius* Silvester, and predator, *Acletoxenus* sp. have been recorded from north India (Smith, 1950). *Encarsia* sp., *Encarsia clypealis* Silvester, *E. citrina* (Craw), *Amitus* sp., *Pullus xerampelinus* Mulsant, *Chrysopa* sp., *Verticillium lecanii, Aschersonia* sp. and *Cladosporium* sp. were recorded on *A. woglumi* in Kodagu, Karnataka (Singh, 1980, 1987). At Pune, 80-90% parasitism by *E. clypealis* was recorded but *A. hesperidum* and *E. clypealis* caused 80% parasitism in Bombay and Bangalore. *A. hesperidum* Linnaeus, *E. opulenta* Silvester, *E. clypealis* and *E. smithi* Silvester, were sent from India and Pakistan to Mexico during 1948-50 for suppression of citrus blackfly (Rao *et al.*, 1971). *Encarsia* sp. nr. *bennetti* Hayat was described from the blackfly (Hayat, 1984). In Nagpur region, *Encarsia* spp., *Eretmocerus indicus* Hayat and *Dialeurolonga elongata* (Dozier) were recorded on the citrus blackfly (Rao *et al.*, 1996). The blackfly was found attacked by *Serangium parcestosum* Sicard, *Mallada boninensis, E. indicus* and *Encarsia* spp. in citrus orchards at Nagpur (Satpute *et al.*, 1986). Releases of *Encarsia* sp. have considerably brought down the blackfly population from 68 to 35% at Nagpur (Anonymous, 1987). On the spiny blackfly *Aleurocanthus spiniferus, Chrysopa fulvolineata* was found feeding (Pruthi and Mani, 1945). The Hussain blackfly, *A. husaini* was found parasitised by *Eretmocerus* sp. (Nair, 1975). The blackfly could be kept under check by naturally occurring parasitoids in India. Release of *M. boniensis* was found effective in reducing the blackfly population (Ingole *et al.*, 2005).

Citrus Whitefly, *Dialeurodes citri* (Ashmead) Riley & Howard (Hemiptera : Aleyrodidae)

It is widely distributed in north India. *Encarsia lahorensis* (Howard) was recorded as the key parasitoid besides *Aphelinus fuscipennis* Howard. The predators often associated with this pest were *Brumus suturalis* Fabricius, *Cryptognatha flavescans* Motschulsky, *Micraspis cardoni* (Weise) and *Chrysopa* sp. (Hussain and Khan, 1945; Pruthi and Mani, 1945). The entomopathogenic fungi, *Aegerita webberi* Fawcatt from Saharanpur (Woglum, 1913), *Cladosporium* sp. from Punjab (Hussain and Khan, 1945; Saini, 1964) and *Aschersonia papillata* Petch from Kumaun hills (Bose, 1953) were observed on citrus whitefly. *Aschersonia papillata* was evaluated and recommended for use in south India (Dharmaraju and Reddy, 1975). Sharma *et al.* (2003) reported *Coccinella septempunctata* (75%), *Coccinella repanda* [*Coccinella transversalis*] (6%), *Cheilomenes sexmaculata* (13%) and *Brumoides suturalis* (6%) as the main predators of the nymphs of citrus whitefly, *D. citri* on Kinnow mandarin. The mean population of coccinellids in March was 11.50 adults, compared to 10.7 nymphs in *D. citri*. The population of coccinellids peaked during April, then declined to zero from June to October. Higher coccinellid population was observed in 2000 than in 2001. Chrysopids (Chrysopidae) were also observed feeding on *D. citri*.

Black Aphid, *Toxoptera aurantii* (Boyer *de* Fonscolombe) (Hemiptera: Aphididae)

About 10 general predators were known to feed on *T. aurantii* in India. *Paralasatomus cephalotus* Ashmead, *Lipolexis scutellaris* Mackauer, *Coelophora biplagiata* Swartz, *Coccinella septempunctata, Cheilomenes sexmaculata, C. transversalis, Scymnus xerampelinus, Nephus* sp., *Chrysopa* sp., *Paragus serratus* (Fabricius), *Dideopsis aegrota* (Fabricius), *Ischiodon scutellaris* (Fabricius) and *Cardiastethus pygmaeus* Poppius were recorded in Kodagu, Karnataka (Singh, 1980, 1987). *Cheilomenes sexmaculata* was observed in considerable numbers clearing the aphid clonies (Singh, 1993a; Mani and Krishnamoorthy, 2005). In Assam, the aphid population was regulated in March by *C. sexmaculata* and *Coccinella transversalis* Fabricius (Bhattacharya and Dutta, 1998).

Leaf Miner, *Phyllocnistis citrella* (Stainton) (Lepidoptera: Phyllocnistidae)

Many parasitoids were recorded on the eggs and larvae of the leaf miner (Singh, 1993a). *Sympiesis* sp., *Kratoysma* sp., *Bracon* sp., *Pediobius* sp., *Tetrastichus* sp. (*miser* group), *Tetrastichus phyllocnistoides* (Narayanan), *Cirrospilus* sp. and *Cirrospilus quadristriatus* Subba Rao & Ramamani were recorded on *P. citrella* in Kodagu, Karnataka (Singh, 1980, 1987). Three larval parasitoids, viz. *Amatellon* sp., *Tetrastichus* sp. and *Elasmus* sp. were earlier recorded parasitising only upto 33% (Narayanan *et al.*, 1957). Among them, *Tetrastichus phyllocnistoides* and *Ageniaspis* sp. were important, the latter one causing upto 80% parasitism (Atwal, 1964). In Punjab, *C.quadristriatus* and *T. phyllocnistoides* caused 30-45% parasitism during July-September (Batra and Sandhu, 1981).

Lemon Butterfly, *Papilio demoleus* Linnaeus, and *P. polites* Linnaeus (Lepidoptera:Papilionidae)

Eggs of lemon butterfly were found parasitised by *Trichogramma evanescens* Westwood, *T. chilonis* Ishii, *Telenomus incommodus* Nixon and *Ooencyrtus papilionis* Ashmead to an extent of 78% in the field (Pruthi and Mani, 1945; Krishnamoorthy and Singh, 1988; Jalali and Singh, 1990). *Dusona* sp., *Melalophacharops tamanukii* (Uchida), *Pteromalus puparum* (Linnaeus), *Brachymeria jambolana* Gahan, *Holcojoppa coelopyga* Morley, *Blepharipa* sp., *Distatrix* (=*Apanteles*) *papilionis* Viereck, *Serratia marcescens* Bizio and *Fusarium* sp. were recorded from caterpillars and pupae in Kodagu, Kannataka (Singh, 1980, 1987). Earlier pupae of *P. demoleus* were attacked by *P. puparium* and *Brachymeria* sp. (Singh, 1967; Misra and Pandey, 1965) and *Erycia nymphalidephaga* Bar (Modawal, 1940). A tachinid, *Ugimyia sericaiae* Cornalia was reported from *P. demoleus* (Thompson, 1946). *Distatrix papilionis, H. coelopyga, Melalophacharops* sp., *B. jambolana* and *P. puparum* were recorded on *Papilo* spp. in Tamil Nadu (Singh, 1991b). Most of the parasitoids of *P. demoleus* also attacked another common species of butterfly *P. polites* (Singh, 1991b). *Distatrix papilionis* was the dominant parasitoid of caterpillars at Bangalore, and *T. chilonis, T. incommodus* and *D. papilionis* caused a cumulative parasitism of 88% (Krishnamoorthy and Singh, 1988). *Trichogramma chilonis, Melalophacharops* sp. and *D. papilionis* could be utilized for the biological suppression of butterflies attacking citrus. *D. papilionis* could be conserved with the use of plant origin pesticides (Mani, 1996; Mani and Krishnamoorthy, 1997a).

At Bangalore, *Nosema* sp. (Anonymous, 1980) and a cytoplasmic polyhedrosis virus (Narayanan and Gopalakrishnan, 1987) were isolated. *Bacillus thuringiensis* Berlin (Dipel) has been found to be very effective against the citrus butterfly caterpillars at 0.05% concentration (Gopalakrishnan and Gangavisalakshi, 2005). It also effectively killed *Helicoverpa armigera* (Hubner) and *Prays citri* Milliere feeding on citrus inflorescence (Singh, 1991a). Earlier, the suscep-tibility of citrus butterfly caterpillar, to 0.2% Bactospeine (Sundara Babu, 1969) and Thuricide 90 TS Flowable (containing 15 billion viable spores/gram) (Narayanan and Jayaraj, 1974) had been reported. Application of *Bacillus thuringiensis* var. *kurstaki* 1 ml/litre + Spreader (Sandovit/Labolene/Teepol 0.01%) was found very effective in controlling the citrus butterfly.

Steinernema (*Neoaplectana*) *carpocapsae* Dutky applied at 600,000 nematodes per tree resulted in 63.33% kill of third instar butterfly caterpillars and 67.67% mortality of fourth and fifth instar caterpillars (Srivastava, 1978). Indigenous entomophilic nematode, *Steinernema* sp. isolated from cutworms was highly effective against the larvae of *Papilio demoleus* (Singh, 1993b).

Leaf Roller, *Posorosticha* (=*Tonica*) *zizyphi* Stainton (Lepidoptera: Oecophoridae)

It is an important pest in Madya Pradesh, Punjab and Maharashtra. *Apanteles* sp., *Pristomerus* sp. and *Brachymeria euploeae* (Westwood) were recorded in Madya Pradesh (Gupta, 1954). In Punjab, the parasitism by *Pristomerus* sp. was higher in July and by *Apanteles machaeralis* Wilkinson during June, however, overall parasitism ranged from 25 to 36% (Sharma and Batra, 1989). In Andhra Pradesh, *Ophion triangularemaculatus* Mots, *Goniozus* sp and *Apanteles* sp., were recorded. During 1979-80, *O. triangularemaculatus* caused 48% parasitism, followed by *Apanteles* sp. with 6-11% and *Goniozus* sp. with 4% parasitism (Nagalingam, 1988).

Fruit Sucking Moths, *Achaea janata* Linnaeus, *Othreis fullonica* Linnaeus and *O. materna* Linnaeus (Lepidoptera: Noctuidae)

Among two dozen parasitoids recorded on *A. janata, Trichogramma chilonis* and *Microplitis maculipennis* Szepare were found important (Rao, 1969).The eggs were successfully parasitised by *T. chilonis and Telenomus* sp. suggesting the possibility of utilising *T. chilonis* for the control of *Othreis* spp. (Dodia *et al.*, 1986; Bhumannavar and Viraktamath, 2001). *Tetrastichus ophiusae* Crawford and *Euplectrus melanocephalus* Girault (*E. lenocos)* were known to attack the larvae of *Othreis* spp. (Nayar *et al.*, 1976). *Euplectrus maternus* Bhatnagar was found parasitising *Othreis* spp. upto 50% in Karnataka, Andhra Pradesh and Maharastra (Bhumannavar and Viraktamath, 2000).

Oriental Mite, *Eutetranychus orientalis* (Klein) (Acari: Tetranychidae)

Heavy infestations of the mite were observed on citrus, guava, banana, fig, etc. in India. *Amblyseius* sp., *Chrysopa* sp., *Scymnus xerampelinus* Mulsant, *Scolothrips* sp. and *Stethorus* sp. were recorded on *E. orientalis* in Kodagu, Karnataka (Singh, 1980, 1987). Among nine insect and mite predators, *Amblyseius alstoniae* Gupta and *Scolothrips indicus* Priesn appeared to be important in regulating the mite populations (Dhooria, 1983).

GRAPEVINE

Pink Hibiscus or Grape Mealybug, *Maconellicoccus hirsutus* (Green) (Hemiptera : Pseudococcidae)

It has become a major pest of grapevine in peninsular India (Manjunath, 1985a; Mani, 1994c; Mani and Thontadarya, 1987c). Six parasitoids namely, *Anagyrus dactylopii* (Howard), *Allotropa* sp. nr.

japonica Ashmead *Gyranusoidea mirzai* (Agarwal), *Alamella flava* Agarwal, *Leptopilinia* sp. and *Chartocerus* sp. nr. *walkeri* Hayat, and seven predators, viz. *Scymnus* sp. *Scymnus coccivora* Ayyar, *Cryptolaemus montrouzieri*, *Chrysopa* sp, *Spalgis epius* Westwood, *Cacoxenus perspicax* (Knab) and *Triommata coccidivora* (Felt) were recorded. Among them, *A. dactylopii* and *S. coccivora* were of considerable importance. *Anagyrus dactylopii* caused upto 70% parasitism in nature (Mani *et al.*, 1987). *Anagyrus dactylopii* male and female completed their development in 14.75 and 22 days, respectively and a progeny of 39.3 was produced (Mani and Thontadarya, 1988c). Third instar nymph and adult female of *M. hirsutus* were found suitable for breeding of *A. dactylopii* (Mani and Thontadarya, 1989). A maximum of 182 offspring was produced at 30°C (Mani and Krishnamurthy, 1992). The parasitoid could be conserved by the application of dichlorvos, diazinon, phosalone, fish oil rosin soap and the commonly used fungicides (Mani and Thontadarya, 1988a). Under field conditions, a positive and significant relationship between the population of *A. dactylopii* and *M. hirsutus* was observed (Mani and Thontadarya, 1987b). *Allotropa japonica* was bred very well on 15-20 days old *M. hirsutus*. It completed its life cycle in 25.5 days at 25.5°C. *Scymnus coccivora* completed its development on *M. hirsutus* in about 20-23 days and, adult males and females lived for 60.5 and 68.7 days, respectively. Each female produced 46.5 eggs. A grub of *S. coccivora* consumed 308 eggs or 62 nymphs or 6.55 adult females (Mani and Thontadarya, 1987d). *Scymnus gratiosus* Wiese consumed about 600 eggs of *M. hirsutus* in its larval development (Ramamurthy, 1982). A single larva of *M. boninensis* was found consuming about 240 mealybug nymphs of *M. hirsutus* (Mani and Krishnamoorthy, 1989b). Fenvalerate, endosulfan and oxydemeton methyl were less toxic to the larvae of *M. boninensis* (Mani, 1993b). Effect of soil application of systemic granular insecticides on the population of grape mealybug and its natural enemies has been studied (Mani and Thontadarya, 1991).

The coccinellid predator *C. montrouzieri* was known to feed about 1000 eggs or 300-500 mealybug nymphs (Mani and Thontadarya, 1987a). Its life cycle was completed in about 30 days. The optimum temperature for the maximal development of *C. montrouzieri* was found to be 30°C (Babu, 1986). A simple method of mass culturing *C. montrouzieri* on the mealybug infested ripe pumpkin was standardised. The performance of the beetle was evaluated in Maharashtra, Andhra Pradesh, Karnataka and Tamil Nadu. Adult beetles of 10 days old were recommended for release in the evening time to control the grape mealybug in 60-75 days of release. Besides the commonly used fungicides, dichlorvos, chlorpyriphos and fish oil rosin soap could be used along with *C. montrouzieri* in the integrated pest management programme in vineyards in India (Mani, 1987, 1988; Mani and Krishnamoorthy, 1990a, 1993a; Mani and Thontadarya, 1988b; Babu, 1986; Manjunath, 1986; Srinivasan, 1987; Shirke and Salunkhe, 1996; Satyanarayana *et al.,* 1991). According to Kamal (1951), complete control of *M. hirsutus* was achieved in Egypt by introducing *Anagyrus kamali* Moursi from Java. Introductions of efficient exotic strains of *A. kamali* should be made and tried against *M. hirsutus* in India.

Citrus Mealybug, *Planococcus citri* (Risso) (Hemiptera: Pseudococcidae)

It is a major pest of grapevine in some grape growing regions in south India. *Cryptolaemus montrouzieri* proved to be very useful in controlling the mealybug in the field (Anonymous, 1986; Mani and Krishnamoorthy, 2008b) and aslo in polyhouse (Mani, 2008). *L. dacylopii* was also recevored in large numbers from *P. citri* infesting grapes (Krishnamoorthy and Mani, 1989b).

Leaf Roller, *Sylepta lunalis* Guien (Lepidoptera: Crambidae)

Sylepta lunalis was reported to be parasitised by larval parasitoid *Apanteles dita* Nixon and prepupal parasitoid *Cardiochiles fulous* Cameron in India (Odak and Dhamdhere, 1970). *Apanteles pusaensis* Lal parasitised the larvae of *S. lunalis* to the extent of 60-70% in nature (Narayanan, 1941).

GUAVA

Aphid, *Aphis gossypii* Glover (Hemiptera: Aphididae)

Aphis gossypii is minor pest of guava but it causes serious damage at times to tender leaves. Five coccinellid predators, viz. *C. sexmaculata, Scymnus* sp., *Pseudaspidimerus circumflexa* (Motschulsky), *C. nigritus, Cryptolaemus montrouzieri* and two syrphids, *Ischiodon scutellaris* (Fabricius) and *Paragus serratus* Fabricius were observed feeding on *A. gossypii* infesting guava. The coccinellid and syrphid predators were quite effective in controlling the aphids in Karnataka (Mani and Krishnamoorthy, 1989b, 2007).

Green Shield Scale, *Pulvinaria psidii* (Maskell) (Hemiptera: Coccidae)

Pulvinaria psidii attacks a wide range of host plants including guava. *Scymnus coccivora, Cryptolaemus montrouzieri, Chilocorus nigrita, C. sexmaculata, Coccophagus cowperii* Girault, *C. bogoriensis* Koningsberger and *Aneristus* sp. were known to attack *P.psidii* (Anonymous, 1979; Puttarudriah and Channabasavanna,1957; Manjunath, 1986; Bennet and Hughes, 1959). Agarwal *et al.* (1984) reported *Bothriophryne* sp. nr. *pulvinariae* Agarwal, Agarwal & Khan and *B.* sp. nr.*tachikawai* Agarwal, Agarwal & Khan from *Pulvinaria* sp. infesting guava trees in Bermuda. *Cryptolaemus montrouzieri* was found to be a voracious feeder on the ovisacs of *P. psidii* consuming about 4000 eggs. A release rate of 20 adult beetles per tree was found to be very effective against *P. psidii* (Mani and Krishnamoorthy, 1990d).

Striped Mealybug, *Ferrisia virgata* (Cockerell) (Hemiptera: Pseudococcidae)

Aenasius advena Compere and *Blepyrus insularis* (Campere), *S. coccivora, Mallada boninensis, Brumus suturalis* and *Spalgis epius* and *C. sexmaculata* were recorded on *F. virgata* (Mani and Krishnamoorthy, 1989c; Mani and Krishnamoorthy, 1993b). *Chrysopa lacciperda* (Kimmis) and *Chrysoperla carnea* (Stephens) were observed on *F.virgata* and *P. citrii* in guava orchards (Krishnamoorthy and Mani, 1989a). A single larva of *M. boninensis* was able to prey about 345, 490 and 560 nymphs of *F. virgata, P.lilacinus* and *P.citri,* respectively (Mani and Krishnamoorthy, 1990c). *Blepyrus insularis* was bred very well on 5-10 old nymphs of *F. virgata* (Mani and Krishnamoorthy, 1991). The key parasitoid, *A. advena* could be conserved by the applications of diazinon, phosalone and dichlorvos (Mani, 1992). *Cryptolaemus montrouzieri* at 10-15 adults/tree gave excellent control of *F. virgata* in the open field (Mani *et al.*, 1990) and polyhouse (Mani, 2008).

Citrus Mealybug, *Planococcus citri* (Risso) (Hemiptera: Pseudococcidae)

It is a polyphagus pest causing severe damage to guava at times. Releases of the exotic parasitoid, *Leptomastix dactylopii* were found to be highly effective against *P. citri* in guava orchards. The local platygasterid, *Allotropa citri* Muesebeck was also found active against

P.citri (Mani, 1994a). In some guava orchards, *P. citri* was found suppressed by *S. epius, C. lacciperda* and *C. montrouzieri* in nature (Mani and Krishnamoorthy, 1990e).

Pink Hibiscus Mealybug, *Maconellicoccus hirsutus* (Green) (Hemiptera: Pseudococcidae)

Local natural enemy complex was very poor on *M. hirsutus* infesting guava. Releases of *C. montrouzieri* were found to be highly satisfactory in suppressing *M. hirsutus* on guava (Anonymous, 1990). *Scymnus coccivora* was found useful in the management of guava mealybug, *Maconellicoccus hirsutus* (Murali Baskaran *et al.,* 2007b).

Waxy Scale, *Drepanococcus (=Ceroplastodes) chiton* (Green) (Hemiptera: Coccidae)

It was found very serious at times on guava in Karnataka. The parasitoids, *Anicetus ceylonensis* Howard, *Diversinervus elegans* Silvestri, *Metaphycus helvolus* (Compere), *Philosindia* sp. nr. *Longicornis* Noyes and Hayat (all encyritids) and *Cephaleta brunniventris* Motschulsky (Pteromalidae) and four coccinellid predators namely *Chilocors nigritus*, *Cryptolaemus montrouzieri*, *Cheolomenes sexmaculata* and *Scymnus* sp. were known to attack this scale. Among them, *A. ceylonensis* and *C. brunniventris* were highly effective in suppressing the scale population. At times, the local predator *C. nigritus* was also observed in large numbers helping in bringing down the pest population (Mani, 1995c). Dichlorvos (0.10%), endosulfan (0.07%) and oxydemeton methyl (0.05%) showed the least residual toxicity to *A. ceylonensis* (Mani and Krishnamoorthy, 1997d).

Spiralling whitefly, *Aleurodicus dispersus* Russell (Hemiptera: Aleyrodidae)

The spiralling whitefly is causing very serious damage to the leaves of guava and many other plants in peninsular India. The coccinellids, *Axinoscymnus puttarudriahi* Kapur & Munshi, *Cryptolaemus montrouzieri, Chilocorus nigritus, Cheilomenus sexmaculata, Angelcia cardoni* Wiese, *Scymnus nubilis* Mulsant, *Serangium parcestosum* Sicard; chrysopids namely *Apertochrysa* sp., *Mallada astur* Banks, *Mallada boninensis*, *Chrysoperla carnea*, *Nobilinus* sp.; a nitidulid, *Cybocephalus* sp.; a drosophilid *Acletoxenus indicus* Malloch and an hemerobid, *Notobiella* sp. were known to attack the whitefly. Among them, *C. montrouzieri, M. astur* and *A. puttarudriahi* were commonly associated with *A. dispersus* (Mani and Krishnamoorthy, 1999c). A single larva of *M. astur* was found consuming about 235 nymphs of spiralling whitefly during its larval development (Mani and Krishnamoorthy, 1999h). A total of 290 nymphs of spiralling whitefly were consumed by an individual grub of *C. montrouzieri* (Mani and Krishnamoorthy, 1999a). However, the predators did not have any significant relationship with the population of spiralling whitefly in guava orchards. They were unable to keep the pest population under check (Mani and Krishnamoorthy, 1996f, 1997e, 1999b). Chemical control was also found to be ineffective in reducing the whitefly population. The aphelinid parasitoids *Encarsia haitiensis* Dozier and *Encarsia guadeloupae* Viggiani were observed on the spiraling whitefly in southern parts of India during 1999. They could be highly useful in tackling whitefly problem (Mani *et al.*, 2000a, 2001, 2003, 2004a,b). Biointensive management of *A. dispersus* has been discussed (Mani and Krishnamoorthy, 2007a)

MANGO

Hoppers, *Idioscopus* spp. (Homoptera: Cicadellidae) and *Amritodus atkinsoni* (Lethierry) (Hemiptera: Jassidae)

Isyndus heros Fabricius, *Epipyrops fulginosa* Tames, *Pipunculus annulifemur* Brundli and *Halictophagus indicus* Bohart were recorded on mango hoppers (Ramachandra Rao, 1930; Wagle, 1934; Rahman, 1939). Twelve species of spiders were also reported preying the mango hoppers (Anonymous, 1979). *Coccinella septumpunctata*, *C. transversalis*, *Cheilomenes sexmaculata*, *Araneus sinhagadensis* Tikader, *Cyrtophora cicatrosa* Stoliczka, *Cheriracanthium danieli* Tikader, *Stagodyphus sarasinorum* Karsch, *Harsilia savigryi* Lucas, *Oxyopes shweta* Tikader, *Rhene indicus* Tikader, *Phiddipus* sp., *Uloborus* sp., *Linyphia* sp., *Theridion indica* Tikader, *Marpissa* sp., *Tetrastichus* sp., *Beauveria bassiana*, *Verticillium lecanii*, *Aspergillus flavus* Link, *Pleospora infectoria* Group, *Cladosporium* sp., *Hirsutella* sp., and *Isaria* sp. were also known to attack mango hoppers in Malihabad in Uttar Pradesh (Singh, 1993a). The reduvid, *Irantha armiper* Stal was found preying voraciously on *A. atkinsoni* (Singh *et al.*, 1993). *Verticillium lecanii* was known to infect the hoppers in nature and it was suggested for evaluation against mango hoppers in different regions of India. In Taiwan, *Gonatocerus* sp. had the highest average rate of parasitism of 78.6% on *I. clypealis* (Lethierry) and *I. neveosparsus* (Lethierry) (Chou and Chou, 1990). *Metarhizium anisopliae* at 1 × 10^{10} spores/ml, was known to cause 100% mortality of mango hoppers (PDBC, 2003)

Oriental Fruit Fly, *Bactrocera dorsalis* (Hendel) (Diptera: Tephritidae)

In North India, *Bracon fletcheri* Silvester, *Opius incisi* Silvestri, *O. watersi* (Fullaway), *O. longicaudatus* (Ashmead), *O. fletcheri* Silvester, *O. vandenboschi* Fullaway, *Cratospila* sp., *Spalangia afra* Silvester, *S. endius* Walker, *Dirhinus luzonensis* Rohwer, *D. giffardii* Silvester, *Pachycrepoideus vindemmiae* (Rondani), *Cynipidae* sp.,*O. carpomyiae* (Silvester), *Silus* sp., *Trichopria* sp. and *O. oophilus* Fullaway (Webber, 1951), and from South India, *Opius persulcatus* (Silvester), *O. oophilus*, *O. compensans* (Silvestri), *O. incisi, O. manii* Fullaway, *O. carpomyiae, Aceratoneuromyia indicum* (Silvester), *S. endius, S. afra, S. grotiuse* Girault, *D. trybliographa* Weld, *Pseudeucolia* spp., *Cothonaspis* spp., *Psilus* spp., *Halticoptera* sp., *Alysiinae* sp., *Tachinaephagus* sp. and *Cratospila* sp. (Clausen *et al.,* 1965) were recorded. According to Narayanan and Batra (1960), 17 species of parasitoids were reported to be associated with the fruit flies in India. Several of them were sent to different countries for trials against *B. dorsalis. Opius oophilus* Fullaway, *O. longicaudatus* Ashmead and *O. incisi* Silvester were responsible for the substantial control of the *B. dorsalis* on many fruits in Hawaii (Ven den Bosch and Messenger, 1973). But recent surveys indicated that many of the parasitoids reported earlier were not present in India.

Stone Weevils, *Sternochetus gravis* (Fabricius) and *S. mangiferae* (Fabricius) (Coleoptera: Curculionidae)

In Tripura, *S. gravis* was found to be predated by ants, *Oecophylla smaragdina* (Fabricius), *Campontus* sp., *Monomorium* sp. and the mite, *Rhizoglyphus* sp. (De and Pande, 1988). Nuclear polyhedrosis virus was recorded in Bangalore on *S. mangiferae* (Shukla *et al.,* 1984). *Beauveria bassiana* 1 × 10^{10} spores/ml was known to cause 100% mortality of weevils (PDBC, 2004)

Mango Coccid, *Rastrococcus iceryoides* (Green) (Hemiptera: Coccidae)

In addition to *Rastrococcus iceryoides,* three pseudococcids, viz. *Planococcus citri* (Risso) and *Ferrisia virgata* and the margorodids *Drosicha stebbingi* (Green), *Perissopneumon ferox* Newstead were also recorded on mango. Several natural enemies were recorded on *R. iceryoides* in Uttar Pradesh and Karnataka. *Anagyrus pseudococci* Girault, *Gyranusoidea* sp., *Praleurocerus viridis* Agarwal, *Allotropa* sp., *Microterys flavus* (Howard), *Dinocarsis* sp., *Promuscidea unfasciativentris* Girault, *Metastenus concinnus* Walker, *Tetrastichus* sp., *Cybocephalus* sp., *Scymnus coccoivora* Ayyar, *Monomorium floricola* (Jerdon), *Coccophagus* sp. and *Proctolaelaps* sp. were recorded from Malihabad. But *P. unfasciativentris* and *A. pseudococci* were important (Tandon and Lal, 1978). *P. isolitus* (Alam) and *Tetrastichus* sp. were also reported earlier (Rawat and Jakhmola, 1970; Shafee *et al.*, 1975). According to Narasimham and Chacko (1988), parasitoids *Anagyrus* sp. nr. *dactylopii* (Howard), *Anagyrus* sp., *C. sexvittatus* Hayat, *Coccophagus* sp. (*pseudococci* group), *Allotropa* sp. and predators namely *Leucopis* sp., *Cacoxenus perspicax* (Knab), *Spalgis epius* were recorded. Besides predatory ants *Camponotus* sp., *Myrmicaria brunnea* Saunders and *Oecophylla smaragdina* (Fabricius) were also known to attack *R. iceroides*. Upto 42% parasitism was observed in the field (Tandon and Lal, 1978). The parasitoid *Anagyrus pseudococci* Girault and the predator *Cacoxenus perspicax* were important in nature. An individual of coccinellid predator, *C. montrouzieri* consumed about 350 mealybug eggs or 500 nymphs during its larval development. *Cryptolaemus montrouzieri* was found very effective in controlling *R. iceryoides* (Mani *et al.*, 1995).

Mango Mealybug, *Drosicha stebbingi* Green (Hemiptera: Margarodidae)

The mango mealybug was predated by several coccinellids including *Rodolia fumida* Mulsant, the chrysopid *Chrysoperla carnea* and drosophilid *Cacoxenus perspicax* and parasitised by *Phygadeuon* sp. (Rahman and Latiff, 1944; Wadi and Batra, 1964; Verma *et al.*, 1976). *Aulis vestitus* Mulsant, *Coccinella septumpunctata, C. transversalis, Cheilomenes sexmaculata, Rodolia fumida, Scymnus cordoni, Cyrtophora cicatrosa* Stoliczka, *Linyphia* sp., *Phiddipus* sp., *Oxyopes shweta* Tikader, *Hersilia savigryi* Lucas, *Rhene indicus* Tikader, *Stagodyphus sarasinorum*, *Marpissa* sp., *Leptus* sp., *Bochartia* sp., *Tetrabrachys robusta* and *Gitona* sp. nr. *magnipalpis* Duda and *Beauveria bassiana* were recorded from Malihabad, Uttar Pradesh. But none of these natural enemies was found to give adequate control of *D. stebbingi* (Singh, 1993a).

Green Shield Scale, *Chloropulvinaria polygonata* (Cockerell) (Hemiptera: Coccidae)

Green shield scale is a very serious pest of mango in several parts of India. *Aneristus ceroplastae* Howard, *Metaphycus* sp. nr. *hederaceous, Tetrastichus* sp., *Anicetus annulatus* Timberlake, *Nicostratus* sp., and *Anagyrus* sp. were recorded from Malihabad (Singh, 1993a). Three parasitoids namely *Coccophagus bivittatus* Compere, *C. nigricorpus* Shafee, *Metaphycus helvolus* (Compere), and three predators, viz. *Cryptolaemus montrouzieri*, *Mallada astur* and *Spalgis epius* were obsereved on mango green shield scale around Bangalore. *Cryptolaemus montrouzieri* was capable of consuming about 2400 eggs of the scale insect. Release of *C. montrouzieri* was found very effective in suppressing the scale population on mango near Bangalore (Mani and Krishnamoorthy, 1998b).

Coconut Scale, *Aspidiotus destructor* Signoret (Hemiptera: Diaspididae)

Aphytis chrysomphali (Mercet), *Aneristus ceroplastae* Howard, *Chrysonotomia* sp. *Comperiella bifasciata* Howard, *Chartocerus* sp. *Thompsonisca desantisiellus* Shafee and *Chilocorus nigritus* were recorded in Uttar Pradesh. Among them *T. desantiellus* was the enemy most commonly encountered in nature (Tandon and Srivastava, 1980).

Mango Scale, *Aulacaspis tubercularis* Newstead (Hemiptera: Diaspididae)

A total of 10 natural enemies including an aphelinid parasitoid *Pteroptrix koebeli* Howard and the predators *Chilocorus circumdatus*, *C. nigritus*, *Rodolia amabilis* Kapur, *Pseudaspidimerus circumflexa* (Motschulsky), *Scymnus* sp., *Ortalia* sp., *Cybocephalus* sp., *Triommata coccidivora* (Felt) and *Mallada bononensis* were found attacking the mango scale in the field. An aphelinid, *P. koebeli* is known to cause upto 90% parasitism in nature around Bangalore (Mani and Krishnamoorthy, 1998a).

Leaf Webber, *Orthaga eudrusalis* Walker (Lepidoptera: Pyralidae)

Brachymeria lasus (Walker), *Tetrastichus* sp., *Pediobius bruchiada* (Rondanii), *Hormius* sp., *Cathartoides* sp. and *Tetratichus* sp. were recorded in Uttar Pradesh (Tandon and Srivastava, 1980).

Aspergillus flavus, *Beauveria bassiana* and *Serratia marcescens* were recorded from Malihabad (Singh, 1993a). *Goniozus* sp. was recorded as an important larval parasitoid in Kerala (George and Abdurahman, 1985).

Mango Leaf Gall Midge, *Procontarinia matteiana* Kieffer and Cecconi (Diptera: Cecidomyiidae)

Chrysonotomyia pulcherrima (Kerrich), *Chrysonotomyia* spp., *Tetrastichus* spp., *Synopeas procon* Austin, *Inostemma oculare* Austin, *Trichacoides indicus* Jackson, *Ormyrus* sp., *Eupelmus* sp. nr. *Testaceiventris* (Motschulsky), *Gastrancistrus* sp. and *Aphanogmus* sp. were recorded on *P. matteiana*. Out of these *C. pulcherrima*, *Eupelmus tetaceiventris* sp nr., *Inostemna oculare* Austin and *Ormyrus* sp. were important, and also sent to Oman (Sankaran and Mjeni, 1985). *Chrysonotomyia pulcherrima*, *Anicetus* sp., *Megastigmus ameraeus* (Chandy Kurian), *Synopeas* sp., *Tetrastichus* sp. and *Amblydromella plebins* Chaudhri were recorded from Malihabad, Uttar Pradesh (Singh, 1993a).

BANANA

Rhizome Weevil, *Cosmopolites sordidus* (Germar) (Coleoptera: Curculionidae)

It is a serious pest in South East Asia and Pacific region. The histerid predator, *Plaesius javanus* Erichson (Origin: Java) was found established and the pest damage was markedly reduced in Fiji and Tahiti (Rao *et al.*, 1971). Soil application of *Steinernema feltiae* (=*Neoaplectana carpocapsae*) (DD-136 nematode) could also be tried against rhizome weevil.

Armyworm, *Spodoptera litura* (Fabricius) (Lepidoptera: Noctuidae)

Larvae of *S. litura* were found causing considerable damage to the leaves of banana in many countries including India. *Telenomus hawaii* Ashmead (origin: Hawaii) was found established in Guam islands (Swezey, 1940). It was also controlled by the application of Sl NPV at the rate of 1.2×10^{12} PIB/ha in Tamil Nadu (Santharam *et al.*, 1978).

Aphid, *Pentalonia nigronervosa* Coquerel (Hemiptera: Aphididae)

In Kerala, *Pseudaspidimerus circumflexus, Micraspis (=Verania) discolor* (Fabricius), *C. sexmaculata, Scymnus nubilis, Coccinella septunpunctata. Paragus serratus* and *Xanthogramma scutellare* were found preying on *P. nigra* (Johnson, 1983). In Karnataka, *Verticillium* sp. was found infecting *P. nigronervosa* (Deshpande *et al.*, 1972). The fungus *Mucor circinelloides* (Vantieghem), isolated from *P. nigronervosa* was found causing 100% mortality of the banana aphid in the laboratory (Mathai *et al.*, 1990).The parasitoids, *Lysiphlebs testaceipes* (Cresson) and *Aphidius colemani* Viereck were successfully mass reared on *P. nigronervosa* and it was justified to recommend mass release of both species on the South Pacific islands (Volkl *et al.*, 1990).

Hairy Caterpillars, *Eupterote geminata* Walker (Lepidoptera: Eupterotidae) and *Pericallia ricini* Fabricius (Lepidoptera: Arctiidae)

The larvae of *Eupterote geminata* were found parasitised by *Henicospilus rufus* Tsq. and *Sturmia* sp. (Jhonson, 1958). The parasitoids namely *H. rufus, Sturmia* sp., *Apanteles ricini* Bhat, *Meteorus* sp., *Thelaira nigripes* Fabriciuu and *Euplectrus* sp. were recorded on *P. ricini.*

Oriental Yellow Scale, *Aonidiella orientalis* (Newstead) (Homoptera: Diaspididae)

Among the natural enemies, *Aphytis* sp. was found causing upto 4% parasitism. The population of *Scymnus* sp. ranged from 3.1 to 13.2. The infection of the scales by *Cladosporum* sp. was observed upto 46% in April. The biotic agents and weather factors except the minimum temperature did not have any significant influence on the population of the scale in April 1995 (Mani and Krishnamoorthy, 1996h).

SAPOTA

Green Scale, *Coccus viridis* (Green) and Green Shield Scale, *Pulvinaria psidii* Maskell (Hemiptera: Coccidae)

Both the green scale *Coccus viridis* and the green shield scale *Pulvinaria psidii* suck the sap from leaves and tender shoots on young sapota plants. The green scale was suppressed by *Chilocorus nigritus* and *Cryptolaemus montrouzieri* while the green shield scale was controlled by *C. montrouzieri* (Mani and Krishnamoorthy, 1997d).

Oriental Mealybug, Planococcus lilacinus (Cockerell) and Citrus Mealybug, *P. citri* (Risso) (Hemiptera: Pseudococcidae)

The two encyrtids *Leptomastix dactylopii* and *Coccidoxenoides peregrinus* were able to suppress the populations of *P. citri,* while *Cryptolaemus montrouzieri* was highly effective in checking the populations of *P. lilacinus* (Mani and Krishnamoorthy, 1997c) and *M. hirsutus* (PDBC, 2004) and *Rastrococcus invadens* (Mani *et al.*, 2004b).

POMEGRANATE

Fruit Borer, *Deudorix (=Virachola) isocrates* (Fabricius) (Lepidoptera: Lycaenidae)

Eggs were found parasitised by *Telenomus* sp., *Ooencyrtus papilionis* Ashmead and *Trichogramma manii.* The larvae were parasitised by *Apanteles* sp. nr. *sauros* Nixon and pupae by *Brachymeria euploeae* Westwood, *B. lasus* (Walker) and *B. nephantidis* Gham, *Charops obtusus* Morley, *C. brachypterum* Gupta & Maheswari (Halleppanavar, 1957; Narayanan, 1954; Narendran,1989; Karuppuchamy, 1994; Shevale and Khire, 1999; Mani and Krishnamoorthy, 1996c; 1999e). *Ooencyrtus papilionis* and *Telenomus* sp. were known to cause upto 60% parasitism in pomegranate orchards in Karnataka. *Ooencyrtus papilionis* alone parasitised 70% of the eggs during August-October 1995 in Maharashtra (Mani and Krishnamoorthy, 1996c, 2003). Culturing and releasing of *Trichogramma chilonis* @ 2.5 lakh/ha four times at 10 days interval had resulted upto 50% parasitism on the fruit borer in Tamil Nadu (Karuppuchamy, 1994). A new species of *Trichogramma* was recorded (Nagaraja and Gupta, 2007). Application of *Bacillus thuringiensis* var *kurstaki* was found to be effective against *D. isocrates* (PDBC, 2002).

Pomegranate Fruit Borer, *Deudorix epijarbas* Moore (Lepidoptera: Lycaenidae)

It is a very serious pest in the hills of Himachal Pradesh and Jammu and Kashmir. Three egg parasitoids *Anastatus* sp. nr. *kashmirensis* Mathur, *Aphelinus gossypii* Timberlake and two larval parasitoids *Apanteles obliquae* Wilkinson and *Apanteles vitripennis* were reported in India. Maximum egg parasitism went upto 62% and the larval parasitism upto 16% (Kakar and Sharma, 1988). Field parasitism of 48-86% in *D. epijarbas* by *T. chilonis* was recorded in Himachal Pradesh (Rawat and Pawar, 1991).Two parasitoids *Anastatus* sp. nr. *kashmirensis* and *Telenomus cyrus* Nixon were found parasitising about average of 56.43% eggs (Thakur *et al.*, 1995). *T. chilonis* was found parasitising 33-93% of the eggs of *D. epijarbis* in Kullu in 1998 (Thakur *et al.*, 1991). Parasitism by *T. cyrus* was 6-10% while parasitism by *Glyptapanteles* (*Apanteles*) *vitripennis* (Curtis) varied from 0-16% (Rawat *et al.*, 1989). However, these were not reported to be very effective in reducing the borer incidence. Releases of *Trichogramma* spp. were able to reduce the borer incidence slightly (Kakar and Sharma, 1988; Thakur *et al.*, 1993).

Aphid, *Aphis punicae* (Passerini) (Hemiptera: Aphididae)

The aphids cause very severe damage to the tender leaves, flowers and fruits. *A. punicae* was found preyed by *Scymnus castaneus* Sicard, *S. latemaculatus* Motschulsky, *C. sexmaculata*, *Chilocorus nigritus*, *Mallada boninensis*, *Leucopis* sp. and *Paragus serratus* and parasitised by *Aphidius* sp., *Aphelinus* sp., *Trioxys* sp. and *Signiphora* sp. Only the predators played a significant role in suppressing the pomegranate aphids (Mani and Krishnamoorthy, 1995b; 1995d; Sreedevi *et al.*, 2006). Release of *Chrysoperla carnea* resulted in good control of *A. punicae* in Tamil Nadu (Karuppuchamy, 1994).The fungus *Entomophthora aphidis* Hoffman was also isolated from the aphid in Karnataka (Gopalakrishnan and Mani, 2000).

Oriental Mealybug, Planococcus lilacinus (Cockerell) (Hemiptera: Pseudococcidae)

It was predated by *Spalgis epius*, *Hyperaspis maindronii* Sicard, *Scymnus severini* Weise, *Eublemma* sp., *Leucopis lutecornis* Malloch and *Anagyrus* sp. (Nair, 1975). *Triommata coccidivora* (Felt), *Spalgis epius*, *Cryptolaemus montrouzieri*, *Scymnus coccivora* and *Cacoxenus perspicaux* were reared from *P. lilacinus*. Only *S. epius* and *Cryptolaemus montrouzieri* were found very efficient in clearing the mealybug populations (Mani and Krishnamoorthy, 1990b; Mani, 1995b). *Tetracnemoidea indica* (Ayyar) played a significant role in reducing the mealybug population during June 1997 on pomegranate (Mani and Krishnamoorthy, 2000).

Citrus Mealybug, *Planococcus citri* (Risso) (Hemiptera: Pseudococcidae)

It is also a major pest of pomegranate at times. *Leptomastix dactylopii* and *Coccidoxenoides peregrinus* were found to be effective in suppressing the mealybug populations (Mani and Krishnamoorthy, 2000). *Verticillium lecanii* at 2-6 g/litre of water effectively checked the mixed mealybug population buildup (Kulkarni *et al*., 2007).

Whitefly, *Siphoninus phillyreae* Haliday (Hemiptera: Aleyrodidae)

Siphoninus phillyreae occurs in large numbers on pomegranate at times. It was difficult to control with conventional pesticides due to protective waxy coating over the bodies of the immature stages. The aphilinids *Encarsia azimi* Hayat (=*Encarsia inaron* Walker) and *Encarsia* sp. were able to suppress the whitefly population effectively in India (Mani and Krishnamoorthy, 1995b, 2002b). Later, six natural enemies, viz. aphelenid parasitoid *Encarsia azimi* and predators, viz. *Scymnus* sp., *Cryptolaemus montrouzieri*, *Chrysopa* sp., *Cheilomenus sexmaculata* and *Acletoxenus indicus* Malloch were also reported on *S. phyllireae*. Among them, only *E. azimi* occurred in considerable numbers causing 40% parasitism in October, 1997, which went upto 95% in March-April, 1998 (Mani & Krishnamoorthy, 1999b). Neem seed kernel extract, neem oil, dicofol, sulphur and copper oxychloride proved harmless to the adults of *E. azimi* (Mani & Krishnamoorthy, 1999f).

Hairy Caterpillar, *Trabala vishnou* Lefevere (Lepidoptera: Lasiocampidae)

The larvae of *T. vishnou* were found parasitised by *Blepheripa zebina* Walker and infected by nucleopolyhedrovirus (NPV) together causing upto 100% mortality in October-November 1998 (Mani *et al*., 2002). During 1999, the pupae were also attacked by *Charops brachypterum* (Cameron) and the fungus *Paecilomyces farinosus* (Holmsk.) (Mani *et al*., 2000a; Raghunatha, 1999).

BER

Wax Scale Insect, *Drepanococcus* (=*Ceroplastodes*) *chiton* (Green) (Hemiptera : Coccidae)

Wax scale insect caused severe damage to ber in Karnataka. Six parasitoids, viz. *Anicetus ceylonensis* Howard, *Diversinervus elegans*

Silvester, *Metaphycus helvolus* Compere, *Philosindia longicornis* Noyes and Hayat and *Cephaleta brunniventris* Motschulsky, and four coccinellid predators, *Chilocorus nigritus*, *Cryptolaemus montrouzieri*, *C. sexmaculata* and *Scymnus* sp. were known to attack the scale in the field. Of them, *A. ceylonensis* and *C. brunniventris* were able to check the scale population very effectively in India (Mani, 1995c). Dichlorvos and deltamethrin were found to be relatively safe to *C. brunniventris* (Mani and Krishnamoorthy, 1995e).

Spherical Mealybug, *Nipaecoccus viridis* (Newstead) (Hemiptera: Pseudococcidae)

Eight parasitoids *Anagyrus agraensis* Saraswat, A.sp.nr. *almoriensis* Shaffee *et al.*, *A. dactylopii* (Howard), *A. mirzai* Agarwal, *Alamella flava* Agarwal, *Gyranusoidea flava* Shaffee *et al.*, *Coccophagus* sp., *Chartocerus* sp., and three predators, *Caxoxenus perspicax*, *Triommata coccidivora* and *Spalgis epius* were recorded on *N. viridis* infesting ber. Among them, *Anagyrus* spp. and *S. epius* were widespread and frequently collected. Releases of *C. montrouzieri* supplemented the local natural enemies in controlling the mealybug on ber (Mani, 1993a).

Citrus Mealybug, *Planococcus citri* (Risso) (Hemiptera: Pseudococcidae)

Two parasitoids, *Coccidoxenoides peregrinus* and *Allotropa* sp. and the predator, *Cryptolaemus montrouzieri* were recorded on *P. citri* infesting ber. These three natural enemies along with *L. dactylopii* were able to check the mealybug very effectively (Mani, 1993a). *Leptomastix dacylopii* was also recovered in large numbers from *P.citri* infesting grapes (Mani *et al.*, 2007).

Oriental Mealybug, *Planococcus lilacinus* (Cockerell) (Hemiptera: Pseudococcidae)

The parasitoid *Aprostocetus purpureus* (Campere) and the lycaenid predator *S. epius* were recorded on *P. lilacinus* (Mani, 1993a). There was reduction in numbers of *P. lilacinus* from 45.40/shoot in December 1994 to 0.40/shoot in first week of January 1995 due to the predation of *S. epius* in Karnataka (Mani and Krishnamoorthy, 1996d).

Pink Hibiscus Mealybug, *Maconellicoccus hirsutus* (Green) (Hemiptera:Pseudococcidae)

It was problem in the nursery as well as main field at times. *C. montrouzieri* is useful to control *M. hirsutus* on ber (Mani and Krishnamoorthy, 1997b; Mani *et al.*, 2007a). Generally, mealybugs posed serious threat to ber cultivation in Maharashtra, Andhra Pradesh, Karnataka, Tamil Nadu and Gujarat.

Hairy Caterpillar, *Thiacidas postica* Walker (Lepidoptera: Noctuidae)

It occurred in serious form at times causing extensive defoliation of the plant. Two parasitoids, viz. *Apanteles creatonoti* Vireck and *Charops obtusus* Morley caused upto 40% parasitism in nature on *T. postica* in ber orchards. A nuclear polyhedrosis virus was found infecting *T. postica* upto 20% during September and October in Karnataka (Pradeep and Rajagopal, 1999). Two tachinids, *Chaetexorista* sp., *Palexorista sorbillans* (Wiedemann), *Charops obtusus* and NPV together caused upto 40% mortality of *T. postica* (Mani *et al.*, 2000a).

CUSTARD APPLE

Mealybugs

Six species of mealybugs, viz. *F.virgata*, *M.hirsutus*, *P. citri*, *P. lilacinus*, *P. pacificus* Cox and *N. viridis* were known to occur on custard apple. Parasitism did not exceed more than 5% on these mealybugs in custard apple orchards. But the predators especially *Spalgis epius* and *Cryptolaemus montrouzieri* played an important role in clearing the mealybug populations in Karnataka (Mani and Krishnamoorthy, 1989a, 2007).

Scale Insects

Aphytis sp. was known to parasitise upto 30% on *Aonidiella orientalis* (Newstead) (Mani and Krishnamoorthy, 1996a). A pteromalid, *Scutellista cyanea* Motschulsky was causing upto 42% parasitism on *Parasaissetia nigra* (Mani and Krishnamoorthy, 1999d). The lac insect, *Kerria communis* (Mahdihassan) was found sucking the sap from the shoots of custard apple. Two encyrtid parasitoids *Parechthorodryinus clavicornis* (Campere) and *Erencyrtus dewitzi* (Mahdihassan) were found causing upto 40% parasitism on *K. communis* in the field (Mani and Krishnamoorthy, 1998d).

PAPAYA

Spiralling Whitefly, *Aleurodicus dispersus* Russell (Homoptera: Aleyrodidae)

Leaves and fruits were found severely damaged by the spiralling whitefly in peninsular India (Mani and Krishnamoorthy, 1999c). The nymphs of *A. dispersus* were found attacked by *Encarsia guadeloupae* on papaya around Bangalore causing upto 85% parasitism (Mani and Krishnamoorthy, 2001).

Papaya Mealybug, *Paracoccus marginatus* Williams & Granara de Willink (Hemiptera: Pseudococcidae)

Adult females are light yellowish white. Eggs are greenish yellow. Body content when crushed on white paper body is yellow, when preserved in alcohol it turns black. It attacks papaya, brinjal, tapioca, okra, acalypha, annona, etc. It is native of Mexico first observed in 1955 and later found to spread to West Indies, Hawii, Florida, Guam, Palau, Pouerto Rico, Sri Lanka. It was first reported in Tamil Nadu in July 2007 and spread fast to A.P., Karnataka, Maharashtra, Tamil Nadu and Kerala It might have been introduced from Sri Lanaka into India (Suresh *et al.,* 2010).

Acerophagus papayae (Noyes and Schauff), *Anagyrus loecki* (Noyes and Menezes) and *Pseudleptomastix mexicana* (Noyes and Schauff) have given spectacular control of papaya mealybug, *Paracoccus marginatus* in Puerto Rico, the Dominican Republic, Guam and Sri Lanka. The local predator, *S. epeus* though found in larger numbers was unable to suppress the mealybug effectively. All the above three parasitoids were imported from Puerto Rico in August 2010 and releases were made in peninsular India, This method has been successfully implemented against papaya mealybug in A.P., Karnataka, Maharashtra, Tamil Nadu and Kerala (Rabindra, 2010).

STAR GOOSEBERRY

Whitefly, *Lipaleyrodes euphorbiae* David & Subramaniam (Homoptera: Aleyrodidae)

The whitefly infestation was higher during January to June than July to December in 1992–94. *Eretmocerus* sp., *Aceltoxenus indicus* Malloch, *Triommata coccidivora* (Felt), *Mallada boninensis* and *Cheilomenes sexmaculata* were found associated with the whitefly in

Bangalore. However, only *Eretmocerus* sp. and *A. indicus* were collected in large numbers. The activity of *Eretmocerus* sp. was observed from January to March and the predator *A. indicus* was found feeding on the whitefly nymphs during September-November (Mani and Krishnamoorthy, 1995c).

APPLE

San Jose Scale, *Quadraspidiotus perniciosus* (Comstock) (Diaspididae : Homoptera)

About 50 predators and parasitoids have been reported to feed on San Jose scale. The important parasitoids namely *Prospaltella perniciosi* Tower, *Aphytis* sp. *proclia* group, *A. diaspidus* Howard, *A. mystilaspidus* (Leb), *Chiloneurus* sp.; and predators *Chilocrus bijugus* Mulsant, *C. circumdatus*, *Coccinella septumpuncta* and *Chilocorus renipustulatus* (Linnaeus) provide good control. Some earlier attempts made to introduce potential parasitoids and predators of San Jose scale did not achieve anything due to irrational and indiscriminate use of insecticides as these natural enemies were killed instantly and vanished. However, the Russian, Chinese and Japanese strains of *P. perniciosi* introduced in 1959 in Kotgarh, Kumarsain and Kotkhai areas of Himachal Pradesh, respectively, provided 89, 86.5 and 8% success. Predators *Cryptolaemus montrouziei*, *Chilocorus kuwanae* Silvester and *Cybocephalus gibbulus* Errick were also released in the same areas (Jolly, 1962). Thakur *et al.* (1989) reported *A. proclia* group Azotus, *Encarsia perniciosi* (Tower) and *Teleterebratus perversus* Compere & Zinna and; predators, *Chilocorus bijugus, C. septempunctata, Pharoscymnus flexibilis* (Mulsant) and *Sticholotis marginalis* Kapur as effective natural enemies of San Jose scale in Jammu and Kashmir, while in Himachal Pradesh *A. proclia, C. bijugus, C. septempunctata* and *P. flexibilis* were reported to attack the pest. Masoodi *et al.* (1989) released the populations of *E. perniciosi* and found 9.11 cent parasitism after its establishment in apple orchards in Kashmir. A significant increase in parasitism of San Jose scale by *Aphytis* sp. *proclia* group was also observed. Rawat *et al.* (1988) after the release of *C. bijugus* and *Pharaoscymnus flexibilis* in Thanedhar area of Himachal Pradesh reported 10-39.7% predation after 3 years of release. Rawat and Pawar (1992) after the release of one beetle of *C. bijugus* per 8 trees, recorded 8-17% reduction in scale population and the releases made in summer were far better than those in autumn or rainy season.

Woolly Apple Aphid, *Eriosoma lanigerum* (Hausmann) (Homoptera: Pemphigidae)

In 1937, attempts were made to introduce and establish exotic parasitoid *(Aphelinus mali* Haldmann) from U K in Kullu valley where it established successfully within a short time and provided upto 98% control (Rahman and Khan, 1941). After this success, it was also introduced at Mashobra and Kotgarh in Shimla hills, Chaubattia (Uttaranchal), Coimbatore (Tamil Nadu) and Shilong (Meghalaya) where it provided significant control at initial stages but later on failed to establish due to varying climatic conditions (Cherian, 1942; Lal and Singh, 1947; Chacko, 1967). The indiscriminate use of toxic insecticides and changes in agroclimatic conditions were also found responsible for poor establishment in Kullu valley (Adlakha and Hameed, 1972). Rawat *et al.* (1989) released *A. mali* in Kinnaur district and reported its low effectivity due to poor dispersal compared to other apple-growing regions. In Kullu valley, 1.2-5.7% parasitism was reported in May while 21.4-38.7% in July by Thakur *et al.* (1988).

Various predators which have been reported to exercise fairly good control of woolly apple aphid include *C. septempunctata* (Lal and Singh, 1945, 1947; Singh, 1942), *C. bijugus*, *Ballia ancharis* Mulsant, *Coleophora sanzeti* Mulsant, *Syrphus confractor* Weid, *Coniocompsa indica* Wittycombe, *Ancylonteryx punctata* Hay, *Exochomus uropygialis* Mulsant and *Syrphus (Episyrphus) balteatus* (De Gree) (Rahman and Khan, 1941; Lal and Singh, 1947; Zaku-ur-Rab, 1972; Verma and Singh, 1985). It was reported that the larvae of *S. (Eupeodes) confractor* Wiedemann and *Chrysopa (Brinckochrysa) scelestes* Banks were also effective against woolly apple aphid. Chakrabarti *et al.* (1988) reported *Harmonia dimidiata* (Fabricius) as a voracious predator by consuming as many as 853.7 and 710.7 individuals of woolly apple aphid at 24 and 17.6°C, respectively. The combined effect of *A. mali* and predators *(Chrysopa scelestes* and *Syrphus confractor)* on woolly apple aphid population was assessed to a tune of 23.8-34.8% by Thakur *et al.* (1991) in Kullu valley. Gupta and Bhalla (1993) recorded higher population of *C. septempunctata* and *Hippodamia variegata* (Goeze) feeding on *E. lanigerum* during April- May.

Codling Moth, *Cydia pomonella* (Linnaeus) (Lepidoptera: Olethreutidae)

Releases of egg parasitoid, *Trichagramma embryophagum* (Hartig) and *T. cacoeciae pallidum* Meir, at 20,000 adults per 50 apple trees per week beginning first fortnight of June to August-end has provided upto 50% egg parasitism and 25-30% reduction in fruit infestation (Pawar *et al.,* 1980). Mirids and anthocorid bugs also feed on eggs.

ALMOND

Bhagat *et al.* (1988) reported *Scymnus gracilis* Motschulsky preying on mites (*Tetranychus urticae* and *Panonychus ulmi*) and hemipteran pests (*Brachycaudus helichrysi* and *Quadraspidiotus perniciosus*) in almond orchards in Kashmir. Mani *et al.* (2003) investigated the parasitism of spiralling whitefly, *Aleurodicus dispersus* by the parasitoid, *E. guadeloupae* in Hyderabad. The number of healthy (yellowish white) and parasitized nymphs (black) of *A. dispersus* on leaves of almond were counted to determine the level of parasitism. Parasitism by *E. guadeloupae* ranged from 20 to 30% in different locations. The samples yielded only adults of *E. guadaloupae* and the results indicated that *E. guadeloupae* is established on *A. dispersus* in Hyderabad.

DATEPALM

Swaminathan and Verma (1991) reported *Pharoscymnus horni* as most important predator of date palm scale, *Parlatoria blanchardi* (Targioni Tozzetti) in western Rajasthan, whereas it was observed that all stages of *C. nigritus* feed voraciously on *P. blanchardii and* its peak period of activity was the last week of September. In Gujarat, on *P. blanchardii, Hemiberlesia lataniae* (Signoret) and *Phoenicococcus marlatti* Cockerell, including the predators, *Cybocephalus* sp., *M. sex-maculatus, P. horni, P. flexibilis* and *C. nigrita*, and the parasitoid, *Archenomus* sp. were reported (Murlidharan, 1993)

MULBERRY

Mathur (1980) reported large-scale utilisation of the parasitoid *Cedria paradoxa* Walker for biological control of mulberry defoliator, *Glyphodes pyloalis* (Walker). Three kinds of pathogens of *Spilosoma obliqua* (Walker) were reported in Karnataka (Sharma *et al.,* 1989). *Nosema* sp. was found in 7.57% of 3rd-instar larvae but not in other instars. Viral polyhedral bodies were more frequently encountered in all instars, and bacterial infection varied considerably between instars. Double infection with virus and bacteria ranged from 13.15 to 21.14% and 16.67-31% of larvae appeared to be uninfected.

Kumar and Rajagopal (1990) studied the occurrence and activity of the predator, *Omphra pilosa* Klug in mulberries plantations in Karnataka. Adults and larvae of the predator fed exclusively on all castes of Isoptera. Each adult predator consumed an average of 15 workers and 2.1 alates/day. The consumption rate increased with increasing prey density. Larvae of predator also stored termites in their tunnels.

Gowda *et al.* (1996) observed that larvae of the lycaenid, *S.epius* were found feeding voraciously on all stages *M. hirsutus.* A single mature adult could consume an average of 132 eggs, 260 nymphs or 75 adults, indicating the potential of the predator as a biological control agent for *M. hirsutus.* But in a subsequent study, Gowda *et al.* (1997) identified 9 parasitoids and 8 predators of *M. hirsutus*, in Karnataka, Andhra Pradesh and Tamil Nadu, while Kumar and Chakraborty (1997) regarded *Scymnus nubilus* as a principal indigenous predator of *M. hirsutus.*

The egg parasitoid, *Trichogramma chilonis* was observed as a potential biocontrol agent against Bihar hairy caterpillar, *Spilarctia obliqua* (Walker) n the mulberry crop system (Katiyar *et al.,* 1999). Release of *T. chilonis* in mulberry gardens at 1 lakh adults (5 tricho-cards)/acre at an interval of 3 days, resulted in 33.12% of egg parasitism. It was also observed that *T. chilonis* can destroy an appreciable population of *S.obliqua* eggs upto 5 meter from the release point, although it was able to search and parasitize the host eggs upto a 10 meter radius but wind direction significantly altered the parasitism rates (i.e. it was greater downwind).

PLUM

Mishra and Bhalla (1974) found *Eulecanium tiliae* (Linnaeus) for the first time in India in 1970 on plum in Himachal Pradesh. Upto 73% of the coccids were parasitised by a species of *Coccophagus* near *C. ishii* Compere. *Harmonia eucharis* (Mulsant) were found effective predators of *E. tiliae* (Zaka *et al.,* 1982). Sharma and Dogra (1986) observed second-instar nymphs of species of *Eulecanium* thought to be *E. tiliae,* to be parasitized by a species of *Coccophagus* near *C. ishii* and adult females by *Blastothrix sericea* (Dalman). The rate of parasitism of nymphs and adults was 7.2 and 40.7%, respectively. Upto 13.58% mortality was caused by the fungus *Rhinocladiella* in 2nd-instar nymphs.

Mealy plum aphid, *Hyalopterus arundinis* Fabricius was found attacked by 10 predators and one parasitoid in a plum orchard in Kashmir (Bhagat and Masoodi, 1988). The appearance of the predators was synchronized with the growth in the aphid population in mid-April. The activity of the natural enemies was greatest in the first week of June when the aphid population was as high as 925 nymphs and 21 alatae per 10 leaves. *Coccinella undecimpunctata, Episyrphus balteatus* (De Geer) and *Sphaerophoria indiana* Bigot were the most abundant.

CONCLUSIONS

Outstanding success in the control of crop pests has been achieved with exotic parasitoids and predators. In India too, excellent control of the pests like cottony cushion scale, apple wooly aphid and citrus mealybug has been obtained with introduced natural enemies. So far only a few natural enemies have been introduced into India for the control of fruit crop pests. Success in biological control is commensurate with the number of efforts made. So importation of natural enemies should be the major approach for the suppression of the fruit crop pests. Inundative releases of indigenous natural enemies, where there is a feasibility of large scale multiplication, should be attempted to control important pests of fruit crops. Attack by the hyper parasitoids is a major limiting factor in the efficacy of indigenous natural enemies. Hence it is difficult to achieve spectacular success with the local parasitoids and predators. However, the important natural enemies can be conserved by the selective use of pesticides. All the commonly used pesticides have to be screened for their safety to the key parasitoids and predators and the information should be made available.

Biological control is the backbone of biointensive integrated pest management. Proven or promising natural enemies have to be included in BIPM programmes. The integration of biocontrol agents with other methods of pest control such as the use of selective chemicals, use of resistant cultivars, trap crop, etc. has to be worked out. The use of pheramones to determine the time of release of parasitoids and predators and kairomones to increase the efficiency of parasitoids should receive the attention. Though the natural enemies were recorded on different fruit crop pests, still exploratory surveys in different parts of our country are needed to facilitate the import and export of parasitoids and predators. Research institutes and other government organisations are unable to supply natural enemies to the farmers due to various limitations. At present only few commercial insectaries has been established in India. Large-scale multiplication units should be started in all the states to ensure the timely supply of proven natural enemies to the farmer.

REFERENCES

Adlakha, R.L. and Hameed, S.F. (1972) Wooly apple aphid *Eriosoma langigerum* (Hausmann), in Kullu valley. *Pesticides,* **6**, 13–22.

Agarwal, M.M., Agarwal, S. and Khan, M.A. (1984) Some new chalcid parasites (Hymenoptera: Encyrtidae) recorded from India. *J. Entomol. Res.,* 8, 61–69.

Ali, M.S. (1957) Some biological studies on *Pseudococcus vastator* (Maskell). *Indian J. Entomol.,* **19**, 54–58.

Anonymous (1979) *Final Technical Report.* All India Co-ordinated Research Project on Biological Control of Crop Pests and Weeds. IIHR, Bangalore.

Anonymous (1980) *Annual Report.* All India Co-ordinate Research Project on Biological Control of Crop Pests and Weeds, IIHR, Bangalore.

Anonymous (1986) *Annual Report.* All India Co-ordinate Research Project on Biological Control of Crop Pests and Weeds, IIHR, Bangalore.

Anonymous (1987) *Annual Report of Citrus Blackfly Scheme.* Nagpur University, Nagpur.

Anonymous (1990) *Annual Report.* All India Co-ordinate Research Project on Biological Control of Crop Pests and Weeds, PDBC, Bangalore.

Atwal, A.S. (1962) Insect pests of citrus in the Punjab. I. Biology and control of citrus psylla, *Diaphorina citri* Kuwayana (Hemi., Psyllidae). *Punjab Hort. J.,* **2**, 104–108.

Atwal, A.S. (1964) Insect pests of citrus in Punjab. VI. Biology and control of citrus leaf miner, *Phyllocnistis citrella* Stainton (Lep. Phyllocnistidae). *Punjab Hort. J.,* **4**, 100–103.

Babu, T.R. (1986) Population density of grape mealybug Maconellicoccus hirsutus (Green) and its control by a predator Cryptolaemus montrouzieri Muls. Ph.D. Thesis, APAU, Hyderabad.

Bartlett, B.R. (1978) Margorodidae in introduced parasites and predators of arthropod pests and weeds-A world review. In: C.P. Clausen (ed.), *USDA Agricultural Hand Book*, pp.132–136.

Batra, R.C. and Sandhu, G.S. (1981) Differential population of citrus leafminer and its parasites on some commercial citrus cultivars. *J. Res. Punjab Agric. Uni.,* **18,** 170–178.

Batra, R.C., Sandhu, G.S. and Sohi, A.S. (1987) Outbreak of California red scale on citrus and suppression through coccinellid predators in the Punjab. *Bull. Entomol.,* **20**, 161–162.

Beingolea, G.O. (1969) Biological control of citrus pests in Peru. *Proceedings 1st International Citrus Symposium,* Riverside, Califorina **2**, 827–838.

Bennet, F.D, David Rosen, P., Cochereau, P. and Wood, B.J. (1976) Biological control of pests of tropical fruits and nuts. In: C.B. Huffakar and P.S. Messenger (eds.), *Theory and Practice of Biological Control.* Academic Press, New York, pp. 359–387.

Bennet, F.D. and Hughes, I.W. (1959) Biological control of insect pests in Bermuda. *Bull. Entomol. Res.,* **50**, 423–436.

Bhagat, K.C. and Masoodi, M.A. (1988) Natural enemies of mealy plum aphid, *Hyalopterus arundinis* Fabricius (Aphididae:Homoptera) in Kashmir. *J. Adv. Zool.,* **9**, 145–147.

Bhagat, K.C., Masoodi, M.A. and Koul, V.V. (1988) A note on the occurrence of a coccinellid predator in Kashmir. *Curr. Res. Uni. Agric. Sci.,* Bangalore, **17**, 49–51

Bhattacharya, B. and Dutta, S.K. (1998) Black citrus aphid, *Toxoptera aurantii* Boyer (Aphididae: Homoptera) in Assam. *Insect Environ.,* **3**, 109.

Bhumannavar, B.S. and Viratamath, C.A. (2000) Biology and behaviour of *Euplectrus maternus* Bhatnagar (Hymenoptera, Eulopidae), an ectoparasitoid of *Othreis* spp. (Lepidoptera Noctuidae) from South India. *Pest Manage. Hort. Ecosys.,* **6**, 1-14.

Bhumannavar, B.S. and Viraktamath, C.A. (2001) Seasonal incidence and extent of parasitisation of fruit piercing moths of the genus *Othreis* (Lepidoptera: Noctuidae). *J. Biocontrol,* **15**, 31–38.

Bindra, O.S. (1969) Insect pests of citrus. Review of work done, lacunae in our knowledge and future lines of work. *ICAR, Workshop, Fruit Research,* Punjab Agricultural University, Ludhiana, April 28–30, 1969.

Bindra, O.S. (1970) *Citrus Decline in India-Causes and Control: IX Insects.* Joint Publication FAO/DSU/USAID, Ludhiana, Punjab, India, pp. 64–78.

Bose, S.K. (1953) An entomogenous fungus on citrus whitefly, *Dialeurodes citri* Ashm. *Proceedings 40th Indian Science Congress*, pp. 79.

Chacko, M.J. (1967) Establishment of *Aphelinus mali* (Haldeman) at Shilong, Assam. *Curr. Sci.,* **36**, 298–299.

Chacko, M.J., Bhat, P.K., Rao, L.V.A., Singh, M.B.D., Ramanarayan, E.P. and Sreedharan, K. (1978) The use of ladybird beetle, *Cryptolaemus montrouzieri* for the control of coffee mealybugs. *J Coffee Res.*, **8**, 14–19.

Chakrabarti, S., Ghosh, D. and Debnath, N. (1988) Developmental rate and larval voracity in *Harmonia* (Leis) *dimidiata)* (Col., Coccinellidae), a predator of *Eriosoma lanigerum* (Hom., Aphididae) in western Himalaya. *Acta Entomol. Bohemslov.*, **85**, 335–339.

Cherian, M.C. (1942) Our present position with regard to the control of fruit pests. *Madras Agric. J.,* **30,** 14–17.

Chou, K.C. and Chou, L.Y. (1990) Survey of the natural enemies of mango leafhoppers in Taiwan. *J. Agric. Res. China,* **39**, 70–75.

Chowdhury, S. and Majid, S. (1954) *Hand Book of Plant Protection.* Department of Agriculture, Assam Publication, Shillong.

CIBC (1984) Survey for natural enemies of diaspine scale insects in South India. *Final Technical Report of US, PL 480 Project, Commonwealth Institute of Biological Control,* Indian Station, Bangalore.

Clausen, C.P., Clancy, D.W. and Chock, Q.C. (1965) Biological control of oriental fruit fly *Dacus dorsalis* (Hendel) and other fruit flies in Hawaii. *Tech. Bull. USDA*, **1322**, pp.102.

Compere, H. (1961) The red scale and its natural enemies. *Hilgardia,* **31**, 173–278.

De, K. and Pande,Y. D. (1988) Bionomics and some behavioural aspects of the mango stone weevil, *Steronochetus gravis* (Fabricus) (Coleoptera, curculionidae). *Entomon,***13**, 17–24.

DeBach, P. (1959) New species and strains of *Aphytis* (Hymenoptera: Eulophididae) parasitic on the California red scale, *Aonidiella auranttii* (Mask) in the Orient. *Ann. Entomol. Soc. Am.,* **52,** 354–362.

Deshpande, R.S., Viswanath, S. and Rahman, M.U. (1972) A new entomogenous fungus on banana aphid (*Pentalonia nigronervosa* Coq.) vector of 'Katte' disease of cardamom (*Elettaria cardamomum* Maton). *Mysore J. Agric. Sci.,* **6**, 54.

Dharmaraju, E. and Reddy, C.S. (1975) Insect pest associated with citrus decline in Andhra Pradesh. Andhara Pradesh Agricultural University, Hyderabad, pp. 46–54.

Dhooria, M.S. (1983) An outbreak of citrus mite, *Eutetranychus orientalis* (Klein) in Delhi. *Pesticides,* **17,** 11 & 36.

Dodia, J.F., Yadav, D.N. and Patel, R.C. (1986) Management of fruit sucking moth, *Othreis fullonica* L. (Lepidoptera: Noctuidae) in citrus orchards at Anand (Gujarat). *Gujarat Agric. Uni. Res. J.,* **11**, 72–73.

Etienne, J. (1978) Biological control of Diaphorina citri, vector of citrus greening disease in Reunion. Rapport annuel 1978, Institut de Recherches Agronomiques Tropicales et des cultures Vivrieres, Reunion. pp. 17–19.

George, S.A. and Abdurahman, U.C. (1985) Some aspects of the reproductive biology of *Goniozus* sp. (Hym., Bethylidae) on external parasite of the mango leaf webber, *Lamida moncusalis* Walker. In: K.J. Joseph and U.C. Abdurahiman (eds.), *Advances in Biological Control Research in India,* Proceedings National Seminar on Entomophagus Insects, Calicut University, Kerala, India, pp. 110–115.

Glover, P.M. (1935) An account of the occurrence of *Chrysomphalus aurantii* Mask and *Laccifera lacca* Kerr. on grape fruit in Ranchi district, Chota

Nagpur with a note on the chalcidoid parasites of *Aspidiotus orientalis* News. *J. Bombay Nat. Hist. Soc.,* **38**, 151–153.

Gopalakrishnan, C. and Gangavisalakhi, P.N. (2005) Field efficacy of commercial formulations of *Bacillu thuriengiensis* var *kurstaki* against *Papilio demoleus* L. on citrus. *Entomon,* **30**, 93–96.

Gopalakrishnan, C. and Mani, M. (2000) Epizootics of *Entomophthora aphidis* Hoffman (Zygomycetes: Entomophthorales) on pomegranate aphid, *Aphis punicae* (Homoptera: Aphididae). *National Symposium on Aphidology*, March, 11-13, 2000, Gorakhpur.

Gowda, D.K.S., Manjunath, D., Kumar, P. and Datta, R.K. (1996) *Spalgis epius* Westwood (Lepidoptera: Lycaenidae)- a potential predator of mulberry mealy bug, *Maconellicoccus hirsutus. Insect Environ.,* **2**, 87–88

Gowda, D.K.S., Manjunath, D., Prasad, K.S., Katiyar, R.L., Ram, K. and Datta, R.K. (1997) Natural enemy complex of the pink mealy bug, *Maconellicoccus hirsutus* (Green) in mulberry crop system. *Indian J. Seric.*, **36**, 55–56

Gupta, P.R. and Bhalla, O.P. (1993) Conservation of biotic agents in apple orchards of Himachal Pradesh. *J. Insect Sci.,* **6,** 204–206.

Gupta, R.L. (1954) Life history of *Tonica zizyphi* Stn. the citrus leaf roller. *Indian J. Entomol.,* **16**, 20–23.

Halleppanavar, N.L. (1957) Two new egg parasites of *Virachola isocratus* Fb. *Sci. Cult.,* **23**, 253.

Hayat, M. (1984) *Encarsia bennetti* sp.n. (Hymernoptera: Aphelinidae) from India, associated with the citrus black fly, *Aleurocanthus woglumi* Ashby (Homoptera: Aleyrodidae). *Bull. Entomol. Res.,* **74**, 399–401.

Hussain, M.A. and Khan, A.W. (1945) The citrus aleurodidae (Hamoptera) in the Punjab and their control. *Memoirs Entomol. Soc. India,* **1**, 1–41.

Hussain, M.A. and Mathur, U.B. (1923) A preliminary list of parasites of economic importance bred in the Punjab. *Proceedings 5th Entomological Meeting*, Pusa, pp. 119–121.

Hussain, M.A. and Nath, D. (1927) The citrus psylla *Diaphorina citri* (Psyllidae: Homoptera) *Memoirs Agric. India, Entomol. Ser.,* **10**, 5–27.

Ingole, D.A., Deotale, V.Y., Deotale, R.O. and Ramkumar, J. (2005) Preliminary study on the performance of *Mallada boniensis* (Okamoto) against *Aleurocanthus woglumi* Ashby on citrus. *J. Biocontrol,* **19**, 77–79.

Jalali, S.K. and Singh, S.P. (1990) A new record of *Ooencyrtus papilionis* (Hymenoptera: Encyrtidae) on the eggs of *Papilio demoleus* (Linn.) from India. *J. Biol. Control,* **4**, 59–60.

Johnson, J. (1983) A note on some common aphidivorous insects of Kerala. *Pranikee,* **4**, 415-418.

Jolly, C.L. (1962) Biological control of San Jose scale in Himachal Pradesh-Retrospect and prospect. *Himachal Hortic.* **2&3,** 163–168.

Kakar, D.K. and Sharma, J.P. (1988) Feasibility of *Trichogramma* spp. for the biological suppression of fruit borers of pomegranate. *National Symposium on Tree Protection*, Dr.Yashwant Singh Parmar University of Horticulture and Forestry, Solan, March, 10-11, 1988.

Kamal, M. (1951) Biological control projects in Egypt with a list of introduced parasites and predators. *Bull. Soc. Fouad Entomol.*, **35**, 205–220.

Karuppuchamy, P. (1994) Studies on the management of pests of pomegranate with special reference to fruit borer, *Virachola isocrates* Fabr. Ph. D Thesis, TNAU, Coimbatore, pp. 181.

Katiyar, R.L., Yogananda, M.C., Manjunath, D., Sen, A.K., Ahsan, M.M. and Datta, R.K. (1999) Role of *Trichogramma chilonis* Ishii in the suppression of field populations of the Bihar hairy caterpillar, *Spilarctia obliqua* Walker. *Indian J. Seric.*, **38**, 40–43

Krishnamoorthy, A. (1989a) Host range, development and sex ratio of *Leptomastix dactylopii* How. on different stages of *Planococcus citri*. *J. Biol. Control,* **2**, 8–11

Krishnamoorthy, A. (1989b) Effect of cold storage of on the emergnce and survival of the adult exotic parasitoid *Leptomastix dactylopii* How. (Hym: Encyrtidae). *Entomon,* **14**, 313–318

Krishnamoorthy, A. (1990) Evaluation of permanent establishment of *Leptomastix dactylopii* How. against *Planococcus citri* (Risso) in citrus orchards in India. *Fruits* (*Paris*), **45**, 29–32.

Krishnamoorthy, A. (1993) Bionomics and development of integrated management strategy for California red scale, *Aonidiella aurantii* (Maskell). Ph.D Thesis, University of Agric. Sci., Bangalore, India.

Krishnamoorthy, A. and Mani, M. (1989) *Coccidoxenoides peregrina*: a new parasitoid of *Planococcus citri* in India. *Curr. Sci.*, **58**, 466.

Krishnamoorthy, A. and Mani, M. (1989a) Records of green lacewings preying on mealybugs in India. *Curr. Sci.*, **58**, 155–156.

Krishnamoorthy, A. and Mani, M. (1989b) Recovery of an exotic parasitoid, *Leptomastix dactylopii* How. from *Planococcus citri* (Risso) infesting some horticultural crops. *J. Biol. Control,* **3**, 125.

Krishnamoorthy, A. and Rajagopal, D. (1995) Comparative toxicity of pesticides to natural enemies of California red scale, *Aonidiella aurantii* (Maskell). *Pest Manage. Hort. Ecosyst.*, **1**, 71–79.

Krishnamoorthy, A. and Singh, S.P. (1987) Biological control of citrus mealybug with an introduced parasite, *Leptomastix dactylopii* in India. *Entomophaga*, **32**, 143–148.

Krishnamoorthy, A. and Singh, S.P. (1988) Observational studies on the occurrence of parasitoids of *Papilio* spp. in citrus. *Indian J. Plant Prot.*, **16**, 79–81.

Kulkarni, S.R., Kadam, J.R. and Chavan, A.P. (2007) Preliminary study on the efficacy of *Verticillium lecanii* (Z.) against mixed populations of mealybugs on pomegranate. *Pest Manage. Hort. Ecosyst.*, **13**, 63–64.

Kumar, B., Mistry, N. C., Singh, B. and Gandhi, C. P. (2010) *Indian Horticulture Data Base-2009.* National Horticulture Board, Gurgaon, India.

Kumar, M.V.S. and Chakraborty, N. (1997) Biology of *Scymnus nubilus* Mulsant (Coleoptera: Coccinellidae), a native predator of pink mealy bug, *Maconellicoccus hirsutus* (Green) (Hemiptera: Pseudococcidae). *J. Entomol. Res.,* **21**, 329–334

Kumar, P. and Rajagopal, D. (1990) Carabid beetle, *Omphra pilosa* Klug (Coleoptera: Carabidae) a potential predator on termites. *J. Biol. Control,* **4**, 105–108

Lal, K.B. and Singh, R.N. (1945) Control of wooly aphid by *Coccinella septempunctata* Linn. *Indian Frmg.,* **6**, 24–25.

Lal, K.B. and Singh, R.N. (1947) Seasonal history and field ecology of the wooly aphid in the Kumaun Hills. *Indian J. Agric. Sci.,* **17**, 211–218.

Mani, M. (1987) A review of the pink mealybug *Maconellicoccus hirsutus* (Green). *Insect Sci. Applic.,* **10**, 157–167.

Mani, M. (1988) Bioecology and management of grape mealybug. *Indian Inst. Hort. Res. Tech. Bull.*, **5**, pp. 32.

Mani, M. (1992) Contact toxicity of different pesticides to the encyrtid parasitoids *Aenasius advena* and *Blepyrus insularis* of the striped mealybug *Ferrisia virgata. Trop. Pest Manage.,* **38**, 386–390.

Mani, M. (1993a) Studies on mealybugs and their natural enemies in ber orchards. *J. Biol. Control,* **7**, 75–80.

Mani, M. (1993b) Evaluation of toxicity of different pesticides to the green lacewing, *Mallada boninensis* (Chrysopidae, Neuroptera). *J. Biol. Control,* **7**, 61–68.

Mani, M. (1994a) Effectiveness of the exotic encyrtid parasitoid, *Leptomastix dactylopii* How. in the control of mealybug, *Planococcus citri* (Risso) in guava orchards. *J. Entomol. Res.,* **18**, 351–355.

Mani, M. (1994b) Recovery of the indigenous *Coccidoxenoides peregrinus* and the exotic *Leptomastix dactylopii* in lemon and acid lime orchards. *Biocontrol Sci. Technol.*, **4**, 49–52.

Mani, M. (1994c) Mealybug problem in Maharashtra on grapevine. *Prog. Hort.*, **26**, 106–108.

Mani, M. (1995a) Comparative development, progeny production and sex ratio of the exotic parasitoid *Leptomastix dactylopii* Howard (Hymenoptera: Encyrtidae) on *Planococcus lilacinus* and *Planococcus citri* (Homoptera: Pseudococcidae). *Entomon,* **20**, 23–26.

Mani, M. (1995b) Studies on the natural enemies of oriental mealybug *Planococcus lilacinus* (Ckll.) (Homoptera: Pseudococcidae) in India. *J. Entomol. Res.,* **19**, 61–70.

Mani, M. (1995c) Studies on the natural enemies of the wax scale, *Drepanococcus chiton* (Green) on ber and guava. *Entomon,* **20**, 55–58.

Mani, M. (1996) Safety of neem and other plant products to parasitoids and predators of citrus insect pests. *J. Insect Sci.,* **9**, 89–90.

Mani, M. (2001) Biological control of fruit crops. In: P. Parvatha Reddy, Abraham Verghese and N.K. Krishna Kumar (eds.), *Integrated Pest Management in Horticultural Ecosystems*, Capital Publishsing Company, New Delhi, pp. 93–107.

Mani, M. (2008) Polyhouse efficacy of *Cryptolaemus montrouzieri* Mulsant for the suppression of *Planococcus citri* (Risso) on grapes and *Ferrisia virgata* (Cockerell) on guava. *J. Insect Sci.*, **21**, 202–204.

Mani, M. and Krishnamoorthy, A. (1989a) Occurrence of mealybugs and their natural enemies on custard apple around Bangalore, South India. *J. Biol. Control,* **3**, 77.

Mani, M. and Krishnamoorthy, A. (1989b) Feeding potential and development of green lacewing, *Mallada boninensis* (Okamoto) on the grape mealybug *Maconellicoccus hirsustus* (Green). *Entomon,* **14**, 19–20.

Mani, M. and Krishnamoorthy, A. (1989c) Record of *Blepyrus insularis* (Cam.) on *Ferrisia virgata* (Ckll.) in India. *Curr. Sci.,* **58**, 644.

Mani, M. and Krishnamoorthy, A. (1990a) Further studies on the safety of pesticides against *Anagyrus dactylopii* (Hym., Encyrtidae) and *Cryptolaemus montrouzieri* (Coleoptera : Coccinellidae). *J. Biol. Control,* **4**, 80–85.

Mani, M. and Krishnamoorthy, A. (1990b) Outbreak of mealybugs and record of their natural enemies on pomegranate. *J. Biol. Control,* **4**, 61–62.

Mani, M. and Krishnamoorthy, A. (1990c) Predation of *Mallada boninensis* on *Ferrisia virgata, Planococcus citri* and *P. lilacinus. J. Biol. Control,* **4**, 122–123.

Mani, M. and Krishnamoorthy, A. (1990d) Evaluation of the exotic predator *Cryptolaemus montrouzieri* Muls. (Coccinellidae, Coleoptera) in the suppression of green shield scale *Chloropulvinaria psidii* (Maskell) (Coccidae, Homoptera) on guava. *Entomon,* **15**, 45–48.

Mani, M. and Krishnamoorthy, A. (1990e) Natural suppression of mealybugs in guava orchards. *Entomon,* **15**, 245–247.

Mani, M. and Krishnamoorthy, A. (1991) Breeding of *Blepyrus insularis* (Hym., Encyrtidae) on *Ferrisia virgata* (Hemip., Pseudococcidae). *Entomon,* **16**, 275–277.

Mani, M. and Krishnamoorthy, A. (1992) Influence of constant temperatures on the developmental rate, progeny production, sex ratio and adult longevity of the grape mealybug parasitoid, *Anagyrus dactylopii* (How.) (Hymenoptera: Encyrtidae). *Insect Sci. Applic.*, **13**, 697–703.

Mani, M. and Krishnamoorthy, A. (1993a) Status of mealybug species and their biological control in vineyards - A global view. In: *Proceedings of the International Symposium on Recent Advances in Viticulture and Oenology*, February 14-17, 1992, Hyderabad, India, pp. 330–342.

Mani, M. and Krishnamoorthy, A. (1993b) Bionomics and management of the striped mealybug, *Ferrisia virgata* (Ckll.) - A world review. *Agric. Rev.*, **14**, 22–43.

Mani, M. and Krishnamoorthy, A. (1995a) Influence of different stages of oriental mealybug, *Planococcus lilacinus* (Ckll.) on the development, progeny production and sex ratio of the parasitoid, *Tetracnemoidea indica* Ayyar. *J. Insect Sci.*, **8**, 192–193.

Mani, M. and Krishnamoorthy, A. (1995b) Natural enemies of *Siphoninus phillyreae* (Homoptera: Aleyrodidae) and *Aphis punicae* (Homoptera: Aphididae) on pomegranate. *Entomon,* **20**, 31–34.

Mani, M. and Krishnamoorthy, A. (1995d) Outbreak of *Aphis punicae* and record of its natural enemies on pomegranate in India. *Indian J. Plant Prot.,* **23**, 89–90.

Mani, M. and Krishnamoorthy, A. (1995e) Toxicity of pesticides to the wax scale parasitoid, *Cephaleta brunniventris. Indian J. Plant Prot.*, **22**, 108–109.

Mani, M. and Krishnamoorthy, A. (1996a) *Aonidiella orientalis* (Newstead) (Diaspididae, Homoptera) and its natural enemies found on sapota, ber, custard apple and banana. *Entomon,* **21**, 273–274.

Mani, M. and Krishnamoorthy, A. (1996b) Discovery of the coccinellid predator, *Chilocorus circumdatus* on the green scale, *Coccus viridis. Entomon,* **21**, 295–296.

Mani, M. and Krishnamoorthy, A. (1996c) First record of *Ooenecyrtus papilionis* on pomegranate butterfly, *Deudorix isocrates* Fabr. (Lycaenidae, Lepidoptera). *Entomon,* **21**, 275–276.

Mani, M. and Krishnamoorthy, A. (1996d) Biological suppression of oriental mealybug *Planococcus lilacinus* on ber. *Pest Manage. Hort. Ecosyst.,* **2**, 49–50.

Mani, M. and Krishnamoorthy, A. (1996e) Response of the encyrtid parasitoid *Tetracnemoidea indica* of the oriental mealybug, *Planococcus lilacinus* to different pesticides. *Indian J. Plant Prot.*, **24**, 80–85

Mani, M. and Krishnamoorthy, A. (1996f) Spiralling whitefly and its natural enemies on guava in Karnataka. *Insect Environ.*, **2**, 12–13.

Mani, M. and Krishnamoorthy, A. (1996g) Record of two insect pests and their natural enemies on phalsa. *J. Insect Sci.,* **9**, 182.

Mani, M. and Krishnamoorthy, A. (1996h) Impact of natural enemies on the oriental yellow scale, *Aonidiella orientalis* (Newstead) on banana. In:

Conference for Banana Production and Utilisation in 21st Century. 24-25 September 1996, Trichy, India.

Mani, M. and Krishnamoorthy, A. (1997a) Safety of plant products and conventional pesticides to *Distatrix papilionis* (Vireck) (Hymenoptera, Braconidae), a parasitoid of citrus butterfly. *National Symposium on Citriculture,* held at Nagpur, November 17-19, 1997.

Mani, M. and Krishnamoorthy, A. (1997b) Australian ladybird beetle, *Cryptolaemus montrouzieri. Madras Agric. J.,* **84**, 237–249.

Mani, M. and Krishnamoorthy, A. (1997c) Suppression of *Planococcus citri* (Risso) on sapota. *Pest Manange. Hort. Ecosyst.,* **3**, 45–47.

Mani, M. and Krishnamoorthy, A. (1997d) Effects of different pesticides upon the wax scale parasitoid, *Anicetus ceylonensis* (Hym.: Encyrtidae). *Int. J. Pest Manage.,* **43**, 123–126.

Mani, M. and Krishnamoorthy, A. (1997e) Suppression of spherical mealybug, *Nipaecoccus viridis* (Newstead) (Homoptera : Pseudococcidae) on jackfruit. *Entomon,* **22**, 161–163.

Mani, M. and Krishnamoorthy, A. (1998a) Studies on the natural enemies of mango scale *Aulacaspis tubercularis* Newstead (Diaspididae, Homoptera). *National Symposium on Mango Production and Export,* June, 25–27, 1998, Lucknow.

Mani, M. and Krishnamoorthy, A. (1998b) Biological control studies on the mango green shield scale, *Chloropulvinaria polygonata* (Ckll.) (Homoptera: Coccidae) in India. *Entomon,* **23**, 105–110.

Mani, M. and Krishnamoorthy, A. (1998c) Suppression of the soft green scale *Coccus viridis* (Green) on acid lime in India. *Proceedings First National Symposium in Pest Management in Horticultural Crops,* Bangalore, 1998, pp. 210–212

Mani, M. and Krishnamoorthy, A. (1998d) Kerria *communis* (Mahdn.) and its natural enemies on custard apple in Karnataka. *Insect Environ.,* **4**, 38–39.

Mani, M. and Krishnamoorthy, A. (1999a) Predatory potential and development of Australian ladybird beetle, *Cryptolaemus montrouzieri* Muls. on the spiralling whitefly Russell. *Entomon,* **24**, 166–171.

Mani, M. and Krishnamoorthy, A. (1999b) Outbreak of ash whitefly *Siphoninus phyllireae* and record of its natural enemies on pomegranate. *Entomon,* **24**, 197–198.

Mani, M. and Krishnamoorthy, A. (1999c) Natural enemies and host plants of spiralling whitefly, *Aleurodicus dispersus* Russell (Homoptera, Aleyrodidae) in Bangalore, Karnataka. *Entomon,* **24**, 75–80.

Mani, M. and Krishnamoorthy, A. (1999d) The nigra scale *Parasaissetia nigra* and its natural enemy on custard apple in Karnataka. *Insect Environ.,* **5,** 20.

Mani, M. and Krishnamoorthy, A. (1999e) First record of *Trichogramma chilotraeae* Nagaraja and Nagarkatti on pomegranate butterfly, *Deudorix isocrates* (Fabr.), *Entomon,* **24**, 399–400.

Mani, M. and Krishnamoorthy, A. (1999f) Selective toxicity of botanicals and conventional pesticides to the pomegranate whitefly parasitoid *Encarsia azimi* (Hayat). Paper presented in the *International Seminar on Integrated Pest Mangement,* Hyderabad, October 8-9, 1999.

Mani, M. and Krishnamoorthy, A. (1999g) Biological control of spherical mealybug, *Nipaecoccus viridis* on acid lime in India. *International Seminar on Citriculture,* Nagpur, 23–26, November 1999.

Mani, M. and Krishnamoorthy, A. (1999h) Development and predatory potential of the green lacewing, *Mallada astur* (Banks) (Neuroptera: Chrysopidae) on the spiralling whitefly, *Aleurodicus dispersus* Russell (Homoptera: Aleyrodidae). *J. Biol. Control,* **13**, 45–49.

Mani, M. and Krishnamoorthy, A. (2000) Biological suppression of the mealybugs *Planococcus citri* (Risso) and *Planococcus lilacinus* (Ckll.) on pomegranate in India. *Indian J. Plant Prot.,* **28**, 187–189.

Mani, M. and Krishnamoorthy, A. (2001) Managing scale insects on fruit crops. *Indian Hort.,* **46**, 4–7.

Mani, M. and Krishnamoorthy, A. (2002) Selectivity of different pesticides to citrus mealybug parasitoid *Coccidoxynoides peregrinus* (Timberlake). *J. Insect Sci.,* **15**, 49–52

Mani, M. and Krishnamoorthy, A. (2002a) Biological suppression of spherical mealybug *Nipaecoccus viridis* (Newstead) (Hemiptera, Pseudococcidae) on acid lime in India. *Entomon,* **27**, 423–424

Mani, M. and Krishnamoorthy, A. (2002b) Role of *Encarsia azimi* Hayat (Aphelinidae, Hymenoptera) in regulating ash whitefly *Siphoninus phyllireae* Haliday (Aleyrodidae, Hemiptera) on pomegranate in India. *Indian J. Plant Prot.,* **30**, 144–148.

Mani, M. and Krishnamoorthy, A. (2005) Impact of the coccinellid predator *Cheilomenes sexmaculata* (Fab.) on black citrus aphid *Toxoptera aurantii* (Boyer) infesting acid lime. *Entomon,* **30**, 91–92

Mani, M. and Krishnamoorthy, A. (2007) Recent trends in the biological suppression of guava pests in India. *Acta Hort.,***735**, 469–481.

Mani, M. and Krishnamoorthy, A. (2007a) Biointensive management of the exotic spiralling whitefly *Aleurodicus dispersus* Russell in India. *J. Insect Sci.,* **20**, 129–142.

Mani, M. and Krishnamoorthy, A. (2008a) Biological suppression of the mealybugs *Planococcus citri* (Risso), *Ferrisia virgata* (Cockerell) and *Nipaecoccus viridis* (Newstead) on pummelo with *Cryptolaemus montrouzieri* Mulsant in India. *J. Biol. Control,* **22**, 169–172.

Mani, M. and Krishnamoorthy, A. (2008b) Biological control of *Planococcus citri* (Risso) on grapevine with *Cryptolaemus montrouzieri* in India. *Indian J. Plant Prot.,* **36**, 125–127.

Mani, M. and Thontadarya, T.S. (1987a) Development and feeding potential of the coccinellid, *Cryptolaemus montrouzieri* Muls. on grape mealybug, *Maconellicoccus hirsutus* (Green). *J. Biol. Control,* **1**, 19–22.

Mani, M. and Thontadarya, T.S. (1987b) Population dynamics of the mealybug *Maconellicoccus hirsutus* (Green) and its natural enemies in the grapevine ecosystem. *J. Biol. Control,* **1**, 93–97.

Mani, M. and Thontadarya, T.S. (1987c) Record of mealybug species on grapevine in Karnataka. *Curr. Sci.,* **56**, 1192.

Mani, M. and Thontadarya, T.S. (1987d) Biological studies on the grape mealybug predator *Scymnus coccivora* (Ayyar). *J. Biol. Control,* **1**, 89–92.

Mani, M. and Thontadarya, T.S. (1988a) Response of *Cryptolaemus montrouzieri* Muls. (Coccinellidae, Coleoptera) to commonly used pesticides in vineyards. *J. Biol. Control,* **2**, 17–20.

Mani, M. and Thontadarya, T.S. (1988b) Field evaluation of *Cryptolaemus montrouzieri* Muls. in the suppression of *Maconellicoccus hirsutus* (Green) on grapevine. *J. Biol. Control,* **2**: 14–16.

Mani, M. and Thontadarya, T.S. (1988c) Biology of the grape mealybug parasitoid *Anagyrus dactylopii* (How.) (Encyrtidae, Hymenoptera). *Entomon,* **13**, 211–214.

Mani, M. and Thontadarya, T.S. (1989) Development of the *Anagyrus dactylopii* (How.) on the grape mealybug *Maconellicoccus hirsutus* (Green). *Entomon,* **14**, 49–51.

Mani, M. and Thontadarya, T.S. (1991) Effect of soil application of systemic granular insecticides on the population of grape mealybug and its natural enemies. *Pestology,* **15,** 24–30.

Mani, M., Dinesh, M.S. and Krishnamoorthy, A. (2000a) Presence of *Encarsia* spp on the spiraling whitefly, *Aleurodicus dispersus* Russell in peninsular India. *Insect Environ.,* **6**, 100.

Mani, M., Dinesh, M.S. and Krishnamoorthy, A. (2003) Establishment of *Encarsia guadeloupae* Viggiani on the spiralling whitefly at Hyderabad. *Insect Environ.,* **9**, 34

Mani, M., Krishnamoorthy, A and Gangavisalakshi, P.N. (2007) Natural parasitisation by the exotic parasitoid, *Leptomastix dactylopii* Howard on *Planococcus citri* on custard apple. *J. Biol. Control,* **21**, 157–158.

Mani, M., Krishnamoorthy, A. and Dinesh, M.S. (2001) Biological control studies on the spiralling whitefly *Aleurodicus dispersus* Russel (Hemiptera: Aleyrodidae). *Entomon* (Special Issue), **26**, 62–69

Mani, M., Krishnamoorthy, A. and Dinesh, M.S. (2004a) Biocontrol of exotic spiralling whitefly. *Indian Hort.,* **49**, 6–7.

Mani, M., Krishnamoorthy, A. and Pattar, G.L. (2007a) Suppression of pink mealybug, *Maconellicoccus hirsutus* on ber, *Zizhyphus mauritiana*. *Indian J. Agric. Sci.*, **77**, 135.

Mani, M., Krishnamoorthy, A. and Singh, S.P. (1990) The impact of the predator *Cryptolaemus montrouzieri* Mulsant on pesticide resistant populations of the striped mealybug, *Ferrisia virgata* (Ckll.) on guava in India. *Insect Sci. Applic.*, **11**, 167–170.

Mani, M., Krishnamoorthy, A. and Srinivasa Rao, M. (2002) Selective toxicity of different pesticides to citrus mealybug prasitoid, *Coccidoxenoides peregrinus* (Timberlake). *J. Insect Sci.*,**15**, 49–52

Mani, M., Sushil, S. N. and Krishnamoorthy, A. (1995) Influence of some selective pesticides on the longevity and progeny production of *Leptomastix dactylopii* How., a parasitoid of citrus mealybug *Planococcus citri* (Risso). *Pest Manage. Hort. Ecosyst.*, **1**, 81–86

Mani, M., Thondarya, T.S. and Singh, S.P. (1987) Record of natural enemies of the grape mealybug *Maconellicoccus hirsutus* (Green). *Curr. Sci.*, **56**, 624–625.

Mani, M., Krishnamoorthy, A., Venugopalan, R. and Pattar, G.L. (2004b) Biological control of exotic spiralling whitefly *Aleurodicus dispersus* Russell on guava by *Encarsia haitiensis* Dozier and *Encarsia guadeloupae* Viggiani in India. *Pest Manage. Hort. Ecosyst.*,**10**, 29–39.

Manjunath, T.M. (1985a) Maconellicoccus hirsutus on grapevine. *FAO Plant Prot. Bull.*, **33**, 74.

Manjunath, T.M. (1985b) Leptomastix dactylopii in India. *Biocontrol News Inform.*, **6**, 297.

Manjunath, T.M. (1986) Recent outbreaks of mealybugs and their biological control. In: S. Jayaraj (ed.) *Resurgence of Sucking Pests-Proceedings National Symposium*, Tamil Nadu Agricultural University, Coimbatore, pp. 249–253.

Masoodi, M.A., Trali, A.R. and Bhat, A.M. (1989) Establishment of *Aphytis sp. Proclia* group on San Jose scale in Kashmir. *Indian J. Plant Prot.*, **17**, 71–73.

Mathai, S., Kumar, G.P., Kumari, T.N. and Hebsybai. (1990) *Mucor circinelloides* Van Tieghem, an entomogenous fungus of the banana aphid *Pentalonia nigronervosa* Coq. (Aphididae: Hemiptera). *J. Biol. Control*, **4**, 124.

Mathur, R.N. (1980) Biology of the mulberry defoliator-*Glyphodes pyloalis* (Wlk.) (Lepidoptera: Pyralidae). *Indian Forest Bull.*, No. **273**, 6 pp.

Mishra, R.C. and Bhalla, O.P. (1974) Occurrence of *Eulecanium sp. tiliae* (L.) (Homoptera: Coccidae) on plum and apple in Himachal Pradesh. *Curr. Sci.*, **43**, 428.

Misra, S.C. and Pandey, N.D. (1965) Some observations on the biology of *Papilio demoleus* Linn. (Papilionidae: Lepidoptera). *Labdev J. Sci. Technol.,* **3**, 142–143.

Modawal, C.N. (1940) The tachnid *Ercia nymphalidephaga* Bar. parasite of *Pappilio demoleus* Linn. *Indian J. Entomol.,* **2**, 242–243.

Murali Baskaran, R.K., Mahendran, G. and Suresh, K. (2007a) Field efficacy of *Scymnus coccivora* Ramkrishna Ayyar for the management of lime mealybug, *Planococcus citri* (Risso). *J. Entomol. Res.,* **31**, 291–294.

Murali Baskaran, R.K., Mahendran, G. and Suresh, K. (2007b) Field evaluation of *Scymnus coccivora* Ayyar for the management of guava mealybug *Maconellicoccus hirsutus* (Green). *J. Entomol. Res.,* **31**, 137–140.

Murlidharan, C.M. (1993) Scale insects of date palm (Phoenix dactylifera) and their natural enemies in the date groves of Kachchh (Gujarat). *Plant Prot. Bull.,* **45**, 31–33

Nagalingam, B. (1988) Studies on the biology of *Psorosticha zizyphi* Stn. on acid lime (*Citrus aurantifolia). Mysore J. Agric. Sci.,* **22**, 48–52.

Nagaraja, H. and Gupta, A. (2007) A new species of *Trichogramma* (Hymenoptera: Trichogrammitidae) parasitic on eggs of pomegranate fruit borer *Deudorix isocrates* (F.) (Lepidoptera: Lycaenidae). *J. Biocontrol,* **21**, 291–293.

Nagarkatti, S., Singh, S.P., Jayanth, K.P. and Bhumannavar, B.S. (1992) Introduction and establishment of *Leptomastix dactylopii* How. against *Planococcus* spp. in India. *Indian J. Plant Prot.,* **19**, 102–104.

Nair, M.R.G.K. (1975) *Insects and Mites of Crops in India.* Indian Council of Agricultural Research, New Delhi.

Narasimham, A.U. and Chacko, M.J. (1988) *Rastrococcus* spp. (Hemiptera, Pseudococcidae) and their natural enemies in India as potential biocontrol agents for *R. invadens* Williams. *Bull. Entomol. Res.*, **78**, 703–708.

Narayanan, E.S. (1941) Notes on some Indian parasitic hymenoptera with a description of new cynipid. *Indian J. Entomol.,* **3**, 59–63.

Narayanan, E.S. (1954) Seasonal pests of crops. The anar butterfly, *Virachola isocrates* Fab. *Indian Farming,* **4**, 8–9.

Narayanan, E.S. and Batra, H.N. (1960) *Fruit Flies and Their Control.* Indian Council of Agricultural Research, New Delhi.

Narayanan, E.S., Subba Rao, B.R. and Kaur, R.B. (1957) Notes on the biology of the parasites of the leaf miner of citrus plants. *Proceedings 44th Indian Sci. Cong.,* **111**, 396–397.

Narayanan, K. and Gopalakrishnan, C. (1987) Occurrence of cytoplastmic polyhedrosis virus in citrus leaf caterpillar, *Papilio demoleus* L. (Papilionidae: Lepidoptera). *J. Biol. Control,* **1**, 73–74.

Narayanan, K. and Jayaraj, S. (1974) *Bacillus thuringiensis* endotoxin on haemolymphacation levels in the citrus leaf caterpillar, *Papilio demoleus. J. Invertebr. Pathol.,* **23**, 125–126.

Narendran, T.C. (1989) *Oriental Chalcidoidea.* Zoological monograph, Department of Zoology, University of Calicut.

Nayar, K.K., Ananthakrishnan, T.N. and David, B.V. (1976) *General and Applied Entomology.* Tata McGrow Hill Publishing Company Limited, New Delhi.

Odak, S.C. and Dhamdhere, S.V. (1970) New record of hymenopterus parasites of *Sylepta lunalis* Guien. (Lepidoptera, Pyralidae). *Indian J. Entomol.,* **32**, 395.

Pawar, A.D.; Tuhan, N.C.; Balasubramanian, S. and Parry, M. (1980) Biological control of codling moth in Ladakh. *Indian J. Plant Prot.,* **8**, 189–191.

PDBC (2002) Annual Report of Project Directorate of Biological Control, Bangalore.

PDBC (2003) Annual Report of Project Directorate of Biological Control, Bangalore.

PDBC (2004) Annual Report of Project Directorate of Biological Control, Bangalore.

Pradeep, S., and Rajagopal, D. (1999) NPV on ber hairy caterpillar *Thiacidas postica* Walker (Nootuidae:Lepidoptera). *Insect Environ.,* **4**, 120.

Pruthi, H.S. and Mani, M.S. (1945) *Our Knowledge of the Insects and Mites Pests of Citrus in India and their Control.* Monograph No. **16**, Indian Council of Agricultural Research, New Delhi.

Puttarudriah, M. and Channabasavanna, G.P. (1957) Beneficial Coccinellidae of Mysore II. *Mysore Agric. J.,* **32**, 4–19.

Rabindra, R.J. (2010) NBAII pioneers successful classical biological control of papaya mealybug. *NBAII News.,* **11**, 1

Raghunatha, R. (1999) Studies on Insect Pests of Pomegranate with Special Reference to the Fruit Borer, *Deudorix isocrates* (Fab.) (Lepidoptera, Lycaenidae) and the Leaf Eating Caterpillar, *Trabala vishnou* (Lefevere) (Lepidoptera, Lasiocampidae). M.Sc. (Ag) thesis, UAS, Bangalore.

Rahman, K.A. (1939) Short notes and exhibits- success of *Aphelimus mali* in the Punjab Hills. *Indian J. Entomol.,* **1**, 97.

Rahman, K.A. and Khan, A.W. (1941) Biology and control of woolly aphids, *Eriosoma lanigerum* (Hausmann). *Indian J. Agric. Sci.,* **11**, 265–278.

Rahman, K.A. and Latiff, A.M. (1944) Description, bionomics and control of the giant mealybug *Drosicha stebbingi* (Homoptera, Coccidae). *Bull, Entomol. Res.,* **35**, 197–209.

Ramachandra Rao, Y. (1930) The mango hoppers in South India. *Agric. J. India,* **25**, 17–25.

Ramachandra Rao, Y. and Cherian, M.C. (1944) The fluted scale, *Icerya purchasi* Mask, as a pest of wattle in South India and its control by biological method. *Madras Agric. J.*, **32**, 20.

Ramachandran, T.S. (1954) Insect pests of orange in Madras state. *Bull. Madras Agric.*, 16 pp.

Ramamurthy, T.V. (1982) Biology and feeding potential of the coccinellid predator, *Scymnus graiosus* Wiese. M.Sc. Thesis, APAU, Hyderabad.

Rao, N.G.V, Lokhande, R.K. and Mohan, P. (1996) Parasitoid of citrus aleyrodids: their behaviour and seasonal occurrence. *Adv. Plant Sci.*, **9**, 147–149.

Rao, V.P. (1969) India as a source of natural enemies of pests of citrus. *Proceedings 1st Int. Citrus Symp.*, **2**, 785–792.

Rao, V.P. and Kamath, S.M. (1966) Record of cottony cushion scale *Icerya purchasi* Maskell in the cannore district of Kerala and its control. *Indian J. Entomol.*, **7**, 552.

Rao, V.P., Ghani, M.A., Sankaran, T. and Mathur, K.C. (1971) *A Review of the Biological Control of Insects and other Pests in South East Asia and the Pacific Region*. Commonwealth Institute of Biological Control Technical Communication, No. **6**.

Rawat, R.R. and Jakhmola, S.S. (1970) Bionomics of the mango coccid, *Rastrococcus iceryoides* Green (Homoptera: Coccidae). *Indian J. Agric. Sci.*, **40**, 140.

Rawat, U.S. and Pawar, A.D. (1991) Field recovery of *Trichogramma chilonis* Ishii (Hymenoptera : Trichogrammatidae) from *Deudorix epijarbas* Moore (Lepidoptera : Lycaenidae) in Himachal Pradesh, India. *Entomon*, **16**, 49–52.

Rawat, U.S. and Pawar, A.D. (1992) Biocontrol of San Jose scale, *Quadraspidiotus perniciosus* (Comstock) by predatory beetle, *Chilocorus bijugus* Mulsant in Himachal Pradesh. *Plant Prot. Bull.*, **44**, 7–10.

Rawat, U.S., Tkahur, J.N. and Pawar, A.D. (1988) Introduction and establishment of *Chilocorus bijugus* Mulsant and *Pharaoscymnus flexibillis* Mulsant, predatory beetles of San Jose scale at Thanedhar area in Himachal Pradesh. *Curr. Sci.*, **57**, 1250–1251.

Saini, B.S. (1964) Bionomics and Control of *Dialeurodes citri* (Riley and Howard) (Homoptera: Aleurodidae). M.Sc. (Agric.) thesis, PAU, Ludhiana.

Sankaran, T. and Mjeni, A.M. (1989) Recent studies on the mango leaf-gall midge *Procontarinia matteiana* Kieffer and Cecconi (Diptera: Cecidomyiidae) and its parasites in India and on prospects for biological control of the pest in Oman. *Acta-Hort.*, **231**, 587–592.

Santharam, G., Regupathy, A., Easwaramoorthy, S. and Jayaraj, S. (1978) Effectiveness of nuclear polyhedrosis virus against field populations of

Spodoptera litura Fab. on banana, *Musa paradisiaca*. *Indian J. Agric. Sci.*, **48**, 676–678.

Satpute, P.P.; Taley, Y.M.; Nimbalkar, S.A. and Kadu, N.R. (1986) Natural enemies of citrus black fly, *Aleurocanthus woglumi* Ashby, in Nagpur region. *Bull. Entomol.*, **27**, 128–131.

Satyanarayana, J., Murthy, M.S. and Srinivasa, N. (1991) Impact of pesticides on grapevine mealybug predator, *Cryptolaemus montrouzieri* Mulsant 2. Residual toxicity and safety of pesticidal sprays to predatory adults. *Indian J. Entomol.*, **53**, 581–586.

Shafee, S.A.; Alam, S.M. and Agarwal, M.M. (1975) *Indian Insect Type*. Aligarh Muslim University Publication (Zoology Series), Aligarh.

Sharga, U.S. (1948) Biological control of some hemipterous insect pests of crops in India *Curr. Sci.*, **17**, 302–303.

Sharma, D.R. (2007) Ecofriendly approaches for the management of fruit pests. In: D. Prasad (ed.), *Sustainable Pest Management*, Daya Publishers, New Delhi, pp. 137–155.

Sharma, D.R. and Batra, R.C. (1989) Parasitoids of citrus leaf roller, *Psorosticha zizyphi* Stain. and influence of abiotic factors on parasitisation. *J. Insect Sci.*, **2**, 162–164.

Sharma, D.R., Kumar, R. and Batra, R.C. (2003) Influence of abiotic and biotic factors on the spatial distribution of *Dialeurodes citri* (Ashmead). *Ann. Plant Prot. Sci.*, **11**, 264–268.

Sharma, J.P. and Dogra, G.S. 1986. Studies on the control of the plum scale *Eulecanium sp? tiliae* (L.) (Homoptera: Coccidae) through chemical and natural elements. *Indian J. Entomol.*,**48**, 258–263.

Sharma, S.D., Balavenkatasubbaiah, M. and Baig, M. A. (1989) Report of the presence of various pathogenic microbes in a wild population of Bihar hairy caterpillar, *Diacrisia obliqua*. *Curr. Sci.*, **58**, 762–763

Shevale, B.S. and Khaire, V.M. (1999) Seasonal abundance of pomegranate butterfly, *Deudorix isocrates* Fabricius. *Entomon,* **24**, 27–31.

Shirke, M.S. and Salunkhe, G.N. (1996) Relative residual toxicity of some insecticides to *Cryptolaemus montrouzieri* Muls., a predator of mealybugs. *J. Maharashtra Agric. Univ.*, **21**, 370–371.

Shukla, R.P., Tandon, P.L. and Singh, S.J. (1984) *Baculovirus*, a new pathogon of mango nut weevil, *Sternochetus mangiferae* (Fabr.) Coleoptera: Curculionidae). *Curr. Sci.*, **53**, 593–594.

Singh, R.N. (1942) Control of woolly aphids (*Eriosoma lanigerum* Hsm.) by spraying and other methods. *Indian J. Agric. Sci.*, **12**, 588–602.

Singh, S.P. (1978) Propagation of coccinellid beetle for the biological control of citrus and coffee mealybugs. *Proc. Scientific Conference on Coffee and Cardamom*, CPA, Madikeri, Karnataka.

Singh, S.P. (1980) Biological control of insect pests of citrus. In: Proceedings of the Third Workshop of AICRP on Biological Control of Insect Pests and Weeds, Bangalore.

Singh, S.P. (1987) Biological control of insect pests of horticulture, In: K.J. Joseph and U.C. Abdurahiman (eds.), *Advances in Biological Control Research in India, Proc. Nat. Sem. Entomophagus Insects,* Calicut University, Kerala, India, pp. 221-231.

Singh, S.P. (1991a) Potential of two commercial formulations of *Bacillus thuringiensis* against citrus butterfly caterpillar, *Papilio demoleus* Linnaeus (Lepidoptera: Papilionidae). *Indian J. Hort.,* **48**, 135–138.

Singh, S.P. (1991b) Natural enemies of *Papilio* spp. *Indian J. Hort.,* **48**, 237–242.

Singh, S.P. (1993a) Biological control of insect pests. In: K.L. Chadha and O.P. Pareek (eds.), *Advances in Horticulture,* Vol. 3, *Fruit Crops.* Malhotra Publishing House, New Delhi, pp. 1591–1615.

Singh, S.P. (1995) Use of entomopathogenic fungus, *Verticillium lecanii* (Zimm.) Viegas, insecticides and their combination for the control of green scale, *Coccus viridis* (Green). *Indian J. Hort.,* **52**, 259–266.

Singh, S.P. (2003) Biological suppression of pests of citrus. In: Shyam Singh, V.J. Shivankar, A.K. Srivastava and I.P. Singh (eds.), *Advances in Citriculture,* Jagmander Book Agency, Karol Bagh, New Delhi, pp. 503–523.

Singh, S.P., Rao, N.S. and Henneberry, T.J. (1993) *Leaf Hopper and Their Natural Enemies.* Technical Bulletin No. 6 of Project Directorate of Biological control, Bangalore.

Singh, Y. (1967) New record of hymenopterus parasites of *Papilo demoleus* Linnaeus, a serious pests of citrus spp. *Bull. Entomol.,* **8**, 163.

Smith, H.D. (1950) A survey of the parasites and predators of citrus blackfly (*Aleurocanthus woglumi* Ash.) in India. *Plant Prot. Bull.,* **2**, 8–10.

Sreedevi, K., Verghese, A., Vasudev, V. and Sudhadevi, K. (2006) Species composition and abundence of predators with reference to the pomegranate aphid *Aphis punicae* P. *Pest Manage. Hort. Ecosyst.,* **12**, 93–97.

Srinivasan, T.R. (1987) Studies on Biology and Control of Grape Mealybug, Maconellicoccus hirsutus (Green) (Homoptera: Pseudococcidae). M.Sc. (Ag.) Thesis, TNAU, Madurai.

Srivastava, R.P. (1978) Studies on the feasibility of the use of microbial pesticides for the integrated control of *Papilio demoleus. XXth International Horticultural Congress,* Sydney, Australia, 15-23 August, 1978.

Subba Rao, B.R., Sangwar, H.S., Abbasi, O.A., Singh, Y. and Ksheer Sagar, A.M. (1965) New records of hymenopterous parasites of *Nipaecoccus vastator* (Maskell) (Homoptera : Coccidae), a serious pest of *citrus* spp from India. *Indian J. Entomol.,* **27**, 109–110.

Subramanian, V.K. (1954) Control of the fluted scale in peninsular India. *Indian J. Entomol.,* **16**, 391–415.

Subramanian, V.K. (1955) Control of the flutted scale in peninsular India. *Indian J. Entomol.,* **17**, 103–120.

Sundara Babu, P.C. (1969) Note on the possible use of *Bacillus thuringiensis* Berliner for the control of certain leppidopterous pests. *Madras Agric. J.,* **56**, 824–826.

Suresh, S., Jothimani, R., Sivasubramaniam, P., Karuppuchamy. P., Samiayyapan, R. and Jonanthan, E.R. (2010) Invasive mealybugs of Tamil Nadu and their management. *Karnataka J. Agric. Sci.,* **23**, 6–9.

Swaminathan, R. and Verma, S.K. (1991) Studies on the incidence of date palm scale, *Parlatoria blanchardi* (Targ.) in western Rajasthan. *Entomon,* **16**, 217–221

Swezey, O.H. (1940) A survey of the insect pests of cultivated plants in Guam. *Hawaiian Planters' Record,* **44**, 151–182.

Tandon, P.L. (1985) Spatial Distribution Sampling Technique and Population Dynamics of Citrus Green Scale *Coccus viridis* (Green) on mandarin. Ph.D. Thesis, UAS, Bangalore.

Tandon, P.L. and Lal, B. (1978) The mango coccid, *Rastrococcus iceryoides* Green (Homoptera, Coccidae) and its natural enemies. *Curr. Sci.,* **47**, 46–48.

Tandon, P.L. and Srivastava, R.P. (1980) New records of parasites and predators of important insect pests of mango. *Entomon,* **5**, 243–244.

Thakur, J.N., Pawar, A.D. and Rawat, A.D. (1988) Observations on the correlation between population density of apple woolly aphid and its natural enemies and their effectiveness in Kullu Vellay. *Plant Prot. Bull.,* **40**, 13–15.

Thakur, J.N., Pawar, A.D. and Rawat, A.D. (1989) Investigations on the occurrence of natural enemies of San Jose scale, *Quadraspidiotus permiciosus* Comstock (Hemiptera: Coccidae) in Jammu & Kashmir and Himachal Pradesh. *Entomon,* **14**, 143–146.

Thakur, J.N., Pawar, A.D. and Rawat, U.S. (1993) Observations on the post release impact of *Trichogramma chilonis* Ishii (Hymenoptera: Trichogrammatidae), an egg parasitoid of anar butterfly on fruit infestation in Kullu valley (H.P.). *Plant Prot. Bull.,* **45**, 54–55

Thakur, J.N., Rawat, U.S. and Pawar, A.D. (1991) Successful introduction of *Trichogramma chilonis* Ishii, an egg parasitoid of anar fruit butterfly, *Deudorix epijarbas* Moore in Kullu valley. *J. Insect Sci.,* **4**, 163–164.

Thakur, J.N., Verma, O.P., Singh, J.P. and Pawar, A.D. (1995) Incidence of *Deudorix epijarbas* Moore (Lepidoptera: Lycaenidae) and its parasitoids on pomegranate in Jammu region. *J. Biol. Control,* **9**, 116–118.

Thompson, W.R. (1946) *Catalogue of Parasites and Predators of Insect Pests. Sec. 1, Parasite Catalogue.* Part 8, Commwealth Agriculural Bureaux, Ottawa.

Tirumala Rao, V. and David, L.A. (1958) The biological control of a coccid pest in South India by the use of beetle *Cryptolaemus montrouzieri* Muls. *Indian J. Agric. Sci.,* **28**, 545–552.

Ven den Bosch, R. and Messenger, P.S. (1973) *Biological Control.* Intext, New York.

Verghese A., Kamala Jayanthi, P.D. and Anil Kumar, H.R. 2001. Foraging strategies of predator *Cryptolaemus montrouzieri* Mulsant in relation to prey *Planococcus lilacinus* (Cockrell). *Pest Manage. Hort. Ecosyst.* **7**, 45–53.

Verma, G.C., Bindra, O.S., Singh, G., Sukhija, H.S. and Sandhu, M.S. (1976) A note on *Rodolia fumida* Muls. a predator of mango mealybug, *Drosicha stebbingi* (Green) *Indian J. Hort.,* **33**, 243.

Verma, K.L. and Singh, M. (1985) Some biological agents associated with the suppression in apple pests in Himachal Pradesh. *Pesticides,* **19**, 54.

Volkl, W., Stechmann, D.H. and Stary, P. (1990) Suitability of five species of aphididae (Hymenoptera) for the biological control of the banana aphid *Pentalonia nigro nervosa* Coq. (Homoptera, Aphididae) in the South Pacific. *Trop. Pest Manage.,* **36**, 249–257.

Wadi, S.R. and Batra, H.N. (1964) Pests of tropical and subtropical fruit trees. In: N.C. Pant (ed.), *Entomology in India,* Entomological Society of India, New Delhi, pp. 464–466.

Wagle, P.V. (1934) The mango hoppers and their control in Karkan Bombay Presidency. *Agric. Livestock India,* **4**, 176–188.

Webber, P.W. (1951) Recent liberations of beneficial insects. *Proc. Hawaii, Entomol. Soc.,* **14**, 327–330.

Woglum, R.S. (1913) Report of a trip to India and the orient in search of the natural enemies of the citrus whitefly. *Bull. US Agric. Buruex Entomol.,* **120**, 58.

Zaka, Ur Rab, M., Rab, M. Zaka Ur and Ur Rab M. Zaka (1982) Predation of the plum scale, *Eulecanium coryli* (L.) (Homoptera: Coccidae), by *Ballia bayaderae* Mulsant (Coleoptera: Coccinellidae), in Kashmir. *J. Bombay Nat. Hist. Soc.,* **79**, 436.

Zaku-ur-Rab, M. (1972) Record of *Syrphus confrater* Wiedmann as predaceous on woolly aphid, *Eriosoma lanigerum* (Hausm). *Indian J. Entomol.,* **34**, 348–349.

10

Biopesticides in Tobacco Pest Management: Problems and Prospects

U. SREEDHAR, S. GUNNESWARA RAO AND V. KRISHNAMURTHY

Central Tobacco Research Institute, Rajahmundry- 533105, India
E-mail: usreedharctri@yahoo.com

INTRODUCTION

Tobacco is one of the important commercial crops in India and plays a prominent role in country's agricultural and industrial economy. India is the 2nd largest producer and exporter of tobacco in the world. Tobacco is cultivated in an area of 4.5 lakh ha in the country with an annual production of 750 million kg. About thirteen types of tobaccos are grown in the country. It is the life line for 6 million farmers and supports 30 million workers. The annual revenue earnings from tobacco are more than ₹14,000 crores and foreign exchange earnings are about ₹4000 crores (CTRI, 2011). Of several constraints that affect production insect pests are the most damaging and cause significant loss of yield and quality. About nineteen insect pests attack this crop in the nursery, planted crop and in storage. Out of these, about six are major pests and economically important in the crop. The major insect pests are tobacco caterpillar, *Spodoptera litura* (nursery and main field), stem borer, *Scrobipalpa heliopa* (nursery and main field), whitefly, *Bemisia tabaci* (nursery and main field) vector of leaf curl virus disease, tobacco aphid, *Myzus nicotianae*, bud worm and seed capsule borer *Helicoverpa armigera* (main field) are the major pests, where as ground beetles, *Mesomorphus villiger* and beet army worm, *Spodoptera exigua* are minor and sporadic in nature

(Chari *et al.*, 1999). These pests cause substantial damage to yield and quality of tobacco under congenial conditions. Sole dependence on synthetic insecticides for pest control is beset with adverse effects such as irreversible damage to the public health, development of insecticide resistance in insect pests, annulations of beneficial organisms, insect pest resurgence, secondary pest outbreaks and environmental pollution. The problem of pesticide residues is one of the important limiting factors in the export of tobacco. As a result of importer and consumer concerns, pesticide residues have become a major problem for tobacco industry (Sreedhar, 2011a,b). In this context use of biopesticides may help to overcome the insect pest problems as well as the problems due to indiscriminate use of pesticides and the subsequent residues in cured tobaccos. Biopesticides usually are inherently less harmful than conventional pesticides. They generally affect only the target pest and closely related organisms, in contrast to broad-spectrum conventional pesticides that may affect organisms as different as birds, insects, reptiles and mammals. Biopesticides often are effective in very small quantities and decompose quickly, thereby resulting in lower exposures and largely avoiding the pollution problems caused by conventional pesticides. When used as a component of integrated pest management (IPM) programs, biopesticides can greatly decrease the use of conventional pesticides which helps in production of residue-free tobacco for export purpose.

BIOPESTICIDES IN TOBACCO

Neem, *Azadirachta indica*

The antifeedant and antiovipositional properties of neem (*Azadirachta indica*) have been thoroughly investigated against *S. litura* at Central Tobacco Research Institute, Rajahmundry. Neem seed kernel suspension (NSKS) has been found effective as an antifeedant and reduced feeding in all the five instars of *S. litura* on tobacco (Joshi and Ramaprasad, 1975). The tobacco nurseries could be effectively protected from *S. litura* damage by using 2% NSKS (Joshi *et al.*, 1978a). A spray schedule of NSKS, first spray with 1% NSKS on 3 week old seedlings and subsequently 2% NSKS sprays at 10 days interval thrice was recommended for tobacco nurseries (Joshi *et al.*, 1978b). The ovipositional repellent property of neem against *S. litura* moths on tobacco was reported by Joshi and Sitaramaiah (1979). NSKS 2% had no adverse effect on *Trichogramma remus*, the egg parasitoid of *S. litura* and NSKS could be integrated with *T. remus* for control of the pest in tobacco nurseries (Joshi *et al.*, 1986). Compatibility of NSKS and the parasitoids, *T. remus* and *Apanteles africanus* on *S. litura* was reported in tobacco (Chari *et al.*, 1996a,b).

Field experiments with NSKS against *S. litura* on FCV tobacco planted crop revealed that NSKS 0.5% gave significant protection and was economical as compared to the recommended insecticides (Joshi *et al.*, 1984). Studies also proved that NSKS 2% in nursery and 0.5% in planted Virginia tobacco gave equal protection compared to that of synthetic insecticides (Ramaprasad *et al.*, 1986, 1987). Integrated schedule for management of *S. litura* in *bidi* tobacco nurseries comprising NSKS, NPV, diflubenzuron and castor as trap crop was found to be at par with chlorpyriphos spray (Chari *et al.*, 1985). Integrated management of *S. litura* in tobacco nurseries with NSKS, egg parasitoid, *T. remus*, larval parasitoid *A. africanus*, NPV and castor trap crop was developed and demonstrated in farmers nurseries and planted crop with a favourable cost benefit ratio (Chari *et al.*, 1995,1996). Based on the work at the institute several neem formulations were developed, evaluated and recommended for use against *S. litura* (Chari *et al.*, 1990; Nagarajan *et al.*, 1998). NSKS 0.5% was found to enhance the trapping of budworm, *H. armigera* on *Tagetes erecta* and *Nicotiana rustica* trap crops in Virginia tobacco (Sreedhar, 2012). Several factors appear to limit the success of botanicals, most notably regulatory barriers and the availability of competing products (newer synthetics, fermentation products, microbials) that are cost-effective and relatively safe compared with their predecessors. In the context of tobacco pest management, botanical insecticides are best suited for use in production of pesticide residue free tobacco for export purpose.

Nuclear Polyhedrosis Virus (NPV)

NPV, a key mortality factor of *S. litura* and *H. armigera,* effectively protects tobacco seedlings and planted crop from these pests. The effectiveness of *Sl*NPV at the rate of 250 LE/ha (1.5×10^{12} PIB/ha) in tobacco nurseries and in tobacco field crop was demonstrated (Ramakrishnan *et al.*, 1981). The seasonal natural incidence of *Sl*NPV was examined in tobacco nurseries and it was found that, the least incidence was observed at 28 to 32°C and highest at 31.2 to 37°C. Residues of *Sl*NPV were not detected in cured leaf in bioassay tests (CTRI, 1987). Water with pH ranging from 6–8 for SlNPV and 7–8 for *Bacillus thuringiensis kurstaki* (*Btk*) were suitable for spraying.

Similarly the EC range of water can be 0.5–6.0 dS/m for both the bio pesticides (Gunneswara Rao *et al.*, 2009).

The effectiveness of *Sl*NPV was potentiated with biochemical adjuvant like boric acid 0.25% and tannic acid 0.25% in a schedule of 4-5 fortnightly sprays in tobacco nursery (Rao *et al.*, 1987). *Sl*NPV retained its efficacy in aqueous solutions ranging from pH 6–8 and electrical conductivity at 0.5–6 dS/m in tobacco and in the field crop at 30–40 days after transplanting. Addition of adjuvant like laundry blue, crude sugar, liquid soap, tannic acid and boric acid had little effect in improving the efficacy over crude preparation of *Sl*NPV (Gunneswara Rao *et al.*, 2009). The flavonoids of tender mango and cashew leaves were found to prevent inactivation of *Sl*NPV on tobacco leaf for three days in sunlight and three hours on exposure to germicidal UV light at 500 and 100 ppm respectively in pot culture experiments (Prasad, 2007).

Integrated management modules involving NPV were found highly effective in the management of *S. litura* and *H. armigera* in tobacco (Sitaramaiah *et al.*, 2001, 2002; Sreedhar *et al.*, 2005). Integrating natural enemies and biopesticides (*Sl*NPV & NSKS) along with castor trap crop, the cost benefit ratios were superior over chemical control (Joshi *et al.*, 1986; Chari *et al.*, 1996a,b). The main advantage of BIPM was trapping of more population of *S. litura* by castor (3- 4 times more population of egg masses and larvae were recovered on castor compared to tobacco). The repellence of NSKS also contributed to effective diversion of ovipostition of *S. litura* to castor rather to tobacco. Application of *Sl* NPV early in the season effectively suppressed *S. litura* population.

In a biopesticides based protection strategy, four sequential sprays of bipesticides viz., *Sl* NPV 250 LE (1.5 X 10^{12} PIB/ha) and *Btk* at the rate of 1 kg/ha gave significantly better protection against *S. litura* in tobacco nursery compared to 4 to 7 sprays of insecticides. The cost:benefit ratio (mean) worked out were 1:1.35 with a net profit of ₹57,366/ha for biopesticides as against 1:1.23 with a net profit of ₹37,313 for chemical control (Farmers' method) (Gunneswara Rao and Sitaramaiah, 2001). During 1994-98 under the mission mode project (DBT) on development, production and demonstration of biological control agents under integrated pest management, a total area of 764.5 ha of tobacco (both nursery and field crop) was covered which included BIPM of *S. litura* (656. 50/ha) and *H. armigera* (108.0 ha). Adopters of BIPM in nursery and field crop realised net profits ranging from ₹ 21,444 to ₹ 3,07,073 and ₹ 27,498–84,938 where as non-BIPM implementers obtained (–) ₹ 577–2,93,693 and ₹ 16,461–80,670, respectively (Sitaramaiah *et al.*, 2002a). Farmer field schools and demonstration of BIPM had a lasting impact on tobacco farmers of East and West Godavari districts of Andhra Pradesh where *S. litura* and *H. armigera* are serious pests. The problems of chemical residues in exportable tobaccos on the other hand also motivated the farmers for adoption of biological control.

Experiments on management of tobacco caterpillar, *S. litura* in tobacco nurseries showed that foliar sprays of NSKS 2% and Sl NPV at the rate of 1.5×10^{12} PIBs/ha or *Btk* at the rate of 1 kg/ha or *N. rileyi* at the rate of 1×10^{8} spores/ml in sequence was promising and comparable to chlorpyrifos 0.05% as shown by reduction in seedlings damaged (Sreedhar *et al.*, 2004). Similarly, sequential spray of profenofos 0.15% and leufenuron 0.005% or Ha NPV 1.5×10^{12} PIB's/ha or *Btk* 1.5 kg/ha were highly effective and equal to two sprays of insecticide against budworm, *H. armigera* in FCV tobacco as shown by less number of leaves damaged, less leaf area damaged and more tobacco yields (Sreedhar *et al.,* 2008). Use of insecticides and biopesticides in sequence is a viable approach as it has several desirable features like selectivity, minimal environmental impact and possibly prolongs the efficacy of biopesticides by avoiding or delaying the development of resistance in target pest.

In the absence of cheaper cell culture techniques, dependence on host insect for multiplication poses a major hurdle. Reasons for limited expansion of baculovirus use include narrow host range, slow killing speed, technical and economical difficulties for *in vitro* commercial production, timing of application based on frequent host population monitoring, variability of field efficacy due to climatic conditions, and farmers' attitudes toward pest control, which have been based on application of fast-killing chemical insecticides. Availability of quality NPVs when required, farmer education regarding use of biological insecticides and their characteristics are considered the major actions necessary for increased use of baculoviruses. Strategies to counteract some of the limitations of baculoviruses, especially their slow killing activity, low persistency are to be addressed on priority. These include the use of chemical or biological substances added to virus formulations and genetic engineering of the viruses themselves fpr increased virulence and to express insect toxins or hormones. Such strategies can enhance viral activity and increase speed of kill as well as reduce larval feeding activity. The efficacy of the presently available semi purified liquid preparations of NPVs of *S. litura* and *H. armigera* can be improved by identifying virulent strains in the country through RFLP (Restricted Fragment Length Polymorphism) mapping and investigating the cross infectivity of NPV of a number of lepidopteran insects may be useful to identify a type of NPV like *Autographa calfornica* NPV with wider host range. Research efforts on formulation technology for biopesticides like NPVs should be intensified to increase the storage life, field efficacy and effective delivery of the virus. Live host insect or insect tissue are essential for multiplication of virus. Research programmes on development of high efficiency rearing systems need to be continued. Commercial scale NPV production becomes viable only when the production process gets automated, host tissue culture production becomes cheaper and formulation and quality control technologies are improved. Until then only small scale production is possible.

Bacteria

Bacillus thuringiensis var. *kurstaki* (*Btk*) was found effective and recommended against *S. litura* and *H. armigera* in tobacco. Commercial formulations Dipel (*B.t. kurstaki* H D 1 strain) and Delfin (*B.t.kurstaki* serotype a, b) at the rate of 1 kg/ha was found effective against *S.litura* in tobacco nurseries and *H. armigera* in tobacco field crop (Gunneswararao *et al.*, 1996; Sreedhar *et al.*, 2003). *Btk* (serotype 3a, 3b 30000IU/mg) was found highly effective at the rate of 2 kg/ha against *S. litura* in tobacco nurseries (Ramaprasad *et al.*, 2000) and *H. armigera* in the field crop (Sreedhar *et al.*, 2003). BARC *Bt* strain (dust formulation) at the rate of 7.5×10^{12} spores/ha was found effective against tobacco capsule borer *H. armigera* on the seed crop of tobacco (Gunneswara Rao and Sitaramaiah, 2001). Though the efficacy of *Bt* insecticides is highest among all the biopesticides available in the market, they did not become very popular among the farmers due to the following constraints (i) the crop coverage has to be better than a contact insecticide as the *Bt* toxin has to be ingested by the pest to be effective, (ii) as the toxin crystals are particulate, they are effective only against surface feeders and not effective against tissue borers as compared to soluble material, (iii) they degrade rapidly and require frequent repeated applications and thus costly and (iv) as the shelf life is less and the markets are uncertain for biopesticides the availability in time is a problem. Due to the increasing concern about pesticide residues in tobacco *Bt* based biopesticides have the potential to emerge as a strong component of integrated pest management in tobacco.

Fungi

Fungal pathogens viz., *Nomuraea rileyi, Beauveria bassiana* and *Verticillium lecani* can be deployed in irrigated and monsoon tobacco growing tracts. Sequential sprays of NSKS 2% and *N. releyi* at the rate of 1×10^8 spores/ml, NSKS 2% and *B. bassiana* at the rate of 1×10^9 spores/ml were found effective against tobacco caterpillar, *S. litrua* in tobacco nurseries (Sreedhar *et al.*, 2004. *Nomuraea rileyi* at the rate of 10^{12} spores/ha was found effective against *H. armigera* in irrigated tobacco crop (Gunneswara Rao *et al.*, 2004). *N. rileyi* at the rate of 10^{13} spores/ha and NSKS 2% were effective against *S. exigua* and comparable with chlorpyriphos in tobacco nurseries (PDBC, 2007). It was found that foliar spray of *Verticillium lecani* at the rate of 3×10^{11} CFUs/ha on FCV tobacco bordered by pearl millet as barrier crop reduced tobacco aphid infestation by 76.39% over control with no adverse effect on the natural enemy population and recorded better yield (CTRI, 2010).

Though mycopesticides are valuable tools in pest management programmes, the full potential of entomogenous fungi as mycopesticides for practical insect pest management has not been realized. Problems with development of effective products include specificity, mass production, stability in storage and efficacy under adverse natural conditions. Success will depend upon formulation and delivery systems as they are crucial besides the biological characteristics of the fungus. The use of genetically improved fungus to overcome the problems offers good scope of the mycopesticides in integrated management. EPN (entomopathogenic nematode) talc based formulation was found effective at the rate of 1×10^{12} IJ/m^2 in succession with *Sl*NPV at the rate of 1.5×10^{12} PIB/ha. EPN individually was also effective at the rate of 4×10^5 IJ/m^2 in containing the damage by *S. litura* by 24 and 22%, respectively over control (Sitaramaiah *et al.*, 2002b). Among the biopesticides though neem, NPVs, *Bt* and myco-insecticides are in the forefront, the importance of other microorganism like entomophathogenic nematodes, protozoa and mycoplasma like organism need not be underscored.

BIOPESTICIDES IN TOBACCO IPM

Mass production of NPVs under Mission Mode Project of DBT resulted in area wide demonstrations covering an area of over 4000 ha in coastal A.P. where *S. litura* and *H. armigera* are key pests in crops like cotton, groundnut, chill and pigeonpea along with tobacco (Gunneswara Rao *et al.*, 2002; Sreedhar *et al.*, 2002). The concept of

IPM with biopesticides was also popularized as IPM kit. BIPM kit was developed under DBT (R&D) project. The kit given as incentive to BIPM farmers provided information on how farmers can prepare *Sl*NPV and neem seed kernel suspension on their own and use it along with sowing of trap crop castor and preparing a seed stock of their own, so that farmers could learn and participate in BIPM process, develop new skills and support scientists in continuously evolving new technologies for the field problems. The BIPM kit contained: 500 ml NPV solution (250 LE), neem seed kernel powder (100 g), castor seed (100 g) and 'How to use' pamphlets in vernacular language -Telugu (CTRI, 1995). The impact of popularization efforts has yielded dividends and due to the meager availability of biocontrol agents in the market, several farmers (particularly tobacco farmers) visit CTRI in the crop season for supply of biocontrol agents especially for *Sl*NPV and *Ha*NPV which were conclusively demonstrated as effective agents along with other potential natural enemies against these pests in tobacco, chillies, groundnut, cotton and pigeonpea under the mission mode project of DBT (Sitaramaiah *et al.*, 2002a).

As tobacco and its products are exported, consumed directly and its alternate uses like seed oil, proteins, organic acids and pharmaceuticals are on the anvil, the chemical pesticide use will be highly restricted and the paradigm shift would be towards organic farming, plant resistance and biological control where biopesticides will play a major role. We have to mould our strategies in the above perspective to produce quality tobacco.

PARASITOIDS AND PREDATORS

Spodoptera litura

Eighteen species of parasitoids were recorded from tobacco caterpillar, *S. litura,* which included one trichogrammatid and one scelionid egg parasitoids, two tachinids, two sarcophagids, eight braconids, three chalcids and one ichneumonid parasitoid (Joshi *et al.*, 1979a; Rao *et al.*, 1981). Many of insect predators of *S. litura* are polyhphagous, have long life cycles and their feeding potential and fecundities were limited including that of spiders (Table 10.1). Generally outstanding results were reported by rather highly prey/ host specific natural enemies. However these polyphagous predators may serve in community balance unlike highly prey specific species or forms which serve in single species control and balance. *Nesidiocoris tennuis*, a mirid predator preferred older eggs and neonate larvae of *S.*

litura and *S. exigua* in laboratory conditions. The activity of the predator in field was high at flowering stage of tobacco (Prasad *et al.*, 2010). On rustica tobacco the predator preferred eggs of *H. armigera* (Prasara and Yadav, 1995) *Trichogramma chilonis* effectively parasitised eggs of *S. litura* when offered without scaly covering usually found on egg masses. The parasitoid failed to paras-itise multilayered egg mass of *S. litura* (Joshi *et al.*, 1980). *Telenomus nawaii* Ashmead was also recorded on *S. litura* with 10-15% parasitisation of egg masses recovered from tobacco nurseries and vegetable gardens (Rao *et al.*, 1981).

Among the braconid group of hymenopteran parasitoids *Gylpapanteles africanus* a solitary early larval parasitoid of *S. litura* was predominant both in nursery and field crop environment. *Nesolynx phaesoma* Waterston, *Eurytoma* sp. and *Aphanogmus fijiensis* Ferrier were recorded as hyperparasitoids from the of cocoon *G. africanus* collected from East Godavari district of A.P. (Rao *et al.*, 1990). In nursery experiments the parasitoid when released @ 100 adults (1: 4 female-male ratio) at 500 neonate larval density on a tobacco seedling bed area of 2 sq m, could reduce damage to seedling up to 50% compared to untreated check. The parasitisation rate was up to 37% (PDBC, 1994).

Cotesia ruficrus a gregarious species was also recovered from field collections. The gregarious parasitoid was less useful for mass rearing because it required late 3rd or early 4th instar larva for oviposition. At this stage *S. litura* larvae disperse and farmers start chemical control at this stage, as the damage to leaves gets visible. Larvae parasitised by gregarious *Cotesia* species continued to feed till 5th instar stage and caused economic damage. In contrast the solitary *C. africanus* completes gestation in host before it reaches third instar stage.

Among the tachinid group of dipteran parasitoids, *Peribaea orbata* is predominant on *S. litura*. Parasitisation rates of *P. orbata* were up to 24% in collections from abandoned tobacco nurseries where as that of larval-pupal parasitoids ranged from 5–10%. Tachinids preferred larvae from third instar onwards. *P. orbata* is a gregarious koinobiont endoparasitoid of *S. litura*. The life cycle of the parasitoid on nursery seedlings of tobacco revealed that from a single parasitised host (*S. litura*) larva yielded 5.50 ± 2.54 puparia. The parasitoid life cycle form egg to adult was completed in 15 days. Even after parasitisation late instars continued to feed and completed metamorphosis. The activity of this tachinid was greatly influenced by physical environment. Warm weather and bright sunlight were essential for activity of the

Table 10.1. Natural enemies of leaf-eating caterpillar, *Spodoptera litura*

Natural enemy	Stage of host	Distribution	Remarks: Potential of parasitism and predation (Reference)
Parasitoids			
Trichogramma chilonis (= *confusum*) (Hymenoptera: Trichogrammatidae)	Egg	East Godavari, Guntur (A.P.)	Joshi *et al.* (1979a)
Telenomus nawaii (Hymenoptera: Scelionidae)	Egg	East Godavari, Guntur, (A.P.)	Rao *et al.* (1990)
Chelonus formosanus (Hymenoptera: Braconidae)	Egg larva	East Godavari, Guntur, Prakasam, Karimnagar (A.P.)	Joshi *et al.* (1979a)
Peribaea orbata (Diptera: Tachinidae)	Larva	East Godavari, Guntur Prakasam (A.P.)	10–56% (IInd to Vth instars), 3–17% (Natural field occurrence). Rao *et al.* (1989)
Apanteles sp. (Octonarius Group) (Hymenoptera: Braconidae)	Early larval instars	East Godavari, Guntur Prakasam (A.P.)	Joshi *et al.* (1979a) Rao *et al.* (1981)
Cotesia (*Glypapanteles*) *africanus* (Hymenoptera: Braconidae)	Early larval instars	East Godavari, Guntur Prakasam (A.P.)	80% lab and 33% field Parasitism. PDBC (1994), Sitaramaiah *et al.* (2002a)
Cotesia (*Apanteles*) *ruficrus* (Hymenoptera: Braconidae)	Early larval instars	East Godavari, Guntur Prakasam (A.P.)	Sitaramaiah *et al.* (2002a)
Charops obtusus (Hymenoptera: Ichneumonidae)	Early larval instars	East Godavari, Guntur Prakasam (A.P.)	5–10% Rao *et al.* (1981)
Zele chloropthalma (Hymenoptera: Braconidae)	Larva	East Godavari (A.P.)	Rao *et al.* (1990)
Chelonus carbonator (Hymenoptera: Braconidae)	Egg larva	East Godavari (A.P.)	Rao *et al.* 1981
Blepherella setigera (Diptera: Tachinidae)	Pupa	East Godavari, Guntur (A.P.)	Joshi *et al.* (1979a)
Chelonus heliopae (Hymenoptera: Braconidae)	Egg Larva	Gujarat	Patel *et al.* (1971a)
Chelonus formosanus (Hymenoptera: Braconidae)	Egg larva	East Godavari (A.P.), Gujarat	Patel *et al.* (1971b) Sitaramaiah *et al.* (2002a)

Natural enemy	Stage of host	Distribution	Remarks: Potential of parasitism and predation (Reference)
Sarcophaga dux (Diptera: Sarcophagidae)	Pupa	East Godavari, Guntur (A.P.)	Joshi *et al.* (1979a)
Sarcophaga albicesps (Diptera: Sarcophagidae)	Pupa	Prakasam (A.P.)	Rao *et al.* (1990)
Brachymeria lasus (Hymenoptera: Chalcididae)	Pupa	East Godavari, Guntur (A.P.)	Chari *et al.* (1989)
Lasiochalcidia erythropoda (Hymenoptera: Chalcididae)	Pupa	East Godavari (A.P.)	Chari *et al.* (1989)
Hybothoracini sp. (Hymenoptera: Chalcididae)	Pupa	East Godavari (A.P.)	Chari *et al.* (1989)
Predators			
Ropalidia sp. (Hymenoptera: Vespidae)	Larva	East Godavari, Prakasam, Guntur, Karimnagar (A.P.)	Chari *et al.*, (1989)
Sphedanolestes sp. (Hymenoptera: Reduviidae)	Larva	Karimnagar (A.P.)	Chari *et al.* (1989)
Polistes stigma (Hymenoptera: Vespidae)	Larva	Prakasam (A.P.)	Chari *et al.* (1989)
Dieuches sp. (Hymenoptera: Lygaeidae)	Larva	Karimnagar (A.P.)	*Chari et al.* (1989)
Icaria marginata (Hymenoptera: Vespidae)	Larva	Prakasam (A.P.)	Chari *et al.* (1989)
Coranus spiniscutis (Hymenoptera: Reduviidae)	Larva	East Godavari (A.P.)	720 + 170.83 Ist and IInd instars of *S. litura*. Sitaramaiah and Ramaprasad (1982)
Coranus sp. (?) *atricapillus* (Heteroptera: Reduviidae)	Larva	East Godavari (A.P.)	790 + 81.23 Ist and IInd instars of *S. litura*. Sitaramaiah and Ramaprasad (1982)
Andrallus spinidens (Heteroptera: Pentatomidae)	Larva	East Godavari (A.P.)	
Rhinocoris squalis (Heteroptera: Reduviidae)	Larva	Guntur (A.P.)	Nymphal stage 104.3 + 3.98. *S. litura* larvae Adult stage: 206.0 + 11.20 *S. litura* larvae. Krishnannda and Satyanarayana (1984)
Chrysopa crassinervis (Neuroptera: Chrysopidae)	Egg larva	Guntur (A.P.)	

Natural enemy	Stage of host	Distribution	Remarks: Potential of parasitism and predation (Reference)
Harpactor costalis (Heteroptera: Reduviidae)	Larva	East Godavari Guntur, Prakasam, Karimnagar (A.P.)	63 *S. litura* larvae in its life time. Sitaramaiah & Satyanarayana (1976)
Rhynocoris fuscipes (Hemiptera: Reduviidae)	Larva	Anand Gujarat	BTRS (1990)
Spiders			
Thomisus lobosus (Araneae: Thomisidae)	Egg larva	East Godavari (A.P.)	Sitaramaiah *et al.* (1980)
Thomisus projectus (Araneae: Thomisidae)	Egg larva	East Godavari (A.P.)	Sitaramaiah *et al.* (1980)
Oxyopes wroughtoni (Araneae: Oxyopidae)	Egg larva	East Godavari (A.P.)	Sitaramaiah *et al.* (1980)
Phidippus punjabensis (Araneae: Salticidae)	Egg larva	East Godavari (A.P.)	Sitaramaiah *et al.* (1980)
Maripissa mandali (Araneae: Lyssomanidae)	Egg larva	East Godavari (A.P.)	Sitaramaiah *et al.* (1980)
Olios punctipes (Araneae: Sparassidae)	Egg larva	East Godavari (A.P.)	Sitaramaiah *et al.* (1980)
Cheiracanthium danieli (Araneae: Miturgidae)	Egg larva	East Godavari (A.P.)	Sitaramaiah *et al.* (1980)
Clubiona sp. (Araneae: Clubionidae)	Egg larva	East Godavari (A.P.)	Sitaramaiah *et al.* (1980)

parasitoid. Late instars of *S. litura* are nocturnal feeders and hide during day time in the thick canopy of tobacco nurseries and fields and escape parasitisation. Early instars are most likely to get exposed to this parasitoid (Rao and Satyanarayana, 1989). However, the parasitoid needs to be conserved. In Shimoga area of Karnataka, *Campoletes chlorideae* is a key mortality factor of *S. litura* and *H. armigera*. Studies on the importance of weather factors in influencing the activity of natural enemies and their impact on succeeding generation of the pests in monsoon tobacco at Shimoga (Karnataka) revealed that during the nursery season copious rainfall and fair distribution of rainy days helped *S. litura* build up as evidenced by moth catches in pheromone traps and consequent high incidence of parasitisation by *C. chlorideae* and NPV which suppressed the incidence of the pest in the subsequent field crop season to a greater extent (Shivayogeswara *et al.*, 2005).

Blepherella setigera, Sarcophaga dux, Sarcophaga albicesps, Brachymeria lassus, Lasiochalcidia erythropoda and *Hybothoracini sp.* were recorded as pupal parasitiods of *S. litura.* The impact of pupal parasitoids is yet to be evaluated. Deep ploughing of nursery and main crop fields after the harvest may damage pupal parasitoid fauna and minimum tillage operations would conserve these cryptic natural enemies.

Myzus nicotianae

Surveys of natural enemies of tobacco aphids yielded two encyrtid parasitoids and one aphelinid parasitoid. Though laboratory parasitisation rates were 60–100% in case of *Aphelinus* sp. field parasitisation was very low (Rao *et al.*, 1981). *Aphidius* sp was recorded from bidi tobacco in Anand but was found highly susceptible to hyperparasitoids (Sharma and Yadav, 1995). Nine species of coccinellids, three species of syrphids, four species of chrysopids and one species of hemerobiid were recorded on *M. nicotianae/M. persicae* (Joshi *et al.*, 1979a; Rao *et al.*, 1981; Rao *et al.*, 1984). The coccinellid and syrphid predators dominated the predatory complex of sucking pests *M. nicotianae/M. persicae* as well as *Bemisia tabaci. M. persicae* has several crop plants and weed hosts which serve as alternate hosts in non-tobacco season such as *Gynandropsis pentaphylla*, a principal weed host on which *M. persicae* survives (Rao *et al.*, 1982). Coccinellid predators were abundantly found feeding the aphids on this weed. Sevaral weed host were identified as reservoirs of aphidophagous insects in tobacco ecosystem. In tobacco crop, early in the season, natural control by syrphids was more pronounced than by coccinellids (Rao and Chandra, 1985). Feeding potentiality of the coccinellid predators on *M. nicotianae* was studied by Jayaramaiah *et al.*, 1996. The larvae of *Coccinella transversalis*, consumed more in the first two instars and *Cheilomenes (Menochilus) sexmaculata*, in later instars. In adult stage *C. transversalis* was better predator than *C. sexmaculata.* On insect species like aphids in unstable habitats, natural enemies have poor impact on population growth. The impact of natural control by coccinellids or syrphids was negligible on tobacco aphids (Rao and Chandra, 1985). *Diplazoan orientalis* parasitised syrphid larvae and pupae in East Godavari district (Rao *et al.*, 1981). The biotic check on initial infestation is poor and intervention of a promising natural enemy through inundative releases along with manipulations of crop environment by identifying suitable associate crops favouring increased natural enemy activity is required. In a study on the influence of border crops on the incidence of insect pests and natural enemies in FCV tobacco crops such as marigold, castor, mustard, safflower and sorghum were evaluated. The percent reduction in infested plants due to border crops over control was 13 in case of aphids and 5% in case of *H. armigera.* Cooccinellids 2.5 to 5.0 and spiders 1.5% were more when ICSV 745 grain sorghum was used as border crop (Gunneswara Rao *et al.*, 2005). In bidi tobacco with local sorghum variety similar findings were reported and tobacco with border crop required only one spray of insecticide where as in border free controls repeated sprays were required (Prasada Rao *et al.*, 2007). Jayaramaiah *et al.* (1995) studied the biology and feeding potential of *Chrysoperla carnea* on *M. nicotianae. Micromus timidus*, a hemerobiid (Neuroptera) was also recorded feeding on *M. persicae.* It laid stalkless eggs and is amenable for mass production a (Gunneswara Rao and Rao, 1986).

Bemisia tabaci

Bemisia tabaci has number of natural enemies on tobacco crop (Table 10.2). On *B. tabaci* five species of coccinellids and two species of syrphids were recorded (Chari *et al.*, 1989; Rao *et al.*, 1990). Two platigastrids, two aphelinids and one scelionid were recorded from *B. tabaci* pupae. *Encarsia* sp. and *Eretmocerus mundus* were most predominant. The others being *Amitus aleurolobi*, *Platygaster* sp. and *Trimorus sp.* (Rao *et al.*, 1990). *B. tabaci* is primarily a vector of leaf curl disease in tobacco and rarely appears as a direct pest. The biological control porgrammes are to be aimed at suppressing the pest in other host crops in the tobacco ecosystem. Sorghum which stood as a stable non host crop during whitefly epidemic was utilized as a barrier crop. Sorghum not only impeded whitefly migration but also contributed to reduction in leaf curl incidence. The abundance of natural enemy fauna is another bonus for utilizing sorghum as barrier crop.

Helicoverpa armigera

Two braconids, one ichneumonid and two tachinid parasitoids were recorded on *H. armigera.* All the predators recorded on *S. litura* were also predating on *H. armigera* which comprise the families of reduvidae, vespidae, chrysopidae, pentatomidae and lygaeidae. The larval parasitoid complex of lepidopterous pests (*S. litura* and *H. armigera*) of tobacco comprised mostly braconids, tachinids and ichneumonids, along with their hyperparasitoids from tobacco nurseries and tobacco fields. *Campoletes chlorideae* an Ichneumonid parasitoid preferred early instars of *H. armigera.* Field parasitisation rates were

Table 10.2. Natural enemies of whitefly, *Bemisia tabaci*

Natural enemies	Stage of host attacked	Distribution in Andhra Pradesh, Gujarat and Karnataka	Remarks
			Potential of parasitism and predation (Reference)
Parasitoids			
Encarsia sp. (Hymenoptera: Aphelinidae)	Pupae	East Godavari, West Godavari, Krishna, Guntur, Prakasam in A.P.	Tobacco areas Rao *et al.* (1990)
Eretmocerus mundus (Hymenoptera: Aphelinidae)	Pupae	East Godavari, West Godavari, Krishna, Guntur, Prakasam in A.P.	Tobacco areas Rao *et al.* (1990)
Amitus aleurolobi (Hymenoptera: Platygastridae)	Pupae	East Godavari, West Godavari, Krishna, Guntur, Prakasam in A.P.	Tobacco areas Rao *et al.* (1990)
Platygaster sp. (Hymenoptera: Platygastridae)	Pupae	East Godavari, West Godavari, Krishna, Guntur, Prakasam in A.P.	Tobacco areas Rao *et al.* (1990)
Trimorus sp. (Hymenoptera: Scelionidae)	Pupae	East Godavari, West Godavari, Krishna, Guntur, Prakasam in A.P.	Tobacco areas Rao *et al.* (1990)
Predators			
Cheilomenes sexmaculata Menochilus sexmaculatus (Coleoptera: Coccinellidae)	Pupae and adults	East Godavari, West Godavari, Krishna, Guntur, Prakasam in A.P. Anand (Gujarat), Nipani Hunsur (Karnataka)	Larval stage 10–15 nymphs/day Chari *et al.* (1989)
Verania discolor (Coleoptera: Coccinellidae)	Pupae and adults	East Godavari, West Godavari, Krishna Guntur, Prakasam in A.P. Anand (Gujarat), Nipani Hunsur, (Karnataka)	Adults feed on 12–50 nymphs/day Chari *et al.* (1989)
Cocinella septumpunctata (Coleoptera: Coccinellidae)	Pupae and adults	East Godavari, West Godavari, Krishna, Guntur, Prakasam in A.P. Anand (Gujarat), Nipani Hunsur, (Karnataka)	Adults feed on 12-50 nymphs/day (Chari *et al.*, 1989)
Cryptolaemus montrouzieri (Coleoptera: Coccinellidae)	Pupae and adults	East Godavari, West Godavari, Krishna, Guntur, Prakasam in A.P. Anand (Gujarat), Nipani Hunsur, (Karnataka)	Adults feed on 12-50 nymphs/day (Chari *et al.*, 1989)

Natural enemies	Stage of host attacked	Distribution in Andhra Pradesh, Gujarat and Karnataka	Remarks Potential of parasitism and predation (Reference)
Coccinella transversalis *Coccinella repanda* (Coleoptera: Coccinellidae)	Pupae and adults	East Godavari, West Godavari, Krishna, Guntur, Prakasam in A.P. Anand (Gujarat), Nipani Hunsur, (Karnataka)	Adults feed on 12-50 nymphs/day (Chari *et al.*, 1989)
Syrphus confrator (Diptera: Syrphidae)	Pupae and adults	East Godavari, West Godavari, Krishna, Guntur, Prakasam in A.P. Anand (Gujarat), Nipani Hunsur, (Karnataka)	Adults feed on 12-50 nymphs/day (Chari *et al.*, 1989)

up to 10–15%. In Gujarat *C. chlorideae*, *Chelonus heliopae* and *Carcelia illota* together accounted for 64% parasitisation of *H. armigera* (Chari *et al.*, 1990). Out of these *C. chlorideae* showed promise for mass multiplication and release against *H. armigera* as this was successfully reared on *S. litura* larvae in laboratory. Rearing of this parasitoid on *Corcyra cephalonica* larvae was successful (Venkatesan *et al.*, 1995) which gave hope for its large scale production and utilization. *Carcelia illota* and *Goniopthalmus sp.* were recorded as larval-pupal parasitoids on *H. armigera.* On *H. armigera* apart from the general predators of *S. litura*, a mirid bug *Cyrtopeltis (Nesidiocoris) tenuis* was reported feeding selectively on *H. armigera* eggs in this crop environment (Chari *et al.*, 1982).

Transgenic Tobacco

Binary vectors suitable for the plant transformation containing Cry 1A and *cry 1* genes that are toxic to the tobacco caterpillar and budworm have been developed at C.T.R.I, Rajahmundry in collaboration with a Dutch Institute. These genes were fused in frame with Kanamycin resistant gene to facilitate selection of transformants and to quantitate toxic gene expression levels. Two flu cured Virginia tobacco cultivars (FCV) Hema and Jayasri have been transformed with insecticidal protein genes *cry1A(b)* and *cry1(c)* from the bacterium *Bacillus thuringiensis*. Genes *cry1A(b)* and *cry1c* were fused in binary vectors pBIN 19 and pPCV. The binary vectors were mobilised into *Agrobactrium tumefaciens* strain, GV 3101 harbouring disarmed helper plasmid MP 90 RK. Transformation was done using co-cultivation method at Tata Energy Research Institute, New Delhi.

Transgenic nature confirmed on Km+ medium and by southern blotting. Expression levels of protein were determined by northern blotting, ELISA and immunoblotting techniques. These transgenic cultivars have been evaluated against the larvae of *S. litura* and *H. armigera* and the plants have been identified to possess combined resistance to these pests. The crystal proteins synthesised by these genes, when ingested by susceptible insects, interact with and lyse their midgut epithelial cells, eventually causing death. These transgenic tobacco cultivars synthesise δ-endotoxins in adequate quantity to kill the larvae of both the pests, the dead larvae show typical *Bt* symptoms; blackening and hardening of body and moult failure. Transgenic *Bt* tobacco cultivars possess resistance to leaf eating caterpillar, *S. litura* and budworm, *H. armigera* (Sitaramaiah *et al*., 2001; Narayanan *et al*., 2003, Sarala *et al*., 2012). These are the first transgenics that were tested under limited field trials in India. However, these lines were not released for commercial cultivation, keeping in view the resistance shown by tobacco importing countries. Production of transgenics will give scope for transferring desirable characters from different species and genera.

CONCLUSIONS

Biopesticides have proved effective in management of inset pests particularly the leaf feeders, viz. *S. litura* and *H. armigera* in tobacco crop. A large number of field trials have been conducted with biopesticides against *S. litura* and *H. armigera* on tobacco resulting in effective control of these pests. However, their large scale use is restricted due to the absence of cheaper mass production techniques, slow process of kill, poor shelf life and stability under field conditions, specific storage requirements unlike chemical pesticides, inadequate quality control norms and their implementation, lack of awareness about the appropriate delivery systems. There is an enormous scope in tobacco pest management in which biopesticides would provide a viable commercial option, where conventional chemical control may result in undesirable pesticide residues in the cured leaf particularly the tobaccos that are meant for export purpose. And also when the chemical control options do not give sufficient control or where there is a case of insecticide resistance. This is mainly because of general disillusionment as to the efficacy of biocontrol methods rather than lack of awareness as far as hazards of chemical pesticides are concerned. Further research and development of biopesticides must be given high priority and people in general and agriculturists in particular must be educated about the dangers posed by handling and use of chemical pesticides. This will lead to a general enlightenment about the benefits of biopesticides and will force governments to make policy decisions reducing the use of chemical pesticides and increasing the use of a green alternative. Biopesticides are set to play an important role in future IPM programmes in tobacco. The large scale use of biopesticides will help tobacco farmers to produce exportable tobaccos that will meet the international customer/consumer preferences. The search for new biopesticides with greater efficacy, persistence and desirable host specificity should continue. Biopesticide formulations with consistent quality, greater efficacy and persistence need to be developed and made available in time for their increased use in agricultural crops particularly tobacco. The large scale utilization of biopesticides will definitely help in production of tobacco that suits international quality requirements and help in boosting the nation's economy.

REFERENCES

BTRS (1990) *Annual Report AICRP on Tobacco.* Bidi Tobacco Research Station, Gujarat Agricultural University, Anand.

Chari, M.S., Patel, S.N. and Bharpoda, T.M. (1982) The predatory habits of *Nesidiocoris tenuis* Reuter. *Tobacco Res.*, **8**, 192–196.

Chari, M.S., Bharpoda, T.M. and Patel, N.M. (1985) Studies on integrated management of *Spodoptera litura* in tobacco nursery. *Tobacco Res.*, **11**, 93–97.

Chari, M.S., Rao, R.S.N. and Rao, S.G. (1989) Progress of biological control of tobacco pests. Proccedings of the Seminar cum workshop held at IISR. Lucknow, October 23-25.1989. Tech. Doc. No. 26. 191 Pp. All India Coordinated Research Project on Biological Control of Crop Pests and Weeds, Bangalore,

Chari, M.S., Nagarajan, K. and Ramaprasad, G. (1999) Pest and disease management in tobacco. In: R.K. Upadhyay, K.G. Mukherji and O.P. Dubey (eds.), *IPM Systems in Agriculture: Cash Crops*, Aditya Books Pvt. Ltd., New Delhi, pp. 227–251.

Chari, M.S. and Ramaprasad, G. (1990) Neem as an anti-feedant and ovipositional repellent for *Spodoptera litura.* In: World Neem Conference, Souvenir, ISTS, Rajahmundry, pp. 81–90.

Chari, M.S., Ramaprasad, G., Rao, R.S.N. and Krishnananda, N. (1990) Present status of *Helicoverpa armigera* in tobacco and strategies for its management in India. In: J.N. Sachan (ed.), *Helicoverpa Management Current Status and Future Strategies,* Indian Institute of Pulses Research, Kanpur, India, pp. 125–133.

Chari, M.S., Rao, R.S.N. and Sreedhar, U. (1995) Integrated management of tobacco leaf eating caterpillar, *Spodoptera litura* F. in India. Tobacco Symposium, Indian Society of Tobacco Science, Rajahmundry, India, pp.70–77.

Chari, M.S., Rao, R.S.N., Sreedhar, U., Naidu S.A. and Naik, P.K. (1996a) Bio-intensive IPM of *Spodoptera litura* F. in field crop of tobacco in Andhra Pradesh. In: S.C. Goel (ed.), *IPM and Sustainable Agriculture-An Entomological Approach*, UPZS, Muzaffarnagar, India. pp. 51–58.

Chari, M.S., Sreedhar, U., Rao, R.S.N. and Reddy, S.A.N. (1996b) Studies on compatibility of botanical and microbial pesticides to the natural enemies of *Spodoptera litura* F. *Tobacco Res.*, **22**, 32–35.

CTRI (1987) *Annual Report.* Central Tobacco Research Institute, Rajahmundry, India.

CTRI (1995) *Mass Production of Natural Enemies of Spodoptera litura.* Final Report of DBT R and D project, Central Tobacco Research Institue, Rajahmundry.

CTRI (2010) *Annual Report 2009-10.* Central Tobacco Research Institute, Rajahmundry, India.

CTRI (2011) *Central Tobacco Research Institute-Vision 2030.* Central Tobacco Research Institute, Rajahmundry.

Gunneswara Rao, S. and Rao, R.S.N. (1986) *Micromus timidus* Hazen (Neuroptera : Hemirobiidae) on *Myzus persicae* on tobacco, In: Souvenir of the VI National Symposium on Tobacco. Indian Society of Tobacco Science, Rajahmundry, Andhra Pradesh, India, Abstracts, 64 pp.

Gunneswara Rao, S and Sitaramaiah, S. (2001) Application strategies for biopesticides in the management of *Spodoptera litura* and *Helicoverpa armigera* in tobacco. *Tobacco Res.*, **27**, 1–6.

Gunneswara Rao,S., Venkateswarlu, P. and Chandra. I.J. (2005) Influence of tobacco types on the occurrence and effectiveness of natural enemies of *Spodoptera exigua* in nurseries. National Conference on Tobacco. October 3–5, 2005. Sientific Strategies for sustainable growth of farm economy and export (Abstracts), Acharya N.G. Ranga Agricultural University, Rajendranagar, Hyderabad.

Gunneswara Rao, S., Prasad, J.V. and Venkateswarlu, P. (2009) Influence of water quality and adjuvants on the efficacy of *Spli* NPV against *Spodoptera litura* in tobacco. *National Symposium on Tobacco. Challenges and Opportnities*, November 24-25, CTRI Rajahmundry.

Gunneswara Rao, S., Chari, M.S., Rao, R.S.N. and Sreedhar, U., (1996) Bio efficacy of *Bacillus thuringiensis* formulations against *Heliothis armigera* and *Spodoptera litura* in tobacco. *J. Appl. Zool. Res.*, **7**, 53–54.

Gunneswara Rao, S., Sitaramaiah, S., Sreedhar, U and Ramaprasad.G. (2002) A mission mode approach for bio-agents production, populariza-tion and human resource development for IPM of *Spodoptera litura* F.

and *Helicoverpa armigera* Hb.in coastal Andhra Pradesh. In: Opender Koul, G.S. Dhaliwal, G.S. Marwaha and J.K. Arora (eds.), *Biopesticides and Pest Management*, Vol. 1, Campus Books International, New Delhi, pp. 370–385.

Gunneswara Rao. S., Venkateswarlu, P., Prasad, J.V. and Chandra, I.J. (2004) *Biopesticide Based Protection Strategy for Management of Helicoverpa armigera in Tobacco.* Biopesticides Emerging Trends, Y.S.Parmar University of Horticulture and Forestry, Nauni, Solan.

Jayaramaiah, M., Jagadish, K.S., Shivayogeswara, B. and Jones Nirmalnath, P. (1995) A studies on the biology of *Chrysoperla carnea* Stephen (Neuroptera: Chrysopidae) on tobacco aphid *Myzus nicotianae* Blackman (Homoptera: Aphididae) and its feeding potential in different larval instars (Abstracts), *Proceedings of Tobacco Symposium- Search for High Quality*, Indian Society Tobacco Science, Rajahmundry, Andhra Pradesh.

Jayaramaiah, M., Jagadish, K.S., Shivayogeswara, B and Jones Nirmala Nath, P. (1996) Feeding potentiality of two coccinellid predators of *Myzus nicotianae. Tobacco Res.*, **22**, 69–75.

Joshi, B.G. and Ramaprasad, G. (1975) Neem kernel as an anti-feedant against the tobacco caterpillar, *Spodoptera litura. Phytoparasitica*, **3**, 59–61.

Joshi, B.G. and Sitaramaiah, S. (1979) Neem kernel as an ovipositional repellent for *Spodoptera litura* moths. *Phytoparasitica*, **7**, 199–202.

Joshi, B.G., Ramaprasad, G. and Sitaramaiah, S. (1978a) Neem kernel suspension protects tobacco nurseries. *Indian Fmg.*, **28**, 17–18.

Joshi, B.G., Ramaprasad, G. and Satyanarayanaa, S.V.V. (1978b) Relative efficacy of neem kernel, fentin acetate and fentin hydroxide as antifeedants against tobacco caterpillar, *Spodoptera litura* in the nursery. *Indian J. Agric. Sci.*, **48**, 19–22.

Joshi, B.G., Sitaramaiah, S. and Ramaprasad, G. (1980) Note on *Trichogramma chilonis* Ishii as a biological control agent against tobacco caterpillar, *Spodoptera litura* F. *Tobacco Res.*, **6**, 131–133.

Joshi, B.G., Ramaprasad, G. and Nageswararao, S. (1984) Neem seed kernel suspension as an antifeedant for *Spodoptera litura* in planted flue cured Virginia tobacco crop. *Phytoparasitica*, **12**, 3–12.

Joshi, B.G., Sitaramaiah, S., Satyanarayana, S.V.V. and Ramaprasad, G. (1979) Note on natural enemies of *Spodoptera litura* F. and *Myzus persicae* Sulz. on flue cured tobacco in Andhra Pradesh. *Sci. Cult.*, **45**, 251–252.

Joshi, B.G., Sitaramaiah, S., Ramaprasad, G. and Rao, R.S.N. (1986) Management of *Spodoptera litura* in tobacco nurseries by integrating *Telenomus remus*, *Curiosa celestas* and neem seed kernel suspension. *Tobacco Res.*, **12**, 16–21

Krishnananda, N. and Satyanarayana, S.V.V. (1984) Biology of *Rhinicoris squalis* Dis. (Heteroptera: Reduvidae) a new predator on tobacco caterpillar, *Spodoptra litura* F. *Tobacco Res.*, **10**, 25–29.

Nagarajan, K., Murthy, P.S.N., Ramaprasad, G., Sitaramaiah, S., Rao, C.V.N., Prabhu, S.R., Sreedhar, U., Nageswara Rao, S., Gnanakumar, D. and Sarojini Sinha (1998) Vision on natural pesticides for tobacco pests. In: *Neem applications in Agriculture, health care and Environment.* Proceedings of the International Conference on Neem- Setting goals for a global vision, Neem foundation, Mumbai, India, pp. 20–26.

Narayanan, A.I., Murthy, T.G.K., Sarala, K., Rao, R.V.S. and. Venugopala Rao, P.V. (2003) Division of crop improvement. In: K. Deo Singh (ed.), Central Tobacco Research Institute, Rajahmundry, pp. 1–16.

Parasara, U.A. and Yadav, D.N. (1995) Predatory potential of *Nesidiocoris tenuis* Reuter (Hemiptera: Miridae) and its role in natural suppression of *Heliothis armigera* in rustica tobacco (Abstract), Tobacco Symposium, Indian Society of Tobacco Science, Rajahmundry.

Patel, R.C., Patel, J.C. and Patel, J.K. (1971a) Studies on the biology of *Chelonus heliopae* Gupta egg larval parasite of *Spodoptera litura* F. *Indian J. Entomol.*, **33**, 50–54.

Patel, R.C., Patel, J.C. and Patel, J.K. (1971b) New records on parasites of *Heliothis armigera* Hb. and *Heliothis peltigera* Schiff from Gujarat. *Indian J. Entomol.*, **33**, 233–234.

PDBC (1994).*Annual Report.* All India Coordinated Research Project on Biological Control of Crop Pests and Weeds, Bangalore, India.

PDBC (2007) *Annual Report.* All India Coordinated Research Project on Biological Control of Crop Pests and Weeds, Bangalore, India.

Prasad, J.V. (2007) Potential of flavonoids as UV protectants against inactivation nuclearpolyhedrosis virus (NPV) of *Spodoptera litura. Tobacco Res.*, **33**, 30–35.

Prasad, J.V., Sreedhar, U. and Gunneswara Rao, S. (2010) Ecological role of *Nesidiocoris tenuis* (Reuter) (Hemiptera: Miridae), an omnivorous mirid bug in tobacco crop ecosystem. *Tobacco Res.*, **36**, 12–16.

Prasada Rao, G.M.V., Venkateswarlu, P., Raju, K. and Sivaramakrishna, M. (2007) Role of tall growing millet crops in the biological suppression of aphid, *Myzus nicotianae* Blackman on bidi tobacco. *J. Biol. Control,* **21**, 55–58.

Ramakrishnan, N., Chaudhari, S., Kumar, S., Rao, R.S.N. (1981) Field efficacy of nuclear polyhedrosis virus against the tobacco caterpillar *Spodoptera litura* F. *Tobacco Res.*, **7**, 129–134.

Rao, R.S.N. and Chandra, I.J. (1984) *Brinckochrysa scelestes* Banks as a predators on *Myzus persicae* Sulz on tobacco. *Entomophaga*, **29**, 283–285.

Rao, R.S.N. and Chandra I.J. (1985) Biological control of *Myzus persicae* Sulz. *Indian J. Agric. Sci.*, **55**, 654–655.

Rao, R.S.N. and Satyanarayana, S.V.V. (1989) Studies on *Peribeae orbata* Weid. (Diptera : Tachinidae) a larval parasite of *Spodoptera litura* F. *Tobacco Res.*, **16**, 1–4.

Rao, R.S.N., Satyanarayana, S.V.V. and Soundararajan. (1981) Note on new additions to the natural enemies of *Spodoptera litura* F. and *Myzus persicae* Sulz. on FCV tobacco in Andhra Pradesh. *Sci. Cult.*, **47**, 98–99.

Rao, R.S.N., Joshi, B.G. and Satyanarayana, S.V.V. (1982) Studies on alternate host plants of *Myzus persicae* Sulz. during non-tobacco seasons. *Tobacco Res.*, **8**, 77–79.

Rao, R.S.N., Gunneswarar Rao and Chandra,I.J. (1987) Biochemical potentiation of NPV of *Spodoptera litura. J. Biol. Control*, **1**, 36–39.

Rao, R.S.N., Chari, M.S. and Gunneswara Rao, S. (1990) Further record of natural enemies on the insect pests of tobacco in Andhra Pradesh. *J. Biol. Control*, **4**, 65–66.

Ramaprasad, G., Joshi, B.G. and Nageswararao, S. (1986) Relative efficacy of some insecticides and neem seed kernel suspension against tobacco caterpillar, *Spodoptera litura* on tobacco. *Indian J. Plant Prot.*, **14**, 69–74.

Ramaprasad, G., Sitaramaiah, S. and Sreedhar, U. (2000) Efficacy of *Bacillus thuringiensis* Berliner formulations against tobacco caterpillar, *Spodoptera litura* F. in tobacco nurseries. *Tobacco Res.*, **26**, 1–6.

Ramaprasad, G., Sitaramaiah, S., Joshi, B.G. and Nageswararao, S. (1987) Relative efficacy of neem seed kernel suspension, some synthetic pyrethroids and insecticides against tobacco caterpillar, *Spodoptera litura* in tobacco nurseries. *Indian J. Plant Prot.*, **15**, 190–192.

Sarala, K., Murthy, T.G.K. and Krishnamurthy, V. (2012) Tobacco (*Nicotiana tabacum* L.). In: D.N. Bharadwaj (ed.), *Breeding of Field Crops*, Agrobios (India), Jodhpur, pp. 831–864.

Sharma, U. and Yadav, D.N. (1995) *Arthropod Natural Enemies Recorded on Major Crops from Gujarat and their Brief Research Status*. Biological Control Laboratory, Gujarat Agricultural University, Anand.

Sivayogeswara, B., Thippeswamy, C., Visweswara Gowda, B.L., Sridhara, S. and Naik, M.I. (1995) Influence of rainfall and abundance of bioagents in tobacco nursery on the infestation of *Spodoptera litura* and *Helicoverpa armigera* in the main filed. National Conference on Tobacco. October 3-5. 2005. Scientific Strategies for sustainable growth of farm economy and export. (Abstracts), Acharya N.G. Ranga Agricultural University, Rajendranagar, Hyderabad.

Sitaramaiah, S. and Satyanarayana, S.V.V. (1976) Biology of *Harpactor costalis* Stal on tobacco caterpillar *Spodoptera litura* F. *Tobacco Res.*, **2**, 134–136.

Sitaramaiah, S. and Ramprasad, G. (1982) Studies on the life cycle of *Coranus spiniscutis* Reuter and *Coranus* sp. (?) *atricapillus* Dist. (Heteroptera : Reduviidae) and *Spodoptera litura* F. from India. *Tobacco Res.*, **8**, 81–85.

Sitaramaiah, S., Joshi, B.G. and Ramaprasad, G. (1980) New records of spiders and predators of tobacco caterpillar, *Spodoptera litura* F. *Sci. Cult.*, **46**, 29–30.

Sitaramaiah, S., Gunneswararao, S., Ramaprasad, G. and Sreedhar, U. (2001) Role of biological agents in tobacco insect pest management. In: R.K. Upadhyay (ed.), *Biocontrol Potential in Sustainable Agriculture*, Kluwer Academic/Plenum publishers, New York, pp. 215–234.

Sitaramaiah, S., Sreedhar, U., Gunneswara Rao, S., Ramaprasad, G. and Nageswararao, S. (2002) Bio-intensive management of leaf eating caterpillar, *Spodoptera litura* F. in tobacco nurseries. In: Opender Koul, G.S. Dhaliwal, S.S. Marwaha and J.K. Arora (eds), *Biopesticides and Pest management*, Vol. 2, Campus Books International, New Delhi, pp. 319–327

Sitaramaiah, S., Gunneswara Rao, S. and Sreedhar, U. (2002a) *Biological Control of Tobacco Pests in India.* Central Tobacco Research Institute, Rajahmundry, India.

Sitaramaiah,S., Gunneswara Rao, S., Hussaini, S.S. and Nageswara Rao, S. (2002b) Entomopahogenic nematodes with special reference to *Steinernima carpocapsae* against *Spodoptera litura* in tobacco nursery In: P.L. Tandon, C.R. Ballal, S.K. Jalali and R.J. Rabindra (eds.), *Biological Control of Lepidopteran Pests*, Society for Biocontrol Advancement, Bangalore, pp. 211–354.

Sreedhar, U. (2011a) Influence of trap crop planting geometry on budworm, *Helicoverpa armigera* in flue cured Virginia tobacco. *Indian J. Plant Prot.*, **39**, 196–199.

Sreedhar, U. (2011b) *Integrated Pest Management and Agrochemical Management in Tobacco.* Souvenir-XIV National Symposium on Tobacco, Indian Society of Tobacco Science, Rajahmundry, pp. 89–94.

Sreedhar, U. (2012) Management of bud worm (*Helicoverpa armigera*) in Virginia tobacco through pest diversionary approach. *Indian J. Agric. Sci.*, **8**, 731–734.

Sreedhar, U., Sitaramaiah, S. and Padmaja, K. (2007) Evaluation of trap crops for management of budworm, *Helicoverpa armigera* in flue cured Virginia tobacco. *Indian J. Plant Prot.*, **35**, 213–217.

Sreedhar, U., Sitaramaiah, S., Gunneswara Rao, S. and Ramaprasad, G. (2002) Management of *Spodoptera litura* F. on chillies with NPV in coastal Andhra Pradesh. In: Opender Koul, G.S. Dhaliwal, S.S. Marwaha and J.K. Arora (eds.), *Biopesticides and Pest Management*, Vol. 2., Campus Books International, New Delhi, pp.173–180.

Sreedhar, U., Rao, R.S.N., Sitaramaiah, S. and Ramaprasad, G. (2003) Evaluation of *Bacillus thuringiensis* formulations against budworm, *Helicoverpa armigera* on FCV tobacco. *Tobacco Res.*, **29**, 43–48

Sreedhar, U., Sitaramaiah S., Gunneswararao, S. and Venkateswarlu, P. (2004) Evaluation of sequential sprays of bio-insecticides against *Spodoptera litura* in tobacco nurseries. *Andhra Agric. J.*, **50**, 503–507.

Sreedhar, U., Sitaramaiah, S., Gunneswararao, S., Ramaprasad G. and Deo Singh, K. (2005) Integrated management of *Spodoptera litura* F. in flue cured virginia tobacco. (Abstract), *Sustainable Crop Production in Stress Environments: Management and Genetic Options*, Symposium organized by JNKVV, Jabalpur.

Sreedhar, U., Sitaramaiah, S., Ramaprasad, G. and Nageswararao, S. (2008) Management of tobacco budworm, *Helicoverpa armigera* with sequential application of conventional and bio-rational insecticides. *Tobacco Res.*, **34**, 29–34.

Venkatesan, T., Singh, S.P., Joshi, S. and Rao, N.S. (1995) Breeding of Campoletis chloridae Uchida (Hymenoptera: Ichneumonidae) on *Corcyra cephalonica* Stainton. *J. Biol. Control*, **9**, 128–129.

11

Status of Biological Control of Pest and Disease Management in Spice Crops

Y. R. SARMA[1*], S. DEVASAHAYAM[1], A. KUMAR[2] AND R. ARAVIND[3]

[1]Indian Institute of Spices Research, Marikunnu P. O., Kozhikode-673012, Kerala, India
*Present address: Aramam, M 10/5 KSHB Colony, Malaparamba, Kozhikode (Calicut)-673009, Kerala, India
[2]Indian Agricultural Research Institute, New Delhi-110012, India
[3]Plant Pathology and Microbiology (PLPM), Texas A & M University, 2132 TAMU, College Station, Texas, 77843, USA
*E-Mail: yammanoor@gmail.com

INTRODUCTION

The present status of biological control as a component of integrated pest and disease management in spice crops has been reviewed emphasizing the need for pesticide free produce. The crop losses caused by biotic stresses especially due to the soil borne disease of spice crops received considerable attention which reflected on the quantum of research investigations and consequent publications. The research on microbial biotechnology for plant protection did establish the bioefficacy and its sustainability under field situations. The importance of botanical pesticides in management of some of the insect pests in an IPM module has been brought out. Since microbial agents are proposed to be used, biosafety of these formulations has been emphasized in addition to the existing regulatory procedures. The future programmes on the ecological aspects of this technology needs critical investigations with the help of additional molecular techniques. It is imperative to establish National Guideline on the biosafety of the use of the microbial biopesticides that are recommended for agricultural field applications. This is of great relevance in view of the recent discovery of *Pseudomonas aeruginosa* - endophytes in black pepper that is capable of suppressing *P. capsici*, *R. similis* and *M. incognita* but could not be utilized for crop protection in view of its human pathogenic nature.

Spices are important components of global food industry, besides being rich source for 'Ayurvedic' Indian medicine that aids to alleviate various human ailments. The active principles of these were also found antimicrobial and also bio-enhancers in uptake of certain allopathic preparations. Global spices industry is to the tune of 11 lakhs tonnes in volume accounting to US$ 3.475 billion (Spices Board, 2012) in value. India's share at global level is about 52% in volume and 58.6% in value (Anonymous, 2010, 2012). These crops are location specific and are grown in certain agro climatic conditions that vary according to the spice crops concerned. In India spices are grown over an area of 3.21 million ha contributing to 5.96 million MT per annum (Anonymous DASD, 2012). During 2011-2012, India exported 5,75,270 MT of spices valued at ₹ 9783.42 crore (US$ 2037.76 million) an all time high (Spices Board, 2012). About 90% of the spices produced in India are consumed locally and only about 10% of these alone are exported.

STATUS OF MAJOR DISEASES AND INSECT PEST PROBLEMS

Out of 109 spice crops recognized globally, in India only 52 crops have been officially recognised under Spices Board, Ministry of Commerce, Government of India. While Spices Board deals with the export and trade, the Indian Council of Agricultural Research (ICAR) and State Agriculture Universities (SAUs) are mandated with the research needs of these crops and developed several agro technologies for their cultivation. Severe crop losses due to pests and diseases still remain as a major production constraint. To produce pesticide free spices, Integrated Pest Management (IPM) is the logical strategy where biological control becomes major focus. Biological control as a strategy gained momentum during the last two decades, where field efficacy of the technology for pest and disease management has made an impact. However, there is an urgent need to convince the farming community about the potential of this bio control technology in making pest and disease management a viable strategy. The role of biological control in crop protection need not be reemphasized because of its relevance and potential as a component of IPM, to ensure pesticide free spice produce which is in great demand at global level. Particularly biocontrol agent has a major role in the organic spices productions which are high demand in worldwide. The biocontrol technology for disease management is well set in spice crops protection because of consistent and continuous research both on basic and applied. But lot more need to be done on the ecological aspects of the technology which is important to make this eco-friendly technology more sustainable and farmer friendly. Biocontrol is more relevant in spice crops due to the following reasons.

- Perennial nature of many economically important spice crops.
- Elusive nature of soil borne/seed borne nature of diseases/ insect pests, for an effective management.
- Lack of high degree of host resistance for major biotic stresses.
- Poor targetability of soil borne plant pathogens/ insect pests to agrochemicals.
- Agrochemicals are prone for leaching due to heavy rainfall in certain locations, thus making crop vulnerable to diseases and pests.
- Spices should be free from pesticide residues as a food safety measure.

The subject was reviewed earlier by several workers (Eapen and Ramana, 1997; Sarma and Anandaraj, 1998; Sarma and Saju, 2004; Sarma, 2006). The biological control of insects pests of spices has also been documented in late 90s (Devasahayam, 1996)

CROPPING SYSTEMS AND BIOCONTROL

In general, spices are cultivated in a mixed cropping system alongside several other crops like coconut, areca nut, tea, coffee etc. Mono cropping of spices particularly in cardamom and black pepper in not uncommon in leading spice growing countries. In High Density Multi-species Cropping System (HDMC) occurrence of multiple plant pathogens including plant parasitic nematodes is experienced as there is no spatial segregation of these pathogens in the soil and they over lap. This renders crop protection more complex in these ecosystems. Therefore, suitable management strategies are needed to be evolved to address the holistic health and productivity of crops. Here under this scenario, the biocontrol technology assumes significance as biocontrol agents possess broad spectrum of activity.

Biological Control of Diseases and Insect Pests of Spice Crops

Many of spice crops are affected by 'Killer' soil borne diseases causing heavy crop losses. Foot rot and slow decline in black pepper, clump rot and capsule rot in cardamom, rhizome rot and bacterial wilt in ginger, rhizome rot in turmeric, wilts in cumin, coriander and root rot in fenugreek, are the major disease problems (Table 11.1). Among the insect pests, 'Pollu beetle' (hollow berry) and root mealy bugs in black pepper, thrips, stem borer and root grub in small cardamom, stem borer and scales in ginger, turmeric and aphid complex in seed spices are the important insect pests that cause considerable crop losses. Though the agrochemical usage in spice crops is much less compared to field crops, a consistent effort has been made in recent years on biological control strategy to produce pesticide free spices. Non availability of durable host resistance is another reason which led to the increased interest in this ecofriendly crop protection technology (Sarma, 2006)

The Potential of Vesicular Arbuscular Mycorrhiza (VAM) in Suppression of Root Diseases of Spice Crops

The role of endomycorrhiza or vesicular arbuscular mycorrhiza (VAM) in disease suppression of soil borne plant pathogens has been well established (Schenk, 1981; Bagyaraj, 2006). Role of VAM in uptake of nutrients like phosphorus, zinc etc is well known besides its ability to induce plant defence (Atkinson *et al.*, 1994). Of the several VAM fungi investigated, *Glomous fasiculatum* is found highly effective in suppression of root rot caused by *P. capsici, R. similis* and *M. incognita* (Anandaraj *et al.*, 1996; Sarma *et al.*, 1996). *Glomus fasiculatum* was found to increase, root biomass in black pepper besides reducing root rot that has resulted in increased yield. Techniques have been evolved for mass production of VAM inoculum on roots of graminaceous hosts especially in Sorghum and Maize. Besides, attenuated axenic root cultures of some grasses, have been standardized for large scale multiplication of VAM. However, routine techniques of multiplication of VAM have been standardized in perlite and sand mixtures that a farmer can adopt for VAM inoculum production for their needs. Altered host metabolism resulting in increased phosphorus uptake, increased sugars, amino acids, phenolic compounds and increased chitinase activity in host plant infected with

Table 11.1. Major diseases of spice crops and the potential biocontrol agents (BCAs) for disease management

Host/Disease	Causal organism/ Pathogen	Biocontrol agents reported/ Programmes initiated
1. BLACKPEPPER (*Piper nigrum* L.)		
Foot rot	*Phytophthora capsici* Leonian (Emend A. Alizadeh and P.H. Tsao)	*T. harzianum* Rifai (1969), *T. virens, T. viride* Pers. (1794), *Pseudomonas fluorescens* (Flügge, 1886), Migula, 1895 *Bacillus* sp. Cohn, 1872, *Glomus fasiculatum* (VAM). *Pseudomonas putida*
Slow decline	*Radopholus similis* Thorne 1949 (Cobb 1893)	*T. harzianum* Rifai, (1969), *T. virens; Bacillus megater-ium; Curtobacterium luteum*
	Meloidogyne incognita (Kofoid and White, 1919)	*Paecilomyces lilacinus* (Thom) Luangsaard, Houbraken, Hywel-Jones and Samson (2011), *Pasteuria penetrans*
	Phytophthora capsici Leonian (Emend A. Alizadeh and P.H. Tsao)	*Pochonia chlamydosporia* var. *catenulata:* (Goddard) zare and W. Gams 2001
Anthracnose	*Colletotrichum gloeospo-rioides* (Penz.) Sac, *C. Piperis* Petch, *C. capsici* (Syd.) E.J. Butler and Bisby, (1931)	*P. fluorescens* Flügge, 1886; Migula, 1895
2. CARDAMOM - Small (*Elettaria cardamomum* Maton)		
Capsule rot ('Azhukal')	*P. meadii* McRae	*T. harzianum* Rifai, (1969)
	P. nicotianae var. *nicot-ianae* Breda Hann var. *nicotianae* Waterhouse	*T. hamatum* *T. virens*
Clump rot	*Pythium vexans* de Bary, (1876)	*T. harzianum* Rifai, (1969)
	Rhizoctonia solani J.G. Kühn 1858	
	M. incognita (Kofoid and White 1919)	
3. GINGER (*Zingiber officinale* Rosc)		
Soft rot/Rhizome rot	*Pythium aphanidermatum* (Edson) Fitzp., (1923)	*T. harzianum* Rifai, (1969)

Host/Disease	**Causal organism/ Pathogen**	**Biocontrol agents reported/ Programmes initiated**
	P. myriotylum Drechsler, (1930)	*T. virens*
	Fusarium solani (Mart.) Sacc.	*Pseudomonas fluorescens*
'Yellows'	*F. oxysporium* f. sp. *zingiberi*	
Bacterial wilt	*Ralstonia (Pseudomonas) solanacearum* (Smith, 1896) Yabuuchi *et al.* 1996 biovar III	*Pseudomonas fluorescens* (Flügge, 1886) Migula, 1895 Endophytic bacteria *Stenotrophomonas maltophilia* *Acinetobacter calcoaceticus*
Dry rot Dry rot	*Pratylenchus coffeae* Goodey, 1951 and *Fusarium* complex *Macrophomina phaseolina* (Tassi) Goid, (1947)	*Trichoderma harzianum* Rifai, (1969) *Trichoderma harzianum* Rifai, (1969)
4. TURMERIC (*Curcuma longa* L. (Syn)) *C. domestica* Val.		
Rhizome rot	*P. graminicolum*	*Trichoderma viride* Pers., (1794), *T. harzianum* Rifai, (1969)
	Fusarium sp.	
	P. aphanidermatum (Edson) Fitzp., (1923) *M. incognita* Kofoid and White, 1919) *R. similis* Thorne 1949 (Cobb 1893)	
5. CUMIN (*Cuminum cyminum* L.)		
Wilt	*F. oxysporium* f. sp. *cumini*	*Trichoderma virens* / *Gliocladium virens*
6. CORIANDER (*Coriandrum sativum* L.)		*Trichoderma* spp.
Wilt	*F. oxysporum* f. sp. *corianderi*	*Streptomyces* sp., Waksman and Henrici, 1943
7. FENUGREEK (*Trigonella foenum-graecum* L.)		
Root rot	*Rhizoctonia solani* J.G. Kühn, 1858	*T. viride* Pers. (1794)

Host/Disease	Causal organism/ Pathogen	Biocontrol agents reported/ Programmes initiated
		Pseudomonas fluorescens (Flügge, 1886) Migula, 1895
8. VANILLA (*Vanilla planifolia* Jacks)	*P. meadii* Mc. Rac *Fusarium oxysporum* sp. vanille	*T. harzianum* (Rifai, 1969) *P. fluorescens* (Flügge, 1886) Migula, 1895

VAM are some of the reasons attributed for its protection. Incorporation of VAM inoculums to the nursery mixture provided good protection against the three rot pathogens (Thankamani *et al.*, 2003). *T. harzianum* + *Glomus mosseae* + G4 bacterial strain provided increase growth and minimum of disease incidence in ginger (Meenu, Gupta *et al.* (Personal communication)). This methodology needs to be exploited for productions of healthy planting material of black pepper, cardamom, ginger and turmeric which are prone to diseases caused by soilborne pathogens. Studies are warranted to undertake investigations on the association of VAM fungi and their role in growth promotion and disease suppression in seeds spices like cumin, coriander and fenugreek etc.

Fungal Biocontrol Agents in Management of Soil Born Diseases in Spice Crops

Trichoderma harzianum, T. viride, T. aureoviride, Pochonia chlamydosporia, Paecilomyces lilacinus are very well documented as biocontrol agents against broad range of plant pathogens. Apart from these, potential of *Pythium acanthophora*, a hyperparasite on *Pythium* spp. needs intensive investigations specifically in spice crops were pathogenic *Pythium* spp. viz., *P. aphanidermatum, P. myriotylum* etc. are involve in crop losses (Lodha and Webster, 1990).

Out of all these *Trichoderma* stands tall. The antagonistic potential, rhizosphere competence and competitive saprophytic ability renders *Trichoderma* highly amenable, adoptable and sustainable in diverse ecological soil conditions and makes it an ideal candidate in biocontrol programmes on soil borne disease of spice crops. The potential of *Trichoderma* as an effective biocontrol agent has been well documented (Papavizas, 1985; Chet, 1987; Harman, 1992; Cook, 1985). *Trichoderma* spp. as an effective biocontrol agent against several soil borne diseases of black pepper (Rajan, 1999; Saju, 2004; Manohara and Dono, 2009; Manohara *et al.*, 2002), ginger (Usman, 1997; Beena, 2006) and cardamom (Suseela Bhai, 1998) has been reported and consequently incorporated in the IDM for these crops.

BLACK PEPPER

Foot rot disease caused by *P. capsici* still remains as a major threat to black pepper and it's prevalent in all pepper growing countries as well as in India (Fig. 11.1). Hence detailed investigations were under taken on etiology, epidemiology and management (Anandaraj, 1997; Sarma *et al.*, 1998).

Fig. 11.1. Phytophthora foot root rot of black pepper. (a) healthy vines, (b) affected vines.

Ecological Studies on *Trichoderma* in Relation to *P. capsici* Suppression

Competition for nutrients and space is achieved by fast multiplication and distribution (Saju, 2004). Different soil types viz., clayee, sandy clay and sandy loam did not influence the survival of *Trichoderma* spp. Soil moisture also plays a critical role in the survival of *Trichoderma spp.* Population of *T. harzianum* IISR 1369 and *T. harzianum* IISR 173 declined gradually over a period of 90 days at 80, 70, and 20% moisture levels and CFU (Colony forming units) of *T. virens* IISR 1370, *T. aureoviride* IISR 126 and *T. harzianum* IISR 165 declined at 80 to 70% moisture levels. Similarly soil temperature is also a factor of importance. Monitoring of propagules of *Trichoderma* spp. in soil incubated at different temperature ranges (<20 and 20–26°C and 26–32°C) did not show significant difference in cfu of *T. harzianum* IISR 1369, *T. aureoviride* IISR 126 and *T. harzianum* IISR 165. However, *T. virens* IISR 1370 showed higher propagule density at lower temperature ranges (<20°C and 20–26°C). The propagule density of *T. harzianum* IISR 173 remained higher at higher temperature ranges (20–26°C and 26–32° C). In addition, *Trichoderma* spp. was able to grow at a wide range of pH (Saju, 2004)

Improvement of bioefficacy. It has been clearly demonstrated that the mode of disease suppression by *Trichoderma* is mainly due to lytic enzymes as well as metabolites. Any improvement of these traits of these strains would enhance the suppressive activities of these antagonistic fungi (Ahamed Baker, 1987; Harman and Stasz, 1991). The studies carried out in black pepper clearly indicated that biosuppressive effect of *T. aureoviride* (IISR 140) on *P. capsici* is superior to *T. harzianum* (IISR 1369). The protoplast fusants obtained from *T. aureoviride* and *T. harzianum* showed 100% inhibition of *P. capsici.* The ethyl acetate extracts from the culture filtrate of these were found highly thermo stable which exhibited 100% inhibition of the zoospore germination of *P. capsici* (Saju, 2004). Further chemical characterization of these metabolites leading to the identification would be rewarding.

Formulations. Because of the saprophytic nature of the fungus several agri-wastes/products like coffee pulp, spent, tea waste, rice bran etc were used for large scale multiplication of *Trichoderma* and these were reviewed earlier (Sarma and Anandaraj, 1998; Prakash *et al.*, 1999). The solid state and the liquid state fermentation technologies are more precise and reproducible and hence preferred. Talcum based formulation are more popular as evident from number of reports. Liquid formulations are also available abroad (Batta, 2004) and in some cases in India too (Sriram, 2010). Liquid formulations would be more useful for seed treatments of ginger, turmeric and several seed spices. Shelf life and biosafety regulation are in place and need to be followed by all commercial organizations producing biocontrol agents (Kulakshetra, 2004).

Management of foot rot disease. As part of the disease management, number of *Trichoderma* isolates from rhizosphere was screened both *in vitro, in vivo* and *in planta.* Five shortlisted isolates tested in the field for foot rot suppression were monitored and statistically analyzed for three years. The results clearly indicated that the isolates could significantly reduce disease incidence Isolates *T. virens*-12 and T. *harzianum*-26 were found more effective to control the disease with disease index ranging from 1.6–4.2. Isolate *T. harzianum*-26 with disease index 8.1 was selected for field application due to its adaptability in the rhizosphere of black pepper (Rajan *et al.*, 2002). Soil application of inoculum 1 kg/vine @ 10^6 to 10^7 cfu/g mixed with farm yard manure or neem cake twice in a year is the present recommendation (Sarma and Saju, 2004). Based on the large scale field evaluation trials, the technology has been transferred to the several entrepreneurs and the profitability of the technology has been further confirmed by a survey. The enterprise was highly profitable with benefit: cost ratio of 1.84. The investment returned a net present worth of ₹ 2,42,618 @ 11% interest rate for 10 years period with an internal rate of return of 121% (Madan *et al.,* 2006). Large scale field evaluation trails in coffee-pepper in Kodagu district of Karnataka and also in APFDC plantation in Andhra Pradesh Forest Development Corporation (APFDC) further confirmed these observations (Fig. 11.2)

Fig. 11.2. Black pepper crop. (a) *Trichoderma* treated plot, (b) untreated plot.

SMALL CARDAMOM

Capsule rot, damping off, clump or rhizome rot and *Fusarium* root tip rot are serious soil bone disease of cardamom. Earlier studies on rhizome rot of cardamom carried out at IISR Regional Station for cardamom at Appangala clearly indicated the efficacy of *T. harzianum* in controlling rhizome rot complex caused by *R. solani* and *P. vexans* (Venugopal unpublished). Biological control of soil borne diseases of small cardamom was earlier reviewed by Sarma *et al.* (1996). Damping of cardamom seedlings was completely absent in *Trichoderma* treated beds and was found to be superior to copper oxy chloride (Fig. 11.3).

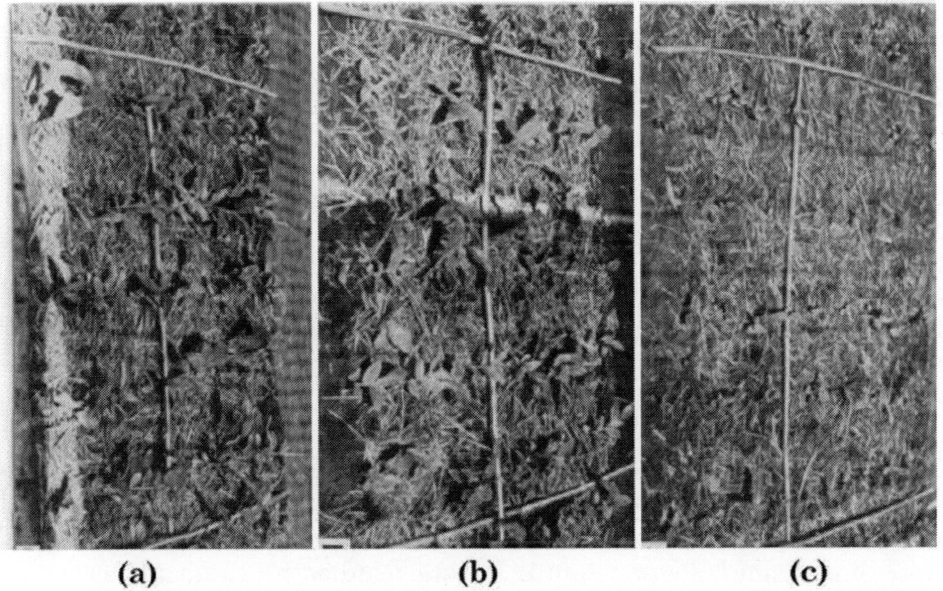

Fig. 11.3. Bioprotection of seedling rot in cardamom nursery. (a) *Trichoderma* treated nursery, (b) COC applied nursery, (c) untreated nursery.

Effective disease management of capsule rot of cardamom with field application of *T. harzianum* in decompose coffee pulp reduce the disease incidence to 83% which was on par with fungicides and its recommended as package to the farming community (Suseela Bhai *et al.*, 1993).

GINGER

P. aphanidermatum and *P. myriotylum* are the two important soil borne/seed borne Oomycete pathogens of ginger causing rhizome rot/ soft rot in majority of the plains in India (Sarma *et al.*, 1994). *P. splendens* has recently been reported from western Himalayas (Shanmugam *et al.*, 2013). Ginger yellows caused by *F. oxysporum* f.sp. *zingiberi* is important in Himachal Pradesh. In addition, dry rot caused by *F. oxysporum/F. solani* in combination with *Pratylenchus coffeae* a plant parasitic nematodes is another important problem in Sikkim as well as in Western Himalayas. These disease complexes need to be trackled through either antagonistic fungi like *T. harzianum* or by rhizobacterial strains or in combination. In Sikkim region rhizome rot associated with *Pythium* spp., *Fusarium oxysporum,* and *Pratylenchus coffeae*. Bacterial wilt caused by *Ralstonia solanacearum* is another disease which is serious. A biocontrol agent, *T. harzianum*, isolated form Sikkim was found effective in control of ginger diseases substantially but large scale field demonstrations are needed (Rajan *et al.*, 2002).

Intensive studies were carried out in Kerala where *P. aphanidermatum* was found as major pathogens of ginger. Seed treatment and soil application of *Trichoderma harzianum* at the time of planting and later after 90 days with a second application was adopted to test the efficacy of *Trichoderma* on rhizome rot suppression in ginger. Studies carried out in sick plots of ginger with, *T. harzianum* under solarized and non on a solarized conditions showed higher yield (Usman *et al.*, 1996). About five species of *Trichoderma harzianum, T. virens, T. hamatum, T. aureoviride* and *T. pseudokoningii* were found in the ginger soils. Among them *T. harzianum* isolate 7 was found to be highly disease suppressive and growth promotive. Thus variability of *Trichoderma* species was noticed in their bioefficacy (Beena, 2006). Recent studies at Indian Institute of Spice research, Calicut showed that *Bacillus amyloliquefaciens* as well as *Serratia marcescens* found effective in disease control of rhizome rot (Anon, 2012). This needs further evaluation both as a seed treatment as well as soil application. Chemical mutagens and irradiation were adopted to induce variation in *Trichoderma harzianum* to increase the bioefficacy for rhizome rot suppression in ginger. Irradiation was found to be more efficient in creating variation. An albino mutant M125 was found to be disease suppressive with higher rhizosphere competence and lytic enzyme (β 1,4 endoglucanase, β 1,3 endoglucanase). The mutant M125 gave 2.8-9% - in disease incident and increase yields

***Fusarium* yellows.** *F. oxysporum* f. sp. *zingiberi* is another serious problem in the Himalayan region and most specifically in western Himalayas. This disease is distinctly different from soft rot caused by *Pythium* spp. Recent studies carried out clearly indicated the involvement of *F. oxysporum* as well as *F. solani* (Shanmugam *et al.,* 2013). The investigations carried out earlier also indicated the involvement of *P. coffeae*, a plant parasitic nematode which causes dry rot in association with *Fusarium* spp. In view of the involvement of more than one pathogen it is important to address this disease using fungal antagonist and rhizobacterial strains. Irrespective of the pathogens involved, it is important to adopt a holistic approach involving healthy seed and biological control strategy especially for North East region of India were organic cultivation is a rule.

VANILLA

In a large scale field evaluation conducted in 3000 vines in Mauritius, the soil application of *T. harzianum* formulation reduced fusarial wilt (Sarma, 2005). *T. lactae* and an avirulent isolates of *F. oxysporum* and were found effective in management of vanilla wilt in Indonesia (Tombe, 2010).

SEED SPICES

Cumin Wilt

Cumin wilt pathogen being soil borne, it was reported as seed borne too (Singh *et al.*, 1972). The reported hot water treatment of the seed at 54°C for 15 min eliminates the seed borne infection by 94% and did not have appreciable effect on germination of the seed. In future it would be appropriate to pellet the hot water treated cumin seed with either *T. harzianum* or *T. viride* to contain the infection as an effective disease management strategy. Incidentally the bioefficacy of *Trichoderma* in cumin wilt suppression has been reported earlier (Vyas and Mathur, 2002), Seed treatment with talc based formulations of *P. fluorescens* closely followed by *T. harzianum* provided excellent control of *Fusarium* wilt under green house condition (Nitin Chawla and Gangopadhyay, 2009). Integratingly soil solarization followed by growing of sorghum in kharif and application of either carbendazim granules @ 10 kg ha^{-1} one month after sowing or application of *T. viride* in organic carrier @ 62.5 kg ha^{-1} was effective for the management of cumin wilt (Jadeja *et al.*, 2008). Similar finding of combination of *T. harzianum* with carbendazim was reported to be highly effective in decreasing the disease incidents to 11.7% with a highest yield of 6.17 q/ha. (Bardia *et al.*, 2007). The recent studies carried out in Rajasthan India showed high disease suppression ensuring 78.5% protection in seed treated with *T. virens* over untreated control plants (Arunkumar and Sharma, 2013). Seed treatment with *Trichoderma* @4g/kg seed followed by 200g of neem cake/kg seed reduced wilt in cumin (Meena *et al.*, 2010).

Significant reduction in wilt incidence was observed with *Trichoderma* applied as seed as compared to control, In *Trichoderma* treated plots, the yield recorded was 6.17 to 6.73 and 6.33 to 6.89 q/ha compared to 4.39 and 4.73 q/ha in untreated control during first and second years of experiment, respectively (Singh, 2007). *T. harzianum, T. hamatum* and *T. koningii* applied along with the peanut haulms compost, was found highly effective in cumin wilt suppression. This has been attributed to antimicrobial metabolites and enhancing the defence system in the plants through increased production of chitinase, protease, Cellulease β-glactosidase and β-13 glucanase (Haggag and Abo Sedera, 2005).

Coriander, Fenugreek, Dill and Ajwain

For coriander wilt management, seed treatment with *Trichoderma* @ 4 g/kg seed has been recommended (Gopal *et al.*, 2010). Root rot of Fenugreek caused by *R. solani* was controlled effectively by seed peletting with talc based *T. harzianum* followed by field application of neem cake @ 150 kg/ha (Kakani *et al.*, 2009). Seed treatment of Dill with *Trichoderma* @ 4 g/kg was reported to reduce root rot caused *Fusarium* sp. (Sharma *et al.*, 2010). Similarly seed treatment of Ajwain with *Trichoderma harzianum* or *T. viride* @ 4 g/kg reduced root rot caused by *R. solani* (Meena *et al.*, 2009).

PLANT GROWTH PROMOTING RHIZOBACTERIA (PGPR) AND ENDOPHYTES

The role of PGPR's and endophytes in growth promotion and disease suppression of crop plants has been well established (Kloepper *et al.,* 1999, Bashan and Holguin, 1988, Hallmann, 2001). The role of rhizobacteria on disease suppression in soil borne diseases spice crops was initiated and the status was reviewed (Sarma *et al.*, 2000).

Black Pepper

Detailed studies were undertaken in black pepper, *P. capsici* pathosystem, amply proved their potential in suppression of root/foot rot and their mechanisms through production of various metabolites mycolytic enzymes and Induced Systemic Resistance (ISR) and growth promotion (Diby Paul, 2004). Extensive *in vitro* screening of bacterial isolates was conducted in dual culture with *P. capsici* to access their antagonistic potential. The potential isolates were tested for their impact on increased root biomass in comparison with root infection in untreated inoculated control (Fig. 11.4, a & b).

Mycolytic enzymes are produced by antagonistic microorganisms are of much importance in biocontrol technology. Under green house conditions β-1, 3 glucanase producing *Burkholderia cepacia* decreased the incidence of disease caused by *Rhizoctonia solani*, *Scleotium rolfsii* and *Pythium ultimum* (Fridlender *et al.*, 1999). Coagulation of cytoplasmic contents in the mycelia of *P. capsici* was noticed in the interaction between *P. fluorescens* and *P. capsici* in addition to production of glucanase and chitinase (Diby Paul *et al.*, 2005). Efforts world-wide which are focused on the mechanism of action of fluorescent Pseudomonades against plant pathogenic fungi stressed on the involvement of production of antifungal metabolites by fluorescent Pseudomonads (Loper *et al.*, 1994, Loper *et al.*, 1997) Phenazine-1-carboxylic acid (PCA), pyoluteorin, pyrrolnitrin, 2-4 diacetyl phloroglucinol and oomycin A are some of the antibiotics which have been attributed for their disease suppressive effects (Bonsall *et al.*, 1997; Howell and Stipanovic, 1979; 1980; Thomashow *et al.*, 1990). The volatile metabolite HCN produced by *P. fluorescens* inhibited the mycelia growth of *P. capsici* over 23.2%. These Pseudomonads strains also produced non-volatile antibiotics viz., pyolutoerin and pyrrolnitrin as evidenced by the thin layer chromatography (Diby Paul, 2004). Mycelial growth inhibition and coagulation of cytoplasm, inhibition of sporangial production, indirect germination of sporangia and zoospores by the PGPR strain are of great relevance in disease suppression (Fig. 11.4). Induction of ISR in black pepper has been clearly established based on the enhanced defence related enzymes (Diby Paul and Sarma, 2005). The role of PGPRs in enhancing the defence system in plants through ISR has been well documented (Kloepper *et al.*, 1993; Kloepper *et al.*, 1980, 1999). PGPRs are able to suppress root rot caused by *P. capsici* in black pepper (Diby Paul and Sarma, 2003; Diby Paul *et al.*, 2000). Increased level of peroxidase (PO), catalase, phenylalanine ammonia lyase (PAL) and polyphenol oxidase (PPO) were induced in leaves apart from the roots of treated plants indicating the systemic protection offered to black pepper by different strains. The induction of defence enzyme was sudden in roots and was gradual in the leaves (Diby Paul and Sarma, 2003). The role of *P. fluorescens* mediated nutrient flux in the soil ecosystem in black pepper growth promotion was confirmed with the higher uptake of nutrients by the bacterized plants after 90 days of treatment and planting. Significant uptake of nitrogen (N) and potassium (K) was noticed in rhizobacteria treated plants as it is related to the dry matter yield. The bacterial population associated with the roots maintained a minimum population of log 6 even up to 90 days after treatment (Diby Paul *et al.*, 2005). PGPRs, especially *P. fluorescens* was found to be highly effective in suppressing root rot caused by *P. capsici* and *R. similis* in black pepper through growth promotion and induced systemic resistance (ISR). These rhizobacteria were found to be compatible with *T. harzianum* and VAM. The production of significantly higher number of leaves, maximum root length and total biomass were noted in this treatment followed by the application of rhizobacteria (Diby Paul *et al.*, 2003; Thankamani *et al.*, 2003). Application of rhizobacteria alone or combination with other biocontrol agents (*T. harzianum* and *G. fasciculatum*) were found effective in growth promotion and disease suppression and production of vigorous black pepper rooted cuttings in the nursery.

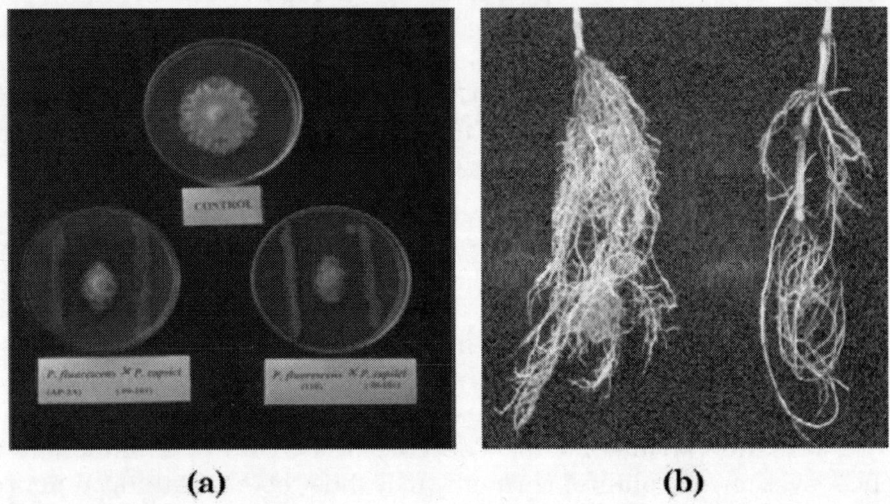

Fig. 11.4. *Phytophthora capsici*-black pepper patho system. (a) In vitro suppression of *P. capsici* by *Pseudomonas fluorescens*, (b) Root rot suppression and enhancement of root biomass in black pepper by *P. fluorescens* in comparison with infected root system.

Cyclic lipopeptide surfactant producing Pseudomonads found to play a major role in *Phytophthora* caused plant diseases (Tran Ha, 2007). Motile zoospores of *P. infestans* were highly sensitive to cyclic lipopeptide (CLP) massetolide A produced by *P. fluorescens* (SS101) followed by mycelium and cysts. This compound caused lysis of zoospores of *P. capsici* at 25 ppm. In Vietnam the black pepper rhizosphere harbored several CLP producing Pseudomonads, particularly of *P. putida* (PCL 1445) strains. These reduced *P. capsici* infection in black pepper cuttings planted in the field. In addition they promoted root and shoot growth in black pepper. However their (CLP) clear role in disease suppression needs further studies. The field evaluations of *P. putida* are in progress in Vietnam (Tran Ha, 2007). The recent investigation carried out in Malaysia on rhizobacterial strains associated with roots of black pepper, showed that *B. amyloliquefaciens*, *B. atrophaeus*, *B. subtillis* and *B. vallismortis* were found to show *in vitro* suppression in dual culture with *P. capsici.* These caused malformation of hyphae and were effective colonizers of black pepper roots. These are being further investigated for their field efficacy (Yap Chin Ann, 2012).

Endophytic bacteria in black pepper, *Pseudomonas* spp., *P. aeruginosa, P. putida* and *B. megaterium* were found to suppress *P. capsici,* to varying in degrees in addition to growth promotion (Aravind *et al.,* 2008; Aravind *et al.,* 2012). However, *P. aeruignosa* cannot be used as a biocontrol agent because it is a human pathogen (Kumar *et al.,* 2012). These endophytes *Curtobacterium luteum* and *B. megaterium* where found suppressive to nematodes. Out of seventy four endophytes of black pepper two isolates namely TC10 followed by BP 17 were found suppressive to *R. similis.* (Aravind *et al.,* 2010). This is an important finding, but this needs further evaluation in large scale nurseries as well as in pepper gardens. A bacterial mixture that would suppressive to *P. capsici* and *R. similis* is an immediate relevance.

Foliar application of *P. fluorescens* in pepper resulted in decrease of anthracnose to 35% disease index compared to 61% in untreated control. Corresponding yields were 2.34 kg/vines in treated compare to 0.78 kg in control (Anon, 2004)

Small Cardamom

A fluorescent pseudomonads isolate from cardamom capsules shown *in vitro* inhibition of *P. meadii* and also found effective in checking panicle infection when spray with bacterial suppression (Suseela Bhai and Sarma, 2003).

Ginger

The *in vitro* suppression of *F. oxysporum* f. sp. *zingiberi* by *B. subtilis* and reduced disease incidence was notice in pot cultures studies (Sharma and Jain, 1979). PGPR strain mixture S2BC-1) (*Bacillus subtilis*) with TEPF sungal (*Burkholderia cepacia*) shown maximum rhizome production (85.2%) reduced *Fusarium* yellows (87.8%) and reduced rhizome rot (88.4%) over control. Defense responses of increased chitinase, β-1,3 glucanase, and polyphenol oxidase were noticed (Shanmugam *et al.,* 2013). In addition *B. atrophaeus* strain S2BC-2 were also reported antagonistic *Fusarium* yellows in Himachal Pradesh and also increased yields. This was attributed to increase chitnolytic activity of antagonist (Shanmugam *et al.,* 2012). In addition *B. atrophaeus* strain S2BC-2 was found antagonistic to *Fusarium* spp. reduced the disease and enhanced the rhizome production by 113.3%. This was attributed to enhance defence system through increased chitinase activity (Shanmugam *et al.*, 2012).

Bacterial wilt of ginger caused by *R. solanacearum* biotype 3 and 4 still remain unresolved for the management of the disease. Studies carried out in other pathosystem with avirulent strains of *R. solanacearum*, *Bacillus* spp. and fluorescent Pseudomonads were not successful. Potential of ginger endophytic bacteria *Stenotrophomonas maltophila*, and *Acinetobacter calcoaceticus* for suppression of *Ralstonia solanacearum* has been observed (Anon, 2009). The recent studies at Indian Institute of Spice Research Calicut indicated *in vitro* suppression of *R. solanacearum* with *Streptomyces* spp., and these need to be further explored for their disease suppressive potential (Anon, 2012). The role of Actinomycetes in spice crop disease management remains unexplored.

Vanilla

As a part of integrated approach in fusarial wilt control *Bacillus patothenticus*, *B. firmus* and *P. fluorescens* were found effective in disease management of vanilla (Tombe, 2010).

Seed Spices

Studies were undertaken on the PGPRs and endophytes of seed spices for their growth promotion. Out of 25 short listed bacterial isolates, *P. putida* from rhizosphere of fenugreek, *Microbacterium paroxidans* an endophyte from fennel were found to increase yield of to varying degrees in coriander and fenugreek. Based on location yields were variable and increased yields ranged from 10–35% (AICRPS, 2009, 2010).

THE ROLE OF BIOCONTROL AGENTS IN PLANT PARASITIC NEMATODE INFECTION / SUPPRESSION IN SPICE CROPS

Suppression of *M. incognita* and increased plant growth in black pepper by *B. pumilus*, *B. macerans* and *B. circulans* was reported (Sheela *et al.*, 1993). Similar effects were reported with *P. fluorescent* in black pepper root rot suppression (Eapen, 1995). The subject was reviewed earlier (Eapen and Ramana, 1997). The disease suppressive effects of *Glomus fasiculatum* arbuscular mycorrhizal fungus in slow decline caused by *R. similis, M. incognita* and *also P. capsici* were recorded earlier (Anandaraj *et al.*, 1991). *Pochonia chlamydosporia, Paecilomyces lilacinus, Trichoderma* spp., *Pasteuria penetrans, P. fluorescens, P. putida* have been reported in suppression of nematode infections in spice crops. Incidentally *T. harzianum, P. fluorescens* and *P. putida* were found to suppress *P. capsici* (Diby Paul, 2004)., Saju, 2004; Tran, 2007). Nematicidal activity of *P. putida* and *B. megaterium* are being pursued (Anon, 2012). At present Indian Institute of Spices Research Calicut, supplies talcum based formulations of *T. harzianum, P. fluorescens* and *P. chlamydosporia,* to farmers.

Plant parasitic nematodes always enhanced the damage in spice crops due to combined infections with plant pathogenic fungi *P. capsici, P. aphanidermatum* in black pepper, ginger and cardamom (Ramana and Eapen, 1996). In homestead cultivation have also high density multi storied cropping systems where coconut-black pepper, Arecanut-black pepper, banana ginger and turmeric are cultivated; the plant parasitic nematodes and plant pathogenic fungi overlap under field situation. The spatial segregation of these pathogens under field situation could never exist and hence the need for a holistic approach is essential to boost up the health and productivity of the crops through microbial bio-suppression. Here microbials with a dual mode of action as well as microbial mixtures would be more relevant (Sarma, 2006).

MICROBIAL MIXTURES/ MICROBIAL CONSORTIA (MM/MC) FOR BIOLOGICAL CONTROL

Black Pepper

In recent years investigations have been intensified to understand the roll of single biocontrol agents as well as their mixtures, and their compatibility on disease management. Studies on compatibility of fungal and bacterial antagonist viz., *Trichoderma harzianum* and *fluorescens* Pseudomonads showed that they are compatible (Jisha *et al.*, 2002). Application of rhizobacteria and *T. harzianum* resulted in significantly enhanced growth of black pepper with increased number of nodes and consequently cuttings (Anandaraj *et al.*, 2003; Saju, 2004; Saju *et al.*, 2003). *P. fluorescens, Bacillus* spp. provided grater protection to cardamom seedlings against *F. oxysporum* and *P. meadii* (Joseph Thomas and Vijayan, 2003). The importance mixtures of plant growth promoting rhizobacteria in suppression of plant diseases are well known (Raupach and Kloepper, 1998). Like in *Trichoderma*, talcum based formulations are extensively used for PGPRs also.

Cardamom

The role of *P. fluorescens* and *Bacillus* in root rot suppression in cardamom seedlings in comparison with *T. harzianum* was studied as seed dressing and soil application adopting challenge inoculation with *P. meadii* and *F. oxysporum*. All the three biocontrol agents provided good protection without any visible root rot (Joseph Thomas and Vijayan, 2002).

Ginger

Recent investigations in Western Himalayas, the talc-based formulations of the PGPR strain XXBC-TN (*Bacillus subtilis*) and mixture of S2BC-1 (*Bacillus subtilis*) with TEPF-Sungal (*Burkholderia cepacia*) which inhibited *F. oxysporum* and *F. solani* were developed for rhizome dressing and soil application in ginger field. Comparatively, the strain mixture recorded maximum rhizome production (84%) with lesser yellows and reduced rhizome rot incidences (84.2 and 79.7%) over the control in polyhouse. Reduction of disease incidence under polyhouse conditions occurred might be through induction of defense gene products such as chitinase, ß-1, 3-glucanase, and polyphenol oxidize. Besides disease suppression, treatment with strain mixture promoted plant growth and enhanced rhizome production by 45.8%. In field experiments, the strain mixture reduced yellows and rhizome rot incidences by about 50.5%, and comparable to that of fungicide mixture (48.6 and 54.0%, respectively) (Shanmugam *et al.*, 2013).

Turmeric

Application of NPK @125:60:90 kg/ha, Farm yard manure @ 10/ha, seed treatment with *T. viride* + *P. fluorescens* @4g/kg seed and soil application of *T. viride* @ 12.5 kg/ha and *P. fluorescens* @ 25 kg/ha resulted in reduction of rhizome rot to 11.7% compared to 37% in untreated control. The yield levels in the treated were @ 28.6 t /ha in treated plots compared to 10.3 t/ha.

Vanilla

Out off four microbial consortia of *P. fluorescens, Enterobacter agglomerans, Bacillus* spp. against *P. meadii, Fusarium oxysporum,* and *Colletotrichum vanillae* shown that three consortia did show maximum disease suppression of 92.9% (Suseela Bhai and Kumar, 2008).

HEALTHY PLANTING MATERIAL PRODUCTION THROUGH BIOCONTROL RESOURCES

In spice crops as mentioned earlier soil borne diseases/ insect pests are seed borne/planting material borne too (Sarma, 2006), Except for the seed spices, many of the spice crops are vegetatively propagated and diseases/insect pests are transmitted to next generation. Even in seed spices, cumin wilt pathogen *F. oxysporum* f. sp. *cumini* is seed borne (Singh *et al.*, 1972). The initial pathogen contamination of seeds leads to initiation of the disease afresh and subsequent spread. Even in areas where the crop was not cultivated earlier, the disease occurrence in the new areas indicates the seed borne nature of the diseases. Specially ginger and turmeric, are typical examples. Protocols of micropropagation (Tissue culture) of black pepper, Small cardamom and Turmeric have been standardized (Nirmal Babu *et al.*, 2006). This technology would be ideal to produce pathogen free (Fungal and Nematodal) planting material. However they will not be free from plant viruses. Until this micro propagation is made simple, cheap and cost effective, the conventional methods of vegetative propagation would continue. Introducing biocontrol agents at hardening phase of the micro propagation would be rewarding (Nowak, 1998).

Black Pepper/ Small Cardamom

The root infection in nurseries caused by *P. capsici, R. similis* and *M. incognita* in black pepper goes unnoticed without any foliar symptoms, even though one can notice infection on the root system. Similarly root infection caused by *F. oxysporum* in small cardamom is another serious problem in healthy planting material production. Distribution of such contaminated planting material is the major source of disease initiation and spread (Sarma and Saju, 2004). Efficacy of *T. harzianum* (Rajan *et al.*, 2002), Vesicular Arbuscular Mycorrhiza (VAM) in suppression root infection caused by *P. capsici, R. similis* and *M. incognita* in black pepper have been reported (Anandaraj *et al.*, 1996). Similarly optimum protection of the root system from these pathogens in nursery management through *P. fluorescens* (Diby *et al.*, 2001), *Bacillus megaterium* (Aravind *et al.*, 2012, *Alcaligenes* (Anith *et al.*, 2002, 2003) has been reported. This technology is now being implemented in majority of the black pepper nurseries.

The combination of *Trichoderma* and fluorescent pseudomonades has been reported effective in suppression of root rot. In small cardamom, the present leads available on bio suppression of *Phytophthora* and *F. oxysporum* need to be persued seriously to produce pathogen free planting material. Once cost effective, micro propagation techniques are evolved it would be most appropriate to introduce the bio control agents at hardening phase of the axenic material, to produce totally pathogen free planting material in black pepper and small cardamom.

Ginger and Turmeric

Apparently normal seed rhizomes of ginger and turmeric collected from farmers fields are often contaminated with *Pythium* spp., *Fusarium* spp, *R. solanacearum* and *P. coffeae*. This initial seed contamination leads to disease initiation and spread. Based on studies carried out in Sikkim, hot water treatment of seed rhizomes at 51° C for ten minutes has been recommended to eliminate the above pathogens except *R.solanacearum*. Hot water treated seed coated with *T.harzianum* provided better protection against rhizome rot ginger in Sikkim (Rajan *et al.,* 2002). Micro tuber technology has been developed recently which is cost effective and totally disease free (Nirmal Babu *et al.,* 2006). Introduction of biocontrol agents (*T. harzianum, P. fluorescens, B. subtilis, B. cepacia* etc.) at this micro tuber stage of the rhizome would be very effective in producing disease free ginger and turmeric seed rhizomes. This need to be popularized.

Seed spices

As mentioned earlier cumin wilt has been reported as seed borne. The seed borne inoculam can be effectively eliminated with hot water treatment at 54°C for fifteen minutes (Singh *et al.,* 1972). Seed pelletting with *T. harzianum* has been already recommended (Meena *et al.,* 2010). Hot water treated seed combined with biocontrol agents would be more rewarding.

NURSERY MANAGEMENT THROUGH FORTIFICATION OF NURSERY MIXTURE WITH BIOCONTROL AGENTS

Solarized or non-solarized nursery mixture fortified (mixed with) formulations of *T. harzianum, T. viride, P. chlamydosporia,* fluorescent pseudomonades like *P. fluorescens, P. putida or B. subtilis, B. megaterium* along with Arbuscular Mycorrhizal *(Glomous fascicul-*

atum) inocula would ensure robust healthy plants free from root infection. The technology would be need based and demand driven for each of the crops based on the compatibility of either microbial mixtures or of a single isolate. Guidance and inputs from the research agencies would become imperative to prove the bioefficacy and cost effectiveness of this ecofriendly technology. Early root colonization of these microbes be with high rhizosphere competence at nursery stage would check root rot infection and trigger defence system in the nursery stock and render them highly robust and healthy.

BIOLOGICAL CONTROL IN INSECT PEST MANAGEMENT

Infestation by insect pests is a major factor responsible for the low productivity of spice crops in India. Spice crops such as black pepper (*Piper nigrum* L.), cardamom (*Elettaria cardamomum* Maton), ginger (*Zingiber officinale* Roscoe), turmeric (*Curcuma longa* L.) and seed spices like coriander (*Coriandrum sativum* L.), cumin (*Cuminus cyminum* L.), fennel (*Foeniculum vulgare* Mill.) and fenugreek (*Trigonella foenum-graecum* L.) are infested by many species of insect pests, causing severe crop losses. Adoption of IPM schedules in which biological control forms a major component is more relevant in these crops since excessive use of insecticides could lead to pesticide residues in the produce and other ecological hazards. There has been a renewed interest during the past decade in developing environment–friendly crop management schedules in sustainable agriculture including spice crops, reflecting the increasing concern over pesticide misuse. The information available on management of major insect pests of spice crops through biological means including plant products and the scope and challenges in its adoption are highlighted in this chapter.

Pollu Beetle

The pollu beetle (*Lanka ramakrishnae* Pratapan and Viraktha-math) (Coleoptera, Chrysomelidae) is the most destructive insect pest of black pepper in India. The pest infestation is higher in plains and midlands in Kerala where 10–32% of berries are reported to be damaged by the pest (Premkumar and Nair, 1988). The adult feeds on tender shoots, leaves and spikes resulting in black patches on shoots and spikes and small irregular holes on tender leaves. The larva (grub) bores into tender berries and feeds on the internal tissues. The infested berries turn black and crumble when pressed (the term pollu denotes the hollow nature of the infested berries in Malayalam). The grub also bores into tender spikes resulting in necrotic patches on them. The pest infestation is more severe in shaded areas in the plantation and on vines trailed on standards (support trees) producing heavy shade (Devasahayam, 2000a).

Natural enemies. Very few general natural enemies such as predatory spider and ant have been recorded on pollu beetle in the field and they may not be of much significance in the control of the pest (Devasahayam and Koya, 1994a; Devasahayam, 1996). Evaluation of entomopathogens such as *Beauvaria* spp. and *Metarhizium anisopliae* (Metsch) Sorokin (Hyphomycetes) from other sources indicated their limited scope for use as biopesticides for the management of the pest (IISR 1996).

Plant products. Evaluation of various commercial neem products in laboratory bioassays against pollu beetle indicated that Repelin, Neemazal-F and neem seed kernel extract were promising (Devasahayam and Anandaraj, 1997). An integrated strategy involving regulation of shade in the plantation and spraying quinalphos (0.05%) during July followed by 3–4 sprays (at 21 day intervals) of neem-based product such as Neemgold 0.6% during August–October, or spraying Neemgold 0.6% during July–October (4–5 sprays at 21 day intervals) has been suggested for controlling the pest (Devasahayam and Koya, 1999). Various plant extracts have been evaluated in the laboratory for their antifeedant activity against pollu beetle. Among them, root bark extracts of *Uvaria narum* Blume and *U. hookeri* King (Annonaceae) (Babu *et al.,* 1996), leaf extracts of *Strychnos nux-vomica* L. (Loganiaceae), *Chromolaena odorata* L. (Asteraceae) and seed extract of custard apple (*Annona squamosa* L.) (Annonaceae) (Devasahayam and Leela, 1997) possessed significant antifeedant activity indicating their potential in utilizing them in IPM schedules.

Scale Insects

Various species of scale insects have been recorded on black pepper in India among which, mussel scale *Lepidosaphes piperis* Green and coconut scale *Aspidiotus destructor* Sign. (Homoptera, Diaspididae) are more common and widely distributed especially at higher altitudes (above 600 m MSL) in Kerala, Karnataka and Tamil Nadu. The mussel scale encrusts stems, mature leaves and berries resulting in chlorotic patches, yellowing and drying of and lateral branches, and mortality of young vines. In the nursery, the pest infestation is generally seen on older cuttings. The coconut scale infests mature leaves resulting in chlorotic patches and sometimes stems of lateral branches and berries (Koya *et al.*, 1996).

Natural enemies. Twenty-one species of natural enemies have been recorded on scale insects in the field among which *Chilocorus circumdatus* (Gyllen.) and *C. nigrita* (Fabr.) were identified to be potential biocontrol agents of *A. destructor* and *L. Piperis*, respectively (Devasahayam, 1996; Selvakumaran *et al.*, 1996). Techniques were also standardized for mass rearing of *C. circumdatus* and *C. nigrita* on *Aonidiella orientalis* (Newst.) (host scale) raised on pumpkins (*Cucurbita moschata* Duch. ex Poir.) in the laboratory (Devasahayam, 1998).

Plant products. Plant and organic products such as neem oil 0.3%, Neemgold 0.3% and fish oil rosin 3% were promising in controlling mussel scale and coconut scale and were also safe to the coccinellid predators indicating their potential use in IPM schedules (Devasahayam, 1998). Spraying of Neemgold also reduced coconut scale infestations on black pepper in the field (Selvan *et al.,* 1996). Various neem and organic products were evaluated at Mudigere (Karnataka) against scale insects among which neem oil @ 5 ml/l and fish oil @ 3 g/l were promising and were as effective as insecticides (Naik *et al.,* 2009).

Top Shoot Borer

The top shoot borer (*Cydia hemidoxa* Meyr.) (Lepidoptera, Tortricidae) is widely distributed in Kerala and is a serious insect pest in younger plantations. The larva infests tender terminal shoots that adhere to the standards and the earlier instars scrapes and feeds on the epidermis of tender terminal shoots and sometimes on tender leaves. The later instars bore into tender shoots and feeds on the internal contents resulting in decay and drying of infested shoots. Repeated infestation of tender terminal shoots affects the growth of young vines (Devasahayam and Koya, 1993a; Devasahayam, 2000a).

Natural enemies. Among the natural enemies recorded on larva of top shoot borer, *Hexamermis* sp. (Mermithidae) and *Apanteles cypris* Nixon (Braconidae) are more common and were identified as potential biocontrol agents (Devasahayam and Koya, 1994a; Devasahayam, 1996). The mermithid nematode was more common during the monsoon period (June to September) at Peruvannamuzhi (Kerala) with a peak parasitism of 66.6–76.7% during July–August (Devasahayam and Koya, 1994b). The hymenopterous parasitoid was more common during the post monsoon period (September–November) with a peak parasitism of 20% during October (Devasahayam and Koya, 1993b).

Leaf Gall Thrips

The leaf gall thrips (*Liothrips karnyi* Bagn.) (Thysanoptera, Phlaeothripidae) is a widespread insect pest on younger vines at higher altitudes and also on rooted cuttings in nurseries in the plains in Kerala, Karnataka and Tamil Nadu. Leaf gall thrips infest tender leaves causing the leaf margins to curl downwards and inwards resulting in the formation of marginal, tubular, hypophyllous galls. As the infested leaves grow they become crinkled, malformed and reduced in size; the leaf margins also turn chlorotic later. Severe infestations of the pest affect the growth of younger vines and rooted cuttings in the nursery (Devasahayam, 1992).

Natural enemies. Among the predators recorded on leaf gall thrips in the field *Montandoniola moraguesi* (Puton) (Anthocoridae) and *Andothrips flavipes* Schmutz (Phlaeothripidae) are widely distributed and were identified as potential biocontrol agents of the pest. *M. moraguesi* predates on all stages of leaf gall thrips whereas, *A. flavipes* generally prefers eggs and juvenile stages of leaf gall thrips (Devasahayam, 1992).

Root Mealybugs

Colonies of root mealybugs (*Planococcus* spp.) (Pseudococcidae) are observed at the basal portion of the stem under the soil and on roots causing yellowing, wilting and mortality of younger vines. The pest infestation is more common at higher altitudes in Wayanad (Kerala) and Kodagu (Karnataka) districts and is more severe on vines affected with *Phytophthora* sp. and nematodes. Root mealybugs are oval, soft bodied insects measuring about 1.5 mm x 1.0 mm in size and the body is covered with white waxy filaments (Devasahayam *et al.*, 2009).

Plant products. Drenching tobacco extract 3% was effective for the management of mild infestations of root mealybugs in the field (Devasahayam, 2006).

Cardamom Thrips

The cardamom thrips (*Sciothrips cardamomi* Ramk.) (Thysanoptera, Thripidae) is the most destructive pest occurring in all cardamom areas. The adults and larvae lacerate the tissues of leaves, shoots, panicles, flowers and immature capsules and feed on the exuding sap. The pest infestation results in shedding of flowers and immature capsules and the formation of corky, scab-like encrustation on the capsules. Severely affected capsules are malformed, shriveled and without seeds. The pest is reported to cause 42–47% yield loss in various areas (Varadarasan 1995; Gopakumar and Chandrasekhar, 2002).

Natural enemies. Very few natural enemies such as chrysopid and anthocorid bug have been recorded on cardamom thrips in the field and they may not be of much significance in the control of the pest (Varadarasan, 1995).

Plant products. Various neem-based products were evaluated against cardamom thrips at Mudigere (Karnataka) in comparison with monocrotophos and the results were not consistent. The mean damage by thrips in moncrotophos 0.05% treated plots was 17.9% when compared to 26.9% in plots treated with neem oil 0.03% (Naik *et al.*, 2006). At Pampadumpara (Kerala), evaluation of plant and other organic products for the management of thrips indicated that among the organic products FOIS (Na) 2.5% + tobacco extract 2.5% significantly reduced the damage caused by thrips; however, spraying quinalphos 0.05% was the most effective for the management of the pest (Rajkumar *et al.*, 2002).

Cardamom Shoot and Capsule Borer

The shoot and capsule borer (*Conogethes punctiferalis* Guen.) (Lepidoptera, Pyralidae) is a serious pest of cardamom in the nursery and main field. The early instars of larvae bore into unopened leaf buds, panicles and immature capsules. The later instars of larvae bore into pseudostems and feed on the central inner portion resulting in drying of the terminal leaf and producing the characteristic 'dead heart' symptom of pest attack. Yield losses of 70–80% have been recorded in severely infested plantations (Varadarasan, 1995; Gopakumar and Chandrasekhar, 2002).

Natural enemies. Twenty species of hymenopterous parasitoids have been recorded on the shoot and capsule borer (Varadarasan, 1995). Among them, *Eriborus trochanteratus* (Morley), *Xanthopimpla* *australis* Kr., *Friona* sp. and *Agrypon* sp. (Ichneumonidae) have been identified as potential biocontrol agents. Parasitization by *E. trochanteratus* and *X. australis* ranged between 10.5–11.1% at Appangala (Karnataka) (CPCRI 1985). *Friona* sp. and *Agrypon* sp. parasitized 8–10% and 16–18% of larvae/pupae, respectively, in the field at Idukki (Kerala) (Varadarasan, 1995).

Plant products. Various neem based products were evaluated at Mudigere (Karnataka) against shoot and capsule borer; however, the results were not promising. The mean incidence of shoot and capsule borer was lowest (2.2%) in plants treated with monocrotophos 0.05% compared to 5.1% in plants treated with neem oil 0.05% (Naik *et al.*, 2006). At Pampadumpara (Kerala), evaluation of plant and other organic products for the management of the pest indicated that none of the treatments were effective (Rajkumar *et al.*, 2002).

Root Grubs of Cardamom

The root grub (*Basilepta fulvicorne* Jacoby) (Coleoptera, Chrysomelidae) is a serious insect pest of cardamom, being distributed in all cardamom areas. The larvae feed on roots and rhizomes and in severe cases of infestation, the entire root system is eaten away. The infested plants turn yellow and become stunted; severely infested plants, especially seedlings, succumb to the pest attack. Seedlings damaged by root grubs are subsequently infected by secondary pathogens resulting in rotting. Crop losses of 29–66% have been reported in various cardamom areas (Varadarasan 1995; Gopakumar and Chandrasekhar, 2002).

Natural enemies. The natural enemies recorded on the root grub include the green muscardine fungus *Metarrhizium anisopliae* (Metsch) Sorokin (Hyphomycetes) and the entomogeneous nematode *Heterorhabditis* sp. (Heterorhabditidae) infesting grubs, and the white muscardine fungus *Beauvaria bassiana* (Bals.) Vuill. (Hyphomycetes) infesting adults in the field. These pathogens are highly pathogenic to the host under laboratory conditions and also in the field indicating their potential as biocontrol agents (Vardarasan *et al.,* 1993; Varadarasan 1995; Varadarasan and Sivasubramonian, 1996; Varadarasan *et al.,* 2010).

Cardamom Aphid

Colonies of cardamom aphid (*Pentalonia caladii* van der Goot) (Homoptera, Aphididae) occur within leaf sheaths of pseudostems feeding on plant sap. Though the direct damage caused by them is not very serious, they are vectors of *katte* and *kokke kandu* diseases of cardamom.

Natural enemies. The natural enemies recorded on cardamom aphid include predatory chrysopids, coleopterans and dipterans and entomopathogenic fungi (Devasahayam, 1996). Laboratory bioassays indicated that entomopathogens such as *B. bassiana* and *Pochonia chlamydosporia* (Hyphomycetes) were promising for the management of the pest (Mathew *et al.*, 1997).

Plant products. Laboratory bioassays of plant extracts and commercial neem products indicated that seed extract of *Annona squamosa* L. (Annonaceae), leaf extracts of *Lawsonia inermis* L. (Lythraceae) and *Leucas aspera* Sprengel (Lamiaceae), rhizome extract of *Acorus calamus* L. (Acoraceae) and Bioneem and Margolin significantly affected the settling and colonization behaviour and multiplication of cardamom aphid. Trials laid out in the field also indicated that neem products such as Margolin and Bioneem (1.6% each) and extracts of *A. squamosa* and *A. calamus* (5% each) were promising for the management of the pest (Venugopal, 1999).

Ginger and Turmeric Shoot Borer

The shoot borer (*Conogethes punctiferalis*) (Lepidoptera, Pyralidae) is the most serious insect pest of ginger and turmeric being widely distributed in India. The larvae bore into pseudostems and feed on the central growing shoot resulting in yellowing and drying of infested shoots. The presence of a bore-hole in the pseudostem through which frays is extruded and the withered central shoot is a characteristic symptom of pest infestation. In ginger significant reduction in yield occurs when more than 50% of the pseudostems in a clump are infested by the pest (Koya *et al.,* 1986; Devasahayam and Koya, 2005; 2007).

Natural enemies. Seventeen species of natural enemies have been recorded on shoot borer in the field. At Peruvannamuzhi (Kerala), *Hexamermis* sp. (Mermithidae) and *Apanteles taragammae* (Braconidae) were more common and were identified as a potential biocontrol agent of the pest (Devasahayam, 1996). Commercial formulations of *Bacillus thuringiensis* such as Dipel 0.3% were effective in controlling the pest infestation on ginger and turmeric in the field when sprayed at 21 day intervals during July–October (Devasahayam, 2000b; 2002). Evaluation of eight native entomopathogenic nematode isolates against shoot borer indicated that *Heterorhabditis* sp., (Heterorhabditidae), *Steinernema* sp. (Steinernematidae) and *Oscheius* sp. (Rhabditidae), caused 100% moratality to larvae in laboratory assays indicating their potential in the management of the pest (Pervez *et al.*, 2011).

Rhizome Scale of Ginger and Turmeric

The rhizome scale (*Aspidiella hartii* Ckll.) (Homoptera, Diaspididae) infests rhizomes of ginger and turmeric both in the field (especially during the later stages of the crop) and in storage. The pest infestation is generally seen in the form of encrustation on the rhizomes during the later stages of the crop and severely infested plants wither and dry. In storage, the pest infestation results in shriveling of buds and rhizomes affecting their sprouting (Devasahayam and Koya, 2005, 2007).

Natural enemies. A few hymenopterous parasitoids and general predators have been recorded as natural enemies of rhizome scale among which *Physcus comperei* Hayat (Aphelenidae) and *Adelencyrtus moderatus* Howard (Encyrtidae) were more common and identified as a potential biocontrol agents of the pest (Devasahayam, 1996).

Plant products. Discarding of severely infested rhizomes and storage in dried leaves of *Strychnos nux-vomina* L. (Loganiaceae) and saw dust in 1:1 proportion was effective in reducing the damage caused by rhizome scale to ginger during storage (IISR, 2003).

Aphids of Seed Spices

Various species of aphids infest seed spices among which the coriander aphid (*Hyadaphis coriandri* (Das) (Homoptera: Aphididae) and green peach aphid (*Myzus persicae* Sulzer) (Homoptera: Aphididae), are more common. Adults and nymphs of coriander aphid suck the sap from leaves, flowers and fruits and exuding honey dew which favors the growth of sooty mould and resulting in poor growth of plants, drying of flowers and shriveling of fruits. The damage by the green peach aphid is generally confined to the umbels (Mittal and Butani, 1994; Singh, 2004).

Natural enemies. Various species of coccinellid beetles predate on aphids in the field among which *Coccinella septumpunctata*, *Bromoides suturalis*, *Menochilus sexmaculatus* and *Adonia variegata* are common. The other common predators include *Chrysoperla carnea* (Chrysopidae), *Episyrphus balteatus* and *Ischiodon scutellaris*

(Syrphidae) and *Aphelinus* sp. (Aphelinidae) also commonly parasitizes the aphids (Mittal and Butani, 1994; Singh, 2004).

Lepidoperans of Seed Spices

Among the various leaf feeding caterpillars infesting seed spice crops, *Spodoptera litura* (Lepidoptera: Noctuidae) and *Helicoverpa armigera* Hubner (Lepidoptera: Noctuidae) are common. The earlier instar larvae scrape and feed on leaf tissues and the later instar larvae feed on the entire leaf and also flowers and fruits. The adult moths of *S. litura* are brown with wavy white markings on the forewings and brown patches on hind wings. The larvae are greenish brown with spots and longitudinal grey bands when fully grown. In *H. armigera*, the forewings are yellowish-orange in females and greenish-grey in males, with a faint broad transversal band. The hind wings are a pale yellow with a narrow brown band at the external edge. The larvae resemble the larvae of *S. litura*, but do not possess the black spots on the thoracic region (Mittal and Butani, 1994; Singh, 2004).

Natural enemies. A number of predators and parasitoids have been recorded on *S. litura* and *H. armigera* that check the field populations of the pest. Nuclear polyhedrosis viruses and *Bacillus thuriengiensis* have also been exploited for the management of the pest in the field. Release of *Trichoderma* sp. at fortnightly intervals thrice has been suggested against S. *litura*. This could be supplemented with release of *Chelonus blackburni* (Braconidae) and *Chrysopela* sp. for controlling the larvae. Against *H. armigera* release of *Trichogramma* spp. (Trichogrammatidae) 4–6 times supplemented with release of parasitoids such as *C. blackburni*, *Bracon brevicornis* (Braconidae), *Telenomus heliothinae* (Scelionidae), *Carcelia ilota* (Tachinidae), or *Campoletis chloridae* (Ichneumonidae) or *Brachymeria* sp. (Chalcididae) and predators such as *C. carnea* and *Scymnus* sp. (Coccinellidae) have been suggested. Erection of bird perches at about 50 perches/ha of cropped field helps in predation of larvae by birds (Mittal and Butani, 1994; Singh, 2004).

Plant products. Plant products, such as neem and *Melia azadirachta* L. have been found effective in control of leaf feeding caterpillars due to their antifeedant activity (Mittal and Butani, 1994; Singh, 2004).

Seed Spice Wasp

The seed wasp (*Systole albipennis* Walk.) (Hymenoptera: Eurytomidae) is a polyphagous species and the female lays its eggs in-between the pericarp and ovule and the emerging larvae feeds on the developing embryo and endosperm of the fruit (Mittal and Butani, 1994; Singh, 2004).

Natural enemies. The parasitoids recorded on seed midge include *Liodontomerus indicus* Rao and Bhatia (Torymidae) and *Dinarmus basalis* (Rondani) (Pteromalidae) (Mittal and Butani, 1994; Singh, 2004).

STRATEGIES FOR THE FUTURE

A number of natural enemies have been recorded on most of the major insect pests of spice crops and potential biocontrol agents have also been identified. However, techniques are to be standardized for the mass culture of many of the biocontrol agents and information generated on their efficacy of these species in the field. The perennial nature or long duration of most of the spice crops and the availability of other crops and plant species along with various insect species in the ecosystem on which the natural enemies can multiply, are ideal factors for incorporation of biocontrol strategies in the integrated management of insect pests of spice crops. Improvement of crop habitat and alternate food sources by cultural manipulations and selective use of safer insecticides for the control of insect pests would help in conservation of natural enemies.

Very few microbial pathogens have been documented on insect pests of spice crops in the field. The humid conditions under most of the spice crops are grown is an ideal factor for the use of microbial pathogens in the management of insect pests of spices. Various aspects of insect-microbe interactions especially variations in epizootic potential, sporulation and dispersal of various strains in the ecosystem are to be studied. The feasibility of multiplication of microbial pathogens in cheap substrates such as agricultural wastes commonly available in plantation crop ecosystems are to be studied.

Spice crop ecosystems contain many other crops, and indigenous plants and trees that are used as shade and support trees. The role of plant volatiles of various plant species and kairomones of insect pests in these ecosystems and their effect on the behaviour of natural enemies in relation to their searching and feeding potential needs to be investigated. A wide variation in the susceptibility of black pepper, cardamom, ginger and turmeric cultivars to insect pests has also been documented. The tritrophic relationships between the resistant crop plants, insect pests and their natural enemies also needs to be investigated for increasing the effectiveness of natural enemies.

Plant products are known to be more effective when used in IPM programmes. The tropical regions where many of the spice crops are grown have a rich biodiversity of plant species with anti-feedant properties. Many of these species have also been screened against insect pests of spice crops, but at the field level many of them are not promising. Plant products degrade rapidly in sunlight and moisture, and though result in reduced persistence and risk to non-target organisms, require precise and frequent applications increasing the cost of cultivation. Some plant products are also quickly metabolized by enzymes in the target pests. Hence future research needs to focus on evaluation of stickers and synergists to enhance the effectiveness of plant products in IPM programmes.

BIOSAFETY

In spite of greater advantage of harnessing microbial diversity to exploit their potential for biological control of plant diseases as well as insect pests, greater consideration need to be given for bio safety as a social security measure. While the regulatory procedures are in place to take care of this aspect its strict implementation is the deciding factor (Kulakshetra, 2004). The recent discovery of *Pseudomonas aeruginosa,* PaBP35 as an endophyte in black pepper (*Piper nigrum* L) from Indian Institute of Spice Research Calicut for its potential to suppress *P.capisici* and *R. similis,* the causal organisms of foot rot and slow decline, though important but could not be utilized for due to its hazardous nature as a human pathogen (Kumar *et al.,* 2012). This implies that there is strict and absolute need to establish National Guidelines on Bio safety on the use of microbials that are recommended for field application. This is especially important in view of the above mentioned *P. aeruginosa.* In such situations alternate microbials like *Trichoderma* spp., *Bacullus* spp. need to be utilized in lieu of organisms like *P. aeruginosa.* As such greater scrutiny of microbials is warranted at regulatory level to ensure bio safety in crop protection programmes where microbials are utilized.

CONCLUSIONS

Lot of basic information on biology, nature of antagonism, hyperparasitasation and the role of lytic enzymes, the rhizosphere competition (RC) and Competitive Saprophytic ability (CST), the scaling up of production of the inocula in various carrier media, the solid and liquid state formulation technologies have been well developed and documented, The formulations/products, their shelf life, and toxicological studies with target biocontrol agents/products have been generated. The improvement of bio efficacy of biocontrol strains need to be intensified for optimum suppression of pathogens/pest to increase the crop production and productivity. It is important to develop single isolate with broad spectrum activity or microbial consortia that would address the holistic crop protection cutting across the crop barriers. This needs critical studies. The large scale field demonstrations and the population of the target biocontrol agents/ products vs populations levels of pathogens in relation to the crop health, specially at the field level are warranted. The recent information on diagnostics generated on detection of *P. capsici* and *R. simillis* in black pepper needs to be effectively utilized for ecological and epidemiological investigations (Dinu *et al.,* 2007, Aravind *et al.,* 2011, Aravind *et al.,* 2009). While the bioefficacy of the organisms both in pot culture studies and micro plots level are the basic, it is equally important and of high priority to demonstrate the field efficacy of the products through field demonstrations in Participatory Technology Development (PTD) mode and convince the farming community on benefit cost ratio of technology to make it attractive, acceptable and popular.

Bio control needs to be considered holistically as major components of IPM/ IDM and GAP, not in isolation. Some critical considerations are emphasized as under:

- Critical ecological and epidemiological studies are warranted on the interaction of both soil borne plant pathogens/insect pests, bio control agents and their population dynamics in relation to crop health and productivity. These are possible with available molecular diagnostics.
- Biocontrol agents with multiple modes of action (broad spectrum activity) that could suppress not only fungal and bacterial plant pathogens but also plant parasitic nematodes are relevant in view of the varied complex cropping systems adopted. This is much more important under perennial cropping systems where spatial segregation of plant pathogens never prevails.
- For the improvement of the existing strains of biocontrol agents (microbials) towards higher level of bio efficacy in

terms of increased production of lytic enzymes, metabolites and higher rhizonsphere competence, the existing molecular techniques need to be invoked ultimately to develop novel biocontrol agents which can fit in to diverse ecosystems

- Field demonstrations need to be set up to convince the farming community about its bio efficacy and economic viability of biocontrol agents (microbials/parasites/predators, microbial inset pathogens)
- An appraisal is needed to convince the farmers that biocontrol will be effective only under effective integrated nutrient management strictly adhering to Good Agricultural Practices (GAP) like phyto sanitation and good drainage and IPM package.
- Quality control of the products need to be strictly implemented based on the regulatory guide lines with a high priority for bio safety.

REFERENCES

Ahmed, J.S. and Baker, R. (1987) Rhizosphere competence of *Trichoderma harzianum*. *Phytopathology*, **77**, 182–189.

AICRPS (2009) *All India Coordinated Research Project on Spices*. Annual Report, 2007-08, Indian Institute of Spices Research, Calicut, Kerala, 96 pp.

AICRPS (2010) *All India Coordinated Research Project on Spices*. Annual Report, 2000-10, Indian Institute of Spices Research, Calicut, Kerala, 131 pp.

Anandaraj, M. (1997) Ecology of *Phytophthora capsici*. In: L. A. Alizadeh and P.H. Tsao (eds.), *Causal Organism of Foot Rot Disease of Black Pepper* (*Piper nigrum* L.), University of Calicut, Calicut, Kerala, India.

Anandaraj, M., Ramana, K.V. and Sarma, Y.R. (1991) Interaction between vesicular arbuscular mycorrhizal fungi and *Meloidogyne incognita* in black pepper. In: D.J. Bagyaraj and A. Manjunath (eds.), *Mycorrhizal Symbiosis and Plant Growth*, Mycorrhiza Network Asia and University of Agricultural Sciences, Bangalore, pp. 110–112.

Anandaraj, M., Raman, K.V. and Sarma, Y.R. (1996) Suppressive effects of VAM to root damage caused by *Phytophthora capsici, Radopholus similis* and *Meloidogyne incognita* in black pepper. In: K.S.S. Nair, J.K. Sharma and R.V. Varma (eds.), *IUFRO Symposium on Important Disease and Insect Pests in Tropical Forests*, Kerala Forest Research Institute, Peechi, Trichur, Kerala, pp. 232–238.

Anandaraj, M., Diby Paul, Jisha, P.J., Kumar, A., Saju, K.A., Thankamani, C.K. and Sarma, Y. R. (2003) Potential of a consortium of plant growth promoting rhizobacteria and *Trichoderma* for effective nursery management in black pepper (*Piper nigrum* L.). In: M.S. Reddy, M. Anandaraj, S.J. Eapen, Y.R. Sarma and J.W. Kloepper (eds.), Indian Institute of Spices Research, Calicut, India, pp. 8–11.

Anith, K.N., Radhakrishnan, N.V. and Manomohandas, T.P. (2002) Management of nursery wilt of black pepper (*Piper nigrum* L.) with antagonistic bacteria. *Curr. Sci.*, **83**, 561–562.

Anith, K.N., Radhakrishnan, N.V. and Manomohandas, T.P. (2003) Screening of antagonistic bacteria for biological control of nursery wilt of black pepper (*Piper nigrum* L.). *Microbiol. Res.*, **158**, 91–97.

Anonymous (2009) *IISR-Annual Report.* IISR-Calicut, Indian Council of Agricultural Research, New Delhi.

Anonymous (2010) *Spices Board.* Ministry of Commerce and Industry, Government of India, Sugandha Bhavan, Cochin

Anonymous (2012) *Spices Board.* Ministry of Commerce and Industry, Government of India, Sugandha Bhavan, Cochin

Anonymous (2012) *Directorate of Arecanut and Spices.* Development, Department of Agriculture and Cooperation, Ministry of Agriculture, Government of India, Calicut.

Aravind, R., Kumar, A., Eapen, S.J. and Ramana, K.V. (2008) Endophytic bacterial flora in root and stem tissues of black pepper (*Piper nigrum* L.) genotype: Isolation, identification and evaluation against *Phytophthora capsici. Lett. Appl. Microbiol.*, **48**, 58–64.

Aravind, R., Kumar, A., Eapen, S.J., Dinu, A. and Ramana, K.V. (2010) Screening of endophytic bacteria and evaluation of selected isolates for suppression of burrowing nematode (*Radopholus similis* Thorne) using three varieties of black pepper (*Piper nigrum* L.), *Crop Prot.*, **29**, 318–324

Aravind, R., Kumar, A., Dinu, A. and Eapen, S. J. (2011) Single tube duplex PCR for simultaneous detection of *Phyotphthora capsici* and *Radopholus similis* infecting black pepper (*Piper nigrum* L.). *Indian Phytopathol*, **64,** 353–357

Aravind, R., Kumar, A. and Eapen, S.J. (2012) Pre-plant bacterisation: A strategy for delivery of beneficial endophytic bacteria and production of disease-free plantlets of black pepper (*Piper nigrum* L.), *Arch. Phytopathol. Plant Prot.*, DOI:10.1080/03235408.2012.657894

Arunkumar and Sharma, A.K.. (2013) Biological control of cumin wilt disease- An innovative approach. *Spice India*, **26**, 8–11.

Atkinson, D., Berta, G. and Hooker, J.E. (1994) Impact of mycorrhizal colonization on root longevity and the formation of growth regulators. In: A.K. Sharma and B.N. Johri (eds.), *Impact of Arbuscular Mycorrhizas on Sustainable Agriculture and Natural Ecosystems*, Birkhauser, Basel, pp. 89–99.

Babu, P.B.S., Padmaja, V. and Hishan, A. (1996) Isodesacetyluvaricin-insect antifeedant against *Longitarsus nigripennis* Mots. *Indian J. Expt. Biol.*, **34**, 377–379.

Bagyaraj, D.J. (2006) VA mycorrhizal fungi and their role in plant disease control. In: B. Ramanujam, and R.J. Rabindra (eds.), *Current Status of Biological Control of Plant Diseases using Antagonistic Organisms in India*, Project Directorate of Biological Control, Bangalore, pp. 125– 134.

Bardia, P.K. and Rai, P.K. (2007) *In vitro* and field evaluation of biocontrol agents and fungicides against wilt of cumin caused by *Fusarium oxysporum* f. sp. *cumini*. *J. Spices Arom. Crops*, **16**, 88–92.

Bashan, Y. and Hoguin, G. (1988) Proposal for the division of plant growth-promoting rhizobacteria into two classifications: biocontrol-PGPB (Plant Growth-Promoting Bacteria) and PGPB. *Soil Biol. Biochem.*, **30**, 1225–1228

Batta, Y.A. (2004) Effect of treatment with *Trichoderma harzianum* Rifai formulated in invert-emulsion on postharvest decay of apple blue mold. *Int. J. Food Microbiol.*, **96**, 281–288

Beena, N. (2006) Potentiating Bioefficacy of Biocontrol Agents and Development of Integrated Disease Management of Rhizome Rot (*Pythium aphanidermatum*) of Ginger (*Zingiber officinale*). University of Calicut, Kerala, India

Bonsall, R.F., Weller, D.M. and Thomashow, L.S. (1997) Quantification of 2,4-diacetylphloroglucinol produced by fluorescent *Pseudomonas* spp. *in vitro* and in the rhizosphere of wheat. *Appl. Environ. Microbiol.*, **63**, 951–955.

CPCRI (1985) *Annual Report 1983*. Central Plantation Crops Research Institute, Kasaragod, India

Chawla, N. and Gangopadhyay, S. (2009) Integration of organic amendments and bioagents in suppressing cumin wilt caused by *Fusarium oxysporum* f.sp. *cumini*. *Indian Phytopathol.*, **62**, 209–216.

Chet, I. (1987) *Trichoderma* – application, mode of action and potential as biocontrol agent of soil borne plant pathogenic fungi. In: I. Chet (ed.), *Innovative Approaches to Plant Disease Control*, Wiley, New York, pp. 137–160.

Cook, R.J. (1985) Biological control of plant pathogens: Theory to applications. *Phytopathology*, **75**, 25–29

Devasahayam, S. (1992) *Bioecology and Control of Leaf Gall Thrips (Liothrips karnyi Bagnall) Infesting Black Pepper*. Final Report of Rsearch Project, National Research Centre for Spices, Calicut, Kerala.

Devasahayam, S. (1996) Biological control of insect pests of spices. In: M. Anandaraj and K.V. Peter (eds.), *Biological Control in Spices*, Indian Institute of Spices Research, Calicut, pp. 33–45.

Devasahayam, S. (1998) *Biological Control of Scale Insects Infesting Black Pepper.* Final Report of Ad-hoc Research Scheme, Indian Institute of Spices Research, Calicut, Kerala.

Devasahayam, S. (2000a) Insect pests of black pepper. In: P.N. Ravindran (ed.), *Black Pepper, Piper nigrum.* Harwood Academic Publishers, Amsterdam, pp. 309–334.

Devasahayam, S. (2000b) Evaluation of biopesticides for the management of shoot borer (*Conogethes punctiferalis* Guen.) on ginger (*Zingiber officinale* Rosc.). In: K.V. Ramana, S.J. Eapen, K.N. Babu, K.S. Krishnamurthy and Kumar, A. (eds.), *Spices and Aromatic Plants, Challenges in the New Millennium*, Indian Society for Spices, Calicut, Kerala, pp. 276–277.

Devasahayam, S. (2002c) Evaluation of biopesticides for the management of shoot borer (*Conogethes punctiferalis* Guen.) on turmeric (*Curcuma longa* L.). In: P. Rethinam, H.H. Khan, V.M. Reddy, P.K. Mandal and K. Suresh (eds.), *Plantation Crops Research and Development in the New Millennium*, Coconut Development Board, Kochi, pp. 489–490.

Devasahayam, S. (2006) *Bioecology and Integrated Management of Root Mealybug* (*Planococcus* sp.) *Infesting Black Pepper.* Final report of ad-hoc research scheme, Indian Institute of Spices Research, Calicut, Kerala.

Devasahayam, S. and Anandaraj, M. (1997) Black pepper (*Piper nigrum* L.) research. In: S.S. Narwal, P. Tauro and S.S. Bisla (eds.), *Neem in Sustainable Agriculture,* Scientific Publishers, Jodhpur, pp. 117–122.

Devasahayam, S. and Koya, K.M.A. (1993a) Effect of top shoot borer (*Cydia hemidoxa*) infestation on young vines of black pepper (*Piper nigrum*). *Indian J. Agric. Sci.*, **63**,762–763.

Devasahayam, S. and Koya, K.M.A. (1993b) Seasonal incidence of hymenopterous parasites of top shoot borer *Cydia hemidoxa* (Lepidoptera: Tortricidae) on black pepper. *J. Entomol. Res.*, **17**, 205–208.

Devasahayam, S. and Koya, K.M.A. (1994a) Natural enemies of major insect pests of black pepper (*Piper nigrum* L) in India. *J. Spices Arom. Crops*, **3**, 50–55.

Devasahayam, S. and Leela, N.K. (1997) Evaluation of plant products for antifeedant activity against pollu beetle (*Longitarsus nigripennis* Motschulsky), a major pest of black pepper. In: P.P. Reddy, N.K.K. Kumar and A. Verghese (eds.), *Advances in IPM for Horticultural Crops: Environmental Implications and Thrusts*, Association for Advancement of Pest Management in Horticultural Ecosystems, Bangalore, pp. 172–174.

Devasahayam, S. and Koya, K.M.A. (1999) Integrated management of insect pests of spices. *Indian J. Arecanut Spices Med. Plants*, **1**, 19–23.

Devasahayam, S. and Koya, K.M.A. (2005) Insect pests of ginger. In: P.N. Ravindran and K.N. Babu (eds.), *Ginger. The Genus Zingiber*, CRC Press, Washington, pp. 367–389.

Devasahayam, S. and Koya, K.M.A. (2007) Insect pests of turmeric. In: P.N. Ravindran, K.N. Babu and V. Sivaraman (eds.), *Turmeric. The Genus Curcuma*, CRC Press, Washington, pp. 169–192.

Devasahayam, S., Koya, K.M.A., Anandaraj, M., Tresa Thomas and Preethi, N. (2009) Distribution and bio-ecology of root mealybugs associated with black pepper (*Piper nigrum* Linnaeus) in Karnataka and Kerala, India. *Entomon*, **34**, 147–154.

Diby Paul (2004) Physiological, Biochemical and Molecular studies on the root rot (caused by *Phytophthora capsici*) suppression in black pepper (*Piper nigrum* L.) by Rhizosphere Bacteria".Ph.D Thesis submitted to University of Calicut, 259 pp.

Diby Paul, Kumar, A., Anandaraj, M. and Sarma, Y R. (2000) Studies on the suppressive action of fluorescent pseudomonads on *Phytophthora capsici*, the foot rot pathogen of black pepper. In IPS southern zone annual meeting and symposium on emerging trends in plant disease management, 7-8 December, Indian Institute of Horticultural Research, Bangalore, India, p. 62.

Diby Paul, Kumar, A., Anandaraj, M. and Sarma, Y.R. (2001) Studies on the suppressive action of fluorescent pseudomonads on *Phytophthora capsici*, the foot rot pathogen of Black pepper. *Indian Phytopathol.*, **54**, 515.

Diby Paul., Srinivasan, V. Anandaraj, M. and Sarma, Y.R. (2003) *Pseudomonas fluorescens* mediated nutrient flux in the black pepper rhizosphere microcosm and enhanced plant growth. In: M.S. Reddy, M. Anandaraj, S.J. Eapen, Y.R. Sarma and J.W. Kloepper (eds.), *Proceedings 6th International Workshop on PGPR*, Indian Institute of Spices Research, Calicut, Kerala, India, pp. 18–24.

Diby Paul and Sarma Y.R. (2003) Induction of defence related enzymes with Plant Growth Promoting Rhizobacteria (PGPRs) in black pepper (*Piper nigrum* L.) and *Phytophthora* patho system. In: M.S. Reddy, M. Anandaraj, S.J. Eapen, Y.R. Sarma and J.W. Kloepper (eds.), *Proceedings 6th International Workshop on PGPR*, Indian Institute of Spices Research, Calicut, Kerala, India, pp. 473–477.

Diby Paul and Sarma Y.R. (2005) *Pseudomonas fluorescens* mediated systemic resistance in black pepper (*Piper nigrum* L.) is driven through an elevated synthesis of defence enzymes. *Arch. Phytopathol. Plant Prot.*, **38**, 139–149.

Diby Paul., Saju, K.A., Jisha, P.J., Sarma, Y.R., Kumar, A. and Anandaraj, M. (2005) Mycolytic enzymes produced by *Pseudomonas fluorescens* and

Trichoderma spp. against *Phytophthora capsici*, the foot rot pathogen of black pepper (*Piper nigrum* L.). *Ann. Microbiol.*, **55**, 129–133.

Dinu A, Kumar A, Aravind R and Eapen S J. (2007) An improved method for selection of antagonistic bacteria against *Phytophthora capsici* Leonian Infections in black pepper (*Piper nigrum* L.). *J. Spices Arom. Crops*, **16**, 1–7.

Eapen, S.J. (1995) *Investigation on Plant Parasitic Nematodes Associated with Cardamom.* Final Report, Indian Institute of Spices Research, Calicut, India.

Eapen, S.J. and Ramana, K.V. (1997) Biological control of plant parasitic nematodes of spices. In: M. Anandaraj and K.V. Peter (eds.), *Biological Control in Spices*, Indian Institute of Spices Research (ICAR). Calicut, Kerala, pp. 20–32.

Eapen, S.J., Ramana, K.V. and Sarma, Y.R. (1997) Evaluation of *Pseudomonas fluorescens* isolates for control of *Meloidogyne incognita* in black pepper (*Piper nigrum* L.) In: S. Edison, KV. Ramana, B. Sasikumar, K. Nirmal Babu and S.J. Eapen (eds.), *Biotechnology of Spices, Medicinal and Aromatic Plants*, Indian Society for Spices, Calicut, Kerala, pp. 129–133.

Fridlender, M., Inbar, J. and Chet, I. (1999) Biological control of soil borne plant pathogens by a β-1,3 glucanase producing *Pseudomonas cepacia. Soil Biol. Biochem.*, **25**, 1211–1221.

Gopakumar, B. and Chandrasekhar, R. (2002) Insect pests of cardamom. In: P.N. Ravindran and K.J. Madhusoodanan (eds.), *Cardamom. The Genus Elettaria*, Taylor and Francis, London, pp. 180– 206.

Gopal, Lal, Meena, S.S, Anwer, M.M, Mehta, R.S and Maheria, S.P. (2010) *Advance Production Technology of Coriander.* National Research Centre on Seed Spices, ICAR, Tabiji, Ajmer, Rajasthan, India.

Haggag, W.M and Abo Sedera, S.A. (2005) Characteristics of three *Trichoderma* species in peanut haulms compost involved in biocontrol of cumin wilt disease. *Int. J. Agric. Biol.*, **7**, 222–229.

Hallmann, J. (2001) Plant interactions with endophytic bacteria. In: M.J. Jegger and N.J. Spence (eds.), *Biotic Interaction in Plant- Pathogen Associations*, CAB International, Wallingford, UK, pp. 87– 120.

Harman, G.E. (1992) Development and benefits of rhizosphere competent fungi for biological control of plant pathogens. *J. Plant Nutr.*, **15**, 835–843.

Harman, G.E. and Stasz, T.E. (1991) Protoplast fusion of the production of superior biocontrol fungi. In: D.O. Tebeest (ed.), *Microbial Control of Weeds*, Chapman and Hall, New York, pp. 171–186.

Howell, C.R. and Stipanovic, R.D. (1979) Control of *Rhizoctonia solani* on cotton seedlings with *Pseudomonas fluorescens* and with an antibiotic produced by the bacterium. *Phytopathology,* **69**, 480–482.

IISR (1996) *Annual Report 1995–96.* Indian Institute of Spices Research, Calicut, Kerala.

IISR (2003) *Annual Report 2002–03.* Indian Institute of Spices Research, Calicut, Kerala.

Jadeja, K.B and Nandoliya, D.M. (2008) Integrated management of wilt of cumin (*Cuminum cyminum* L.). *J. Spices Arom. Crops*, **17**, 223–229.

Joseph Thomas and Vijayan, A.K. (2003) *Fusarium oxysporum.* A new threat to cardamom cultivation. In *Proceedings of PLACROSYM*-XV, pp. 535–540

Jisha, P.J., Diby Paul, Kumar, A., Anandaraj, M. and Sarma, Y.R. (2002) Biocontrol consortium for a cropping system involving black pepper, ginger and cardamom (Ab). *Indian Phytopathol.*, **55**, 374–376.

Kakani, R.K, Anwer, M.M, Meena, S.S and Saxena, S.N. (2009) *Advance Production Technology of Fenugreek.* National Research Centre on Seed Spices, ICAR, Tabiji, Ajmer, Rajasthan, India.

Kloepper, J.W., Schroth, M.N. and Miller, T.D. (1980) Effects of rhizosphere colonization by plant growth promoting rhizobacteria on potato yield and development. *Phytopathology,* **70**, 1078– 1082.

Kloepper, J.W., Tuzun. S., Liu. L. and Wei. G. (1993) Plant growth promoting rhizobacteria as inducers of systemic resistance. In: R.D. Lumsden and J.L. Vaughn(rds.), *Pest Management, Biologically Based Technologies,* American Chemical Society, Washington DC, pp. 156–165.

Kloepper, J.W., Rodriguez Kabana, R., Zehnder, G.W., Murphy, J.F., Sikora, E. and Fernandez, C. (1999) Plant root-bacterial interactions in biological control of soilborne diseases and potential extension to systemic and foliar diseases. *Aust. J. Plant Pathol.*, **28**, 21–26.

Koya, K.M.A., Balakrishnan, R., Devasahayam, S. and Banerjee, S.K. (1986) A sequential sampling strategy for the control of shoot borer (*Dichocrocis punctiferalis* Guen.) in ginger (*Zingiber officinale* Rosc.) in India. *Trop. Pest Manage.*, **32**, 343–346.

Koya, K.M.A., Devasahayam, S., Selvakumaran, S. and Kallil, M. (1996) Distribution and damage caused by scale insects and mealybugs associated with black pepper (*Piper nigrum* Linnaeus) in India. *J. Entomol. Res.*, **20**, 129–136.

Kulakshetra, S. (2004) Status of regulatory norms for biopesticides in India. In: Nutan Kaushik (ed.), *Biopesticides for Sustainable Agriculture*, Tata Energy Research Institute, New Delhi, India, pp. 67–71.

Kumar, A., Munder, A., Aravind, R., Eapen, S.J., Tűemmler, B. and Raaijmakers, M. (2012) Friend or foe: genetic and functional characterization of plant endophytic *Pseudomonas aeruginosa. Environ. Microbiol.* doi:10.1111/1462- 2920.12031

Lodha, B.C. and Webster, J. (1990) *Pythium acanthophorum*, mycoparasite rediscovered in India and Britain. *Mycol. Res.*, **94**, 1006–1008.

Loper, J.E., Kraus, J. and Henkels, M. (1994) Antagonism of soil borne plant pathogens by rhizosphere pseudomonads. In: D.D. Bills and S. Kung (eds.), *Biotechnology and Plant Protection*, World Scientific Publishing Company, Singapore, pp. 73–84.

Loper, J.E., Nowak-Thompson, B., Whistler, C.A., Hagen, M.L., Corbell, N.A., Henkels, M.D. and Stockwell, V.O. (1997) Biological control mediated by antifungal metabolite production and resource competition: an overview, In: A. Ogoshi, A. Kobayashi, Y. Homma, F. Kodama, N. Kondo and S. Akino (eds.), *Proceedings of the Fourth International Workshop on Plant Growth Promoting Rhizobacteria*, Sapporo, Japan, pp. 73–79.

Madan, M.S., Ramana, K.V., Manoj, K.A., Anandaraj, M., Suseela, Bhai R. and Meera, I.S. (2006) Economic viability of large-scale production of the biocontrol agent *Trichoderma harzianum* Rifai. *J. Spices Arom. Crops*, **151,** 48–51.

Manohara, D. and Wahyuno, D. (2009) *Foot Rot Disease of Black Pepper and its Control Strategies.* Food and Agriculture Organization, Indonesian Centre for Estate Crops Research and Development, Indonesia.

Manohara, D., Mulya, K., Wahyuno, D. and Noveriza, R. (2002) Viabilitas *Trichoderma* harzianum pada beberapa furmulasi dan efikasinya terhadap *Phytophthora* capsici,. Risalah Simposium Nasional. Penelitian PHT Perkebunan Rakyat. Pengembangan dan Implementasi PHT Perkebunan Rakyat Berbasis Agribisnis. Bogor, Indonesia, pp. 199–206 (with English summary).

Mathew, M.J., Saju, K.A. and Venugopal, M.N. (1997) Efficacy of enotmogenous fungi on biological suppression of *Pentalonia nigronervosa* f.sp *caladii* van der Goot of cardamom (*Elettaria cardamomum* Maton). *J. Spices Arom. Crops*, **7**, 43–46.

Meena, S.S, Mehta, R.S, Anwer, M.M, Gopla, Lal, Sharma, Y.K., Kakani, R.K. and Saxena, S.N. (2009) *Advance Production Technology of Ajwain.* National Research Centre on Seed Spices, ICAR, Tabiji, Ajmer, Rajasthan, India.

Meena, R.S, Anwer, M.M, Gopal, Lal, Krishna Kant and Mehta, R.S. (2010) *Advance Production Technology of Cumin.* National Research Centre on Seed Spices, ICAR, Tabiji, Ajmer, Rajasthan, India.

Mittal, V.P. and Butani, P.G. (1994) Pests of seed spices. In: K.L. Chadha and P. Rethinam (eds.), *Advances in Horticulture, Plantation and Spice Crops,* Vol. 10, Part 2, Malhotra Publishing House, New Delhi, pp. 825–855.

Naik, D.J., Belavadi, V.V., Thippesha, D., Kumar, M.D. and Madaiah, D. (2006) Field efficacy of neem products against thrips and capsule borer of small cardamom. *Karnataka J. Agric. Sci.*, **19**, 144– 145.

Naik, D.J., Thippesha, D., Belavadi, V.V., Devaraju, K.M., Lakshmana, D., Rangaswamy, S.D. and Venkatesha, J. (2009) Evaluation of insecticides and organic products against pepper mussel scale, *Lepidosaphes piperis*

Gr. and its correlation with weather factors. *Karnataka J. Agric. Sci.*, **22**, 582–584.

Nirmal Babu, K., Minoo, D., Geetha, S.P. and Jayakumar, V.M. (2006) Progress of spices biotechnology in India. In: P. N. Raveendran, K. Nirmal Babu, K.N. Siva and A.K. Jhony, *Advances in Spices Research, History and Achievement of Spices in India Since Indipendance*, Agro Bios –India, pp. 169–194.

Nowak, J. (1998) Benefits of *in vitro* 'biotization' of plant tissue culture with microbial inoculants. *In Vitro Cell Dev. Biol. Plant*, **34**, 122–130.

Papavizas, G.C. (1985) *Trichoderma* and *Gliocladium:* biology, ecology and potential for biocontrol. *Annu. Rev. Phytopathol.*, **23**, 23–54.

Pervez, R., Eapen, S.J., Devasahayam, S. and Jacob, T.K. (2011) Efficacy of entomopathogenic nematodes against insect pests of ginger and their mass multiplication. *In Souvenir, SYMSAC VI, Exploiting Spices Production Potential of the Deccan Region,* 8-10 December 2011, Dharwad. Indian Society for Spices, Kozhikode, p. 258.

Prakash, M.G., Vinayagopal, K., Anandaraj, M. and Sarma, Y.R. (1999) Evaluation of substrates for mass multiplication of fungal biocontrol agents *Trichoderma harzianum* and *T. virens. J. Spices Arom. Crops,* **8**, 207–210.

Premkumar, T. and Nair, M.R.G.K. (1988) Effect of some planting conditions on infestation of black pepper by *Longitarsus nigripennis* Mots. *Indian Cocoa Arecanut Spices J.*, **10**, 83–84.

Rajan, P.P. (1999) Approaches towards the integrated disease management of *Phytophthora* infection of black pepper (*Piper nigrum* L.). Ph.D Thesis, University of Calicut, Calicut, 190 pp

Rajan, P.P., Sarma, Y.R. and Anandaraj, M. (2002) Management of foot rot disease of black pepper with *Trichoderma* spp. *Indian Phytopathol.,* **55**, 34–38.

Rajan, P.P., Gupta, S.R., Sarma, Y.R. and Jackson, G.V.H. (2002) Diseases of ginger and their control with *Trichoderma harzianum. Indian Phytopathol.*, **55**, 173–177.

Rajkumar, A.J., Kurian, P.S., Backiyarani, S. and Murugan, M. (2002) Evaluation of biorationals against thrips (*Sciothrips cardamomi* Ramk.) and shoot and capsule borer (*Conogethes punctiferalis* Guen.) in cardamom. *J. Spices Arom. Crops*, **11**, 132–134.

Ramana, K.V. and Eapen, S.J. (1995) Parasitic nematodes and their management in spice crops. *J. Spices Arom. Crops,* **4**, 1–16.

Raupach, G.S. and Kloepper, J.W. (1998) Mixtures of plant growth promoting rhizobacteria enhance biological control of multiple cucumber pathogens. *Phytopathology*, **88**, 1158–1164

Saju K.A (2004) Factors Affecting the Biological Control of Phytophthora capsici Infection in Black Pepper (*Piper nigrum* L.). University of Calicut, Kerala.

Saju, K.A., Anandaraj, M. and Sarma, Y.R. (2003) Evaluation of *Trichoderma* spp and *Pseudomonas fluorescens* for suppression of *Phytophthora capsici* infecting black pepper. In: M.S. Reddy, M. Anandraj, S.J. Eapen, Y.R. Sarma and J.W. Koloepper (eds.), *Proceedings 6th International PGPR Workshop*, University of Calicut, Kerala, India, pp. 52–58.

Sarma, Y.R. (20060 Recent trends in the use of antagonistic organisms for disease management in spice crops. In: B. Ramanujam and R.J. Rabindra (eds.), *Current Status of Biological Control of Plant diseases using antagonistic Organisms in India,* Project Directorate of Biological Control, Bangalore, pp. 49–73.

Sarma, Y.R. 2005. Approaches towards biological control of stem and root rot of vanilla in Mauritius-A Report, pp 1–30.

Sarma, Y. R., Anandaraj, M. and Venugopal, M. N. 1994. Diseases of spice crops In: K.L. Chadha and P. Rethinam (eds.), *Advances in Horticulture,* Vol 10, Malhotra Publishing House, New Delhi, Plantation and Spice Crops Part 2, pp. 1015–1057

Sarma, Y.R., Anandaraj, M. and Venugopal, M.N. (1996) Biological control of diseases of spices. In: M. Anandaraj and K.V. Peter (eds.), *Biological Control in Spices*, Indian Institute of Spices Research, Calicut, Kerala, India, pp. 1–19.

Sarma, Y. R. and Anandaraj, M. (1996) *Phytophthora* foot rot of black pepper. In: V.P. Agnihotri, A.K. Sarabhoy and D.V. Singh (eds.), *Management of Threatening Plant Diseases of National Importance*, Malhotra Publishing House, New Delhi, pp. 237–248.

Sarma, Y.R. and Anandaraj M. (1998) Biological suppression of disease of plantation crops and spices. In: S.P. Singh and H.H. Hussaini (eds.), *Biological Suppression of Disease of Plants, Phytoparasitic Nematodes and Weeds*, Project Directorate of Biological Control, Bangalore, India, pp. 24–47.

Sarma, Y.R., Anandaraj, M. and Ramana, K.V. (1998) Major diseases of black pepper and cardamom and their management. In: S.M.P. Khurana (ed.), *Pathological Problem of Economic Crop Plants and Their Management*, Scientific Publishers, Jodhpur, India, pp. 282–293.

Sarma, Y.R., Rajan, P.P., Diby, Paul, Beena, N. and Anandaraj, M. (2000) Role of rhizobacteria on disease suppression on spice crops and future prospects. Abstract in *Seminar on Biological Control with Plant Growth Promoting Rhizobacteria to Sustainable Agriculture,* April 3-4, University of Hyderabad, Hyderabad, India.

Sarma, Y.R. and Saju, K. (2004) Biological control for the management of foot rot and slow decline diseases of black pepper. In: *Focus on Pepper* (*Piper nigrum* L.), Vol. 1, International Pepper Community, Jakarta, Indonesia, pp. 25–51.

Schenk, N.C. (1981) Can mycorrhizae control root disease? *Plant Dis.*, **65**, 230–234.

Selvakumaran, S., Kallil, M. and Devasahayam, S. (1996) Natural enemies of two major species of scale insects infesting black pepper (*Piper nigrum* L.) in India. *Pest Manage. Hort. Ecosyst.*, **2**, 79–83.

Selvan, M.T., Gopakumar, B., Chandrasekhar, S.S., Singh, J., Kumaresan, D. and Varadarasan, S. (1996) Survey on the occurrence of coccids infesting black pepper in higher altitudes of Idukki District, Kerala and adaptive trial for their management. *Pestology*, **20**, 19–23.

Shanmugam, V., Thakur, H. and Gupta, S. (2012) Use of chitinolytic *Bacillus atropheaus* strain S2BC-2 antagonistic to *Fusarium* spp. for control of rhizome rot of ginger. *Ann. Microbiol.* DOI 10.1007/s13213-012-0552-2

Shanmugam, V., Gupta, S. and Dohroo, N.P. (2013) Selection of a compatible biocontrol strain mixture based on co-cultivation to control rhizome rot of ginger. *Crop Prot.*, **43**, 119–127

Shanmugam, V., Thakur, H., Kaur, J., Gupta, S., Rajkumar, S. and Dohroo, N.P. (2013) Genetic Diversity of *Fusarium* spp. Inciting Rhizome Rot of Ginger and its Management by PGPR Consortium in the Western Himalayas. Institute of Himalayan Bioresource Technology, CSIR, Palampur, India.

Sharma, N.D. and Jain, A.C. (1979) Studies on the biological control of *Fusarium oxysporum* f.sp. *zingiberi*, the causal organism of yellows disease of ginger. *Indian Phytopathol.*, **31**, 260–261.

Sharma, Y.K., Anwer, M.M., Meena, S.S. and Saxena, S.N. (2010) *Crop Production Technology of Dill* (*Sowa*). National Research Centre on Seed Spices, ICAR, Tabiji, Ajmer, India.

Sheela, M.S., Venkitesan, T.S. and Mohandas, N. (1993) Status of *Bacillus* spp. as biocontrol agents of root knot nematode (*Meloidogyne incognita*) infesting black pepper (*Piper nigrum* L.). *J. Plantation Crops*, **21** (Suppl), 218–222.

Singh, M.P. (2007) Integrated pest management in seed spices. In: S.K. Malhotra and B.B. Vashishtha (eds.), *Production, Development, Quality and Export of Seed Spices*, National Research Centre on Seed Spices, Ajmer, pp. 139–148.

Singh, R.D., Choudhary, S.L. and Patel, K.G. (1972) Seed transmission and control of *Fusarium* wilt of cumin. *Phytopathol. Medit.*, **11**, 19–24.

Sriram, S. (2010) Invert-emulsion formulation of a fungal antagonist for biological management of plant diseases. Indian Patent Application No. 3603/CHE/2010 filed at Indian Patent Office.

Suseela, Bhai R. (1998) *Studies on Capsule Rot (Azhukal) Disease of Cardamom*. University of Calicut, Kerala, India.

Suseela, Bhai R. and Kumar, A. (2008) Effect of rhizobacteria on *Phytophthora meadii, Fusarium oxysporum* f.sp. *vanillae* and *Colletotrichum vanillae* infecting vanilla. *J. Biol. Control*, **22**, 33–41.

Suseela, Bhai R., Joseph, Thomas and Naidu, R. (1993) Biological control of Azhukal disease of cardamom caused by *Phytophthora meadii* McRae. *J. Plantation Crops*, **21** (Suppl.), 134–139.

Suseela, Bhai R. and Sarma, Y.R. (2003) In vitro Effect of *Pseudomonas fluorescens* on capsule rot of cardamom (*Elettaria Cardamomum* Maton) caused by *Phytophthora Meadii.* In: M.S. Reddy, M. Anandaraj, S.J. Eapen, Y.R. Sarma and J.W. Kloepper (eds.), *Proceedings 6th International PGPR Workshop*, University of Calicut, Kerala, pp. 74–78.

Thankamani, C.K., Anandaraj, M., Diby, Paul and Kandiannan, K. (2003) Compatibility of *Pseudomonas fluorescens,* VAM (*Glomus fasciculatum*), *Trichoderma harzianum,* pesticides on establishment and growth of black pepper under nursery conditions. In: M.S. Reddy, M. Anandaraj, S.J. Eapen, Y.R. Sarma and J.W. Kloepper (eds.), *Proceedings 6th International PGPR Workshop*, University of Calicut, Kerala, pp. 82–88.

Thomashow, L.S., Weller, D.M., Bonsall, R.F. and Pierson, L.S. (1990) Production of antibiotic phenazine-l-carboxylic acid by fluorescent *Pseudomonas* species in the rhizosphere of wheat. *Appl. Environ. Microbiol.*, **56**, 908–912.

Tombe, M. and Adwerd, C.Y.L. (2010) Fungal diseases of vanilla. In: Eric Odoux and M. Grisoni (eds.), *Vanilla*, CRC Press, Boca Raton, Fl., pp. 125–140.

Tran Thi Thu Ha (2007) Interactions between biosurfactant-producing *Pseudomonas* and *Phytophthora* species. PhD thesis Wageningen University, The Netherlands with summaries in English and Dutch.

Usman, N.M. (1997) Biological control of rhizome rot of ginger. Ph.D. thesis, University of Calicut, Kerala.

Usman, N.M., Balakrishnan, P. and Sarma, Y.R. (1996) Biocontrol of rhizome rot of ginger. *J. Plantation Crops*, **24** (Suppl), 184–191.

Varadarasan, S. (1995) Biological control of insect pests of cardamom. In: T.N. Ananthakrishnan (ed.), *Biological Control of Social Forest and Plantation Crops Insects*, Oxford and IBH Publishing Company Private Limited, New Delhi, pp. 109–119.

Varadarasan, S., Sivasubramonian, T., Manimegalai, R. and Naidu, R. (1993) Integrated management of cardamom root grub *Basilepta fulvicorne* Jacoby (Eumolpinae: Chrysomelidae: Coleoptera). *J. Plantation Crops*, **21** (Suppl.), 191–194.

Varadarasan, S. and Sivasubramonian, T. (1996) Studies on natural occurrence of bioagents and their bioassay on cardamom root grub, *Basilepta fulvicorne* Jacoby (Coleoptera: Chrysomelidae). *J. Plantation Crops*, **24** (Suppl.), 286–290.

Venugopal, M.N. (1999) Characterization, early detection and management of 'kokke kandu' disease of ardamom. *Final Report of Ad-hoc Research Scheme*. Indian Institute of Spices Research, Calicut, Kerala.

Vyas, R. K. and Mathur, K. (2002) Introduction of *Trichoderma* spp. in Cumin rhizosphere and their potential in suppressing of wilt. *Indian Phytopathol.*, **55**, 451–457.

Yap Chin Ann (2012) Rhizobacteria of pepper (*Piper nigrum*) and their antifungal activities. *Afr. J. Microbiol. Res.*, **6**, 4185–4193.

12

Biological Control of Insect Pests of Tuber Crops

C. A. JAYAPRAKAS*, R. S. SREERAG, E. R. HARISH AND L. RAGESH

Central Tuber Crops Research Institute,
Sreekariyam - 695017, Thiruvananthapuram
*E-mail : cajprakash@yahoo.com

INTRODUCTION

Tuber crops, which include cassava, sweet potato, yams and aroids, are the third most important food crops after cereals and grain legumes. These crops are unique in their inherent qualities like high dry matter production, capacity to withstand the vagaries of climate change, and ensure food security to millions of people in the tropical and subtropical countries (Ram, 2008). The world's total harvested area of tubers is nearly 51 million ha with one third found in Africa and one third in Asia and Pacific regions. Tuber crops are generally rain fed in nature, and these store starch in their vegetative parts, mostly inside roots or subterranean stem modifications (tuber, corms, stolons). Tuber crops have a higher biological efficiency as food producers and show the highest rate of dry matter production per day per unit area among all the crops. These crops are recognized as the most efficient converters of solar energy. Apart from that these tuber crops are known to supply cheap source of energy especially for the weaker sections of the population. These attributes make tuber crops ideal for cultivation in the less developed countries of the world. Among the tropical tuber crops of economic importance, cassava (*Manihot esculenta*), *sweet potato* (*Ipomoea batatas*), yams (*Dioscorea* species) and aroids namely taro (*Colocasia esculenta*), tannia

(*Xanthosoma sagittifolium*) and elephant foot yam (*Amorphophallus paeoniifolius)* are the major tropical tuber crops cultivated in many countries including India. These tubers are popularly used as vegetables in various delicious cuisines, possess medicinal properties and also serve as byproducts for industries (Babu *et al.*, 2012). With the use of new varieties, yields of some of the tuber crops have spurted handsomely. The average productivity of cassava in India, for instance, has shot up to around 35 tonnes/ha, which is nearly three times the global average of 11 to 12 tonnes/ha. At present, the cultivation of tuber crops, other than potato, is confined largely to the peninsular and north-eastern states of India. The Central Tuber Crop Research Institute in India is making efforts to introduce these crops elsewhere as well, especially in the agriculturally-progressive areas in the north. It is collaborating with institutions in Punjab which can yield positive results, paving the way for the introduction of cassava as a summer crop for harvest before planting wheat in winter.

However, insect pests and diseases are some of the limiting factors in the production of tropical tuber crops. Climate change has paved the way for transformation of hitherto known minor pests to the status of major pests in tuber crops. The occurrence of papaya mealy bugs, for instance, has not been recorded till recently, but now it has been havoc in cassava and other standing crops (Hedge and Jayaprakas, 2011). Invasion of insect pest on various tuber crops creates serious economic loss for the marginal farmers. Potyviruses cause significant losses in a wide range of crop plants and are transmitted by aphids in a non-persistent manner and tuber crops are no exception. The entire aforementioned tuber crops are known to be infected by one or the other species of potyvirus. Cassava brown streak virus (*Ipomovirus* in *Potyviridae* family) in cassava, *sweet potato feathery mottle virus* (SPFMV) in sweet potato, *Dioscorea alata virus* (DAV) and *yam mosaic virus* (YMV) in yams and *Dasheen mosaic virus* (DsMV) in edible aroids are the major potyviruses causing economic loss in tuber crops (Babu *et al.*, 2012). However, management of pests and disease by chemical measure is highly discouraged due to its proven adversaries to man and environment. A bio-intensive integrated pest and disease management approach is the rational solution for a sustainable development and ecofriendly management of important insect pests (Table 12.1) of tuber crops is discussed as under.

Table 12.1. Some insect pests of tuber crops and their natural enemies

	Crops	Pest	Major Bio-control organisms
1	Cassava	1 *Paracoccus marginatus*	*Cryptolaemus montrouzieri, Anagyrus loecki, Pseudleptomastix mexicana, Acerophagous papayae*
		2 . *Bemisia tabci*	*Prospaltella flava, Coccinella septempunctata*
		3. *Aleurodicus dispersus*	*Nephaspis oculatus, N. bicolor, Encarsia haitiensis, Axinoscymnus puttarudriahi* and *Cheilomenes sexmaculata*
		4. *Aonidomytilus albus*	*Chilocorus nigritus, Chilomenes sexmaculata, Scolothrips indicus*
		5.*Rhyzopertha dominica*	*Anisopteromalus calandrae, Lyctocoris campestris, Acarophenax assanovi*
		6. *Sitophilus oryzae*	*Anis, Lariophagus distinguendus*
2	Sweet potato	1. *Cylas formicarius*	*Rhaconotus* sp. and *Bracon* sp
		2 *Agrius convolvuli*	*Trichogramma australicum, T. achaea, T. agriae.*
		3 *Aspidomorpha miliaris*	*Cassidocida aspidiomorphae* and *Tetrastichus colemani*
3	Yams	1. *Aspidiella hartii*	*Physcus comperi, Adelencyrtus moderatus*
		2. *Ferrisia virgata*	*Allograpta javana, Triommata coccidivora*
4	Aroids	1. *Aphis gossypii*	*Aphelinus mali, Coccophagus cowperi Aphidencyrtus aphidivorus, Menochilus sexmaculatus, Verania discolor, Pseudospidimerus cicrumflexus*
		2. *Spodoptera litura*	*Charops hersei*
		3. *Monolepta signata*	*Eocanthecona furcellata*
5	Chinese potato	*Spoladea recurvalis*	*Microgaster psarae* and *Leptobatopsis indica*

Source: Palaniswami, M.S. (1994).

CASSAVA

Cassava (*Manihot esculenta*) is a woody shrub belongs to Euphorbiaceae family. Cassava feeds half-a-billion people in Africa every day, and is also grown by millions of smallholder farmers in Southeast Asia and Latin America. It is an important tuber crop widely cultivated in tropical and subtropical regions with total production of 8.06 million tonnes from India alone (Faostat, 2011). Globally only about 11% of total cassava production is traded. The largest trade comes from Indonesia, Thailand and Vietnam, who ship chips and starch to China. In 2007, about 18% of chips were exported to Western Europe. Total traded cassava jumped 56% from 25 million tons in 2007 to 39 million tons in 2009. Chinese demand for cassava continues to increase well beyond what Indonesia, Thailand and Viet Nam can supply, given their own growing internal demands (Kueneman *et al.*, 2012). Crop management practices coupled with the use of improved varieties have resulted in steady, albeit slow, increases in productivity in Africa from average yields of about 5 tonnes/ha in the 1960s to about 10 tons/ha today. The crop is relatively new in most parts of Asia, but average yields of the 1970s of less than 10 tons/ha have been about doubled (20 tonnes/ha). In Latin America yields have stayed on the average rather stagnant around 12–14 tonnes/ha. However, there is considerable loss of the crop due to insect pests and diseases. Root yield losses have been recorded for several pests, including mites, mealybugs, whiteflies, hornworm, lacebugs, thrips and burrower bugs (Bellotti *et al.*, 2012) that have been controlled by number of natural enemies (Table 12.2). Therefore, pest management is essential for good production depending on the timing and severity of the attack. Disease control has largely been through the release of resistant varieties. Viruses continue to be a challenge in Africa (cassava mosaic and the more recent brown streak viruses) and lower production output on susceptible varieties. Use of disease-free planting materials is also promoted along with new varieties, but it is generally difficult to provide 'clean' planting stakes each season in any routine way to hundreds of thousands of small-holder growers. Of the various arthropod pests and rodents which reduce the yield or production, insect pests are most devastating (Bellotti and Schoonhoven, 1978; Lal and Pillai, 1981). The major insect pests being sucking pests which de-sap the plant and also function as predisposing factors for various virus diseases. Some major insect pests of cassava are discussed as under:

Paracoccus marginatus (Papaya Mealybug)

Paracoccus marginatus Williams and Granara de Willink is a polyphagous pest native of Central America/Mexico infesting more than 60 species of plants spread over 50 countries. In India, it was first reported from Tamil Nadu during 2008 (Muniappan *et al.*, 2008) infesting papaya, cassava and a wide list of agricultural and horticultural crops. In the cassava growing area, particularly in Tamil Nadu and Andhra Pradesh, this mealy bug is a severe problem and its management by synthetic insecticides poses a challenge to the farmers.

Table 12.2. Number of natural enemies of some of the principal pests of cassava

Pests	Parasitoids	Predators	Pathogens
Cassava green mite		60	2
Cassava hornworm	18	15	15
Whiteflies	17	5	6
Mealybugs	25	46	2
Bores			
Chilomina clarkei	5	2	5
Lagochirus sp.	2		
Burrower bugs		1	5
White grubs	2	1	3
Lacebugs		1	
Thrips		1	
Scales	4	9	2
Total	**73**	**141**	**40**

Source: Melo (2002)

From Biological control point of view, natural enemies of the papaya mealybug include the mealybug destroyer (*Cryptolaemus montrouzieri*), other lady beetles, lacewings and hover flies. The papaya mealy bug was successfully controlled by the parasitoids, *Anagyrus loecki*, *Pseudleptomastix mexicana* and *Acerophagous papayae* (Meyerdirk *et al.*, 2004) which were mass released to various infested areas. *Anagyrus loecki* and *A. papayae* appear to be promising biological control agents and no field recovery of *P. mexicana* was made in spite of several field releases (Muniappan *et al.*, 2006). Following the introduction of these exotic parasitoids mealy bug population density level reduced below detectable levels in a six month period.

Phenacoccus manihoti (Cassava mealy bug)

Phenacoccus manihoti Matile-Ferrero is one of the most serious pests of cassava worldwide, has recently reached Asia, raising significant concern over its potential spread throughout the region.

Model predictions suggest *P. manihoti* imposes an important, yet differential, threat to cassava production in Asia. Though this bug does not exist in India, predicted risk is most acute in the southern end of Karnataka in India, the eastern end of the Ninh Thuan province in Vietnam, and in most of West Timor in Indonesia (Parsa *et al.*, 2012). However, biological control of this mealybug in Africa by introduction of parasitoid, *Epidinocarsis lopezi* from South America (Herren and Neuenschwander (1991) represents one of the largest and most successful biological control programmes.

Bemisia tabci (Cassava/Sweet Potato Whitefly)

Bemisia tabci Gennadius was first described in 1889 by Gennadius in Greece as a pest of tobacco (*Nicotiana tabacum*). There is a cassava specific population of *B. tabaci* christened as 'the Indian cassava biotype' that acts as a vector for Indian cassava mosaic disease (ICMV), but the sweet potato biotype is a non vector of mosaic disease in cassava (Lisha *et al.*, 2003). More over high reproductive potential, broad host range and resistance to insecticides are some of the constraints facing for their management. Whereas natural enemies provide satisfactory control over this noxious pest and several potential parasitoids have been reported suppressing its population. Among them, *Encarsia* sp. and *Erectmocerus* sp. are very common in India. It also parasitized by an aphelinid parasitoid *Prospaltella flava* (Palaniswami *et al.*, 1995). Lal (1980) reported this parasitoid in 1976-77 on cassava in Kerala, India and the level of parasitism varied from 6.81% in December to 29.50% in October.

Cassava is an important staple food from which over 200 million people derive over 50% of their carbohydrate intake in Africa. *Bemisia tabaci,* the whitefly has played havoc with cassava production in some parts of Africa. The International Institute of Tropical Agricu-lture (IITA) researchers estimate in their disease survey conducted in 2009 that the whitefly problem, which originated from south-eastern Uganda before entering Tanzania, causes cassava plants to lose 90% of yields. Use of parasitoids and predators has been one of the major options to control this pest (Gerling *et al.*, 2001). In Uganda, Two species of aphelinid parasitoids, *Eretmocerus mundus* Mercet and *Encarsia sophia* Girault and Dodd have been identified to control *Bemisia tabaci* on cassava where trends in the buildup in numbers of both parasitoids species and apparent parasitism were similar but differed significantly in response to seasonal situation. It was, however, noted that percent parasitism decreased with nymph number (Otim *et al.*, 2006). Simmons and Abd-Rabou (2007) recorded fifteen species of natural enemies of *B. tabaci*, including 5 species of predators and 10 species of parasitoids. *Coccinella septempunctata* was the most commonly found predator, and it was found in four of the crops. These results help define the diversity of natural enemies of *B. tabaci* among vegetable crops. According to a report of IITA (2010) the institute is investing in projects that identify the most effective natural enemies of the whitefly that can be deployed to reduce its population. It also explores cassava varieties, including wild relatives, with resistance to the pest.

Field trials for the management of whitefly in the cassava are also known using botanical insecticides. During 2006-2007 and 2007-2008 at farmer's holdings of Attur Taluk, Salem District in India, pooled data revealed that application of sweetflag rhizome *Acorus calamus* 10 D @ 25 kg/ha and neem seed kernel extract (NSKE) 5% at 15 days interval induced significant reduction in whitefly population. This treatment also showed reduction in mosaic incidence of 9.10% with a tuber yield of 15.5 t/ha and benefit cost ratio of 2.80 compared to the untreated control where a tuber yield of 11.92 t/ha was recorded (Venketasan and Palinasamy (2010).

Aleurodicus dispersus (Spiraling Whitefly)

Aleurodicus dispersus Russell is a sap feeding insect is a native of Caribbean islands and Central America (Russel, 1965) and the incidence of spiraling whitefly in India was first reported in 1994 at Thiruvananthapuram (Palaniswami *et al.*, 1994) and subsequently from five states in peninsular India on a variety of crops (David and Regu, 1995; Mani and Krishnamoorthy, 1999; Rajamma *et al.*, 2001; Aishwariya *et al.*, 2007b). Coccinellid predators. *Nephaspis oculatus, N. bicolor, Encarsia haitiensis* Dozier and *Encarsia* sp. and two aphelinid parasitoids were found effective in suppressing the whitefly population (Waterhouse and Norris, 1989). Aishwariya *et al.* (2007a) opined that among the predators, *Axinoscymnus puttarudriahi, Cryptolaemus montruozier* and *Cheilomenes sexmaculata* were found more frequent than others. Also, *Paecilomyces farinosis* (Holms.) is known to infect the nymphs of spiralling whitefly in India (Mani *et al.*, 2001). *Fusarium semitectum* and *V. lecanii* are also known to kill spiralling whitefly (Aiswariya *et al.*, 2007b).

Overall survey has revealed the presence of 45 predators and two parasitoids namely *Encarsia guadeloupae* Viggiani and *E. haitiensis.* Both these seem likely to cover all the spiralling whitefly areas and cause remarkable reduction in the population of *A. despersus* in India as witnessed in other countries (Mani, 2010).

Various botanicals like tobacco extract (4%,) was found effective in minimising the spiralling whitefly. Spraying of neem oil (2%), fish oil rosin soap (4%) and detergent soap solution (5%) also reduces the whitefly population (Mani, 2010).

In African regions also niche separation between *Encarsia dispersa* and *Encarsia guadeloupae*, two biological control agents of the spiraling whitefly *Aleurodicus dispersus* has been established, more specifically in Benin, West Africa (Obinna *et al.*, 2011). It is shown that the parasitoids established their geographical niches, with *E. dispersa* being more abundant (up to 89%) in the coastal south bellow latitude 7°30′N and *E. guadeloupae* (up to 84%) in the north between latitudes 7°30′ and 10°30′N. *Stenthonus* species (a small dark beetle) has also been found to prey on spiralling whiteflies, nymphs and pupae as predator which is more or less a natural biological control of *A. disperses* in Nigeria on cassava (Banjo *et al.*, 2004).

Aonidomytilus albus (Cassava Scale/White Mussel Scale)

Though *Aonidomytilus albus* Cockerell is not a serious problem in the field, but cassava stems under storage are severely infested by this pest. It was first recorded in India by Ananthanarayanan *et al.* (1957) from Salem. Infestation is common on stem, branch or even on leaf petiole and it occurs in dry seasons and aggravates during prolonged moisture stress (Lal and Pillai, 1981). Natural enemies of scale insects commonly seen in India are *Chilocorus nigritus*, *Chilomenus sexmaculata*, *Scymnus* and *Stethorus* spp. and predatory thrips *Scolothrips indicus.* These scale insects was found heavily preyed upon by coccenellids (*Chilochorus nigritus, Scymnus guimeti, Menochilus sexmaculatas* and *Stethorus pariepunctatus*) and parasitized by two hymenopteran parasitoids *Aspidiophagus* sp. and *Signiphora* sp. (Anon., 1983).

Mononychellus tanajoa (Cassava green mite)

The cassava green mite, *Mononychellus tanajoa* is the most important species, especially in lowland tropical regions with prolonged (3 to 6 months) dry seasons. It is native to the Americas, possibly from northern South America or Northeast Brazil, where it was reported for the first time in 1938. The mite attacks young leaves and meristems, preferably feeding on the underside of the leaves, which develop a mottled to bronzed appearance in the form of a mosaic with chlorotic spots until the leaves become deformed. *Mononychellus tanajoa* was introduced accidentally to the African continent during the 1970's, where it caused 13-80% yield loss (Yaninek and Herren 1988). Subsequently, it was responsible for a yield loss in cassava in Africa of between 30 and 50% until a natural enemy of the pest helped to contain the devastation. Several natural enemies and biocontrol agents of cassava green mite, *Mononychellus tanajoa* have been rreported (Table 12.3). In 1993, scientists from IITA and partners identified *Typhlodromalus aripo* as one of the most efficient enemies against cassava green mite. The introduction of *T. aripo* reduced pest populations by as much as 90% in the dry season when pest populations are usually high; in the wet season, pest attacks are not as severe. *T. aripo* from Brazil was first released on cassava farms in Bénin and, subsequently, in 11 countries; it is now confirmed as established in all of them, except Zambia. *T. aripo* has also spread into Togo and Côte d'Ivoire from neighboring countries. It spread to about 12 km in the first year, and as much as 200 km in the second year. Today, the predator of the cassava green mite has been established on more than 400,000 km^2 of Africa's cassava-growing areas. Overall, in cassava 66 species of Phytoseiidae were collected, of which 25 were new for science and 13 were very common in other crops. *Typhlodromalus manihoti* (Moraes) was collected most frequently, being found in over 50% of the fields sampled. It is followed by *Neoseiulus anonymus* Chant and Baker *T. aripo* De Leon, *Galendromus annectens* (De Leon), *G. helveolus* (Chant) and *Amblyseius aerialis*, (Muma) among others. *Typhlodromalus aripo, T. manihoti* and *N. idaeus* play an important role in the control of *M. tanajoa* in Africa. *Typhlodromalus aripo* has proven to be the most promising species. Field evaluations in Africa indicated that *T. aripo* can reduce the mite population from 30-90%, bringing about a 30-37% increase in cassava production (Yaninek *et al.* 1993).

It is observed that chemical control of the pest was ruled out because of possible adverse effects of chemicals on illiterate farmers and the environment. Also, disease pathogens and pests tend to develop gradual resistance to chemical pesticides over time. Moreover, most chemical pesticides are not selective and might destroy the natural enemies and the pests together. Since the release of *T. aripo*, benefits in Nigeria have been estimated at $1.367 billion, followed by Ghana $305 million, and Bénin $54 million (http://r4dreview.org/2011/11/ biocontrol-offers-benefits-to-africa/). In India the cassava is infested by *Tetranychus telarius* L., *T. neocaledonicus* Andre,

Eutetranychus orientalis Klein, and *Oligonychus biharensis* (Hirst) but is not a major problem.

Table 12.3. Natural enemies and biocontrol agents of cassava green mite, *Mononychellus tanajoa*

Predators	Pathogens
Neuroptera	Fungi
Chrysopa sp.	*Neozygites floridana*
Coleoptera	*Hirsutella thompsonii*
Stethorus sp.	Virus
Oligota spp.	Not identified (found in Colombia, unpublished information)
Acari (114 strains)	
Typhlodromalus manihoti	
T. aripo	
Neoseiulus idaeus	
Others	

Source: Bellotti *et al.*, 2005

Other natural enemies of mites are the pathogenic fungi belonging to the genera *Neozygites* (Zygomycetes: Entomophthora) and *Hirsutella* (Hyphomycetes: Monilia) (Table 12.3). The former is a pathogenic fungus that appears sporadically in Colombia and Northeast Brazil (*Neozygites cf floridana*), causing up to 100% mortality of the cassava green mite in 1-2 wk. Some strains are specific to the genus *Mononychellus*. In evaluations done in Africa, *Hirsutella* sp. has proven to be very effective controlling mite populations (Bellotti *et al.*, 2005).

PESTS OF STORED CASSAVA CHIPS

High perishability is the major concern of fresh cassava tubers for storage. Raw tubers cannot be stored for more than 2–6 days after harvest without quality deterioration. Chips can be used as a form in which cassava is usually stored for longer duration after proper drying. The major insect pests infesting the cassava chips are as follows

Rhyzopertha dominica (Lesser Grain Borer)

Rhyzopertha dominica Fabricius is an important internal and external feeder of plain dried and parboiled cassava chips. Parker

(1981) recognized this as the most destructive pest on stored cassava chips.

The species of arthropods found to attack the beetle, either as parasites or as predators are *Anisopteromalus calandrae, Rhabdepyris* sp., *Cephalonomia* sp., *Lyctocoris campestris, Acarophenax assanovi, Cheyletus eruditus* and *Pyemotes ventricosus (Pediculoides ventricosus). C. eruditus* appeared to be the most widely distributed, being found in all the infested grain products studied. According to tests, it devours 3-5 eggs of the beetle per day. Anisopteromalus was found in a few areas and parasitised up to 13.3-17.1% of the beetle population. *Cephalonomia* and *Rhabdepyris* parasitised 1.9–3.1% in certain areas. The mite *Acarophenax* occurred on pupae (rarely) and adult beetles. It is a small species, and is stated to have been only recently recognised. Lyctocoris seemed to be the most successful predator of all, destroying 8-12 eggs/day. It is polyphagous and also attacked other stored product pests.

Sitophilus oryzae (Rice Weevil)

Sitophilus oryzae Linnaeus is a polyphagous pest infesting on a number of stored-products, including plain dried and parboiled cassava chips (Premkumar *et al.*, 1996; Jayaprakas and Karnavar 1999), and distributed throughout the tropical and warm temperate areas of the world. Among natural enemies, *Anisopteromalus calandrae* and *Lariophagus distinguendus* are ectoparasitic wasps parasitizing larvae of rice weevil and their potential for controlling *S. oryzae* has already been demonstrated (Arbogast and Mullen, 1990).

Prostephanus truncatus (Large Grain Borer)

Prostephanys truncatus (Horn) is a devastating pest of dried cassava. it has spread to 16 countries in which it is one of the m ost serious pests of stored products constituting a continental problem in Africa. The most promising aspect of integrated pest management is the use of *Teretrius nigrescens,* the natural enemy of *P. truncatus,* which was released in Kenya in 1992. The release recorded positive impact on the control of *P. truncatus* (Giles *et al.*, 1996). Despite the release, areas where releases were done still shows presence of damage as a result of *P. truncatus* infestation (Meikle *et al.*, 2002). Consequently, there is need for additional control measures in order to reduce the dependant of the farmers on synthetic insecticides which are expensive with high toxic residues for protecting stored foods. The biological activity of *Azardirachta indica* (neem leaves), *Tetrapleura tetraptera* (seeds), *Piper guinenses* (alligator pepper seeds) and *Xylopia aethiopica* (black pepper seeds) extracts surface treatment to dried cassava chips (*Esculenta manihot*) against *Prostephanus truncatus* (Horn) to determine infestation rate via mortality and emergence of the pest and weight loss in the dried cassava chips have been documented. Out of the four plants extract *A. indica* has the best insecticidal properties, followed by *P. guinensee*, *T. tetraptera* and *X. aethiopica* in a descending order of efficacy (Popoola and Hassan, 2011).

SWEET POTATO

The sweet potato (*Ipomoea batatas*) is a dicotyledonous plant that belongs to the family Convolvulaceae. Its large, starchy, sweet-tasting, tuberous roots are an important root vegetable. Many insect species attack sweet potato and the importance of different species varies between agroecological zones. The crop is widely cooked by deep frying and consumed in forms of French fries and chips (Farinu and Baik, 2007).

Cylas formicarius (Sweet Potato Weevil)

Sweet potato weevil, *Cylas formicarius* Fabricius is often considered to be the most serious pest of sweet potato, with reports of losses ranging from five to 97% in areas where the weevil occurs. Sweet potato weevil *Cylas formicarius* (Fab.) is a serious insect pest of sweet potato worldwide. The weevil can be managed by Integrated Pest Management (IPM) strategy, which includes host plant resistance, biotechnological approaches, cultural methods, biological control, use of semiochemicals, botanical and chemical pesticides etc. Weevil infestation varied among genotypes based on the physical characters of tubers. After the discovery of sex pheromone for *C. formicarius*, the management strategy for this pest has changed significantly. The female sex pheromone (Z)-3-doden-1-ol (E)-2-butenoate occupied an inclusive component in the management of *C. formicarius* worldwide, and the pheromone trap is modified to accommodate entomopathogenic fungi to hasten the process of its mortality (Korada *et al.*, 2010).

Maeto and Uesato (2007) reported a new species of braconid, *Bracon yasudai* from the south-west islands of Japan. It is a solitary idiobiont ectoparasitoid of the larvae of the sweet potato weevil.

Palaniswami and Rajamma (1986) reported the braconids *Rhaconotus* sp. and *Bracon* sp. and an unidentified hymenopterous parasitoid on the larvae of these weevils. Jansson and Lecrone (1991) reported *Euderus purpureas*, an eulophid parasitizing on *C. formicarius* in Southern Florida. Nevertheless, the success of all these parasitoids at field conditions is doubtful since they are recorded in a very few numbers.

Weevils were also infected by the bacteria *Bacillus thuringiensis* in the laboratory (Sundara Babu *et al.*, 1970). Palaniswami *et al*, (1988) found that the green muscardine fungus *Metarrhizium anisopliae* was infecting the sweet potato weevil. A nuclear polyhedrosis virus (NPV) and a bacterium were observed in the grubs (KAU, 1986). Two hymenopteran parasites and three hymenoptera predators were also collected and identified (Palaniswami *et al.,* 1991). The hymenopteran parasites include *Trichopria* sp (Diaprididae) and *Spathius dido* (Branconidae), and three hymenopteran predators include *Tetramorium bicarinatus, Monomorium* sp. and *Monomorium* sp. (Salmonis group).

Among botanicals, several plant materials have been evaluated against this weevil. An extract of the gum from leaves and twigs of *Croton linearis*, which is used medicinally in Jamaica, contain a diterpene compound, which has marked insecticidal activity to adult *C. formicarius* (Alexander *et al.*, 1991). Sesquiterpene furan, and epingaione isolated by guided isolation of the crude ethanol extract of the leaves and stems of *Bontia daphnoides* are also insecticidal (Williams and Williams, 1997). NeemAzal 5 (Trifolio-M, Lahnau, Germany), at a concentration of 0.05% a.i. v/v, is an oviposition inhibitor for this weevil and reduced feeding (Jayaprakash *et al.*, 2000). The essential oil of leaves, stem, and fruits of *Hyptis capitata* are dominated by oxygenated sesquiterpenes (>50%), with viridiflorol being the major component (Facey *et al.*, 2006). Viridiflorol is insecticidal against *C. formicarius.*

Agrius convolvuli (Convolvulus Hawk Moth)

Yield losses can occur due to *Agrius convolvuli* Linnaeus, if heavy defoliation takes place when the crop is young. A large caterpillar can defoliate a plant and a large population of late instar larvae can defoliate a field overnight. The larvae feed on the leaf blades, causing irregular holes, and may eat the entire blade, leaving only the petiole.

In India, *A. convolvuli* is an occasional pest of sweet potato, *Vigna mungo* and *Vigna radiata*. Eggs are also laid on the leaves of *Colocasia antiquorum* and *Clerodendrum chinense*, but no significant feeding occurs on these latter plants. Four species of parasitoid, *Trichogramma australicum*, *T. achaeae*, *T. agriae* and a species of *Telenomus* (Eulophidae) attack the eggs of *A. convolvuli*. The abundance of each parasitoid varied with the plant species on which the eggs were laid. *T. agriae* was the commonest species in eggs collected on *Colocasia*, followed by *T. achaeae* and *Telenomus* sp., up to a total of 43.6%. *T. australicum*, followed by *Telenomus* sp., were the main species emerging from eggs on *Clerodendrum*, up to a total of 63.9% (Nagarkatti 1973). *Trichogramma confusum* has also been recorded from the eggs of *A. convolvuli* on *Clerodendrum chinense* (Nagarkatti and Nagaraja 1978).

In China, *A. convolvuli* larvae were parasitised by *Charops bicolor* (Ichneumonidae), *Apanteles* spp. (Braconidae) and eulophid wasps (Wu, 1983). *A. convolvuli* is a common pest in the eastern part of South Africa. Although it is generally not abundant, from time to time large areas of sweet potatoes have been almost completely defoliated by it. There are 3 generations a year and overwintering occurs as the pupa. Natural enemies include the whitebellied stork (*Ciconia nigra*) which, on occasion, destroys large numbers of larvae (Anon. 1927). In Zimbabwe *A. convolvuli* larvae on sweet potato were parasitised by the tachinids *Zygobothria atropivora* and *Sturmia dilabida*, both of which are widely distributed in South Africa. The latter parasitoid also attacks larvae of *Spodoptera exigua* (Cuthbertson, 1934).

Aspidomorpha miliaris (Spotted Tortoise Beetle)

Both adults and larvae of *Aspidomorpha miliaris* Fabricius eat large round holes in the leaves. Attacks are sometimes sufficiently severe to completely skeletonize the leaves and peel the stems.

The tortoise beetle *(Aspidomorpha miliaris)* of sweet potato is often parasitized by *Cassidocida aspidiomorphae* and *Tetrastichus colemani* in India (Nair, 1975). Sweet potato ecosystem in Philippines has rich community of natural enemies. A total of 40 species of beneficial arthropods and 2 entomopathogens have been recorded (Amalin and Vasquez, 1993). Among spiders, *Oxyopes javanus* is common and among predatory beetles; *Harmonia octomaculata, Menochilus sexmaculata* and *Micraspis hirashimai* are well known. Over all promising biological control agents for sweetpotato weevils appear to be the fungi *B. bassiana* and *Metarrhizium anisopliae* and the nematodes *Heterorhabditis* spp. and *Steinernema* spp.The fungi attack and kill adult weevils, whereas the nematodes kill the larvae (Ames *et al.*, 1996).

AROIDS

Aroids belong to the family Araceae and comprises of 110 genera and more than 2000 species, which are generally grown in homesteads. The most edible aroids are taro, tannia and elephant foot yam. Tubers of taro and tannia contain about 15 to 25% starch and fairly good amount of protein and minerals (Moorthy and Padmaja, 1991). Taro (*Colocasia esculenta*) is an important vegetable crop in Bangladesh. Taro was once an important crop in north Mauritius, particularly in the humid regions of Crève Cour, Montagne Longue, Les Marianes and Congomah. It was also found in marshy regions of Bambous, Clementia, Schoenfeld and Careau Acacia. But this is no longer so; taro leaf blight came in 1995 and damaged the crop substantially. Unfortunately, the local varieties in Mauritius do not have tolerance to the disease, and control measures using fungicides are very expensive. The only practical measure is to introduce resistant varieties. Another important aroid is elephant foot yam (EFY), often referred to as the "King of tubers" due to its palatability, cooking quality, medical utility and therapeutic values (Sengupta *et al.*, 2008). These tubers are also attacked by number of insects.

Aphis gossypii (Aphid)

Aphis gossypii Glover is one of the important sucking pests of taro which reduce the chlorophyll content of leaves causing yellowish, speckling and curling. They attack both sides of leaves and petioles. Several biocontrol agents such as *Aphelinus mali*, *Aphelinus sp.*, *Coccophagus cowperi* and *Aphidencyrtus aphidivorus* have been reported. Palaniswami *et al*, (1991) reported aphidophagous predators *Menochilus sexmaculatus*, *Verania discolor*, *Pseudospidimerus cicrumflexus*, *V. inops*, *Ischiodon scutellaris* in India. The aphidophagus predators observed on these crops were coccenellids (*Menochilus sexmaculatus, Verania discolor, V. inops* and *Pseudospidimerus circumflexus*) and syrphid (*Ischiodon scutellaris*). The predation was higher during larval stage while it was less in the adult stage. The mean aphid consumption per day by a grub of *M. sexmaculatus* ranged from 15 to 18 while it was 10 to 18 in the case of *V. discolor*. Among the aphidophagous predators, *I. scutellaris* is a voracious feeder. The mean consumption per day ranged from 20 to 28 aphids. In general, coccenellids and syrphids were effective against aphid. Four species of parasitoids, viz., *Aphelinus mali*, *Aphelinus* sp., *Cocophagus cowperi* (Aphelinidae) and *Aphidencyrtus aphidivorus* (Encyrtidae) were active on *Aphis gossypii* in the fields of taro, tannia and elephant foot yam (edible aroids). The total parasitization in the field ranged from 2–14%. *A. maliand, A. aphidivorus* were predominant among others (Palaniswami and Pillai, 1980).

Spodoptera litura (Hornworm)

Spodoptera litura Fabricius are leaf feeders which are stout and greenish to brown in color with stipes. Often they are regarded as minor pests of sporadic occurrence. They were parasitized by a parasitoid *Charops hersei* (Palaniswami and Pillai, 1983). Leaf eating caterpillars (*Spodoptera litura* and *Hippotion* spp.) of edible aroids in the field were found preyed upon by a wasp (*Polistes* sp.). The caterpillars of *Hippotion* spp. (horn worm) were usually parasitized (6–10%) by a hymenopteran *Charops hersei* under field conditions (Palaniswami and Pillai, 1983). In Western Samoa, Stechmann and Semisi (1984) collected information on *S. litura* damage levels in relation to natural populations of *Apanteles* spp. in taro fields. They found that this pest was more severe on taro where insecticides and herbicides were widely used, which perhaps created imbalance between the pest and its natural enemies. Barrion and Litsinger (1987) reported the presence of *Peribaea orbata* as a gregarious larval parasitoid on *S. litura*.

Monolepta signata (White Spotted Beetle)

Monolepta signata Olivier is an important pest feeding on leaves of taro. Besides the feeding it also predisposes to sclerotium rot. The grubs are preyed upon by *Eocanthecona furcellata* in field (Palaniswami, 1983).

YAMS

Yams (the *Dioscorea* spp.) are starchy staples in the form of large tubers produced by annual and perennial vines grown in Africa, the Americas, the Caribbean, South Pacific and Asia. There are hundreds of wild and domesticated *Dioscorea* species. White Guinea yam, *D. rotundata,* is the most important species especially in the dominant yam production zone in West and Central Africa. It is indigenous to West Africa, as is the Yellow yam, *D. cayenensis*. Water yam, *D. alata,* the second most cultivated species, originated from Asia and is the most widely distributed species in the world. Yam tubers consist of about 21% dietary fiber and are rich in carbohydrates, vitamin C and essential minerals. Worldwide annual consumption of yams is 18 million tons, with 15 million in West Africa. Annual consumption in West Africa is 61 kilograms per capita. Yams are boiled, roasted, baked or fried. In Africa they are also mashed into a sticky paste or dough after boiling (http://www.iita.org/yam).

Worldwide yam production in 2007 amounted to 52 million tons, of which Africa produced 96%. Most of the world's production comes from West Africa representing 94%, with Nigeria alone producing 71%, equaling more than 37 million tons. African countries imported more than 2,000 tons in 2002, and exported 15,500, of which Nigeria exported 12%. Yam production is declining in some traditional producing areas due to declining soil fertility, increasing pest pressures and the high cost of labor. Small holders, therefore, need access to innovations to reduce labor and improve productivity. Yams are grown by planting pieces of tuber, or small whole tubers ('seed yams') saved from the previous season. Small-scale farmers, the majority of producers, often intercrop yams with cereals and vegetables.

The major pests that affect yams include insects such as leaf and tuber beetles, mealy bugs, and scales; parasitic nematodes; fungi causing anthracnose, leaf spot, leaf blight, and tuber rot; and viruses, especially the yam mosaic virus (YMV).

Aspidiella hartii (Yam Scale Insect)

Yam scale, *Aspidiella hartii* attacks stored yams in Africa, Asia and the Pacific; in the Caribbean its principal damage is to young vines which may be destroyed completely. *Aspidiella hartii* Cockerell also affects ginger, turmeric, taro and elephant foot yam. The pest problem is perpetuated through seed tubers. In severe infestation, attacked plants show drying due to continuous sucking and desapping (Palaniswami, 1991). The infested tubers shrivel affecting the quality, viability and marketability. The nymphs are parasitized by an aphelinid *Physcus comperi* H. and an encyrtid *Adelencyrtus moderatus* (Palaniswami., 1991). Yam scale insects are frequently checked by two parasitoids, viz., *Coccobius* (*Physus*) *comperi* and *Adelencyrtus moderatus.* Both are solitary endoparasitoids, former being more abundant in the field and storage.

Lema lacordairei (Yam Blue Beetle)

Lema lacordairei Baly is a leaf feeding chrysomelid that occurs on all *Dioscorea* spp. In India, it has been reported as a pest of *Dioscorea allata* in Kerala (Visalakshi and Nair, 1978). It is often attacked by a tachnid parasite, *Pseudalsomyia* sp. The field parasitism ranges from 7-9%. Adult female oviposits on young grubs (usually second instars). The parasite grows along with the host and completes the egg and larval period within 5-6 days. More than one parasitoid developes from single host (Palaniswami and Pillai, 1984).

Ferrisia virgata (Yam Mealy Bug)

Palaniswami *et al.* (1982) reported the incidence of *Ferrisia virgata* Cockerell mealy bug on *D. dumetorum*. They de saps the leaf and the high incidence of this pest causes drying of leaves and withering. *Aenasius advena, Anagyrus qadrii* and *Blepyrus insularis* parasitized 0.4, 1.4 and 25.1%, resp., of the *F. virgata* collected. The predators *Scymnus* spp., a genus near *Scymnus, Allograpta javana, Leucospis* sp., *Gitona* sp., *Triommata coccidivora* and *Mallada* sp. were also recorded. *Acerophagus* sp. was also found to be parasitic on *Ferrisia virgata.*

CHINESE POTATO/COLEUS POTATO

Plectranthus rotundifolius or *Solenostemon rotundifolius* (Chinese Potato) is a perennial herbaceous plant, of the mint family, native to tropical Africa, cultivated in parts of West Africa, South Asia and Southeast Asia for its edible tubers. In English, *P. rotundifolius* is often called "Chinese potato", "Coleus potato" or "Hausa potato". The leaf and shoot folder is an important pest of Chinese potato.

Spoladea recurvalis (Leaf and Shoot Folder)

In Chinese potato, *Spoladea recurvalis* Fabricius is a major leaf and shoot folder, which is controlled by two beneficial endoparasitoids, *Microgaster psarae* on *Pycnarmon cribrata* (Ayyar, 1938) and *Leptobatopsis indica* on *Phostria piasusalis.* The parasitoid showed preference to second instars, total developmental period ranged from 8 to 14 days. The field ranged from 15–50% (Palaniswami, 1979).

CONCLUSIONS

The major tuber crops, like, Cassava, Sweet potato, Aroids, Yams and Chinese potato are affected by different insect pests at various crop stages. They are causing significant damage and yield reductions to the crops. The major insect pests affecting Cassava in field conditions are papaya mealybug, cassava white fly, spiraling whitefly and cassava scale. Also stored cassava chips are affected by storage pests like lesser grain borer and rice weevil. The major insect pests affecting Sweet potato are sweet potato weevil, convolvulus hawk moth and spotted tortoise beetle. Aroids and Yams are affected by various insect pests like aphid, horn worm, white spotted beetle, blue beetle, scale insect and mealy bug. The leaf and shoot folders are major pests affecting chinese potato.

Pests of tuber crops are having lot of natural enemies. There are various predators, parasitoids and microbial control agents, which help to keep the pests population with in checks. Predators mainly belong to insect orders like coleopteran (coccinellids), hemiptera (bugs), neuropteran (lace wings), diptera (hover flies) and hymenoptera (Salmonis group). Hymenoptera, diptera, thysanoptera (thrips) and acari (mites) are main orders of Parasitoids which attack these pests. Main microbial control agents include bacteria (*Bacillus thuringiensis*), fungus (*Metarrhizium anisopliae*) and virus (NPV).

They help to regulate pest populations in natural conditions. But in artificially cultivated field conditions, the ratio of natural enemies to pests will be very low. So they may not be able to keep the pest populations with in checks, and from causing havocs in fields- after attaining big numbers. So it is imperative to use the natural enemies in required sufficient numbers after artificial rearing, for pest management. Proper management and multiplication of these natural enemies and timely intervention with them could be more sustainable and eco-friendly way of pest management in tuber crops, in the coming days.

REFERENCES

Aishwariya, K.K., Manjunatha, M., Mohan and Naik, I. (2007a) Seasonal incidence of spiraling whitefly, *Aleurodicus dispersus* Russell and its natural enemies in relation to whether in Shimoga. *Karnataka J. Agric. Sci.*, **20**, 146–148.

Aishwariya, K.K., Manjunatha, M. and Naik, M. (2007b) Evaluation of fungi *Fusarium semi tectum* Berk and Ravenel and *Verticillium lecanii* (Zimm.) Viegas against spiraling whitefly *Aleurodicus dispersus* Russell on guava, *Karnataka J. Agric. Sci.*, **20**, 283–287.

Alexander, I.C., Pascoe, K.O., Manchard, P. and Williams, L.A.D. (1991) An insecticidal diterpene from *Croton linearis*. *Phytochemistry*, **30**, 1801–1803.

Alvin M. Simmons and Shaaban Abd-Rabou (2007) Survey of natural enemies of the sweet potato whitefly (Hemiptera: Aleyrodidae) in ten vegetable crops in Egypt. *J. Agric. Urban Entomol.*, **24**, 137–145.

Amalin, D.M. and Vasquez, E.A. (1993) *A Handbook on Philippine Sweet Potato Arthropod Pests and Their natural Enemies*. International Potato Center, Los Banos, Philippines.

Ames, T., Smit, N.E.J.M., Braun, A.R., O'Sullivan, J.N. and Skoglund, L.G. (1996) *Sweetpotato: Major Pests, Diseases, and Nutritional Disorders*. International Potato Center (CIP), Lima, Peru.

Ananthanarayanan, K.P., Subramaniam, T.R. and Muthukrishnan, T.S. (1957). A note on tapioca scale, *Aonidomytilus albus* Cockrel. *Madras Agric. J.*, **44**, 281–282.

Anonymous (1927) Miscellaneous entomological notes. *Farming South Afr.*, **1**, 41–46.

Anonymous (1983) *Pest Management. Two Decades of Research*. CTCRI, Thiruvanan-thapuram.

Arbogast, R.T. and Mullen, M.A. (1990) Interaction of maize weevil (Coleoptera: Curculionidae) and parasitoid *Anisopteromalus calandrae* ((Hymenoptera: Pteromalidae) in small bulk of stored corn. *J. Econ. Entomol.*, **83**, 2462–2468.

Ayyar, T.V.R. (1938) Two new caterpillars of economic importance in south India. *J. Bombay Nat. Hist. Soc.*, **40**, 336–338.

Babu, B., Hegde, V., Makeshkumar, T. and Jeeva, M.L. (2011) Rapid and sensitive detection of potyvirus infecting tropical tuber crops using genus specific primers and probes. *Afr. J. Biotechnol.*, **11**, 1023–1027.

Banjo, A.D., Hassan, A.T., Ekanayake, I.J., Dixon, A.G.O., Jackal, L.E.N. (2004). Effect of *Aleurodicus dispersus* Russel (Spiralling whitefly) on growth indices and yield of three genotypes of cassava (*Manihot esculanta* Crantz). *J. Res. Crops*, **5**, 252–260.

Barrion, A.T. and Litsinger, J.A. (1987) A larval parasite of swarming caterpillar and common cutworm in Philippines. *Int. Rice Res. Inst. Newsletter*, **12**, 34–35.

Basant Ram (2008) Role of *Amorphophallus* in food and nutritional security for humankind in Uttar Pradesh. Abstract, *National Seminar on Amorphophallus: Innovative Technologies*, Patna, Bihar, July 19–20, 2008.

Bellotti, A. and Schoonhoven, A. (1978) Mite and insect pests of cassava. *Annu. Rev. Entomol.*, **23**, 639–667.

Bellotti, A., Melo, E.L., Arias, B., Herrera, C.J., Hernandez, M.P., Holguin, C.M., Guerrero, J.M. and Tryjillo, H. (2005) Biological control in the neotropics: A selective review with emphasis on cassava. In: M.S. Hoddle (ed.), *Biological Control of Arthropods*, USDA Forest Service, Washington DC, pp. 206–227.

Bellotti, A., Campo, B.V.H. and Hyman, G. (2012) Cassava production and pest management: Present and potential threats in a changing environment. *Trop. Plant Biol.*, **5**, 39–72.

Cuthbertson, A. (1934) Biological notes on some Diptera in Southern Rhodesia. *Proc. Rhodesian Sci. Assoc.*, **33**, 32–50.

David, B.V. and Regu, K. (1995) *Aleurodicus dispersus* Russel (Aleyrodidae: Homoptera), a whitefly pest new to India. *Pestology*, **19**, 5−7.

Facey, P.C., Porter, R.B.R. and Williams, L.A.D. (2006) The chemical composition and insecticidal activity of the essential oils from Jamaican *Hyptis capitata* Jacq. In: L.A.D. Williams and P.B. Reese (eds.), *Biologically Active Natural Products for the 21st Century*. Research Signpost, Trivandrum, India, pp. 13–19.

FAO, FAOSTAT, Statistical Database (Online). Available: http://faostat.fao. org/ site/567/DesktopDefault.aspx?PageID=567#ancor (10 June 2012)

Farinu, A. and Baik, O.D. (2007) Heat transfer coefficients during deep fat frying of sweet potato: Effects of product size and oil temperature. *Food Res. Int.*, **40**, 989−994.

Gerling, D., Alomar. O. and Arno, J. (2001) Biological control of *Bemisia tabaci* using predators and parasitoids. *Crop Prot.*, **20**, 779-799

Giles, P.H., Hill, H.G., Nang'ayo, F.L.O., Farrel, G. and Kibata, G.N. (1996) Release and establishment of the predator *Teretriosoma nigrescens* LEWIS for the biological control of *Prostephanus truncatus* (Horn) in Kenya. *Afr. Crop Sci. J.*, **4**, 325–337.

Hedge, V. and Jayaprakas, C.A. (2011) Management of pest and diseases in tropical tuber crops by adopting innovative technologies, In: M.S. Sajeev (eds.), *Proceedings of the National Seminar on Climate and Food security- Challenges and Opportunities of Tuber Crops*, Kerala, pp. 349−361.

Herren, H.R. and Neuenschwander, P. (1991) Biological control of cassava pests in Africa. *Annu. Rev. Entomol.*, **36**, 257–284.

Jansson, R.K. and Lecrone, S.H. (1991) *Euderus purpureas* (Hymenoptera: Eulophidae) a parasitoid of sweet potato weevil (Coleoptera: Apionidae) in southern Florida. *Fla. Entomol.*, **74**, 596–598.

Jayaprakas, C.A. and Karnavar, G.K. (1999) The biology and taxonomy of *Sitophilus oryzae* (L.) (Coleoptera: Curculionidae). *J. Root Crop Res.*, **24**, 1−10.

Jayaprakas, C.A., Rajamma, P., Palaniswami, M.S. and Sreekumar, J. (2000) Bioactivity of cassava seed extract on pests of tuber crops. *J. Root Crops*, **26**, 1–3.

Korada, R.R., Naskar, S.K., Palaniswami, M.S. and Ray, R.C. (2010) Management of sweet potato weevil, *Cylas formicarius* (Fabricius): An overview. *J. Root Crops*, **36**, 14–26.

Kueneman, E., Raswant, V., Lutaladio, N. and Cooke, R. (2012) *Global Consultation on Cassava as a Potential Bioenergy Crop*. IFAD, Rome.

Lal, S.S. (1980) A note on *Prospaltella flava* Shafee, an aphelinid parasitoid of *Bemisia tabaci* (Gennadius). *Madras Agric. J.*, **67**, 557–558.

Lal, S.S. and Pillai, K.S. (1981) Cassava pests and their control in Southern India. *Trop. Pest Manage.*, **27**, 349–351.

Lisha, V.S., Antony, B., Palaniswami, M.S. and Henneberry, T.J. (2003) *Bemisia tabaci* (Gennadius) (Homoptera: Aleyrodidae) biotypes in India. *J. Econ. Entomol.*, **96**, 322–327.

Maeto, K. and Uesato, T. (2007) A new species of *Bracon* (Hymenoptera: Braconidae) parasitic on alien sweet potato weevils in the south-west islands of Japan. *Entomol. Sci.*, **10**, 55–63.

Mani, M. (2010) Origin, introduction, distribution and management of the invasive spiralling whitefly *Aleurodicus dispersus* Russell in India. *Karnataka J. Agric. Sci.*, **23**, 59–75.

Mani, M. and Krishnamoorthy, A. (1999) Natural enemies and host plants of spiraling whitefly *Aleurodicus dispersus* Russel in Bangalore, Karnataka. *Entomon,* **24**, 75–80.

Mani, M., Ragunatha, R., Dinesh, M.S. and Krishnamoorthy, A. (2001) Spiralling whitefly *leurodicus dispersus* Russell in Hyderabad. *Insect Environ.*, **7**, 82.

Meikle, W.G., Gutierrez, A.P. and Herren, H.R. (1996) Decision making by smallholder farmers in an oil palm based economy in southern Benin, West Africa. *Trop. Agric. (Trinidad)*, **73**, 217–225.

Melo, E. L. (2002) Potencial de Control Biológico en el Manejo de las Plagas de la Yuca. In: B. Ospina and H. Ceballos (eds.), *La Yuca en el Tercer Milenio. Sistemas Modernos de Producción, Procesamiento, Utilización y Comercialización*, Centro Internacional de Agricultura Tropical, Cali, Colombia, pp. 234–249.

Meyerdirk, D.E., Muniappan, R., Warkentin, R., Bamba, J. and Reddy, G.V.P. (2004) Biological control of the papaya mealybug, *Paracoccus marginatus* (Hemiptera: Pseudococcidae) in Guam. *Plant Prot. Quart.*, **19**, 110–114.

Moorthy, S.N. and Padmaja, G. (1991) Comparative study on digestibility of raw and cooked starches of different tuber crops. *J. Root Crops*, **17**, 255–258.

Muniappan, R., Meyerdirk, D.E., Sengebau, F.M., Berringer, D.D. and Reddy, G.V.P. (2006)Classical biological control of the papaya mealybug, *Paracoccus marginatus* (Hemiptera: Pseudococcidae) in the republic of Palau. *Fl. Entomol.*, **89**, 212–217.

Muniappan, R., Shepard, B.M., Watson, G.W., Carner, G.R., Sartiami, D., Rauf, A. and Hammig, M.D. (2008) First report of the papaya mealybug, *Paracoccus marginatus* (Hemiptera: Pseudococcidae), in Indonesia and India. *J. Agric. Urban Entomol.*, **25**, 37–40.

Nagarkatti, S. (1973) *A Note on Egg Parasitism of Agrius convolvuli in Karnataka*. Technical Bulletin, CIBC, Bangalore.

Nagarkatti, S. and Nagaraja, H. (1978) Experimental comparison of laboratory reared vs. wild-type *Trichogramma confusum* (Hym.: Trichogrammatidae). Fertility, fecundity and longevity. *Entomophaga*, **23**, 129–136

Nair, M.R.G.K. (1975) *Insects and Non-insect Pests of Crops in India*. ICAR, New Delhi. Palaniswami, M.S. (1979) A potential parasite *Leptobatopsis indica* Cameron (Ichneumonidae: Hymenoptera) on coleus top shoot folder *Phostria piasusalis* W. (Pyralidae: Lepidoptera). *J. Root Crops,* **5**, 36–38.

Obinna, A., Peter, N. and Korie, S. (2011) Niche separation between *Encarsia dispersa* and *Encarsia guadeloupae*, two biological control agents of the spiraling whitefly *Aleurodicus dispersus*, in Benin, West Africa. *BioControl*, **56**, 277–282.

Otim, M., Legg, J., Kyamanywa, S., Polaszek, A. and Gerling, D. (2006) Population dynamics of *Bemisia tabaci* (Homoptera: Aleyrodidae) parasitoids on cassava mosaic disease /resistant and susceptible varieties. *Biocontrol Sci. Technol.*, **16**, 205–214.

Palaniswami, M.S. (1991) Yams scale insect *Aspidiella hartii* Ckll and its parasitoids. *J. Root Crops,* **17**, 69–70.

Palaniswami, M.S. (1994) Pests of Edible Aroids, Yams and Chinese Potato. In: K.L. Chadha, and G.G. Nayar (eds.), *Advances in Horticulture-TuberCrops*. Malhotra Publishing House, New Delhi, pp. 490–491.

Palaniswami, M.S. and Pillai, K.S. (1980) Parasites on *Aphis gossypii* G. infesting taro and tannia. *Curr. Sci.,* **21**, 830.

Palaniswami, M.S. and Pillai K.S. (1983). Occurrence of *Charops hersei* G & M as a parasite on taro hornworm *Hippotion oldenlandiae F. Entomon.,* **8**, 103–104.

Palaniswami, M.S. and Pillai, K.S. (1984) *Pseudalsomyia* sp. a new parasite of *Lema lacordairei* B. *Entomon.,* **9**, 75.

Palaniswami, M.S. and Rajamma, P. (1986) Biological control of sweet potato weevil and cassava spider mites. Annual progress report 1986, Central Tuber Crops Research Institute, Trivandrum, India, pp. 90–92

Palaniswami, M.S., Pillai, K.S. and Abraham, K.(1982). Occurance of *Ferrisia virgata* on *Dioscorea dumetorum*. *J. Root Crops*, **8**, 69–70.

Palaniswami, M, S., Rajamma, P. and Pillai, K.S. (1988) Biotic agents of pests of tropical tuber crops- A review. *J. Root Crops,* **14**, 41–46.

Palaniswami, M.S., Pillai, K.S., Nair, R.R. and Mohandas, C. (1995) A new cassava pest in India. *Cassava News Lett.*, **19**, 6–7.

Parsa, S., Kondo, T. and Winotai, A. (2012) The cassava mealybug (*Phenacoccus manihoti*) in Asia: First records, potential distribution, and an identification key. *PLoS ONE* 7(10): e47675. doi:10.1371

Parker, B. L., Booth, R. H. and Haines, C. P. (1981). Arthropods infesting stored cassava (*Manihot esculenta* Crantz.) in peninsular Malaysia. *Prot. Ecol.*, **3**, 141–156.

Pillai, K.S. (1994) Pests of cassava. In: K.L. Chadha, and G.G. Nayar (eds.), *Advances in Horticulture- Tuber Crops*, Malhotra Publishing House, New Delhi, pp. 429–456.

Pillai, K.S. and Hrishi, N. (1975) *Retithrips syriacus*, a new pest on cassava and ceara rubber in Kerala. *J. Root Crops*, **1**, 88.

Pillai, K.S. and Palaniswami, M.S. (1985). The ecology of spider mite, *Tetranychus cinnabarinus* Boisd and *Eutetranychus orientalis* Klein on cassava. In: M.S. Palaniswami (ed.), *Proceedings of the National Symposium on the production and utilization of tropical tuber crops*, CTCRI, Trivandrum, pp. 187–192.

Pillai, K.S. and Palaniswami, M.S. (1988) Evaluation of cassava accessions resistant to spider mites and factors governing resistance in India. *Proceedings of VIII Symposium of the International Society for Tropical Root and Tuber Crops*, Bangkok, Thailand, pp. 341–350.

Pillai, K.S., Rajamma, P. and Palaniswami, M.S. (1990). Integrated pest management in Tapioca, pp. 22-23.

Popoola, K.O.K. and Hassan, A.T. (2011) Insecticidal effects of four plants extract against *Prostephanus truncatus* (Horn) (Coleoptera: Bostrichidae) in infested dried cassava chips. *Curr. Res. J. Biol. Sci.*, **3**, 1–5.

Premkumar, T., Moorthy, S. N., Balagopalan, C., Jayaprakas, C.A. and Rajamma, P. (1996) Quality changes in market cassava chips infested by insects. *J. Stored Prod. Res.*, **22**, 183–186.

Rajamma, P. (1980) On some natural enemies of sweet potato weevil *Cylas formicarius* Fab. (Curculionidae: Coleoptera). *J. Root Crops,* **6**, 59–60.

Rajamma, P., Premkumar, T. and Jayaprakas, C.A. (2001) Incidence of spiraling whitefly, *Aleurodicus dispersus* Russel (Homoptera: Aleyrodidae) in cassava. *J. Root Crops,* **27**, 335–340.

Russel, L.M. (1965). A new species of *Aleurodicus Douglas* and two closely relatives (Homoptera: Aleurodidae). *Fl. Entomol.*, **48**, 47–55.

Sengupta, S, Chowdhary BM, Singh BN and Ray RN (2008). Status of elephant foot yam cultivation in Jharkand, In: M.S. Palaniswami (ed.), *National Seminar on Amorphophallus: Innovative Technologies*, July 19–20, Patna, Bihar-abstract book, status papers and extended summary: pp. 30–34.

Stechmann, D.H. and Semisi, S.T. (1984) Insektenbekämpfung in West-Samoa unter besonderer Berücksichtigung des Standes biologischer und integrierter Verfahren. *Anz. für Schädling. Pflanzen. Umwelts.*, **57**, 65–70.

Venkatesan, S. and Palanisamy, V. (2010) Eco-friendly management of cassava whitefly, *Bemisia tabaci gennadius. Madras Agric. J.*, **97**, 78–80.

Visalakshi, A. and Nair, M.R.G.K. (1978) Biology of *Lema lacordairei*, new record coleoptera chrysomelidae criocerinae, a pest of yam *Dioscorea alata* in Kerala India. *Entomon*, **3**, 129–132

Waterhouse, D.F. and Norris, K.R. (1989) Biological control: Pacific prospects- Supplement1. ACIAR Monograph No. 12, vii + 125 p.

Williams, L.A.D. and Williams, L.C. (1997) Insecticidally active sesquiterpene furan from *Bontia daphnoides* L. *Philipp. J. Sci.*, **126**, 155–162.

Wu, Z. Q. (1983) Identification of parasitized larval bodies of lepidopterous insects in sweetpotato fields. *J. Fujian Agric. College,* **12,** 153–157.

13

Current Status and Potential of Biopesticides in Diseases Management with Special Reference to Coconut Root Wilt

N. SRINIVASAN*

Division of Crop Protection, Central Plantation Crops Research Institute, Regional Station (ICAR), Kayangulam, Krishnapuram- 690533, Kerala, India
E-mail: srinivasannarayanasamy@yahoo.com

INTRODUCTION

Plant diseases have been conventionally controlled by application of chemicals. The contribution of pesticides to agriculture and especially plant diseases control in particular has lead to widespread diseases management/control leading to the possibility of manifold production in quantitative terms. Even as application of pesticides assumed to be the most common practice to control plant pathogens and that too in large quantities these practices lead to widespread environmental pollution (Pimentel and Greiner, 1997). These measures ultimately lead to badly hitting the environment besides a negative impact on various stakeholders in production-consumption chain. Therefore and with the changing scenario, this technology are unsustainable warranting alternative measures for adoption. More over certain diseases such as viral, phytoplasmal etc. are not readily or sufficiently amenable by conventional/traditional measures for effective control. The demand for organic based and pesticide-free products has grown leap and bound as consumers are more sensitive to pesticide-based products (Diaz and Fereres, 2007). A new disease control technology with integrated system involved has become necessary.

A number of parameters are involved in the integrated disease management system and such a system has assumed significant importance in modern day Agriculture. Biological control has emerged as an important component in the plant disease management system. A special importance in this system of biological control is availability of unique phenomenon of induced systemic resistance (ISR) in host plants against pathogens (Ramamoorthy *et al.*, 2001). As such the technology underneath the biological control strategy consequently covers the organic agriculture system and thus non-chemical alternative mechanism for coconut cultivation also comes into effect. Biopesticides of plant/microbial origins are available in formulation state also against several plant diseases. Several studies have been conducted by different workers to evaluate the effect of biocontrol agents against various plant diseases. Efficacies of biocontrol agents as such and in bioformulations state in control of coconut leaf rot fungi have been shown. *Bacillus subtilis*, *Pseudomonas fluorescens*, *Trichoderma* spp. etc. have been used as biological control agents (Srinivasan, 2001, 2002a, 2004a). This chapter elaborates general aspects on biological control of plant diseases and in context of coconut root (wilt) – leaf rot diseases complex.

Coconut (*Cocos nucifera* L.), commonly refereed to as 'Kalpavriksha' is an important plantation crop providing livelihood security to millions of small-marginal farmers in India particularly in southern states, besides in certain other countries - Coconut cultivation in India is concentrated in Kerala state, followed by Tamil Nadu, Karnataka and Andhra Pradesh. Currently the crop is grown in the country on nearly 1.95 million hectares with an estimated annual production of 14.81 billion nuts per year. The crop plays a major social, religious and economical role to millions of rural people in such regions (Rethinam, 2007). Coconut cultivation in India is concentrated primarily in Kerala state, followed by Tamil Nadu, but the crop is affected by a number of pests and diseases. In coconut, pests and diseases are serious constraints in the production and productivity. Rhinoceros beetle, red palm weevil, leaf eating caterpillar, eriophyid mite, root grubs, etc. are important pests of coconut. The important diseases of the palm crop are leaf rot along with root (wilt), bud rot, stem bleeding, basal stem rot and grey blight. Root (wilt) disease occurs predominantly in southern districts of Kerala state, besides its neighboring states – particularly and heavily in Theni district of Tamil Nadu. The root (wilt) affected palm is commonly infected by leaf rot, wherein the leaf rot incidence having a strong interrelationship with root (wilt) firmly established. As such, the root (wilt) in association with leaf rot is economically very important diseases complex in coconut and it is a unique case of phytoplasma–fungal disease complex. Even as leaf rot generally appears on palms affected by root (wilt), the palms having latent infection of root (wilt) might be also infected by leaf rot. The destruction of the photosynthetic area of leaves (due to leaf rot) in the diseased palm is quite clear and conspicuous, and therefore root (wilt) affected palms come to the farmers' attention instantly with the onset of leaf rot. It is viewed that innovation and technology evolving play a contributory role in stabilizing yield potential of the crop and thereby sustainable agricultural activities and reducing risk perceptions.

As need for consolidating the information available on the use of biological control agents in management of plant diseases with environmental consideration in general and especially the advances of such a system with special reference to coconut root (wilt)–leaf rot diseases complex is felt necessary this chapter has been evolved wherein recent trends are conceptually discussed. The use and scope for application of the biological control- ISR measure as a tool in the integrated disease management system is further evolved as the root (wilt)–leaf rot is a unique case of phytoplasma–fungal disease complex.

BIOPESTICIDES IN PLANT DISEASE CONTROL

Mechanism Involved in Biological Control

The biological control of plant diseases can be simply defined as the method of controlling/managing plant diseases using certain living organisms instead of using chemicals-fungicides etc. The biotic agents involved in biological control are often referred to as antagonists. The means by which the antagonists kill the pathogens are parasitization of pathogen, by the secretion of metabolites (antibiotics) which kill the pathogens and by competing with pathogen for food and other needs leading to death of pathogen. Hence, the major mechanisms involved or by which the antagonists function as biological control agents against pathogens can be accounted (Ramamoorthy *et al.*, 2001) under the following: (i) Parasitism (suppression of plant pathogens by simple parasitic invasion), (ii) degradation of cell wall (lytic enzymes produced by antagonists degrade cell walls of plant pathogens (fungi) leading to death of pathogen propagules), (iii) competition (displacement of pathogen by antagonists through intensive competition for food, essential elements, etc. For example, the siderophores (certain chemicals) produced by non-pathogenic microbes remove iron ions from the soil, (iv) limiting the growth/population of pathogens), (v) secretion of harmful metabolites (antibiotics produced by antagonists function as antimicrobial substance directly inhibiting the growth of pathogens) and (vi) induction of defenses in plant system (induced resistance established in the plant) following intentional exposure of plants to non-pathogenic antagonistic organism - plant defenses thus evolved limit the infection by pathogens. This system not necessarily can eliminate the pathogen/disease outright but bring down the pathogen inoculums below economic threshold level. Thus it is a technology of reducing the level of pathogens by using one or more organisms referred as antagonists. Often biocontrol is also considered as an ecological management (Anandaraj *et al.*, 2003) of a community of organisms, which usually occur in the ecosystem such as root and leaf surfaces of plants. Even as more attention has been receiving in the past in control of soil- borne plant diseases using biological control agents as compared to foliar diseases (where relatively more hostile factors - climate etc. - interact with microbes on leaf surfaces) biological control of foliar diseases also started receiving importance as evidences of general inhibition of pathogen or competition on leaf surface prior to invasion of host tissue or suppression of the pathogen after invasion (either by hyper parasitism or by induced systemic resistance) accumulated.

Antagonists Involved in Biological Control

Interesting features of the antagonists are that the antagonists are non-pathogens, compete with pathogenic organisms and exclude them from colonizing plant tissues and these features are accomplished naturally or through manipulation of environment, host or pathogen by mass introduction of one or more antagonists. Certain bacterial (*Pseudomonas*, *Bacillus*, etc.) and fungal (*Trichoderma*, *Gliocladium*, etc.) antagonists have been extensively used against plant diseases (Bharathi, 2001) and by various methods of application of fungal and bacterial antagonists several economically important plant diseases have been effectively managed. Certain species of antagonists are most commonly used in disease control (e.g., Bacteria: *Pseudomonas fluorescens*, *Bacillus subtilis*; Fungi: *Trichoderma viride, T. harzianum*) for having established their effectiveness. Nevertheless, the biocontrol agents should have certain desirable traits in order to achieve a satisfactory degree of disease control and for acceptability (Renuka *et al.*, 2003; Saju *et al.*, 2003; Srinivasan, 2005, 2007a). A candidate bioagent should have relatively a large number of favourable attributes for successful effect against pathogen/disease.

INVOLVEMENT OF INDUCED SYSTEMIC RESISTANCE

Importance of Induced Resistance in Plants

The basis for induced protection of plants against biotic and abiotic stresses was evolved from the term of "acquired physiological immunity". Induced resistance was also described as "systemic acquired resistance", "translocated resistance", "plant immunization", etc. (Ryalls *et al.*, 1996; Sticher *et al.*, 1997). In other words, the induced resistance is an enhanced defensive capacity of plants following stimulation against pathogens and pests. Enhanced resistance after induction by an agent upon pathogen's infection is called as "induced systemic resistance" (ISR) (e.g., by plant growth promoting rhizobacteria, abbreviated as PGPRs) or "systemic acquired resistance" (SAR) (by other agents). SAR is maximally expressed when the inducing agent causes necrosis of tissues; ISR induced by PGPRs do not establish necrotic symptoms in the host. The PGPRs are root-colonizing bacteria (benefits of inducing systemic resistance in host as well as biological control of pathogens availed). However, both SAR and ISR are the expression of latent resistance (Van Loon *et al.*, 1998) following a challenge introduction of a given pathogen of a disease - biotic (virulent pathogens, non-pathogens, elicitors etc.) and abiotic (salicylic acid, ethylene, benzothiadiazole, etc.) inducers of SAR evolved. As experimental evidences has proved PGPRs as effective inducers of ISR. The use of PGPRs against pathogens came into being widely in recent years. PGPRs colonize not only rhizosphere/ rhizoplane of plant roots, but also act as endophytes by residing in intercellular spaces of root epidermal cells and vascular tissues – they move upward and downward from point of application and prevent entry of pathogen into vascular stele (Weller, 1988). Ultimately, by utilizing PGPRs as inducers of defenses, a degree of plant immunization is possible.

Plant Disease Categories in Induced Resistance

The PGPRs mediated ISR is found effective in both dicotyledonous and monocotyledonous plants – a matter of wider applicability – and resistance induction achieved against a vast number of fungal, bacterial and viral diseases of plants besides pests. ISR induction could be evolved against both systemic diseases (e.g., viral infections, fusarial infections) as well as localized infections (spots, blights, rots etc.) of plants (Jeyalaksmi *et al.*, 1998; Kandan *et al.*, 2003) following applications of PGPRs. Experimentally it is proved that mixtures of antagonists - consortium/consortia - are more efficient and synergistic in disease control (Raupach and Kloepper, 1998) as for example, combination of strains of *P. fluorescens* and combination of *P. fluorescens* and *B. subtilis* are found more effective over individually alone against pathogens (Srinivasan *et al.*, 2006). In nature itself it is possible that mixtures of antagonists may function against pathogens. Through further insights, it has also been proved that same strains of PGPRs can induce ISR against multiple pathogens in a given crop and therefore, a broad spectrum PGPRs mediated ISR would be more important – selection of broad-spectrum strains against multiple pathogens/ diseases thus assumes rewarding (Ramamoorthy *et al.*, 2001); development of antagonist consortia assumes an added promise under ISR.

Resistance Induction Mechanism

The ISR is brought about by enhancing the physical strength of cell wall (rapid mechanical changes at sites of pathogen entry delays the onset of infection process that aids time for the host to build up defense reactions to resist the growth of pathogen), changes physiological-biochemical reactions of host resulting in synthesis of defense chemicals (accumulation of PR – proteins, i. e., pathogenesis related proteins such as β–1,3, glucanases, endochitinases, peroxidase, phenyl alanine ammonia lyase, polyphenol oxidase, catalase, etc.) against pathogen, phytoalexins and other secondary metabolites (Chen *et al.*, 2000; Meena *et al.*, 2000; Ramamoorthy *et al.*, 2002; Thangavelu *et al.*, 2003). Understanding these mechanisms of PGPRs bringing about ISR in host provides more insights on the functional system of defense in plants. In short, ISR is achieved by application of PGPRs through changes in cell wall structures followed by biochemical changes of accumulation of PR–proteins and/or phytoalexins/secondary metabolites in the host system. It is also to be understood that certain traits of PGPR bacteria are involved in determining the ISR. These traits are lipopolysaccharides of outer membrane of bacterial cell wall, siderophores (e.g., pseudobactin) produced under iron-limited conditions (both the traits may be contributing in a complementary manner for final ISR) and production of salicylic acid. Although different determinants are involved in ISR (Duffy and Defago, 1999), these determinants vary with iron limiting conditions, bacterial strains, host and cultivars of host species involved.

Utilization of Biological Control Agents

The need for formulation of bioagents and process them into packages in a manner similar to pesticide packing and make them available conveniently is necessary. For field delivery of biological control agent, the biopesticide formulation should preserve the microbial activity for relatively a long period (shelf life) in the chain of production and utilization. Thus mass multiplication of bioagents is inevitable where candidate organism is grown in a relevant liquid medium followed by mixing the culture with a suitable solid (powder) carrier material. The powder based bioformulation of antagonist came into being in 1980s (Kloepper and Schroth, 1981). The beneficial organisms have been found to survive for a long period in such dry formulations (e.g., talc powder) and also suitable for various methods of application. More over, the powder formulations have been found to favour the bioagent to colonize on plant surface and root system better than just the suspension form. The talc-based and peat-based formulations of *P. fluorescens* have been successfully delivered for systemic protection of crop plants against several diseases (Vidyase-karan and Muthamilan, 1995; Nandakumar *et al.*, 2001; Radja Commare *et al.*, 2002); talc-based bioformulations have been evolved (Mukhopayhyay, 1987; Rajan *et al.*, 2002; Chaluvaraju *et al.*, 2003) for *B. subtilis* and also fungal antagonists such as *T. viride*, *T. harzianum*, etc. that have been successfully utilized in field against diseases. These bioformulations can be applied by different methods depending upon the need, nature of disease and crop involved (seed treatment, root treatment, vegetative part treatment, soil application, foliar application, etc.).

The importance lies how best the ISR can be prolonged in the plant system for protection against diseases under field conditions as duration of ISR in plants may vary from crop to crop and also depending upon the strain of bioagent involved. The determination may not be always as easy in perennial crops as that in annuals. Therefore, in perennials, the requirement of more than one method of application and frequent applications could be well envisaged. Further, utilization of superior strain or mixture of bioagents may be necessary for extended field performance of ISR. Nevertheless, irrespective of the lasting duration of ISR in plant system, the PGPRs treatment aids the plants by way of improved growth also besides biocontrol of pathogens. A notable feature as for example, *Bacillus* spp. has been shown to produce growth regulators like IAA, GA etc. favouing improved growth of plants besides biocontrol of pathogens (Velusamy *et al.*, 2003). Accumulated evidences amply prove that PGPRs application benefits plants in a three dimensional manner, *viz.* in plant growth, control of pathogens by biological means and induction of systemic resistance against pathogens (Hammerschmidt,

1999); an immunization like accomplishment is conferred on to plants against infection by pathogens through ISR.

Bioagent Strains and Organics in Biological Control

Biological diversity in biosystem is the rule and thus variations in biocontrol potential among strains of antagonists do occur. Availability of superior strains of bioagents against a single or multiple pathogens, as the case may be, and also synergistic effect with consortium of strains is always a possibility. With available know-how it would be expected that consortia of PGPRs eliciting more broad-spectrum activity that could be exploited particularly against multiple pathogens; benefits of combination of bacterial with fungal antagonists, biofertilizers, etc. are also accessible. Combined effect of fungicides and/or plant products with antagonistic organisms is another important area of knowledge (Raupach and Kloepper, 1998; Amruta and Chandrasekar, 2003; Hazarika and Phookan, 2003; Singh and Sinha, 2004) and such combined applications are shown to be more effective in control of certain diseases. Further, plant derivatives have also proved their importance as insensitivity or low sensitivity of antagonistic organisms to plant products which are inhibitory to plant pathogens (Anishkumar *et al.*, 2004); organics could as well be utilized as low cost carrier materials for the antagonists. The integration of plant derivatives-antagonists would itself function as potential consortium for disease control. Various organic materials have been used as medium for mass production/carrier material/mix base of antagonists (Suseela Bhai *et al.*, 1994; Zaidi and Singh, 2004, Ramanujam *et al.*, 2004, Srinivasan, 2007a).

BIOPESTICIDES AGAINST COCONUT ROOT (WILT) - LEAF ROT DISEASES

Biological Control of Coconut Diseases

Coconut based cropping system provides opportunities for extensive use of biocontrol agents and therein *in situ* recycling of organics in combination with antagonists could be effectively integrated. The basal stem rot and stem bleeding are soil-borne diseases where the effects of particularly fungal antagonists (*Trichoderma* spp., *Gliocladium* spp.) have been discerned (Ramanujam *et al.*, 2000; Srinivasulu *et al.*, 2004). In case of leaf rot and bud rot, use of bacterial antagonists has been brought out (Srinivasan and Rohini Iyer, 2004). In other cases the system of biocontrol is in emerging state. Nevertheless, biological control measures have been accepted as an integral part of coconut protection measures.

Root (Wilt) – Leaf Rot Diseases

Root (wilt)-leaf rot diseases are heavily distributed in Kerala state besides specific regions in Tamil Nadu, bordering Kerala (Srinivasan *et al.*, 2000; Srinivasan and Sasikala 2001). The root (wilt) in coconut is a debilitating malady, caused by phytoplasma and the affected palms suffer systemically (Solomon *et al.*, 1999; Srinivasan, 1999). Further, the root (wilt) affected palms are vulnerable to infection by leaf rot and hence leaf rot is interrelated with root (wilt). Leaf rot is caused by a group of fungal pathogens. Among them *Colletotrichum gloeosporioides*, *Exserohilum rostratum* and *Fusarium solani* are the major pathogens (Srinivasan, 1991, 2004b; Srinivasan and Gunasekaran, 1993, 1996, 2000). Thus leaf rot control forms an integral part in the management of root (wilt) complex (Srinivasan *et al.*, 2001) - with the incidence of leaf rot on root (wilt) affected palms, leaf area lost, the crown disfigured and the palm's vigor/yield declines rapidly (Srinivasan and Gunasekaran, 1999).

Root (wilt). The direct effect of biological control agents *in vitro* against phytoplasma is not feasible. Observations have pointed to the probability of PGPRs playing some role in nature against root (wilt) disease. Microorganisms playing role in the rhizosphere/soils of coconut, in relation to root (wilt) affected palms and in context of fertility/nutrition of palms have been studied. It has been observed that general bacterial population is higher in surface soils (rhizosphere) in the vicinity of root (wilt) diseased palms but very less population in soil from the vicinity of healthy (field tolerant) palms. But specific bacteria like nitrogen fixers and silicate solubilizers, besides actinomycetes, have been observed to be more in soils near healthy (field tolerant) palms as compared to soils from the vicinity of root (wilt) affected palms (Muraligopal *et al.*, 2003).

The study on induced systemic resistance (ISR) against root (wilt) disease was done involving certain chemicals (Digitonin, Salicylic acid, Triazole) and biological control agents (*Pseudomonas fluorescence* and *Bacillus amyloliquefaciens*) that were known to elicit the mechanism of ISR in plant system as reported elsewhere. The experiment was first started in pot culture condition using six months' old coconut seedlings. For this purpose coconut seed nuts were collected from root (wilt) disease-free mother palms and seedlings raised in micro plots. After 2 years of experimentation in pot culture the experimental plants were transplanted in to the main field. In the first two years the treatments were at Quarterly intervals in potted condition (8 rounds); subsequently at six monthly intervals in field condition (6 rounds). Therefore, a total of 14 rounds of treatments were given to these experimental plants. Activities of defense enzymes like Polyphenol oxidase, Peroxidase, etc. were analyzed in the experimental seedlings - the evaluations yielded results of enhancement of certain defense enzymes in the host system following the treatments (CPCRI 2006, 2007). In these measures, *P. fluorescens* and *B. amyloliquefaciens* induced higher activities of peroxidase and polyphenol oxidase in treated seedlings indicating the activation of defense mechanism by the biological control agents (Table 13.1A). While peroxidase activity was steadily increased in *P. fluorescens* treated seedlings from first application itself, higher activities of phenyalanine ammonia lyase and glucanase were also induced by *B. amyloliquefaciens*. Higher activities of peroxidase, polyphenol oxidase, phenyalanine ammonia lyase and glucanase were observed among applications of the aforementioned chemicals also.

Table 13.1A. Study of induced systemic resistance against root (wilt) using chemicals and biological control agents: Activity of selected defense enzymes in treated coconut (units/minute/gram protein)

Treatments	Polyphenol oxidase	Peroxidase
B. amyloliquefaciens	36	113
P. fluorescens	31	84
Triazole	53	165
Salicylic acid	31	82
Digitonin	45	136
Control	15	69

Source: CPCRI (2007)

Periodical visual observation of the experimental plants with respect to appearance of symptoms of root (wilt), leaf rot, population of vectors (*Stephanitis typica* and *Proutista moesta*), major pests etc. provided some useful information of effect of treatments - Disease incidence has been generally less in treated palms as compared to control and also the treated palms found to grow robust as compared to untreated controls (Table 13.1B). As such certain positive influence of biocontrol agents and chemicals in the line of inducing systemic resistance against root (wilt) disease and leaf rot has been pointed through this trial, besides favouring vigour to coconut plants. This experiment, the ISR studies with root (wilt) disease, forms the first of its kind and with accrual of more information and through new trials the effect of biotic and abiotic factors (biocontrol agents/defense elicitors/growth promoting microbes etc.) more insights on induced systemic resistance against the disease are likely to be generated.

Table 13.1B. Visual observations of experimental coconut palms under the trial of induced systemic resistance against root (wilt) disease - 10 palms per treatment

Treatment	**Morphological factors (Mean of all palms in each treatment)**						
	No. of leaves	**No. of palms with Inflore-scence***	**Girth of bole (cm)**	**Ht. of palm/ Stem length (cm)**	**General impression of palm growth**		
					+	**++**	**+++**
Control	16.9	2	158.0	42.4	6	4	0
Digitonin	17.6	0	151.0	44.8	1	4	3
Triazole	17.9	3	147.8	41.9	4	6	0
Salicylic acid	17.0	2	140.0	43.5	2	5	2
P. fluorescens	15.8	4	143.8	47.2	4	5	1
B. amyloliquifaciens	19.5	6	152.9	56.2	2	6	2

+: Poor growth ++: Good growth +++: Very good growth

Source: CPCRI (2012)

Leaf rot. *B. subtilis* capable of inhibiting a pathogen of leaf rot, *Helminthosporium halodes* (= *Exserohilum rostratum*), has been recorded earlier (Lily *et al.*, 1952, 1955). The feasibility of applying the biological control technology, however, has expanded. With leaf rot being established as a disease of fungal complex, more studies were initiated and more knowledge accrued on biological control aspects of leaf rot disease. Instances of using antagonistic bacteria and fungi are common in the control of various plant diseases and some of these organisms bestow beneficial effects such as direct growth promotion of plants, biological control of pathogens and induction of resistance in host plants the technology received attention. Antagonistic bacteria (*Pseudomonas fluorescens*, *Bacillus subtilis*) and fungi (*Trichoderma viride*) have been found to kill or prevent the growth of the major pathogens of leaf rot (Table 13.2). Antagonistic activities of isolates of bacteria from coconut rhizosphere have been evaluated against pathogens of leaf rot and effective isolates (fluorescent *Pseudomonas* and *Bacillus*) evolved (Gupta *et al.*, 2000; Srinivasan and Bharathi,

2006). PGPRs isolated from coconut do show biological activity in terms of biocontrol processes such as production of secondary metabolites (antibiotics, siderophores, hydrogen cyanide, ammonia, etc.) and extra-cellular elaboration of lytic enzymes (Chinase, β 1, 3 glucanase) affecting the pathogens (Gupta, 2002, 2003). *P. fluorescens* and its culture filtrate were found inhibitory to the growth of major pathogens of leaf rot *in vitro* with notable inhibition zones; suspension culture of the bacterium also reduced the development of leaf rot (Gunasekaran *et al.*, 2003; Srinivasan *et al.*, 2006), thus growing cultures, besides cell free culture filtrates of such bacterial isolates also inhibited the growth of leaf rot pathogens.

Table 13.2. Effect of bacterial (*Bacillus subtilis* and *Pseudomonas fluorescens*) and fungal (*Trichoderma viride)* biocontrol agents, individually and in combination, on inhibition in growth of major pathogens of leaf rot *in vitro**

Biocontrol agents and their combinations	Mean colony diameter of pathogens (mm)*			Mean per cent inhibition of pathogens*			Mean inhibition zone of pathogens (mm)*		
	C.g	E.r	F.s	C.g	E.r	F.s	C.g	E.r	F.s
Bacillus subtilis	26.6	37.2	32.8	12.6	04.3	09.5	10.4	04.5	06.9
Pseudomonas fluorescens	23.8	27.2	27.0	22.1	31.1	27.8	14.0	11.1	11.7
Trichoderma viride	08.0	09.6	07.5	67.2	74.4	78.2	02.9	02.8	03.1
B. subtilis + P. fluorescens	23.6	28.6	28.5	23.6	26.8	21.5	14.5	10.3	10.8
B. subtilis + T. viride	09.6	11.1	09.1	63.1	70.7	73.8	05.9	06.4	05.3
P. fluorescens + T. viride	10.0	11.6	08.9	60.3	69.3	74.3	05.8	06.0	05.4
B. subtilis + P. fluorescens + T. viride	10.3	11.9	08.2	61.5	69.0	76.3	05.1	05.9	05.9
Control	30.3	39.1	36.6	-	-	-	-	-	-
C. D. P= 0.05	12.8	14.3	13.4	24.6	26.3	27.4	02.1	02.4	02.2

Source: Srinivasan and Barathi (2006)

*Mean of 3rd, 6th and 9th days of observations – averaged out of 6 replications for each day.

C.g: *Colletotrichum goleosporioides*; E.r: *Exserohilum rostratum*; F.s: *Fusarium solani*.

Prevalence of antagonistic bacteria (*Bacillus subtilis*, *Pseudomonas fluorescens*) and fungi (*Trichoderma viride*, *T. harzianum*) in coconut rhizosphere is evident by consistent inhibitory effect of *B. subtilis, P. fluorescens* and *T. viride* on all three pathogens. Inhibition zones developed by bacterial antagonists were found higher and more effective over the inhibition zones developed by the fungal antagonist. An effective isolate of each antagonist as evaluated for compatibility among them and interaction with pathogens also yielded interesting information, i.e. compatibility of *B. subtilis* with *P. fluorescens* and *T. viride*; non-compatibility of *T. viride* with *P. fluorescens*; consistent inhibition of all pathogens with inhibition zones by bacterial and fungal species; heavy suppression of growth of pathogens by *T.viride* individually (with less inhibition zone) and in combination with bacterial bioagents (with more inhibition zone); synergism in consortia like *B. subtilis* + *P. fluorescens*, *B. subtilis* + *T. viride*, *B. subtilis* + *P. fluorescens* + *T. viride*, etc. have been observed. The role of microbes in coconut phylloplane was also evaluated and there the prevalence of antagonistic and other groups of microorganisms was ascertained through their *in vitro* reactions against major pathogens of leaf rot. Wider variation in biocontrol potential among bacterial (*Bacillus* sp.) and fungal (*Aspergillus* sp., *Penicillium* sp.) isolates was observed (Srinivasan, 2006; Srinivasan *et al.*, 2006; CPCRI, 2007). These results emphasized the significance of biocontrol having antagonistic potential of popular bioagents in coconut rhizosphere and superior isolates as antagonists were evolved for utilization.

Biological control technology could be extended to field level using a talc-based *P. fluorescens* formulation. Amelioration of leaf rot has been found possible (Srinivasan, 2003; Srinivasan *et al.*, 2006). The biopesticide is acceptable to spindle leaf tissues of coconut and in palms treated with *P. fluorescens* that reduced leaf rot index in the emerged leaves/disease considerably (Srinivasan *et al.*, 2006; CPCRI, 2007). Favourable influence of consortium of antagonists like *B. subtilis* and *P. fluorescens* and its utilization in the field management of coconut leaf rot disease was achieved and are shown in Tables 13.3A and B. Response of *B. subtilis* and *P. fluorescens* to fungicides has been observed. *B. subtilis* was relatively tolerant to certain fungicides (Contaf, Indofil M-45) as compared to sensitivity of *P. fluorescens*. *B. subtilis* was able to grow in lower concentration of Tilt or Bavistin but *P. fluorescens* was not compatible with these fungicides. However, viability of *P. fluorescens* in lower concentration of Contaf was possible. *B. subtilis, P. fluorescens* and *T. harzianum* were relatively tolerant to Antracol in lower concentrations. In combinations, *B. subtilis* or *P. fluorescens* with Contaf, notable synergism against pathogens was observed. As such integrated use of bioagents with fungicides at minimal concentrations in leaf rot control is possible (Srinivasan *et al.*, 2006; Srinivasan and Gunasekaran, 2003). Thus scope for use of consortium of antagonists with fungicide in the disease control is evident.

Table 13.3A. Leaf rot disease index levels in coconut palms - in cumulatively emerged leaves - in experiment of treatment application of biological control agent, talc-based preparation of *Pseudomonas fluorescens* (Mean of 20 palms in each category)

S. N.	Category of palms	Pre-treatment Disease Index*	Leaf rot Disease Index (DI) in emerged leaves in different rounds **					Mean of Rounds
			I	II	III	IV	V	
1	Palms treated with *Pseudomonas fluorescens*	14.71	12.65	11.80	08.83	08.25	08.00	9.91
	Improvement over pre-treatment (Decrease in DI)	-	(-) 14.00%	(-) 19.78%	(-) 39.97%	(-) 43.92%	(-) 45.62%	(-) 32.66%
	Improvement over control (Decrease in D I)	-	(-) 32.75%	(-) 41.35%	(-) 63.99%	(-) 67.39%	(-) 70.91%	(-) 55.28%
2	Control palms	15.68	18.81	20.12	24.52	25.30	27.50	23.25
	Deterioration over pre-treatment (Increase in DI)	-	(+) 19.96%	(+) 28.32%	(+) 56.38%	(+) 61.35%	(+) 75.38%	(+) 48.28%
	Deterioration over treatment (Increase in DI)	-	(+) 48.69%	(+) 70.50%	(+) 177.69%	(+) 206.67%	(+) 243.75%	(+) 149.46%

Source: Srinivasan *et al.* (2006)

*For leaves in whole crown, at the time of Experiment initiation.

**For leaves emerged subsequent to pre-treatment – Cumulatively emerged leaves.

Table 13.3B. Effect of application of talc-based preparations of *Bacillus subtils* and *Pseudomonas fluorescens* - individually and consortium mode on leaf rot disease on coconut palms*

S. No.	Treatment	Pre-treatment Disease Index**	Leaf rot disease index in emerged leaves in different rounds ***					Mean of Rounds
			I	II	III	IV	V	
1	Application of *Bacillus subtilis*	6.44	6.10	5.82	5.32	5.10	4.90	5.45
	Improvement over pre-treatment (Decrease in DI) - %		(-) 5.28	(-) 9.63	(-) 17.39	(-) 20.81	(-) 23.91	(-) 15.37
	Improvement over control (Decrease in D I) - %		(-) 2.87	(-) 10.46	(-) 37.85	(-) 52.91	(-) 60.99	(-) 39.11
2	Application of *Pseudomonas fluorescens*	6.76	5.98	5.70	5.10	4.80	4.50	5.22
	Improvement over pre-treatment (Decrease in DI) - %		(-) 11.54	(-) 15.68	(-) 24.56	(-) 28.99	(-) 33.43	(-) 22.78
	Improvement over control (Decrease in D I) - %		(-) 4.78	(-) 12.31	(-) 40.42	(-) 55.68	(-) 64.17	(-) 41.68
3	Combined application of *B. subtilis* and *P. fluorescens*	5.59	4.86	4.80	4.25	3.75	3.63	4.26
	Improvement over pre-treatment (Decrease in DI) - %		(-) 13.06	(-) 14.13	(-) 23.97	(-) 32.92	(-) 35.06	(-) 23.79
	Improvement over control (Decrease in D I) - %		(-) 22.61	(-) 26.15	(-) 50.35	(-) 65.37	(-) 71.10	(-) 52.40
4	Phytosanitation + Combined application of *B. subtilis* and *P. fluorescens**	6.26	4.90	4.75	4.00	3.50	3.35	4.10
	Improvement over pre-treatment (Decrease in DI) - %		(-)21.73	(-) 24.12	(-) 36.10	(-) 44.09	(-) 46.49	(-) 34.50
	Improvement over control (Decrease in D I) - %		(-)21.97	(-) 26.92	(-) 53.27	(-) 67.68	(-) 73.33	(-) 54.19

5	Control palms	4.70	6.28	6.50	8.56	10.83	12.56	8.95
	Deterioration over pre-treatment (Increase in DI) - %		(+)33.62	(+ 38.30	(+)082.13	(+)130.4	(+)167.2	(+)090.43
	Deterioration over treatment* (Increase in DI) - %		(+)28.16	(+)36.84	(+)114.00	(+)209.4	(+)274.9	(+)118.29

Source: Srinivasan *et al.* (2011a)

*Mean of 10 palms in each category.

* *For leaves in whole crown, at the time of Experiment initiation.

***For leaves emerged subsequent to pre-treatment–Cumulatively emerged leaves.

Effect of botanicals on the pathogens of leaf rot has also been evaluated (Anishkumar *et al.*, 2004; CPCRI, 2007, Srinivasan, 2007a). Among various plant species, *Adenocalymma alliceae* and *Lawsonia inermis* inhibited leaf rot pathogens. Between the two antagonists, *B. subtilis* was less affected by plant extracts than *P. fluorescens*. The antagonists were generally less sensitive to various plant extracts (*B. subtilis* more tolerant) in comparison to inhibition of pathogens by plant extracts. For example, the bacterial antagonists were insensitive or relatively less sensitive to neem extract as compared to inhibition of leaf rot pathogens by neem. Thus organics such as neem cake, etc. could be utilized as carriers in the preparation of bioagents (Table 13.4). In this direction the coconut-based organics such as coir pith, vermi compost, etc. are also likely to serve as effective carriers or mix material for antagonistic organisms (Srinivasan *et al.*, 2010).

Mass Production of Biocontrol Agents - Bioformulations

Specific media such as King's 'B' broth medium (*P. fluorescens*), Nutrient broth medium (*B. subtilis*), selective medium/molasses yeast medium (*Trichoderma* spp.), etc. are used for growth/multiplication of biocontrol agents and processed for the preparation of bioformulations using a solid carrier material such as talc powder. During the course of studies with biocontrol of coconut root (wilt)-leaf rot complex, coconut water supported the growth of bacteria and fungal antagonists. Therefore, the suitability of coconut water as a natural growth medium for mass multiplication of bioagents was evolved. Consequently coconut water media could be consistently prepared and the antagonists successfully grown in such media, where their growth was comparable with that in conventional media. The biocontrol agents multiplied in coconut water based media could be effectively converted into their bioformulations using talc powder as a carrier material (Srinivasan, 2007b; Srinivasan *et al.*, 2010a,b, 2011a).

Table 13.4. Viability of biocontrol agents – grown in coconut water and conventional media – in talc powder-various organics mixture carrier (1:1 ratio, w/w) in different months of storage*

Biocontrol agent	Nature of talc-organic mixture carrier	Medium of growth	Initial population in talc-organic mixture carrier (x 10^{12})	Population of bacterial bioagent (Cfu/gm) in talc- organic mixture carrier at different months of storage					
				1st x 10^{12}	2nd x 10^{10}	3rd x 10^{10}	4th x 10^{9}	5th x 10^{8}	6th x 10^{8}
Pseudomonas fluorescens	Neem cake	Coconut water	128	176	178	120	70	65	52
		King's B Broth	130	180	179	142	72	67	53
	Vermi compost	Coconut water	126	174	176	118	68	62	44
		King's B Broth	129	176	177	132	71	65	48
	Coir Pith	Coconut water	125	152	167	108	65	56	39
		King's B Broth	128	161	171	121	69	59	42
			(×10^{8})	Population of fungal bioagent (Cfu/gm) in talc- organic mixture carrier at different months of storage (x 10^{8})					
Trichoderma viride	Neem cake	Coconut water	2.7	3.1	3.1	2.8	2.5	2.3	2.0
		Potato Dextrose Broth	2.9	3.7	3.5	3.3	2.9	2.6	2.2
	Vermi-compost	Coconut water	2.6	2.9	2.9	2.6	2.5	2.3	2.0
		Potato Dextrose Broth	2.7	3.0	2.9	2.8	2.7	2.5	2.2
	Coir Pith	Coconut water	2.5	2.8	2.6	2.4	2.3	2.1	1.9
		Potato Dextrose Broth	2.8	2.9	2.8	2.5	2.5	2.2	2.0

Source: Srinivasan *et al.* (2011a); *Mean of 3 replicates.

Biopesticides on Coconut Seedlings

In recent years seed biopriming of coconut with bacterial antagonists, *B. subtilis* and *P. fluorescens*, individually and in combination, i.e. application of biocontrol agents onto coconut seeds prior to sowing, has been found favourable with a positive impact on their germination trend. Further, PGPR bacterial treatment to roots of coconut seedlings could lead to increased level of their growth (in terms of parameters such as height, girth and shoot and root weights, etc). In related studies the vigour and growth of coconut seedlings raised from seed nuts (treated with formulated *B. subtilis*, *P. fluorescens*, combination of *B. subtilis* and *P. fluorescens* beside *T. viride* as such and each fortified with neem cake) was found to be good and a distinct effect was observed with consortium of *B. subtilis* and *P. fluorescens*. *P. fluorescens* significantly increased the growth of treated coconut seedlings (seedling height, total number of leaves and girth at collar region) and the effect of consortium of *P. fluorescens* and *B. subtilis* among various treatments (*B. subtilis*, *P. fluorescens*, *B. subtilis* + *P. fluorescens*, *T. viride* (each formulated and individually fortified with neem cake) on the growth and vigour of coconut seedlings (transplanted into poly covers after root treatment) was stronger (Tables 13.5A–D). Inoculations of major pathogens of leaf rot (*Colletotrichum gloeosporioides*, *Exserohilum rostratum*, *Fusarium solani*) through trials with leaf samples of seedlings raised through seed biotreatment as well as seedlings subjected to biotreatment showed lesser disease lesions as compared to control. In control seedlings (without biotreatment) the disease development was higher (Srinivasan, 2010a).

Table 13.5A. Mean growth characters of coconut seedlings raised through seed bio priming (Experiment I)

Treatment	No. of seed nuts germinated as on 6th month after sowing per 10 seed nuts	Mean characters of one year old seedlings (Out of germinated seed nuts)					
		Height (cm)	Collar Girth (cm)	No. of leaves	No. of seedlings with split leaf	Fresh Weight (kg.)	No. of fresh roots
Bacillus subtilis	10 (100%)	150.5	14.5	6.3	4	1.765	13.90
Pseudomonas fluorescens	10 (100%)	155.2	13.9	6.2	3	1.825	14.50
B. subtilis + *P. fluorescens*	09 (90%)	162.0	14.8	6.2	3	1.850	14.80
Control	09 (90%)	150.0	13.8	6.2	3	1.750	13.15

Table 13.5B. Mean growth characters of coconut seedlings raised through seed bio priming (Experiment II-A)

Treatment of biocontrol agent with neem cake	No. of seed nuts germinated as on 6th month after sowing (per 20 seed nuts)	Mean characters of 9 month old seedlings (out of germinated seed nuts)				
		Height (cm)	Collar Girth (cm)	No. of leaves	No. of seedlings with split leaf	General robustness of seedlings (No.)
Bacillus subtilis	20	89.9	11.5	5.0	1	6
Pseudomonas fluorescens	17	94.3	11.2	4.6	1	6
B. subtilis + *P. fluorescens*	18	100.3	11.8	4.8	3	9
Trichoderma viride	19	93.9	10.7	4.6	3	5
Control (Neem cake)	20	86.0	10.2	4.5	1	4
Absolute Control	20	83.8	10.1	4.5	0	3

Table 13.5C. Mean growth characters of coconut seedlings raised through seed biopriming (Experiment II-A)

Treatment of biocontrol agent with neem cake	No. of seed nuts germinated as on 6th month after sowing (per20 seed nuts)	Mean characters of one year old seedlings (Out of germinated seed nuts)						
		Height (cm)	Collar Girth (cm)	No. of leaves	No. of seedlings with split leaf	Fresh weight (Kg.)	No. of Fresh roots	General robustness of seedlings (No.)
Bacillus subtilis	20	158.4	15.0	6.3	8	1.795	14.23	10
Pseudomonas fluorescens	17	160.2	14.8	6.3	9	1.867	14.85	10
B. subtilis + *P. fluorescens*	18	168.8	15.6	6.5	11	1.975	16.50	12
Trichoderma viride	19	152.3	13.2	6.0	7	1.826	13.85	7
Control (Neem cake)	20	140.4	12.8	6.0	5	1.725	13.28	7
Absolute Control	20	138.2	12.4	5.5	5	1.652	12.32	5

Table 13.5D. Mean growth characters of bio treated coconut seedlings in poly bags after one year of transplanting 12 months old seedlings (Experiment II-B)

Treatment of biocontrol agent with neem cake	No. of seed-lings treated and trans-plan-ted	Mean characters of two year old seedlings in poly bags						
		Height (cm)	Collar Girth (cm)	No. of leaves	No. of seed-lings with split leaf	Fresh weight (Kg.)	No. of fresh roots	General Robust-ness of seedlings (No.)
Bacillus subtilis	10	185.6	17.5	11.3	7	2.735	24.35	6
Pseudomonas fluorescens	10	188.5	16.8	10.8	8	2.721	24.25	6
B. subtilis + P. fluorescens	10	195.3	18.5	12.8	8	2.855	26.85	8
Trichoderma viride	10	173.4	14.8	10.2	7	2.682	23.65	5
Control (Neem cake)	10	165.5	14.2	9.3	6	2.527	23.54	4
Absolute Control	10	158.2	13.7	8.5	5	2.251	20.15	4
Mean pre treatment characters of 60 seedlings raised in bulk		142.3	11.8	5.5	18/60 (= 3/10)	1.685	12.82	12/60 (=2/10)

Source: Srinivasan (2010a)

Encouraging results were obtained with application of bioagents onto coconut seeds/seedlings. The effect of biological control agents by way of induced systemic strategy against the root (wilt) was more rewarding.

Techniques of Application of Biopesticides

Biopesticides are applicable to coconut in different stages of cultivation. They can be applied from sowing of seed nuts in nursery through adult palms in plantations. Mass treatment of coconut seeds/seedlings with biocontrol agents in nursery stage itself perhaps would serve to establish vigorous seedlings with induced resistance in the plant system against the disease complex (plants in young stage itself require consideration as root (wilt) affected young palms seldom flower). To take care of the root rots, fungal antagonists (*T. viride*, etc.) could be applied. Such bioagents would provide protection to root system, besides aiding enhanced growth of the plant. The talc-based *T. viride* formulation is used normally as dry seed treatment for various crops. It has been reported that *T. viride* treatment also enhances the crop growth appreciably besides diseases control and there by the crop yield is increased. Therefore, *T. viride* treatment to coconut seeds and/or root system of coconut seedlings, and coconut based cropping system is beneficial. As organics are ecofriendly, economical and environmentally safe such materials draw importance in preparation of biopesticides (Srinivasan, 2007c; Srinivasan *et al.*, 2007a, 2010a). The grown up cultures of bioagents can be mixed with locally available organic materials, byproducts from agro industries, etc. (saw dust, coir pith compost, neem cake, groundnut cake, mahuva cake, vermicompost, other organic composts, rice bran, etc.) as carrier materials in formulating the biopesticides. Talc-based preparations can be also mixed with organic materials and then applied onto plants in the field. The organic compounds not only function as carriers of bioagents but would also sustain the population of the organisms besides supplying nutrients to plants. While root (wilt) is a systemic symptom the leaf rot is non-systemic, necrotrophic syndrome. Use of mixtures of antagonistic organisms (consortium) is apparently important due to their synergistic interaction which leads to broad spectrum/enhanced activity against multiple pathogens. Mass introduction of multiple biocontrol agents into coconut system would achieve suppression of two different pathogens/infections/symptoms.

Field Extension of Biocontrol Technology

Mass production and use of biocontrol agents in the integrated management of coconut leaf rot disease was initiated in our institute at the behest of DBT sponsored program in order to provide specific technology to growers, women, rural people and other stakeholders in the disease affected regions. The biocontrol agents were mass-produced and processed into bioformulations and demonstrated to rural people through effective contacts/collaboration. Innovative programmes were conducted like on-campus trainings, off-campus trainings, household/field contacts, awareness measures through lectures, orientations, field schools, group discussions, etc. Thereby the rural people were enlightened on mass production of biocontrol agents and their utilization. Multilocational field demonstrations on technology of utilization of biological control agents against the disease were also conducted in coconut based cropping system by farmers' participatory approach. Promotional literature-materials on the leaf rot complex, guide on biological control agents evolved periodically besides mass-produced biopesticides were widely distributed to the beneficiaries of such programmes, and thereby the technology on biological control agents percolated in to rural masses (Table 13.6). Therefore, utility of biocontrol technology as an ecofriendly strategy in the integrated field management of leaf rot has been advanced (Srinivasan, 2008; Srinivasan *et al.*, 2007, 2011b).

Table 13.6. Consolidated data on number of persons benefited through various programmes conducted through implementation of DBT project, "Mass production and use of biocontrol agents in the integrated management of coconut leaf rot disease" in disease affected regions during different years

S. No.	Type of programme conducted in different years of project implementation	No. of persons availed benefits through participation in the programme		
		Men	Women	Total
I. Year – 2005-06:				
1	On campus trainings	037	045	082
2	Off campus trainings	423	108	531
3	House hold/Field visits	072	077	149
4	Other Awareness programmes	070	129	199
Year Total		602	359	961
II. Year – 2006-07:				
1	On campus trainings	0	43	43
2	Off campus trainings	838	258	1096
3	House hold/Field visits	202	61	263
4	Other Awareness programmes	151	69	220
Year Total		1191	431	1622
III. Year – 2007-08:				
1	On campus trainings	184	087	271
2	Off campus trainings	135	092	227
3	House hold/Field visits	178	099	277
4	Other Awareness programmes	117	125	242
Year Total		614	403	1017
Grand Total for all years		**2407**	**1193**	**3600**

Source: Srinivasan *et al.* (2011b)

CONCLUSIONS

Contribution of pesticides to agriculture in general and pest-diseases control in particular has been a land mark phenomenon that lend in their management/control and production made possible. But with the changing scenario, such a technology is unsustainable and it is imperative a new control technology with integrated system infused in the pest and disease management. In integrated management system a number of parameters are involved. Therefore, integrated management practices against pests and diseases are being intensively researched and in recent years biopesticides drawn immense potential. The biological control strategy covers the organic agriculture system as well, and being a non-chemical alternative mechanism of crop protection the system ensures environmental safety. Biological control as an important component in the plant protection management system has lead to success in management of several pests and diseases of various crop plants including coconut.

Coconut is an important plantation crop providing livelihood security to millions of small-marginal farmers in India particularly in southern states (coconut cultivation in India concentrated in Kerala state, followed by Tamil Nadu, Karnataka and Andhra Pradesh), besides in several other countries. In coconut, pests and diseases are serious constraints in the production and productivity. Rhinoceros beetle, red palm weevil, leaf eating caterpillar, eriophyid mite, root grubs etc. are the very important pests of coconut. The important diseases of the palm crop are leaf rot along with root (wilt), bud rot, stem bleeding, basal stem rot and grey blight. Root (wilt) disease occurs predominantly in southern districts of Kerala state, besides its neighboring states – particularly and heavily in Theni district of Tamil Nadu. The root (wilt) affected palm is commonly infected by leaf rot, wherein the leaf rot incidence is in strong interrelationship with root (wilt).

Efficacy of biocontrol agents as bioformulations for the control of coconut leaf rot fungi has been demonstrated using *Bacillus subtilis*, *Pseudomonas fluorescens* and *Trichoderma* spp. as biological control agents. The use and scope for application of the biological control, ISR has been a major tool in the integrated disease management. The effect of antagonistic organisms would be manifested over a period of time and, therefore, the potential could be exploited on a long term basis.

Innovation low cost technology could play an important role in stabilizing yield potential of the crop and thereby provide sustainable coconut agriculture with least risk perceptions. Correct adoption of the technology would avoid accumulation of adverse factors and assure environmental protection. Biocontrol agents in combination with organics need to be popularized. In India, the technology is utilized at the coconut nursery stage and value added seedlings utilizing the bioagents are raised. Supply of such value added, bioprimed, seedlings for large scale planting in the diseased region is gaining importance as no immune coconut variety is available against the root (wilt)-leaf rot complex. Innovative programmes were conducted to reach rural people through effective contacts/collaboration. Multilocational field demonstrations on utilization of biological control agents against the disease complex were also conducted by farmers' participatory approach along with promotional material on mass-production of antagonists. Therefore, the knowledge of biologicl control technology has been specially and widely popularized in disease-affected regions through implementation of various awareness programmes.

REFERENCES

Amruta, S.B. and Chandrasekar, S.C. (2003) *In vitro* evaluation of antagonistic agents and plant products against *Rhizoctonia bataticola* causing root rot of groundnut. *Indian Phytopathol.,* **56**, 335–338.

Anandaraj, M., Diby Paul, Jisha, P.J., Kumar, A., Saju, K.A., Thankamani, C.K. and Sarma, Y.R. (2003) Potential of a consortium of plant growth promoting rhizobacteria and *Trichoderma* for effective nursery management in black pepper (*Piper nigrum* L.). In: M.S. Reddy, M. Anandaraj, S.J. Eapen, Y.R. Sarma and J.W. Kloepper (eds.), *Abstracts and short papers, 6th International PGPR Workshop,* Indian Institute of Spices Research, Calicut, India, pp. 8–12.

Anishkumar, V.K., Jyothi Rahna, S. and Srinivasan, N. (2004) Evaluation of botanicals against major pathogens of coconut leaf rot disease and their antagonistic organisms. *J. Plantation Crops*, **32** (Suppl.), 321–329.

Beatriz, M.D. and Fereres, A. (2007) Ultraviolet-blocking materials as a physical barrier to control insect pests and plant pathogens in protected crops. *Pest Technol.*, **1**, 85–89.

Bharathi, R. (2001) Development of a rhizobacteria based bioformulation for the management of major pests and diseases in chilli. M. Sc. (Ag,) Thesis, Tamil Nadu Agricultural University, Coimbatore, India, 111 pp.

Chaluvaraju, G., Deepak, S.A., Nandini, P.S., Shekar Shetty, H. and Reddy, M.S. (2003) Comparative efficacy of different seed treatment formulations of plant growth promoting rhizobacteria in pearl millet plant growth promotion and downy mildew disease management. In: M.S. Reddy, M. Anandaraj, S.J. Eapen, Y.R. Sarma and J.W. Kloepper (eds.), *Abstracts and short papers, 6th International PGPR Workshop,* Indian Institute of Spices Research, Calicut, India, pp. 345–353.

Chen, C., Belanger, R.R. Benhamou, N. and Paultiz, T.C. (2000) Defense enzymes induced in cucumber roots by treatment with plant growth promoting rhizobacteria (PGPR). *Physiol. Mol. Plant Pathol.,* **56**, 13–23.

CPCRI. (2006) *Annual Report 2005-06.* Central Plantation Crops Research Institute, Kasaragod, Kerala, India.

CPCRI. (2007) *Annual Report 2006-07.* Central Plantation Crops research Institute, Kasaragod, Kerala, India.

CPCRI. (2012) *Annual Report 2011-12.* Central Plantation Crops research Institute, Kasaragod, Kerala, India.

Duffy, B,K. and Defago, G. (1999) Environmental factors modulating antobiotic and siderophores biosynthesis by *Pseudomonas fluorescens* biocontrol strains. *Appl. Environ. Microbiol.,* **63**, 2147–2154.

Gunasekaran, M., Gupta, A. and Srinivasan, N. (2003) *In vitro* biocontrol of coconut leaf rot pathogens with *Pseudomonas fluorescens. J. Plantation Crops,* **31**, 52–54.

Gupta, A. (2002) Mechanisms involved in biocontrol activity of antagonists of fungal pathogens of leaf rot disease of coconut. Abstract, *National Symposium on Crop Protection and WTO – An Indian Perspective,* Kasaragod.

Gupta, A. (2003) Important biological traits of microbial antagonists of leaf rot disease of coconut. *CORD,* **19**, 27–35.

Gupta, A., Gunasekaran, M. and Srinivasan, N. (2000) Isolation of bacterial antagonists from rhizosphere and their *in vitro* evaluation against pathogens of coconut leaf rot disease. In: Zhou Guang-he and Li Huai-fang (eds.), *Proceedings of the 1st Asian Conference on Plant Pathology (ACPP 2000),* China Agricultural Scientech Press, Beijing, pp. 268–274.

Hammerschmidt, R. (1999) Induced disease resistance: How do induced plants stop pathogens? *Physiol. Mol. Plant Pathol.,* **55**, 77–84.

Hazarika, D.K. and Phookan, A.K. (2003) Combination of *Glomus fasciculatum* with *Pseudomonas fluorescens* and *Trichoderma harzianum*: Effect of biocontrol potential and growth promotion in tea seedlings. In: M.S. Reddy, M. Anandaraj, S.J. Eapen, Y.R. Sarma and J.W. Kloepper (eds.), *Abstracts and short papers, 6th International PGPR Workshop,* Indian Institute of Spices Research, Calicut, India, pp. 289–294.

Jeyalaksmi, L., Durairaj, P., Seetharaman, K. and Sivaprakasam, K. (1998) Biocontrol of fruit rot and die back of chilli using antagonistic microorganisms. *Indian Phytopathol.,* **51**, 180–183.

Kandan, A., Radjacommare, R., Ramiah, M., Ramanathan, A. and Samiyappan, R. (2003) Plant growth promoting rhizobacteria induce systemic resistance in cowpea (*Vigna unguiculata*) against tomato spotted wilt virus by activating defense related enzymes and compound. In: M.S. Reddy, M. Anandaraj, S.J. Eapen, Y.R. Sarma and J.W. Kloepper (eds.), *Abstracts and short papers, 6th International PGPR Workshop,* Indian Institute of Spices Research, Calicut, India, pp. 480–486.

Kloepper, J.W. and Schroth, M.N. (1981) Development of a powdery formulation of rhizobacteria for inoculation of potato seed pieces. *Phytopathology*, **71**, 590–592.

Lily, V.G., Nair, U.K., Pandalai, K.M. and Menon, K.P.V. (1952) Observations on the inhibitory activity of a species of bacterium on some fungi parasitic on the coconut palm. *Indian Coconut J.*, **5**, 162–170.

Lily, V.G., Radha, K. and Menon, K.P.V. (1955) Observations on the inhibitory activity of a bacterium. Activity of the antifungal substance produced by the bacterium *Bacillus subtilis. Indian Coconut J.*, **8**, 137–149.

Meena, B., Radhajeyalakshmi, R., Marimuthu, T., Vidyasekaran, P., Doraisamy, S. and Velazhahan, R. (2000) Induction of pathogenesis-related proteins, phenolics and phenylalanine ammonia lyase in groundnut by *Pseudomonas fluorescens. J. Plant Dis. Prot.,* **107**, 514–527.

Mukhopadhyay, A.N. (1987) Biological control of soil borne plant pathogens by *Trichoderma* spp. *Indian J. Mycol. Plant Pathol.,* **17**, 1–10.

Muraligopal, M., Gupta, A. and Nair, R.V. (2003) Differences in rhizosphere microbial ecology of root (wilt) diseased and field tolerant coconut palms. In: M.S. Reddy, M. Anandaraj, S.J. Eapen, Y.R. Sarma and J.W. Kloepper (eds.), *Abstracts and short papers, 6th International PGPR Workshop,* Indian Institute of Spices Research, Calicut, India, pp. 404–405.

Nandakumar, R., Viswanathan, R., Babu, S., Shella, J., Raghuchander, T. and Samiyappan, R. (2001) A new bioformulation containing plant growth promoting rhizobacterial mixture for the managemenent of sheath blight and enhanced grain yield in rice. *BioControl*, **46**, 493–510.

Pimentel, D. and Greiner, A. (1997) Environmental and socio-economic costs of pesticide use. In: D. Pimentel (ed.), *Techniques for Reducing Pesticides: Environmental and Economic Benefits*, John Wiley and Sons, UK, pp. 51–78.

Radja Commare, R., Nandakumar, R., Kandan, A., Suresh, S., Bharathi, M., Raguchander, T. and Samiyappan, R. (2002) *Pseudomonas fluorescens* based bioformulation for the management of sheath blight and leaf folder in rice. *Crop Prot.*, **21**, 671–677.

Rajan, P.P., Sarma, Y.R. and Anandaraj, M. (2002) Management of foot rot disease of black pepper with *Trichoderma* spp. *Indian Phytopathol.,* **55**, 34–38.

Rajbir Singh and Sinha, A.P. (2004) Comparative efficacy of local bioagents, commercial bioformulations and fungicide for the management of sheath blight of rice, under glass house conditions. *Indian Phytopathol.*, **57**, 494–496

Ramamoorthy, V., Raguchander, T. and Samiyappan, R. (2002) Induction of defense-related proteins in tomato roots treated with *Pseudomonas fluorescens* Pf1 and *Fusarium oxysporum* f. sp. *lycopersici*. *Plant Soil*, **239**, 55–68.

Ramamoorthy, V., Viswanathan, R., Raguchander, T., Prakasam, V. and Samiyappan, R. (2001) Induction of systemic resistance by plant growth promoting rhizobacteria in crop plants against pests and diseases. *Crop Prot.*, **20**, 1–11.

Ramanujam, B., Nambiar, K.K.N. and Iyer, R. (2000) Identification of potential antagonistic fungi for biocontrol of *Thielaviopsis paradoxa*, causing stem bleeding disease of coconut. In: N. Muraleedharan and R. Rajkumar (eds.), *Recent Advances in Plantation Crops Research*, Allied Publishers Ltd., New Delhi, pp. 353–367.

Ramanujam, B., Nambiar, K.K.N., Iyer, R. and Biddappa, C.C. (2004) Mass multiplication of *Trichoderma* sp. on various agricultural and forest products. *J. Plantation Crops*, **32** (Suppl.), 395–398.

Raupach, G.S. and Kloepper, J.W. (1998) Mixtures of plant growth promoting rhizobacteria enhance biological control of multiple cucumber pathogens. *Phytopathology*, **88**, 1158–1164.

Renuka, R., Prakasam, V. and Parthiban, V. K. (2003) Antagonistic effect of *Pseudomonas fluorescens* and *Bacillus subtilis* isolates against *Alternaria chrysanthemi* causing leaf spot of chrysanthemum. In: M.S. Reddy, M. Anandaraj, S. J. Eapen, Y.R. Sarma and J.W. Kloepper (eds.), *Abstracts and short papers, 6th International PGPR Workshop,* Indian Institute of Spices Research, Calicut, India, pp. 219–220.

Rethinam, P. (2007) Acceleration of value addition in coconut. *Indian Coconut J.*, **38**, 2–14.

Ryalls, J., Neuenschwander, U., Willits, M., Molina, A., Steiner, H.Y. and Hunt, M. (1996) Systemic acquired resistance. *Plant Cell*, **8**, 1809–1819.

Saju, K. A., Anandaraj, M. and Sarma, Y.R. (2003) Evaluation of *Trichoderma* spp. and *Pseudomonas fluorescens* for suppression *Phytophthora capsici* infecting black pepper. In: M.S. Reddy, M. Anandaraj, S.J. Eapen, Y.R. Sarma and J.W. Kloepper (eds.), *Abstracts and short papers, 6th International PGPR Workshop,* Indian Institute of Spices Research, Calicut, India, pp 52–58.

Solomon, J.J., Nair, C.P.R., Srinivasan, N., Gunasekaran, M. and Sasikala, M. (1999) Coconut root wilt - the malady and remedy. *J. Plantation Crops*, **27**, 71–92.

Srinivasan, N. (1991) Occurrence of coconut leaf rot in relation to root (wilt) disease. *Indian Coconut J.*, **21**, 14–18.

Srinivasan, N. (1999) Know about phytoplasma as pathogen in relation to coconut diseases and with reference to root wilt. *Indian Coconut J.*, **30**, 3–6.

Srinivasan, N. (2001) Diseases and disorders of coconut and their management. In: P.C .Trivedi (ed.), *Plant Pathology*, Pointer Publishers, Jaipur, India, pp. 194–254.

Srinivasan, N. (2002) Coconut leaf rot complex and perspectives for the disease control - Status report. *Indian Coconut J.*, **32**, 2–9.

Srinivasan, N. (2003) Efficacy of *Pseudomonas fluorescens* against leaf rot in root (wilt) affected coconut palms. *Indian Phytopathol.,* **56**, 210–211.

Srinivasan, N. (2004a) Coconut root (wilt) disease and its management. *Indian Fmg*, **54**, 3–4, 8–9, 11.

Srinivasan, N. (2004b) Incidence of leaf rot disease in relation to root (wilt) in young coconut palms and fungi in petiole and lamina lesions. *CORD*, **20**, 28–36.

Srinivasan, N. (2005) Advances in leaf rot and its management in root (wilt) affected coconut. In: Sampat Nehra (ed.), *Plant Diseases: Biocontrol Management*, Aavishkar Publishers, Distributors, Jaipur, India, pp. 19–37.

Srinivasn, N. (2006) Mass production and use of biocontrol agents in the integrated management of coconut leaf rot disease, Progress report submitted to Department of Biotechnology, Government of India, New Delhi, India, 35 pp.

Srinivasan, N. (2007a) Use of biocontrol agents in integrated management of leaf rot in root (wilt) affected coconut. In: P.C. Trivedi (ed.), *Biocontrol of Plant Diseases*, Aavishkar Publishers, Distributors, Jaipur, India, pp. 83–106.

Srinivasan, N. (2007b) Mass production of biocontrol agents in coconut water based media. *J. Plantation Crops*, **35**, 190–194.

Srinivasan, N. (2007c) Induced systemic resistance-biocontrol technology in plantation crops with special reference to coconut root (wilt) and leaf rot diseases. In: R.V. Nair, V. Krishnakumar, Regi J. Thomas, K. Muralidharan and George V. Thomas (eds.), *Breeding for resistance to diseases and insect pests in plantation crops*, Central Plantation Crops Research Institute, Kasaragod, Kerala, India, pp. 237–243.

Srinivasan, N. (2008) Management of coconut leaf rot disease, *Technical Bulletin No. 58*, Central Plantation Crops Research Institute, Kasaragod, India, 10 pp.

Srinivasan, N. (2010a) Production of value added coconut seedlings through biopriming with biocontrol agents against leaf rot disease. Abstracts, *National Conference on Production of Quality Seeds and Planting*

Material – Health Management in Horticultural Crops", 11-14 March 2010, New Delhi, p. 229.

Srinivasan, N. (2010b) Advances in coconut leaf rot disease and its management measures. In: P.C. Trivedi, (ed.), *Plant Diseases and its management*, Pointer Publishers, Jaipur, India, pp. 53–92.

Srinivasan, N. and Bharathi, R. (2006) Biocontrol agents against pathogens of coconut leaf rot disease. *J. Plantation Crops*, **34**, 494–499.

Srinivasan, N., Bharathi, R., Shanty Issak and Rajeev, G. (2011a) Field management of coconut leaf rot disease with biological control agents, their bioformulation and viability in organics. *J. Plantation Crops*, **39**, 99–104.

Srinivasan, N., Chandramohanan, R., Bharathi, R., Radhika, N. S. and Shanty Issak. (2011b) Development of technology and popularization of biological control agents for integrated management of coconut leaf rot disease. *CORD*, **27**, 48–60.

Srinivasan, N., Chandramohanan, R. and Bharathi, R. (2007) *Biocontrol Agents in Integrated Management of Coconut Leaf Rot Disease – A Guide for Preparation and use of Biopesticides.* Training Folder, Central Plantation Crops Research Institute, Regional Station, Kayangulam, Kerala, India.

Srinivasan, N. and Gunasekaran, M. (1993) Fungi associated with leaf rot disease of coconut. *Indian Coconut J.*, **23**, 2–7.

Srinivasan, N. and Gunasekaran, M. (1996) Incidence of fungal species associated with leaf rot disease of coconut palms in relation to weather and the stage of lesion development. *Ann. Appl. Biol.*, **129**, 433–449.

Srinivasan, N. and Gunasekaran, M. (1999) Coconut leaf rot disease complex - A review. *CORD*, **15**, 33–65.

Srinivasan, N. and Gunasekaran, M. (2000) Etiology and recurrence of coconut leaf rot with special reference to seedlings. In: N. Muraleedharan and R. Raj Kumar (eds.), *Recent Advances in Plantation Crops Research,* Allied Publishers Ltd., New Delhi, India, pp. 400–403.

Srinivasan, N. and Gunasekaran, M. (2003) Curative and prophylactic control of fungal leaf rot complex in root (wilt) affected coconut palms in India. *CORD*, **19**, 17–26.

Srinivasan, N., Jyothi Rahna, S. and Anishkumar, V.K. (2006) Evaluation of fungicides and antagonistic organisms against major pathogens of leaf rot disease of coconut and their eco-friendly management. *CORD*, **22**, 27–50.

Srinivasan, N., Koshy, P.K., Kamalakshy Amma, P.G., Sasikala, M., Gunasekaran, M. and Solomon, J.J. (2000) Appraisal of the distribution of coconut root wilt and heavy incidence of the disease in Cumbum Valley of Tamil Nadu. *Indian Coconut J.*, **31**, 1–5.

Srinivasan, N. and Iyer, R. (2004) Leaf rot and bud rot of coconut. In: P. Santhakumari (ed.), *Advances in the Diseases of Plantation Crops and Spices*, International Book Distributing Co., Lucknow, India, pp. 122–152.

Srinivasan, N. and Sasikala, M. (2001) Spread and distribution of coconut root (wilt) disease in the region of Theni - Dindigul districts of Tamil Nadu. *Indian Coconut J.*, **31**, 7–11.

Srinivasan, N., Solomon, J.J., Sasikala, M., Gunasekaran, M., Geetha, L. and Rajeev, G. (2001) Managing coconut root wilt and leaf rot diseases. *Indian Hortic.*, **46**, 28–29.

Srinivasulu, B., Aruna, K., Vijay Krishna Kumar, K., Sabitha Doraiswamy and Rao, D.V.R. (2004) Biocontrol potential of *Trichoderma viride* against basal stem rot disease of coconut. *J. Plantation Crops*, 32, 28–31.

Sticher, L., Mauch-Mani, B. and Metraux, J.P. (1997) Systemic acquired resistance. *Annu. Rev. Phytopathol.*, **35**, 235–270.

Suseela Bhai, R., Joseph Thomas and Naidu, R. (1994) Evaluation of carrier media for field application of *Trichoderma* spp. in cardamom growing soils. *J. Plantation Crops*, **22**, 50–52.

Thangavelu, R., Velazhahan, R. and Sathiamoorthy, S. (2003) Induction of defense mechanisms in banana by plant growth-promoting microbes, plant activators and inoculation with *Fusarium oxysporum* f. sp. *cubense*. In: M.S. Reddy, M. Anandaraj, S.J. Eapen, Y.R. Sarma and J.W. Kloepper (eds.), *Abstracts and short papers, 6th International PGPR Workshop,* Indian Institute of Spices Research, Calicut, India, pp. 617–626.

Van Loon, L.C., Bakker, P.A.H.M. and Pieterse, C.M.J. (1998) Systemic resistance induced by rhizospheric bacteria. *Annu. Rev. Phytopathol.*, **36**, 453–483.

Velusamy, P., Kavitha, S., Preeti Vasudevan and Gnanamanickam, S.S. (2003) Role of plant-associated bacteria and their antimicrobial products in the suppression of major rice diseases in India. In: M.S. Reddy, M. Anandaraj, S.J. Eapen, Y.R. Sarma and J.W. Kloepper (eds.), *Abstracts and short papers, 6th International PGPR Workshop,* Indian Institute of Spices Research, Calicut, India, pp. 256–259.

Vidhyasekaran, P. and Muthamilan, M. (1995) Development of formulations of *Pseudomonas fluorescens* for control of chickpea wilt. *Plant Dis.*, **79**, 782–786.

Weller, D.M. (1988) Biological control of soil borne plant pathogens in the rhizosphere with bacteria. *Annu. Rev. Phytopathol.*, **26**, 379–407.

Zaidi, N.W. and Singh, U.S. (2004) Mass multiplication of *Trichoderma harzianum* on cow dung. *Indian Phytopathol.*, 57, 189–192.

14

Status of Potential Biocontrol Components for Integrated Management of Forest Insect Pests in India

NITIN KULKARNI

Tropical Forest Research Institute, Jabalpur- 482 021, India
E. mail: kulkarnin@icfre.org, kulkarni_n27@icfre.org

INTRODUCTION

India ranks tenth in the world list of most forested nations with 76.87 million hectare of forest and tree cover, accounting for about 23.4% of the total geographical area of the country (Kishwan *et al.*, 2009). Along with the ever increasing pressure from human encroachments of the forest lands, the forest cover has to cope up with many other biotic and abiotic factors also, including the biological determinants like insect and disease pests. Forest tree species at various life stages; flowering to seed setting, fruits, seeds, seed germination to young seedlings, saplings, standing trees and stored timbers, are infested by various species of insect pests (Beeson, 1941; Pathak *et al.*, 1995; FAO, 2006, 2007). The insect pest outbreaks can contribute either directly or indirectly to economic and environmental losses and have adverse effects on tree growth.

Unlike agricultural counterparts, where losses caused by insect pests can easily be assessed, it is very difficult for the forest entomologists to generate statistics on the overall extent of damage caused by the insect and pathogen pests to forests (FAO, 2005). Inapproachability of many of the forest areas, long gestation period crops with longer rotation period (sometimes 70 to 80 years) are some of the major attributable factors. It is only in the last two decades that the forestry research has been focussed after the formation of Indian Council of Forestry Research, Dehradun with the research institutes representing forest types, in the country. These institutes are not only generating data on status of forest health but also undertake researches on environment friendly management strategy for the management of forest pests. Besides, the Food and Agriculture Organisation (FAO) has been assisting, advising, and supporting countries and national institutes to safeguard the health of forests, forest ecosystems and trees outside forests, with special reference to insects, diseases and other harmful biotic and abiotic agents. Apart from preventive measures and integrated pest management (IPM), FAO has also recommended actions to minimise risks of trans-boundary transfer. It also offers assistance to countries in establishing long-term prevention and forest protection strategies (FAO, 2011).

In India, principal plantation tree species include *Tectona grandis, Eucalyptus* spp. *Eucalyptus globulus*, *E. grandis* and *E. tereticornis;* acacias, *Acacia auriculiformis*, *A. catechu*, *A. mearnsii*, *A. nilotica* and *A. tortalis*. Other commonly planted broadleaf species are *Albizia* spp., *Azadirachta indica*, *Casuarina equisetifolia*, *Dalbergia sissoo*, *Gmelina arborea*, *Populus* spp., *Prosopis* spp., *Shorea robusta* and *Terminalia* spp. Among conifers, *Cedrus deodara* and *Pinus roxburghii* occupy a major area; *Pinus patula* and *P. caribaea* have been planted to a limited extent.

The sporadic incidences of pests occurring in large plantations or natural forests normally do not attract any interventions, except in cases where their incidence exceeds to epidemic form leading to either tremendous losses to the products or leads to tree mortality. More interventions are required in forest nurseries, which simulate an agricultural farm and delicate seedlings are more prone to the attack by the insect pests. Forest entomology has still to generate necessary field data for the development of integrated models against individual major insect pests using the ecofriendly options in the country (Kulkarni *et al.*, 2004a). Otvos (2004) while discussing potential and challenges in IPM in forestry, in his recent review, has discussed many such constraints in success of IPM in forestry. The status and strategies to maintain the forest health for improved productivity from the limited and rather decreasing forest areas needs to be updated times and again for the researchers, academicians and environmentalists.

MAJOR INSECT PESTS

The insect pests of forestry can be grouped into the following major categories, based on stages of the plants they attack and nature of damage; (i) flower and inflorescence feeders, seed borers on standing trees and in seed storage, (ii) insect pests of forest nurseries and (iii) insect pests of plantation and natural forests. Major insect pests damaging root system of the seedlings in forest nurseries are termites (*Macrotermes* spp., *Odontotermes* spp., *Microtermes* spp., *Odontotermes* spp.; wood dwelling termite, *Stylotermes bengalensisi*; carton nest building termite, *Microcerotermes* spp.; white grubs, *Holotrichia consanguinea, H. serrata, H. reynaudi, H. rustica, H. mucida,* and *Schizonycha ruficollis*) (Thakur, 2000; Kulkarni, 2010; Kulkarni *et al.*, 2007a, 2009), defoliators like *Hyblaea puera* and *Eutectona machaeralis, Calopepla leayana, Aulacophora foveicollis, Dasychira grotei, D. mondosa, D. thwaitesi, Euproctis bipartita, E. fraterna, E. guttata, Lymantria lepcha, L. mathura, Trabala vishnou;* Maharukh (*Ailanthus excelsa*) webworm, *Atteva fabriciella,* Sissoo defoliator, *Plecoptera reflexa;* Kullu defoliator *Sterculia urens, Sylepta balteata*; bamboo leaf rollers; *Algedonia* (*Pyrausta, Crypsiptya*) *coclesalis, A. bambucivora, Crocidophora ptyophora, Matapa aria, Piorea flavofimbriata* and *Sylepta derogata* from June to October. Other species of defoliators recorded were *Spodoptera litura* feeding on *Sesbania* spp. *Euproctis divisa* feeding on *Albizia procera,* semilooper *Perciana flavifusa* feeding on *Grewia optiva* and *Clostera fulgurita, C. cupreata* and *Phalanta phalantha* feeding on *Populus deltoides. Moringa* seedlings were sometimes killed by a leaf defoliator and shoot borer, *Noorda blitealis.* Some of the major sap suckers are aphids, *Aphis gossypii, Drosichiella phyllanthi, Tingis beesoni,* psyllid, *Heteropsylla cubana, Eucalyptus* gall maker, *Leptocybe invasa* and *Trioza obsoleta.*

The major seed borers of forest tree species have been reviewed by (Kulkarni and Joshi, 1998c). Seed borers include bruchid borer, *Caryoborus gonagra, Bruchidius pisorum, B. pygoquadrimaculatus, B. saundersi, B. schrodci, B. uberatus, B. bilineatopygus, B. sparsemaculatus* and *Caryedon gonagra* that were observed to bore the seeds of *Albizia* seeds (Joshi *et al.*, 1990). *Sitophilus rugicollis, Pagyda sylvalis* and *Dichocrocis leptalis* occur on *sal* (*Shorea robusta*) seeds in central India (Kulkarni *et al.*, 1998; Choubey *et al.*, 2004, 2008).

Among the insect pests of shoots and young stems of teak and *Gmelina arborea* phassus borer *Sahyadrassus* (*Phassus*) *malabaricus*; shoot borer, *Zeuzera coffeae*; bark eating caterpillars, *Indarbela quadrinotata* on teak, *shisham, sissoo*, *Eucalypts, Albizia* spp., etc. in Chhattisgarh, Madhya Pradesh and Maharashtra. *Sal* heartwood borer, *Hoplocerambyx spinicornis*; root and stem borer, *Coelosterna scabrator* have been recorded as borers of *babul* and *Eucalyptus* in different areas of Madhya Pradesh. Sapling borer, *Cossus cadambae*; *Zeuzera* sp.; *monga* shoot borer, *Noorda blitealis*; *semul* shoot borer, *Tonica niviferana*; *aonla* shoot gall insect, *Betousa stylophora* and the bark eating caterpillar, *Indarbella quadrinotata* have also been recorded at different localities of Madhya Pradesh and Maharashtra. *Cyrtotrachelus longimanus* and *Olethreutes* (*Argyroploce*) *paragramma*, bamboo hispine beetle, *Estigmena chinensis* and green shoot weevil borer, *Myocalandra exarata* on young bamboo culms, *Aeolesthes holosericea* on standing dead/felled *sal, babul, bel,* mango, teak, etc. are also well known. This borer was observed very much similar in shape, size and behaviour with *Hoplocerambyx spinicornis,* which causeshavoc in sal dominating forest areas in the country with repeated epidemics (Kulkarni *et al.,* 2009).

CURRENT STATUS OF MANAGEMENT PRACTICES

The insect pests infesting forest tree species were identified and preserved as reference collection in Forest Research Institute (FRI), Dehradun, since 19th century and later at regional ICFRE institutes in the country. These insect collections also included insect natural enemies (parasitoids and predators) of the insect pests. However, investigations on evaluation of potentiality of these biocontrol agents as independent components of IPM, were initiated one to two decades back. Insect control was not given a due consideration in forestry, especially for large plantation and natural forests, due to the feasibility problem and until the realization that the only way to increase forest productivity is to judiciously manage the available resources in the limited land area (Joshi *et al.*, 2001a,b). However, application of suitable chemical pesticides has long been practiced especially for fruit and seed borers during storage (Kulkarni and Joshi, 1998c), soil inhabiting insects like white ants, white grubs, cut worms, voracious leaf defoliators and borers of seedlings in forest nurseries (Kulkarni, 2010) and young saplings in plantations and borers of wooden logs, planks and furniture, etc. Joshi *et al.* (2001b) have compiled detailed account of the chemical insecticides evaluated against the forest insect pests. The harmful environmental impact of the chemical insecticides needs no discussion and, therefore, recent researches on IPM options of forest insect management (Kulkarni *et al.*, 2004a,b), particularly different biological control components in the country are being discussed herein.

Silvicultural and Mechanical Methods

Beeson (1941) has discussed many silvicultural (cultural) and mechanical methods for the management of forest insect pests, which have been discussed times and again in various technical bulletins and training lectures (Joshi and Namdeo, 1990). Cultural control represents a fundamental integrated pest management tactic even if the adoption of the practices, as driven by biotic factors of the environment, may remain unrecognized by growers (Bajwa and Kogan, 2004). However, this very important component was rarely applied for the management of forestry insect pests. It is possibly because IPM modules for the management of major insect pests are still lacking for the want of detailed field data on bioecology of many of the major forest insect pests.

Recently, detailed bioecological investigations on the white grub species damaging teak seedlings in forest nurseries were carried out. Three new scarab species *Holotrichia rustica, H. mucida* and *Schizonycha ruficollis*, were not only reported for the first time to infest any forestry species besides many other known species (Kulkarni *et al.*, 2007a, 2009), but utilized the detailed bioecological observations to propose an integrated pest management model, including the prediction of the attack (Kulkarni, 2009a,b, 2010), for the management of scarab pests in teak nurseries of Maharashtra for the Forest Development Corporation of Maharashtra (FDCM), Nagpur.

The management sal heartwood borer (*Hoplocerambyx spinicornis*) in natural *sal* (*Shorea robusta*) forests is another most important and widely practiced silvicultural or mechanical method of management of insect pests in forestry. This cerambycid insect pest has annual life cycle, newly hatched grubs of the pest enter the tree trunk during rainy season after mating by the emerging beetles, feed on softwood and then heartwood making large tunnels inside the tree trunk and emerging again as adult during next rainy season. The attacked tree succumbs to the injuries, dies and has merely as fuel wood value. Thus, the dreaded borer has an annual life cycle and has been causing havoc through reoccurring epidemics in *sal* forest areas of Madhya Pradesh and Chhattisgarh (Joshi *et al.*, 2006) in India. The recent epidemic was during 1996 to 2001, which resulted in removal of damaged/ attacked dead *sal* trees to the tune of ten lakhs from the forest floor to maintain forest hygiene. Despite some successful investigations on identification of kairomonal compounds from the *sal* bark (Kulkarni *et al.*, 2004b), the cultural/silvicultural method practiced to manage this pest is through tree trapping operation. In tree trapping measured length of broken logs of *sal* are beaten to allow loosening of bark and oozing out of cell sap during the evenings of rainy season to attract beetles, which are collected every morning before sun rise and killed during epidemic periods (Joshi *et al.*, 2006, Kulkarni *et al.*, 2007b).

Relative Resistance in Trees

After the publication of the classic *Plant Resistance to Insects* by Painter (1951), the plant resistance to insects has been worked out very well in agriculture, extended further to forestry as discussed in several contemporary reviews on host plant resistance in forest trees (Stark, 1965; Gerhold *et al.*, 1966; Emden, 1973; Hanover, 1980). However, long rotation periods of most of the important tropical tree species has been a major hurdle in making any headway.

There are very few examples on application of host tree resistance in forestry in the country. The available reports concentrate on the important timber tree species, teak (*Tectona grandis*). Kaushik (1956) had discussed possible existence of resistance in 'Teli' (meaning 'oily') variety of teak against defoliators, through field study in Karnataka. Kedharnath and Singh (1975) observed relative resistance in eight clones of *T. grandis* and one allied species, *T. hamiltoniana* to larvae of leaf skeletonizer. Of these 'Teli' clone of *T. grandis* and *T. hamistoniana* are highly resistant to larvae of *E. machaeralis*. Later, Ahmad (1991) evaluated the relative resistance of 20 teak clones of southern India planted in clonal orchard at Walayar (Kerala). Clones TNT-11, TNT-6 and KLK-2 were found to be most resistant ones, based on food preference bioassays. Mishra (1992) investigated the natural variation in susceptibility of 16 teak clones maintained in teak germplasm bank at Newforest, Dehradun and based on quantity of leaf consumed by the larvae, Bori and FG-1 clones were reported as the most resistant, followed by Gorakhpur FG-19, Allapalli, Togo and Orissa. Further, more scientific work started with the selection of phenotypically superior trees (plus trees) in different parts of the country (Kedharnath, 1984, Tewari, 1992), with a few studies included to screen clones for resistance to insects (Thakur, 1983). Meshram (1993) and Meshram *et al.* (1994) examined clones of teak for their natural variation in susceptibility to both the teak defoliators (*H. puera* and *E. mechaeralis*) and APT-20, APT-8 and MHSC-J2 were observed to be the most resistant whereas UP-D1 and TNT-11 as the the most susceptible ones, based on leaf area consumption.

All the above fragmented data led to confusion and, therefore, could not be utilized practically, until 167 clones collected from 10 teak growing states of India, planted at National Teak Germplasm Bank, Lohara, Chandrapur, Maharashtra were screened out in an interdisciplinary approach (Roychoudhury, 2001). On the basis of quantitative assessment on the severity of the damage, clone of Orissa origin ORANR$_3$ was assessed as most resistant followed by the remaining 17 highly resistant clones namely APNPL$_2$, APT$_{20}$ (Andhra Pradesh), MHAL-P$_5$, MHAL-A$_9$, MHSC-A$_3$, MHSC-J$_2$, MHWY-K$_1$ (Maharashtra), MYHV$_3$, ST$_{-6}$, ST$_{-14}$, ST$_{-17}$ (Karnataka), TNT-3, TNT-17 (Tamil Nadu), UP$_{-D}$, UP$_{-N}$, (Uttar Pradesh), KLN$_{-1}$ (Kerala), and PT$_{-46}$ (Madhya Pradesh). These observations were later confirmed by the laboratory feeding tests and all the clones, besides most (1 clone) and highly resistant (17 clones), were ranked in to resistant (28 clones), moderately resistant (38 clones), least resistant (30 clones), moderately susceptible (32 clones), susceptible (13 clones), highly susceptible (6 clones) and most susceptible (2 clones) (Roychoudhury and Joshi, 1996; Roychoudhury *et al.*, 1998). To demonstrate the above, 40 clones out of above were planted at Tropical Forest Research Institute, Jabalpur and have been studied for their growth rates and resistance to skeletonizer. Clones APNPL-2 of Andhra Pradesh and UP-N of Uttar Pradesh, have been proven to be resistant and have also exhibited superior growth in terms of height, basal diameter and number of leaves (Roychoudhury *et al.*, 2000). Teak seed orchards grown in different areas of Madhya Pradesh were also screened for resistance against the larvae of leaf skeletonizer *E. machaeralis*. On the basis of damage, CSC-9 clone of Chhindwara was recorded less attacked as compared to the clones of Seoni and Balaghat. Feeding bioassay conducted in laboratory confirmed that CSC-9 of Chhindwara was proved less consumed having lesser leaf moisture percentage and the larvae gained less weight as compared to other clones (Roychoudhury, 2001).

Javaregowda (2005) while investigating some teak clones from Karnataka on similar lines, found clone MYMK-1 as resistant (R2) and MYHD-1 as susceptible (S1). In the same report, clone MYHUT-3 (Hunsur origin), MYSA-1 (Shimoga origin) and MYHD1 (Haliyal origin) were observed as most susceptible (S2).

The existence of resistance in other important multipurpose tree species, *Albizia* spp. was also investigated against *siris* (*Albizia* spp.)

cutworm or foliage feeder, *Spirama retorta*. Relative resistance of *safed siris, Albizia procera* seedlings obtained from seeds sampled from Jabalpur, Bijadandi, Barghat, Khandwa, and Umaria of Madhya Pradesh; Bilaspur, Shakti and Lakhna (Raipur) of Chhattisgarh; Kanpur in Uttar Pradesh; Haldwani in Uttarakhand and Agartala in Tripura were evaluated at the institute. The results showed that seed source of Agartala (Tripura) followed by Shakti and Lakhana- Raipur (Chhattisgarh) were resistant to the attack of this cutworm. There were some fragmentary observations on *Spodoptera litura* on teak in nurseries (Roychoudhury *et al.*, 1995) and *Clostera cupreata*, resistance in poplar, *Populus deltoides* (Ahmad, 1993), which require more data to be generated.

In general, genetic inheritance of resistance in trees is still a controversial subject (Nair *et al.*, 1997) and requires more confirmatory investigations on different tree species in the country. Nair *et al.* (1989; 1997) have discussed that many isolated trees can remain unattacked by defoliator during first round of selection by the insect but were attacked in the second year and thus proving an involvement of only an escape mechanism through early leaf maturity. The resistance has also been attributed to the morphological features of leaf, physiological conditions of host plants, the extent of essential nutrients present in the soil, biochemical composition of phytomass and presence of chemical compounds and their interactions, influencing ultimately the palatability of leaves, which in turn influences the susceptibility of foliage to defoliating insects (Berryman, 1988; Anathakrishnan, 1990). However, these kinds of examples have been termed as ecological resistance by Panda (1979).

Botanical Biopesticides

Plants biosynthesize dazzling array of anti-insect biological activities, some known since more than 300 years in India and Europe (Kulkarni, 2001; Koul and Wahab, 2004) as grain protectants. It was only after the publication of scientific report by Pradhan *et al.* (1962) on feeding deterrence property in neem seed kernels against locusts that the claims of earlier workers (Fletcher and Ghosh, 1919; Chopra, 1928) was established. This led to many good reports against the agricultural insect pests (Koul *et al.*, 2003). However, evaluations with the crude and organic extracts of botanicals against the forestry insect pests were initiated only in nineties (Kulkarni, 1998, 1999; Kulkarni and Joshi, 1997, 1998a,b,c; Kulkarni *et al.*, 1996a,b, 1997a,b, 1999). The researches with botanicals against the forest insect pests in India have been compiled by Kulkarni *et al.* (1995, 1998), Kulkarni (2001) and recently by Gahukar (2010).

Due to the limited scope of application, botanical pesticides in forestry were not investigated prior to nineties before it was realized that forest nurseries offer insects the agriculture-like ecosystem and thus invite more input of chemical insecticides and, therefore, are the places where chemicals can be replaced by ecofriendly botanical pesticides as part of integrated system of insect pest management. Some reports of evaluations of plant products against forest insect pests worthy of detailed discussion are Kulkarni *et al.* (1996a,b; 1997a,b). Kulkarni *et al.* (1996a) have evaluated biological activity of defatted methanolic neem seed extract on the larvae of *E. machaeralis.* Methanolic neem seed extract at 1 and 2% aqueous concentrations were strongly antifeedant by preventing 70.15 to 75.44% feeding of teak leaves. Kulkarni *et al.* (1997a) reported antifeedant property of methanolic extract of a weed *Lantana camara* var. *aculeata* and *Aloe vera* leaves against the larvae of *E. machaeralis* in laboratory. The lowest concentration for an antifeedant activity of the larvae was recorded as 0.5%. No activity was obtained below this concentration. The activity increased to its threshold at the concentration of 2.0% and above, at which feeding was reduced to 8.35% and 6.3% against 34.9% in control. The activity exhibited by *A. vera* and *L. camara* may be due to the active toxic groups like "Aloin" and "Lantadane A" present in the leaves. Further increase in concentration resulted in mortality in some larvae due to starvation. Kulkarni (2001) has discussed anti-insect activities of various plant parts of 97 plant species against 28 insect pests of forestry importance.

Kulkarni *et al.* (2003) have reported field stability of *A. squamosa* leaves extracted directly in methanol when evaluated in field-cum-lab experiment and compared with commercially available neem product Amrutguard, methanolic extract proved at par with Amrutguard. In some recent field-cum-lab investigations on efficacy of botanicals, the results indicated that teak seedlings sprayed prophylactically with 2.0% neem product allowed less egg laying as evident by significantly less mean number of larvae, i.e., 14.66, as compared to 22.33 larvae on seedlings treated with water as control and other lower treatments. Results indicated that planned and time-bound use of biopesticides like neem formulation at/above 2.0%, offer anti-ovipositional effect against teak defoliator and can be used as a prophylactic treatment to inhibit egg laying of teak defoliator as a component of IPM module. Neem products also inhibited feeding in larvae, concentrations at and above 0.5% caused significant ($P < 0.001$) antifeedant effect (95.93%) over other treatments. Other natural botanical products also proved effective. Derisom® at and above 0.15% proved effective with above 90% mortality ($P < 0.05$), followed by 0.10%. The botanical pesticide, Ozomite® at concentrations at and above 0.01% gave best results with mortality observed above 90%. Its lower concentrations of 0.002 and 0.005% caused 50 and 54.17% mortality, respectively (Kulkarni, 2009b).

Javaregowda (2005) in his recent study against teak defoliator, *Hyblaea puera* investigated efficacy of a few plant products (in different concentration of 0.5, 1.0, 1.5 and 2.5%), viz., *Azadirachta indica* A. Juss (seed and leaf), *Melia azedarach* L. (leaf), *Strychnus nux-vomica* L. (leaf, seed and bark), *Jatropha curcas* L. (seed and leaf), *Cassia fistula* L. (leaf and bark), *Cassia siamea* Lam. (leaf), *Gnidia glauca* L. (leaf and bark), *Ricinus communis* L. (leaf), *Lobellia nicotinifolia* L. (leaf), *Amorphophallus campanulatus* Blume (tuber), *Vitex negundo* L. (leaf), *Derris indica* (L.) (leaf), *Clerodendrum inerme* Garten (leaf), and *Lantana caftera* L. (leaf). While all the treatments were effective as antifeedant when compared to control in minimising the consumption, 2.5% fresh kernel extract of *Azadirachta indica* was the best. Evaluation of ovicidal effect of neem concentrations of 0.5, 1.5 and 2.5% on eggs of teak defoliator (*H. puera*) revealed *A. indica* kernal and leaf extracts were effective ovicidal products (Javaregowda, 2005).

Bacterial Biopesticides

Entomopathogenic bacteria, which are ubiquitous prokaryotic organisms, causing diseases through gut epithelium in insect pests (Steinhaus, 1963a,b) need no introduction. The major species of bacteria belong to the genus *Bacillus* (Kawanishi and Held, 1990), like *B. thuringiensis, B. lentimorbus, B. popillae* and *B. cereus*, of which *B. thuringiensis* has only has received much attention because of its high potency and long range of host killing ability. The action of *B. thuringiensis* starts at 10-15 hours after the larval feeding on the sprayed leaves and symptoms were clearly visible after 24 hours of feeding. The larvae steadily become inactive and lethargic, and cease feeding due to the loss of appetite followed with the discoloration of the integument. The larvae at first, develop light brown colour, shrink in size and finally turn dark brown with concomitant larval death (Heimpel and Angus, 1963). Application of *B. thuringiensis* is still the most effective option for managing the insect pests of trees in plantations or clonal seed orchards (CSOs) of some major timber species. However, its large scale application has not been possible due to the higher cost of bacterial products, which inhibits the government agencies, under whom most of the forest areas are there in the country. *Bacillus thuringiensis* var. *kurstaki* products were evaluated against the forest insect pests.

During initial evaluations as early as in 1978, Singh and Mishra (1978) tested Thuricide® against 26 forest insect pests in laboratory and found 0.5% of potency 30X10^9 spores/g killed 80–90% larvae of *Arthroschita hilaralis, Ascotis infixaria, A. selenaria, Atteva eridantis, Diphania laticos, Dirades theclata, Eutectona machaeralis, H. puera, Metanastria latipennis, Plecoptera reflexa, Selepa celtis, Sylepta derogata* and *Tyspanodes linealis.* Mishra and Singh (1993) reported bioassays of Thuricide® against 12 lepidopteran insect pests of forest trees.

Invariably, foliar spraying of *B. thuringiensis* var. *thuringiensis* and *B. thuringiensis* var. *kurstaki* (LDC or Dipel) have been found to be effective to kill the larvae of teak defoliator, *H. puera.* Its 1.0% concentration in water proved highly effective to kill even an average of 93.33 to 98.33% last instar larvae within 72 h of the spray (Roychoudhury *et al.*, 1994; Joshi *et al.*, 2001a.b). Foliar spraying of *Bacillus thuringiensis* var. *kurstaki* 1.0% was recommended against the *sal* defoliator after testing its effectiveness in the laboratory. *Sal* defoliators consisting of more than four species, viz. *Ascotis selenaria, Ophiusa janata, O. tirhaca* and *Paectes subapicali* had caused an epidemic in *sal* dominating forest areas in central India (unpublished). *B. thuringiensis* var. *kurstaki* 1.0% proved highly effective to kill 80.0% full grown larvae of *A. fabriciella* within 72 hrs of the spray when treated in combination with malathion 0.07%. Joshi *et al.* (2001a) has discussed available reports on evaluation of *B.t.* products against major forest insect pests, including teak defoliator and skeletonizer.

Besides *Bacillus*, an actinomycete product spinosad 45 SC @ 0.01% was effective against *E. machaeralis* (Kulkarni, 2009a,b) and *Atteva fabriciella* (Roychoudhury *et al.*, 2009) in lab-cum-evaluation. Recently, some biopesticides including the product of bacillus and actinomycetes were also evaluated against teak defoliator, *H. puera* and skeletonizer, *E. machaeralis* (Kulkarni, 2009). Field-cum-laboratory experiment with Agropest *bt*® caused mean corrected mortality varying from 59.26–77.78%. Javaregowda (2005) tested commercial formulations of *B. thuringiensis* like Delfin®, Biobit®, Dipel® and

Halt® for comparative pathogenicity against *H. puera* at the dose of 0.1, 0.2, 0.5, 1.0 and 1.5 g/l, and showed that higher doses are required for causing mortality in last instar larvae. Maximum mortality in the present study was observed in 1-2 days, perhaps due to the time required for bacterial spore to germinate and colonise the midgut (Briggs, 1963, Dulmage and Rhodes, 1971). While *B.t.* products are an established option for managing insect pests in economically important younger plantations, yet cost of products inhibits the forest managers to go for its large scale application.

Viral Pathogens

Nuclear polyhedrosis viruses (NPV) have also been promising alternatives against many forest insect pests. In India, NPVs have been recorded from the larvae of *Pericallis recini* (Kumar and Jayraj, 1978; Narayan and Jayaraj, 1979; Nair and Jacob, 1980), *Diacrisia obliqua* (Prasad and Chand, 1981), poplar defoliator, *Clostera (Pygaera) fulgurita* (Ahmed and Sen-Sarma, 1983), larvae of *Catopsilia crocale* (Ahmed, 1994), larvae of teak defoliator, *H. puera* (Mohammad Ali *et al.*, 1992) and willow defoliator, *Lymantria obfuscate*. There have been very few investigations on efficacy of viruses against forest insect pests.

The infection of NPV of the teak defoliator, *H. puera* under field conditions was reported from Nilambur, Kerala by Sudheendrakumar *et al.* (1988), suggesting HpNPV as quick acting and causing significant mortality of larvae within 3 days. Early laboratory and field studies showed that the virus is highly pathogenic and there was no cross infectivity to larvae of *E. machaeralis, Eligma narcissus, Atteva fabriciella, Helicoverpa armigera, Spodoptera litura, Amsacta albistriga* and *Bombyx mori* (Rabindra *et al.*, 1997; Sudheendrakumar and Biji, 2009). Disease transmission studies (Nair *et al.*, 1998) indicated intra-host persistence of the virus and transovarian transmission from parental generation up to the second filial generation under controlled laboratory conditions. Preliminary field studies carried out at Nilambur in the year 1993, for HpNPV against the teak defoliator showed that one-time foliar application of a crude preparation of HpNPV at the rate of 105 POBs per ml of the spray fluid, at the earliest sign of each infestation, gave 70 to 76% protection of foliage during the first two infestations. In protected trees, a basal area increment was enhanced by 41%, indicating the efficacy of HpNPV as a biocontrol agent against teak defoliator (Nair and Mohandas, 1996). Out of seven types of formulations prepared,

Sudheendrakumar *et al.*, (2006) reported freeze dried wettable powder formulation with activated charcoal as UV protectant to be the most efficient from the point of view of virulence, sunlight stability and field efficacy and storage life of about 18 months with retention of a high percentage of original activity. Biosafety tests of freeze dried HpNPV formulation and an Indian process patent has been obtained for the development of the freeze dried formulation of HpNPV or now a commercial product named Hybcheck®, which has recommended field dose of 1×10^{10} POBs/ha to target third instar larvae. It may not be economical to target much older larvae as it may incur higher dose of the virus. This biopesticide is a new addition to the repertoire of measures undertaken in tropical forestry to reduce environmental hazards (Sajeev *et al.,* 2007, Sudheendra Kumar and Biji, 2009). Javaregowda (2005) also tested HpNPV having an potency of $5x10^{10}$ POB/g, obtained from KFRI, Kerala in concentrations of $1x10^6$, $1x10^5$, $1x10^4$, $1x10^3$ and $1x10^2$ POBs/ml.

Fungal Pathogens

Fungi are known as insect pathogens since eighteenth century when Bassi (1835) recorded *Beauveria bassiana* on *Bombyx mori.* Patil (1981) also reported its pathogenicity to teak skeletonizer, *E. machaeralis.* Rajak *et al.* (1993) and Sandhu *et al.* (1993) reported, that the fungus *B. bassiana, Metarhizium anisopliae* and *Nomuraea rileyi* are epizootic strains causing epizootics in the larvae of *H. puera.* Detailed history of fungi as microbial control against forest insect pests has been compiled by Joshi *et al.* (2001a,b), who has discussed evaluations of 21 entomopathogenic fungi against around 40 insect pests.

Laboratory and field evaluation of biopesticides Biopro(R) (containing spores of *B. bassiana, Poecilamyes famisorosueus* and *Entomophaga grilli* @ 1.5×10^8 s/g) and Bioseal(R) (containing spores of *M. anisopliae, E. grilli* and *Aschersonia afeyrodis* at the rate of 2×10^9 s/g) in combination with soil insecticide methyl-parathion 2% DP with one week old white grubs proved best to kill 100% grubs of the white grub, *Holotrichia rustica* (Kulkarni and Joshi, 2005). The larvae of teak defoliator, *H. puera* and skeletonizer, *E. machaeralis* were also observed to be infested with entomopathogenic fungi, *B. bassiana, M. anisopliae* and *Acremonium zeylanicum* in the nature during monsoon. They killed 40.65 to 73.15% of the larvae of above defoliators (Khan *et al.*, 1988; Rajak *et al.*, 1993; Sharma *et al.*, 2005).

In all, five species of entomopathogenic fungi, viz. *B. bassiana, M. anisopliae, A. zeylanicum, Fusarium oxysporum* and *F. fusariodes* were identified to attack the grubs of *sal* heartwood borer. The pathogenicity of these fungi was evaluated in the laboratory and it was discovered that *B. bassiana* 1.7 × 10^7 spores were highly effective followed by *M. anisopliae* 1.5 × 10^7 spores to kill 75.19 and 64.08% young grubs of *sal* heartwood borer, respectively (Sharma and Joshi, 2007; Joshi, 2009). Javaregowda (2005) has reported efficacy of monoculture of *B. bassiana, M. anisopliae* and *N. releyi* grown on Subouraud's maltose yeast extract medium containing peptone 1%, maltose 4%, yeast extract 1% and agar in distilled water at pH 6.8 at the dose of 0.5 and 1.0 g/l of distilled water so as to get the concentration of 1 × 10^4 and 1 × 10^5 spores/ml. Susceptibility of first to fifth instars larvae was evaluated by spraying 10 ml of suspension using an atomizer. It was found that maximum mortality of 100% occurred within 3 days in first, second and third instar larvae of *H. puera*. The percentage mortality decreased and the time taken for death increased as the age of the larvae advanced indicating that older larvae were relatively more tolerant to the infection.

Parasitoids and Predators

Plantations or natural forests offer scope only for biological control insect's own natural enemies, which are capable of vertical movement along the height of trees. Collection, identification of native strains of potential insect parasitoids, their multiplication and release in the epidemic areas of some potential parasitoids are the best options (Kulkarni *et al.*, 2004; Yousuf *et al.*, 2004). The trichogrammatid egg parasitoids have shown great promises, for which initial screenings were made through some exotic species. Sudheendrakumar *et al.* (1995) reported a successful parasitisation of *H. puera* eggs by *T. embryophagum and T. dendrolini* under laboratory conditions. Further, Patil and Mohan Naik (1998) recorded higher parasitisation of eggs of *H. puera* by *T. chilonis* under field conditions. At Tropical Forest Research Institute, Jabalpur, Field experiments were conducted with four exotic egg parasitoids, viz. *Trichogramma brasiliensis, T. chilonis, T. japonicum and T. pretiosum* to investigate the lowest highly effective quantity of wasps of egg parasitoid by liberating 0.25–3 lakh wasps/ha to minimise the attack of skeletonizer during the years 2001 and 2002 in teak forests at Udaipur in Bijadandi range of Mandla forest division. The result showed that liberating 1.25 lakh wasps/ha and above of *Trichogramma brasiliensis* was effective to minimise the attack of *E. machaeralis* (Joshi *et al.*, 2001a,b, 2007).

The exotics were though effective but unable to sustain in released teak plantations and forests. Therefore, an indigenous egg parasitoid, was collected from teak forests, identified as *T. raoi*, multiplied, tested in laboratory and finally evaluated for field efficacy by supplementing in 50 ha area of Compartment Nos. 240 and 243 in Kalpi range and 50 ha area of Compartment Nos. 242 and 149 in Bijadandi range of Mandla Forest Division of Madhya Pradesh by releasing 1.25 lakhs/ha in September–October. The quantitative estimation of damaged leaves in 50 ha area of Kalpi and Bijadandi ranges was compared with non- released sites in November before leaf shedding. The result showed that the average percentage of leaf skeletonized by larvae in egg parasitoid released site was only 57.32 and 48.75% as compared to 89.97 and 83.40% skeletonized leaves in non-released sites respectively. The same experiments were repeated in 50 hec teak plantation of compartment no. 396 of Mamla range under West Chanda Forest Product Division, Chandrapur and 50 ha area in Compartment No. 160 of Sonegaon range, Bhandara Forest Project Division. The egg parasitoid, *T. raoi* at the rate of 1.25 lakh/ha was released in 4 instalments between August to September. The observations on percentage of sleletonized leaves were taken in the 3rd week of October. The results showed that the average percentage of skeletonized leaves in parasitoid released site was 50.50 and 34.19% as compared to 93.45 and 89.21% skeletonized leaves in non released sites, respectively. Thus, it was concluded, that the supplementing egg parasitoid *T. raoi* alone in teak plantations or forests at the rate of 1.25 lakh wasps was sufficient to minimise at least half of the skeletonized leaves in released sites resulting in saving of half of the annual loss caused. The cost of releasing egg parasitoid, *T. raoi* is calculated to ₹1000/ha only.

Similar field releases were also made in Teak Seed Orchards (TSOs) during the year 2004-05 in Madhya Pradesh. The institute provided the technical expertise as well as nearly 147 lakh egg parasitoids to be released in 117 ha Teak Seed Orchard (Joshi *et al.*, 2001b, 2007) against teak defoliator, *H. puera* and skeletonizer, *E. machaeralis*. Javaregowda (2005) tested five species of *Trichogramma*; *T. acheae, T. japanicum, T. chilonis, T. pretiosum* and *T. dendrolini* and reported that all accepted the eggs of *H. puera*. However, fairly good parasitization (55.2, 47.0 and 45.0%, respectively) was found with *T. chilonis, T. acheae* and *T. Japanicum* on freshly laid eggs and 22.0, 19.0 and 13.0% on one day old eggs, respectively. So far, there are no reports on such parasitism of these species on *H. puera* eggs under laboratory conditions. Fresh eggs were preferred by all the five species of *Trichogramma* parasitoids.

Entomopathogenic Nematodes (EPNs)

The rhabditid EPNs, as candidate biological control agents were discovered and described prior to 1990. Starting from the first record of EPN *Steinernema* (*Aplectana*) *kraussei*, parasitizing the sawfly, *Cephalcia* (*Lyda*) species (Steiner, 1923, 1929) in nature, few more EPN species had been found parasitizing insect pests (*N. affine* by Bovien, 1937; *S. feltiae* by Filipjev, 1934; *N. carpocapsae* by Weiser, 1955). The biological control potential of steinernematid EPN was first published by Glaser (1932) with *S. glaseri* which he found were safe to useful insects and plants (Glaser, 1940). Another genus *Heterorhabditis* was established by Poinar (1976) when he described *Heterorhabditis bacteriophora* found parasitizing *Helicoverpa punctigera* in Australia. Eleven years later another *Heterorhabditis* species, *H. megidis* was described from the important scarab *Popilla japonica* in Ohio, USA (Poinar *et al.*, 1987). Later, 49 species of *Steinernema* and 12 species of *Heterorhabditis* were described (Nguyen, 2006). Bedding (2006) has excellently reviewed the EPNs research stating from their discovery to utilization.

The work on nematodes as biocontrol agents in India was limited to mermethrids (Iyengar, 1927, 1938) till 1966, which were not considered promising as their multiplication was not possible in laboratory and it was slower. Mathur *et al.* (1966) also reported a few species of the rhabditid nematodes, other than Steinermatids and Heterorhabditis. The investigations in India on rhabditid EPNs were initiated with the identification and potentiality of DD136 strain of *S. glaseri* in 1966 (Rao and Manjunath, 1966). Since then the potentiality of rhabditid EPNs was well recognized and it led to many isolations and evaluations from all over the country (Hussaini *et al.*, 2003). Kulkarni *et al.* (2008) have compiled current status on identification and application of EPNs against insect pests of agriculture and plantation crops in the country. The need to isolate native strains/populations/species for economical mass-production and better efficacy has been stressed in almost all the available reviews and publications relating to the EPNs along with their promising role in the management of insect pests of agricultural and forestry pests (Hussaini *et al.*, 2003; Kulkarni *et al.*, 2008).

The countable number of evaluations against the forest insect pests include; ulmus bark boring beetles, *Scolytus scolytus* (Finney and Walker, 1979), field efficacy of *S. kraussei* against spruce pest *Cephalcia abietis* (Mracek and David, 1986), salix wood boring beetles, *Anoplophora glabripennis* (Qin *et al.*, 1988), cotoneaster and taxus root grubs, *Rhizotrogus majalis* (Wright *et al.*, 1988; Nielsen and Cowles, 1998), tilia, gleditsia, fraxinus, prunus, malus, taxus root-feeding grubs, *Popillia japonica* (Mannion *et al.*, 2001), and vine weevil *Otiorhynchus sulcatus* (Lola-Luz *et al.*, 2005). Besides these, some important laboratory and field evaluations against insect pests of forestry and plantation crops have been summarized as Table 14.1. Notably there have been no reports on the evaluation of entomopathogenic nematodes against forest insect pests in India, except on insect pests of plantation crops like coconut, arecanut, palm and cashew (Sosamma, 2003), teak (*Tectona grandis*) (Kulkarni *et al.*, 2011).

Constraints

Unlike agriculture, forestry is a complicated ecosystem not only by virtue of inapproachability of many forest areas but also by the long gestation period crop. In consonance, many of the insect pests attacking forestry species have complex behaviour patterns and life cycle extending from one to four years. Such conditions offer most suitable platform for integration of the available options for the management of insect pests. In spite of extensive research carried out in the past and ongoing current research programmes, it is a fact that routine insect pest management is not taking place anywhere in the country. While we realize that there are irreversible trend of conversion of forest land into agricultural land and then in due course to urbanization due to population pressure, we also know that the increasing demands of the forestry products can be met only be judiciously increasing the productivity of the limited forest lands. The forest productivity can be enhanced through the effective pest management.

Unlike agriculture, which are private lands, forests are under the direct control of governments and thus many factors govern the management decisions other the than the simple economics. It also includes a natural hesitation to adopt to the newer techniques, which are although, more ecofriendly but might involve cost implications.

The network of forestry extension has not been very powerful in the past. Although it is now being strengthened by the Indian Council Forest Forestry Research by way of opening Van Vigyan Kendra (VVKs) in line with the Krishi Vigyan Kendras (KVKs).

Table 14.1. Entomopathogenic nematodes (EPNs) against the insect pests of forestry and plantation crops in India and abroad (*Source*: Kulkarni *et al.*, 2008)

EPN Species	Host/ Plantations crop/ country	Insect pests
S. feltiae, S. carpocapsae (2 populations), *Heterorhabditis* spp. (2 populations)	Poplar in Australia	Xylophagous insect pests of poplar, *Paranthrene tabaniformis* and *Cryptorhynchus lapathi*
S. bibionis, S. feltiae, S. glaseri, H. bacteriophora	Wood borers and tenthredinids	*Anoplophora glabripennis, A. nobilis,* Cossus, *Pristiphora* sp. and *Arge przhevalskii*
S. bibionis 3 populations.	Forest tree hosts	Cerambycid, *Anoplophora glabripennis* and Cossid, *Holcocerus insularis* in laboratory.
S. carpocapsae, S. bibionis, S. feltiae, S. glaseri, H. heliothidis (10 populations)	Bamboos in southern China	Bamboo weevils, *Cyrtotrachelus longimanus* in field.
Steinernema spp.	Spruce in Canada	Spruce budworm, *Zeiraphera canadensis* in field
S. carpocapsae *H. bacteriophora*	Forest insect pests	*Agelastia coerulea, Pryeria sinica* and *Glyphodes perspectalis* in laboratory
EPN species	Forest nurseries	--
Heterorhabditis sp. *S. carpocapsae* *S. feltiae* *S. krausseri*	Forest Ecosystem of Pecan & Abies in Italy	Spruce web-spining sawfly, *Cephalia arvensis* in field experiment
Heterorhabditis sp.	*Taxus baccata, Rhododendron* spp. and Strawberries in Germany	*Otiorhynechus sulcatus* in open field stands
S. carpocapsae	Forest nurseries in Canada	Black vine weevil, *O. sulcatus*
H. megidis, H. downesi *H. heliothidis*	Strawberry plantation Potted plants	Black vine weevil, *O. sulcatus* Black vine weevil, *O. sulcatus*
S. carpocapsae	Conifer species	*Hylobius congener*
Steinernema spp., *Heterorhabditis* spp.	Forestry crops in Korea	Forest insect pests in laboratory
EPN species	Pines	*Hylobius abietis*
S. carpocapsae *S. glaseri,* *S. feltiae* *H. bacteriophora* EPN species	Lichi trees, China Cranberry plantation	*Comorilis albicapilla* (Yponomeutidae) Cranberry girdler, *Chrysoteuchia topiaria*

S. carpocapsae *H. bacteriophora*	Chest nut insect pests, Korea	*Curculio, sikkimensis, Dichocrocis punctiferalis, Cydia kurokoi* in laboratory
H. megidis, *S. carpocapsae*	Bitka spruce, *Bicea sitchensis*	Large pine weevil, *Hylobius abietis* in laboratory
S. carpocapsae *S. ribobrave* *H. bacteriophora* *H. indica*	--	Subterranean termites, *Reticulitermes flavipes* and *Coptotermes formosanus*
S. feltiae	*Pinus halepensis* reforestation sites in Italy	Pine procersionary larvae, *Thauntopoea pityocampa*
H.bacteriophora. *S. carpocapsae*	Pecan orchards in U.S.	Pecan weevil, *Curculio caryae* in laboratory
S. glaseri, S. rarum and nine species and 15 strains of EPNs *S. carpocapsae,* *S. riobrave* strains	Stone and pome fruit, North America	Plum curculio, *Conotrachelus nenuphar*
S. carpocapsae *S. feltiae* *S. longicadam* *H. bacteriophora*	Eucalyptus in China	*Odontoterms formosanus*
S. feltiae, *S. affinae,* *S. carpocapsae,* *H. megidis,* *H. bacteriophora*	Urban forestry, Poland	Winter moth Geometrids, *Operoptera brumata* and *O. fagata* in laboratory
Heterorhabditis indica 128 populations, *Steinernema* sp. 3 populations	Coconut, Arecanut, Oil Palm and Cashew in India.	Nine Important insect pests

Source: Kulkarni *et al.* (2008)

CONCLUSIONS

The management of insect pests of forest trees is not an easy task, especially for the pests infesting old plantations and natural forests. It is due to various limiting factors of the application of any single method of insect pest management. The necessity for developing biorational and ecofriendly methods of the insect pest management in forestry further makes the task more challenging.

Development of IPM models requires comprehensive knowledge of the reasons for pest outbreaks and, further, an understanding of which process can be manipulated to reduce the severity of any outbreaks. In recent years, many individual IPM components have been evaluated against variety of insect pests. They have certainly helped to generate species-specific data on optimum doses related aspects, but requirement to develop integrated models based on available base-line data on the individual insect pests, to avert their epidemics. The present review reveals that, while preliminary investigations in this directions have already been initiated, despite all odds, more in-depth understanding of insect-host bioecological interrelationship, specific to each economically important insect/host species in forestry is need of the hour. The investigations are particularly essential on the major insect pests of region-specific principal plantation tree species in the country, like *Tectona grandis, Eucalyptus* spp. *Eucalyptus globulus, E. grandis* and *E. tereticornis;* acacias, *Acacia auriculiformis, A. catechu, A. mearnsii, A. nilotica* and *A. tortalis, Albizia* spp., *Azadirachta indica, Casuarina equisetifolia, Dalbergia sissoo, Gmelina arborea, Populus* spp., *Prosopis* spp., *Shorea robusta* and *Terminalia* spp, conifers, *Cedrus deodara* and *Pinus roxburghii.*

While the concepts of IPM are intuitively sound, the practical implementation of these concepts to reach a successful conclusion is not always so easily achieved. In fact, case studies to illustrate IPM success in forestry are relatively few, if a strict definition of 'integrated' is adopted, such that there is a requirement for a multifaceted approach across a range of disciplines. In reality, although there are multiple variables to contend with, management will tend to rely on one or two key elements to achieve the pest reduction. More detailed investigations on the individual major insect pests as a base-line data for planning and modelling the IPM, can be expected in the near future.

REFERENCES

Ahmad, M. (1991) Natural resistance in teak clones to leaf skeletonizer, *Eutectona machaeralis* (Lepidoptera : Pyralidae). *Indian J. Fores.*, **14**, 228–231.

Ahmad, M. (1993) Relative resistance in different clones of *Populus deltoides* to poplar defoliator, *Clostera cupreata* (Lepidoptera : Notodontidae). *Ann. Fores.*, **1**, 1–12.

Ahmed, S.I (1994) Use of entomopathogenic viruses in forestry: potential for their use in developing nations. *Antenna*, **18**, 14–19.

Ahmed, S.I. and Sen-Sarma, P.K. (1983) Investigations on newly recorded nuclear polyhedrosis virus of *Pygaera fulgurita* Walk. (Notodontidae: Lepidoptera). I. Nature of disease and insect virus ineraction. *J. Ent. Res.*, **7**, 150–153.

Ananthakrishnan, T.N. (1990). Facets of chemical ecology in insect plant interactions. An overview. *Proc. Indian Acad. Sci.*, **99**, 177–183.

Atkinson, D.J. (1936) *On the Biology of Hapalia machaeralis Wlk.* Indian Forest Record (N.S.), Govt. of India Publication, New Delhi.

Bajwa, W.I. and Kogan, M. (2004). Cultural Practices: Springboard to IPM. In: O. Koul, G.S. Dhaliwal and G.W. Cuperus (eds.), *Integrated Pest Management: Potential, Challenges and Constraints*, CABI Publishing, Wallington, pp. 21-38.

Bassi, A. (1835) Del mal del sengno, calcinaccio o moscardino, malattia che affligge I bachi de seta e sul modo di liberarne le bigattaeie anche le piu infestate. Part I, teoria. Orcesi, Lodi.

Bedding, R. (2006) Entomopathogenic nematodes from discovery to application. *Biopestic. Int.*, **2**, 87–119.

Beeson, C.F.C. (1941) The Eecology and Control of the Forest Insects of India and Neighbouring Countries. Vasant Press, Dehradun.

Berryman, A.A. (1988) Towards a unified theory of plant defence. In: W.J. Mattson, J. Levieux, and C. Bernard-Dagan (eds.), *Mechanisms of Woody Plant Defence against Insects*, Springer-Verlag, New York, pp. 39–55.

Bovien, P. (1937) Some type of association between nematodes and insects. *Vidensk. Meddel. Dansk. Naturhistor. For.*, **101**, 1–114.

Brigges, J.D. (1963) Commercial production of insect pathogens. In: E.A. Steinhauss (ed.), *Insect Pathology, An Advanced Treatise*, Academic Press, London, pp. 519–548.

Chopra, R.N. (1928) *Annual Report of Department of Agriculture*, 1925-26. Punjab, Lyallpur **1,** 67–125.

Choubey, V., Kulkarni, N. and Bhandari, R. (2004) Incidence of seed insect pests of *sal* (*Shorea robusta* Gaertn. F.). *Indian J. Trop. Biodiver.*, **12,** 48–52.

Choubey, V., Kulkarni, N. and Bhandari, R. (2008) Status of *sal* (*Shorea robusta*) seed insect pests in six sal dominating regions of central Indian state of Madhya Pradesh. *Indian J.* Fores., **31**, 595–598.

Dulmage, H.T. and Rhodes, R.A. (1971) Production of pathogens in artificial media. In: H.O.D. Burges and N.W. Hussey (eds.), *Microbial Control of Insects and Mites*, Academic Press. New York, pp. 507–540.

Emden, H.F. van (1973) *Insect/Plant Relationships*. John Wiley & Sons, New York.

FAO (2005) *Global Forest Resources Assessment 2005 – India – Country Report.* Forestry Department, Forest Resources Assessment 2005, Country Report 001, FAO, Rome.

FAO (2006) Global Forest Resources Assessment 2005 – Progress towards Sustainable Forest Management. Forestry Paper No. 147, FAO, Rome.

FAO (2007) *Overview of the Forest Pests: India,* Forest Health and Biosecurity Working Papers, Working Paper FBS/18E, Forest Management Division FAO, Rome. http://www.fao.org/docrep/012/ al007e/al007e00.pdf

FAO (2011) *Guide to Implementation of Phytosanitary Standards in Forestry,* FAO Forestry Paper 164, Food and Agriculture Organization of United Nations, FAO, Rome. http://www.fao.org/docrep/013/i2080e/i2080e00.htm

Filipjev, I.N. (1934) Miscellanea Nematologica 1. Eine neue Art der Gattung *Neoaplectana* Steiner nebst Bermerkungen über die systematische Stellung der letzteren. *Magazine de Parasitologie du Musée Zoologique de l'Académie des Sciences de l'USSR,* 4, 229–240.

Finney, J.R. and Walker, C. (1979) Assessment of field trial using the DD-136 strain for *Neoplectana* sp. for the control of *Scolytus scolytus. J. Invertebr. Pathol.*, **33,** 234–241.

Fletcher, T.B. and Ghosh, C.C. (1919) Reports of the Proceedings of the 3rd Entomological Meeting held at PUSA, 1919 :14. Borers in Sugarcane rice: Stored grain pests, 712–761.

Gahukar, R.T. (2010) Bioefficacy of indigenous plant products against pests and diseases of Indian forest trees: a review. *J. Fores. Res.*, **21**, 231–238.

Glaser, R.W. (1932) Studies on *Neoaplectana glaseri*, a Nematode Parasite of the Japanese Beetle (*Popillia japonica*). Circular 211, New Jersey Department of Agriculture, Treton, NJ.

Glaser, R.W. (1940) The bacteria-free culture of a nematode parasite. *Proc. Soc. Exp. Biol. Med.,* **43**, 512–514.

Gerhold, H.D., Schreiner, E.J., McDermott, R.E. and Winieski, J.A. (1966) *Breeding Pest-Resistant Trees.* Pergamon Press, London.

Hanover, J. (1980) Breeding forest trees resistant to insects. In: F.G. Maxwell and P.R. Jennings (eds.), *Breeding Plants Resistance to Insects,* John Wiley & Sons, New York, pp. 487–511.

Heimpal, A.M. and Angus, T.A. (1963) Disease caused by certain spore forming bacteria. In: F.A. Steinhause (ed.), *Insect Pathology, An Advanced Treatise.*, Academic press, New York, pp. 21–73.

Hussaini, S.S., Rabindra, R.J. and Nagesh, M. (2003) *Current Status of Research on Entomopathogenic Nematodes in India.* Project Directorate of Biological Control, PDBC, Bangalore, India.

Hussaini, S.S., Singh, S.P. and Shakeela, V. (2001) Influence of temperature on infectivity of entomopathogenic nematodes to black cutworm, *Agrotis ipsilon* (Hufnagel) larvae. *National Congress on Centenary of Nematology in India – Appraisal and Future Plans*, Indian Agricultural Research Institute, New Delhi, 5-7 Dec. 2001, pp.120.

Iyengar, M.O.T. (1927) Parasitic nematodes of *Anopheles* in Bengal. *Eastern Assoc. Trop. Med. Trans.*, **3**, 128–135.

Iyengar, M.O.T. (1938) Natural parasites of mosquitoes in India. *Proc. Nat. Insect Sci.*, **4**, 237–239.

Javaregowda (2005) Studies on the seasonal incidence, biology and management of teak defoliator, *Hyblaea puera* Cramer (Hyblaeidae: Lepidoptera) in Uttara Kannada district of Karanataka. *Ph.D* Thesis in Agricultural Entomology, University of Agricultural Sciences, Dharwad, Karnataka, India, 169p.

Joshi, K.C. (2009) Biology, life history of sal heartwood borer, factors responsible for outbreak and its management – an updated information. *Environ. Fores. Bull.*, **9**, 1–8.

Joshi, K.C. and Namdeo, R.K. (1990) *Common Forest Insect Pests and their Control Measures.* IDF *Technical Bulletin* No. 3, 1990, Tropical Forest Research Institute, Jabalpur, India.

Joshi, K.C., Meshram, P.B, Ushakiran and Bhajan, Ram (1990) A note on the insect pests of economically important stored forest seeds from N.E. and Central India and possibilities of their control measures. *J. Trop. Fores.*, **6**, 261–263.

Joshi, K.C., Roychoudhury, N. and Sharma, N. (2001a) Microbial pesticides for forest insect control. In: P.K. Shukla and K.C. Joshi (eds.), *Recent Trends in Insect Pest Control to Enhance Forest Productivity*, Tropical Forest Research Institute, Jabalpur, pp. 61–84.

Joshi, K.C., Roychoudhury, N., Kulkarni, N. and Sambath, S. (2001b) *Implementation Completion Report of World Bank Forestry Research, Education and Extension Project (FREEP)*, entitled Entomology (04). Tropical Forest Research Institute, Jabalpur.

Joshi, K.C., Roychoudhury, N., Kulkarni, N. and Sambath, S. (2006) *Sal* heartwood borer in Madhya Pradesh. *Indian Forester*, **132**, 799–808.

Joshi, K.C., Sambath, S., Yousuf, M., Chander, S., Roychoudhury, N. and Kulkarni, N. (2007) Evaluation of *Trichogramma* spp. to minimize the attack of teak leaf skeletonizer. *Indian Forester*, **133**, 527–533.

Kaushik, R.C. (1956) Teli variety of *Tectona grandis*. *Proceedings of the Ninth Silvicultural Conference*, Forest Research Institute, Dehradun, Part 2, pp. 92-94.

Kawanashi, and Held, (1990) Viruses and bacteria as sources of insecticides. In: E. Hodgeson and R.J. Khar (eds.), *Safer Insecticides: Development and Use*, Marcel Dekker, New York, pp.351–383.

Kedharnath, S. (1984) Forest tree improvement in India. *Proc. Indian Acad. Sci. (Plant Sci.)*, **93**, 401–412.

Kedharnath, S. and Singh, P. (1975) Studies on natural variation in susceptibility of *Tectona* to leaf skeletonizer, *Pyrausta machaeralis* (Lepidoptera : Pyralidae). Paper presented in FAO/IUFRO Symposium on Forest Diseases and Insects, 7-12 April, 1975, New Delhi.

Khan, A.K., Agrawal, G.P., Rajak, R.C. and Sandhu, S.S. (1988) Microbial control agents of Insect Pests of Forests at Jabalpur. *Proceedings Seminar on Perspectives Plan for Institute of Deciduous Forests and Research Priorities*, November 2-3, 1988, Institute of Deciduous Forests, Jabalpur.

Kishwan, J., Pandey, R. and Dadhwal, V.K. (2009) *India's Forest and Tree Cover: Contribution as a Carbon Sink*. Technical Paper 130, (ICFRE) BL-23, Jabalpur.

Koul, O., and Wahab, S. (2004) *Neem Today and the New Mellenium*. Kluwer Academic Publishers, Netherland.

Koul, O., Dhaliwal, G.S., Marwaha, S.S. and Arora, J.A. (2003) *Biopesticides and Pest Management*, Vol. II, Campus Books International, New Delhi, India.

Kulkarni, N. (1998) Feeding deterrence property of some plant extracts against poplar defolliator, *Clostera cupreata* Butler (Lepidoptera: Notodontidae). *Insect Environ.*, **3**, 98.

Kulkarni, N. (1999) Antifeedant and growth inhibitory effects of *Azadirachta indica* A. Juss, seeds against the larvae of *Spodoptera litura* (Fab.) Boursin (Noctuidae:Lepidoptera). *Nat. Acad. Sci. Lett.*, **22**, 24–26.

Kulkarni, N. (2001) Anti-insect bioactivities of some botanicals: their prospects as component of integrated pest control system. In: P.K. Shukla, and K.C. Joshi (eds.), *Recent Trends in Insect Pest Control to Enhance Forest Productivity*, Tropical Forest Research Institute, Jabalpur, pp. 95–137.

Kulkarni (2009a) Development of Model for the Management of White Grubsin Teak Nursery under the Concept of Integrated Pest Management, Final Technical Report, Tropical Forest Research Institute, Jabalpur, India.

Kulkarni (2009b) Evaluation of Biopesticidal Products against Teak Pests, Teak Defoliator; Hyblaea puera and Teak Skeletonizer, Eutectona machaeralis in Forest Nursery. Research Component Completion Report, Tropical Forest Research Institute, Jabalpur, India.

Kulkarni, N. (2010) Bioecology and management of white grub complex in teak forest nursery in India. In: M. Cram (ed.), Proceedings of the 7th Meeting of IUFRO Working Party 7.03.04 Diseases and Insects in Forest Nurseries, Hilo, Hawaii, USA, July 13to 17, 2009. On-line publication source: http://www.fs.fed.us/r8/ foresthealth/publications/nursery/ IUFRO_7_03_04_HiloProc.pdf , pp. 84–91.

Kulkarni, N. and Joshi, K.C. (1997) Insecticidal action of some plant extracts against Albizia defoliator, *Rhesala imparata* Walk. (Lepidoptera : Noctuidae. *Entomon*, 22, 75–79.

Kulkarni, N. and Joshi, K.C. (1998a) Feeding inhibition property of a common weed, *Ipomoea carnea* (Jacq.) ssp. *fistulosa* (Austin) against bamboo leaf roller, *Crypsiptya coclesalis* (Walk.) (Lepidoptera:Pyralidae). *Allelopathy J.*, 5, 93–96.

Kulkarni, N. and Joshi, K.C. (1998b) Antifeedant property of some botanical extracts against Maharukh webworm, *Atteva fabriciella* Swed. (Lepidoptera : Yponomeutidae). *J. Trop. Forest Prod.*, 4, 11–16.

Kulkarni, N. and Joshi, K.C. (1998c) Insect pests of forest tree seeds: their economic impact and control measures. *J. Trop. Forest Sci.*, 10, 438–455.

Kulkarni, N. and Joshi, K.C. (2005) Efficacy of *Beauveria bassiana* and *Metarhizium anisopliae* in combination with M-parathion against white grubs of *Holotrichia rustica* Burm. (Coleoptera : Scarabaeidae). *Paper presented in Biopesticide Conference, Biopestidies: Emerging Trends BET 2005*, Nov. 11th–13th, 2005, Institute of Himalayan Biotechnology, Palampur, India.

Kulkarni, N., Joshi, K.C. and Meshram, P.B. (1996a) Bioactivity of methanolic neem seed extract against the teak skeletonizer, *Eutectona machaeralis* Walk. (Lepidoptera : Pyralidae). *J. Environ. Biol.*, **17**, 189–195.

Kulkarni, N., Joshi, K.C. and Ramarao, N. (1996b) Screening of some plant extracts for feeding inhibition property against major forest insect pests. *Myforest*, **32**, 118–127.

Kulkarni, N., Joshi, K.C. and Gupta, B.N. (1997a) Antifeedant property of *Lantana camara* var. *aculeata* and *Aloe vera* leaves against teak skeletonizer, *Eutectona machaeralis* Walk. (Lepidoptera : Pyralidae). *Entomon*, **22**, 61–65.

Kulkarni, N., Joshi, K.C. and Kalia, S. (1997b) Feeding inhibition property of some botanical extracts against sissoo defoliator, *Plecoptera reflexa* (Lepidoptera:Noctuidae). *Indian J. Fores.*, **20**, 390–394.

Kulkarni, N., Joshi, K.C. and Lal, R.B. (1998) Plant pesticides: their biological activities against forest insect pests. *Adv. Fores. Res.*, 18, 161–182.

Kulkarni, N., Joshi, K.C. and Shukla, P.K. (2003). Antifeedant activity of *Annona squamosa* Linn. against *Crypsiptya coclesalis* Linn. (Lepidoptera : Pyralidae), *Entomon*, **28**, 389 392.

Kulkarni, N., Joshi, K.C. and Shukla, P.K. (2004a). Integrated Insect Pest Management of forest insect pests. In: G.T. Gujar, (ed.), *Contemporary Trends in Insect Science*, Campus Books, New Delhi, pp. 370–410.

Kulkarni, N., Tripathi, S. and Joshi, K.C. (2004b) Kairomonal activity of compounds isolated from bark of sal (*Shorea robusta* Gaert. f.) for attracting the sal heartwood borer, *Hoplocerambyx spinicornis* Newman (Coleoptera: Cerambycidae). *Indian J. Fores.*, **27**, 321–325

Kulkarni, N., Soni, K.K. and Joshi, K.C. (2007a) Biotic and abiotic factors responsible for outbreak of the sal heartwood borer, *Hoplocerambyx spinicornis* Newman in Achanakmar-Amarkantak biosphere reserve. In: K.C. Joshi and A.K. Mandal (eds.), *Research Needs for Achanakmar-Amarkantak Biosphere Reserve*, Tropical Forest Research Institute, Jabalpur (M.P.), India, pp.115–126.

Kulkarni, N., Chandra, K., Wagh, P.N., Joshi, K.C. and Singh, R.B. (2007b) Incidence and management of white grub, *Schizonycha ruficollis* on seedlings of teak (*Tectona grandis* Linn. f.). *Insect Sci.*, **14**, 411–418.

Kulkarni, N., Singh, N., Joshi, K.C. and Bhargava, A. (1995) Neem tree (*Azadirachta indica* A. Juss) as a potent source of biopesticide: application in forest insect control. *J. Trop. Fores.*, **13**, 166–174.

Kulkarni, N., Singh, N., Joshi, K.C. and Lal, R.B. (1999) Activity of *Lantana camara* var. *aculeata* extract against the larvae of bamboo leaf roller, *Crypsiptya coclesalis* (Walker) (Lepidoptera : Pyralidae). In: S. Ignacimuthu and Alok Sen (eds.), *Biopesticides in Insect Pest Management*, Phoenix Publishing House Pvt. Ltd., New Delhi, India, pp.118–123.

.Kulkarni N., Paunikar S., Hussaini S.S. and Joshi K.C. (2008) Entomopathogenic Nematodes in insect pest management of forestry and plantations crops: An appraisal. *Indian J. Trop. Biodiversity,* **16**, 155–166.

Kulkarni, N., S. Paunikar, K.C. Joshi, and John Rogers (2009) White grubs, *Holotrichia rustica* (Burm.) and *Holotrichia mucida* Gyll. (Coleoptera: Scarabaeidae) as pests of teak (*Tectona grandis* L. f.) seedlings in central India. *Insect Sci.*, **16**, 519–525.

Kulkarni, N., Paunikar, S. and Singh, R.B. (2010) Efficacy and field persistence of some insecticides against the white grub beetle, *Holotrichia rustica* (Burm.) (Scarabaeidae : Melolonthinae). *Indian J. Entomol.* (Accepted : in press).

Kulkarni, N., Paunikar, S. and Hussaini, S.S. (2011) Susceptibility of teak skeletonizer, *Eutectona machaeralis* (Walker) to the EPN, *Heterorhabditis indica* Poinar. *World J. Zool.*, **6**, 33–39.

Kumar, S. and Jayaraj, S. (1978) Pathogenicity and sites of infection of nuclear polyhedrosis virus of the castor hairy caterpillar, *Pericallia recini* Fab. *Indian J. Entomol.*, **40**, 469–470.

Lefroy, H.M. (1906) *Indian Insect Pests*. Reprinted 1990. Today & Tomorrow's Printers & Publishers, New Delhi.

Lefroy, H.M. (1909) *A Manual of the Insects of Plains (Tropical India) 1. Lepidoptera*. Indian Agricultural Research Institute, Pusa, New Delhi.

Lola-Luz, T., Downes, M. and Dunne, R. (2005) Control of Black vine larvae *Otiorhynchus sulcatus* (Fabricius) (Coleoptera :Curculionidae) in grow bags outdoors with nematodes. *Agric. Forest Entomol.*, **7**, 121–126.

Mannion, C.M., McLane, W., Klein, M.G., Moyeseenko, J., Oliver, J.B. and Cowan, D. (2001) Management of early-instar Japanese Beetles (Coleoptera : Scarabaeidae) in field grown nursery crops. *J. Econ. Entomol.*, **94**, 1151–1161.

Mathur, S.B., Srivastava, R.E. and Subba Rao, B.R. (1966) Records of nematodes parasitizing *Chilo zonellus* Stem borer of Maize, *Indian J. Entomol.*, **28**, 415.

Meshram, P.B. (1993) Relative resistance of certain clones of *Tecona grandis* to *Tectona grandis* Walk. *J. Appl. Zool. Res.*, **4**, 87–88.

Meshram, P.B., Joshi, K.C. and Sarkar, A.K. (1994) Relative resistance of certain clones of *Tectona grandis* to leaf skeletonizer, *Eutectona machaeralis* Walk. (Lepidoptera : Pyralidae). *Indian Forester*, **120**, 58–61.

Mishra, S.C. (1992) Comparative natural resistance of different clones of *Tectona grandis* L.f. to teak skeletonizer, *Eutectona machaeralis* Walk. (Lepidoptera : Pyralidae). *Indian Forester*, **118**, 274–278.

Mishra, R.M. and Singh, P. (1993) Bioassay of thuricide – A microbial insecticide against important forest insect pests, II. *Indian Forester*, **119**, 33–35.

Mohammad, Ali, M.I., Varma, R.V. and Sudheendrakumar, V.V. (1992) Microbial pathogens associated with forest insect pests in Kerala. In: T.N. Ananthakrishnan (ed.), *Emerging Trends in Biological Control of Phytophagous Insects*, Oxford and IBH Pubhishing Co. Pvt. Ltd., New Delhi, pp. 131–141.

Mracek, Z and David, L. (1986) Preliminary field control of *Cephalecia abietis* L. (Hynoptera: Pamphilidae) larvae with steinernematid nematodes in Czechoslovakia, *Z. Angew. Entomol.*, **102**, 260–263.

Nair, K.S.S (2007) Tropical Forest Insect Pests: Ecology, Impact and Management. Cambridge University Press, New York.

Nair, K.P.V. and Jacob, A. (1980) Influence of Nuclear Polyhedrosis on moulting in the larvae of *Pericallia recini* F. *Agric. Res. J. Kerala*, **18**, 111–112.

Nair, K.S.S., Kedharnath, S., Koshy, M.P., Sudheendrakumar, V.V., Varma, R.V., Mohanadas, K. and Mathew, G. (1989) *Search for Natural Resistance to the Insect Pest, Hyblaeapuera in Teak.* Kerala Forest Research Institute, Peechi, India.

Nair, K.S.S. and Mohandas, K., (1996). Early events in the outbreak of the teakcaterpillar, *Hyblaea puera. Int. J. Ecol. Environ. Sci.*, **22**, 271–279.

Nair, K.S.S., Sudheendrakumar, V.V., Mohandas, K., Varma, R.V., Mathew, G., Mathew, P., Koshy, M.P. and Kedharnath, S. (1997) Search for teak trees resistant to the defoliator, *Hyblaea puera* Cramer (Lepidoptera : Hyblaeidae). In: A. Raman (ed.), *Ecology and Evolution of Plant Feeding Insects in Natural and Man-made Environments*, International Scientific Publications, New Delhi, pp. 109–122.

Nair, K.S.S., Varma, R.V., Sudheendrakumar, V.V., Mohanadas, K. and Mohamed-Ali, M.I. (1998) *Management of the Teak Defoliator (Hyblaea puera) using Nuclear Polyhedrosis Virus (NPV).* Kerala Forest Research Institute, Peechi, India.

Narayanan, K. and Jayaraj, S. (1979) Occurrence of mixed infection of virus and protozoa in two species of lepidoptera. *Curr. Sci.*, **48**, 825.

Nguyen, K.B. (2006) Online source: Web site: www.steinsp1.htm.

Nielsen, D.G. and Cowles, R.S. (1998) Preventing white grub infestation in container-grown nursery stock. *J. Environ. Hortic.*, **16**, 202–207.

Otvos, I.S. (2004). *IPM in Forestry: Potential and Challenges.* In: O. Koul, G.S. Dhaliwal and G.W. Cuperus (eds.), *Integrated Pest Management: Potential, Challenges and Constraints*, CABI Publishing, U.K., pp. 205–254.

Painter, R.H. (1951) *Insect Resistance in Crop Plants.* The MacMillan, New York.

Panda, N. (1979) *Principles of Host-Plant Resistance to Insect Pests.* Allanheld, Osmum and Universe Books, New York.

Pathak, S.C., Kulkarni, N. and Meshram, P.B. (1995) Forest Insect Control. *Encyclopaedia of Environmental Biology*, Vol. 3, Academic Press, San Diego, Netherlands, pp. 119–130.

Patil, B.V. (1981) Bionomics, life-table and biocontrol of the teak skeletonizer, *Pyrausta machaeralis* Walker (Lepidoptera : Pyralidae). *Ph.D. Thesis,* University of Agricultural Sciences, Bangalore, 56 pp.

Patil, S.U. and Mohan Naik, I. (1998) Natural enemies of teak defoliator, *Hyblaea puera* Cramer and their seasonal incidence. *Indian J. Fores.*, **21**, 253–255.

Poinar, G.O., Jr. (1976) Description and biology of a new insect parasitic rhabditoid *Heterorhabditis bacteriophora* n. gen., n. sp. (Rhabditida: Heterorhabditidae). *Nematologica,* **21,** 463–470.

Poinar, G.O. Jr, Jackson, T. and Klein, M. (1987) *Heterorhabditis megidis* sp. n. (Heterorhabditidae: Rhabditida), parasitic in the Japanese beetle, *Popillia japonica* (Scarabaeidae : Coleoptera), in Ohio. *Proc. Helminthol. Soc. Washington,* **54**, 53–59.

Pradhan, S., Jotwani, M.G. and Rai, B.K. (1962) Neem seed deterrent to locusts. *Indian Fmg.*, **12**, 7–11.

Prasad, D. and Chand, P. (1981) Effect of different dilutions of nuclear polyhedrosis viruses on fourth instar or onward caterpillars of *Diacrisia obliqua* Wlk. *Indian J. Ent.*, **43**, 64–66.

Qin, X.X., Kao, R.T., Yang, H.W., Zhang, G.Y. (1988) Study on application of entomopathogenic nematodes *Steinernema bibionis* and *S. feltiae* to control *Anoplophora glabripennis* and *Holcocerus insularis*. *Forest Res.*, **1**, 179–185.

Rabindra, R.J., Geetha, N., Jayaraj, S., Brown, M. and Grzywacz, D. (1997) Evaluation of cross infectivity of some Baculoviruses to *Helicoverpa armigera* Hbn. *Pest Manage. Hortic. Ecosys.*, **3**, 61–65.

Rajak, R.C., Agrawal, G.P., Khan, A.R. and Sandhu, S.S. (1993) Susceptibility of teak defoliator, *Hyblaea puera* Cramer and teak skeletonizer, *Eutectona machaealis* Walker to *Beauveria bassiana* (Bals) Vuill. *Indian J. Exp. Biol.*, **31**, 80–82.

Rao, V.P. and Manjunath, T.M. (1966) DD-136 nematode that can kill many insect pests. *Indian Fmg.*, **16**, 43–44.

Roychoudhury, N. (2001) Tree resistance to insects: a novel approach of forest insect management. In: P.K. Shukla and K.C. Joshi (eds.), *Recent Trendsin Insect Pest Control to Enhance Forest Productivity*, Tropical Forest Research Institute, Jabalpur, pp. 28–60.

Roychoudhury, N. and Joshi, K.C. (1996) Relative toxicity of certain insecticides against the larvae of teak defoliator, *Hyblaea puera* Cramer. *Indian J. Ent.*, **58**, 253–255.

Roychoudhury, N., Jain, A. and Joshi, K.C. (1998) Identification of resistant teak clones to leaf skeletonizer. *ENVIS Newsletter*, **1**, 40–41.

Roychoudhury, N., Jain, A. and Joshi, K.C. (2000) Insect pests of teak and breeding for insect resistance. In: A.K. Mandal and S.A. Ansari (eds.), *Genetics and Silviculture of Teak*, International Book Distributors, Dehradun, pp. 187–205.

Roychoudhury, N., Joshi, K.C., Sambath, S., Humne, S. and Pandey, D.K. (1994) Effectiveness of three varietal toxins of *Bacillus thuringiensis* against teak leaf skeletonizer, *Eutectona machaeralis* Walker (Pyralidae). *Myforest*, **30**, 57–60.

Roychoudhury, N., Kalia, S. and Joshi, K.C. (1995) Pest status and larval feeding preference in *Spodoptera litura* (Fabricius) Boursin (Lepidoptera : Noctuidae) on teak. *India Forester*, **121**, 581–583.

Roychoudhury, N., Singh, K. and Joshi, K.C. (2009) Insecticidal activity of spinosad against the larvae of Ailanthus webworm, *Atteva fabriciella* Swederus (Lepidoptera: Yponomeutidae). *Pestology*, **33**, 34–36.

Sajeev, T.V., Sudheendrakumar, V.V., Mahiba Helen, S., Meera, C.S., Bindu, T.N. and Bindu, K.J. (2007) Hybcheck- the biopesticide for managing teak defoliator (*Hyblaea puera* Cramer) : an announcement. In : K.M. Bhat, K.V. Balasundaran, E.M. Muralidharan and P.K. Thulasidhas (eds.), *Processing and Marketing of Teak Wood Products of Planted Forests*, Kerala Forest Research Institute, Peechi, India, pp. 173–177.

Sandhu, S., Rajak, R.C. and Agrawal, G.P. (1993) Microbial control agents of forest pests of Jabalpur. *Ann. Fores.*, **1**, 136–140.

Sharma, N., Joshi, K.C. and Harsh, N.S.K. (2005) Entomopathogenic fungi of teak defoliator, *Hyblaea puera* Cramer and leaf skeletonizer, *Eutectona machaeralis*Wlk. *J. Trop. Fores.*, **21**, 70–77.

Sharma, N. and Joshi, (2007) Entomopathogenic fungi attacking sal heartwood borer, *Hoplocerambyx spinicornis* Newman. *Indian J. Fores.*, **27**, 133–140.

Singh, P. and Mishra, R.M. (1978) Bioassay of Thuricide, a microbial insecticide against important forest pests. *Indian Forester*, **104**, 838–842.

Sosamma, V.K. (2003) Utilization of entomopathogenic nematodes in plantation crops. In: S.S. Hussaini, R.J. Rabindra and M. Nagesh (eds.), *Current Status of Research on Entomopathogenic Nematodes in India*, Project Directorate of Biological Control, Bangalore, India, pp. 109–112.

Stark, R.W. (1965) Recent trends in forest entomology. *Annu. Rev. Entomol.*, **10**, 303–324.

Steiner, G. (1923) *Aplectana kraussei* n. sp. Einer in der Blattwespe Lyda sp. Parasitieerende Nematodenform, nebst Bemeerkungen uber das Seitenorgan der parasitischen Nematoden. *Zentralbl. Bakteriol. Parasitenkd. Hyg. Abt.*, 2.

Steiner, G. (1929) *Neoaplectana glaseri* n. g. n. sp. (Oxyuridae), a new nemic parasite of the Japanese beetle (*Popillia japonica* Newm.) *J. Washington Acad. Sci.*, **19**, 436–440.

Steinhaus, E.A. (1963a) *An Advanced Treatise*. Vol. 1, Academic Press, New York.

Steinhaus, E.A. (1963b) *An Advanced Treatise*. Vol. 2, Academic Press, New York.

Sudheendrakumar, V.V. and Biji, C.P. (2009) Biocontrol of the teak defoliator, *Hyblaea puera* : prospects and constraints. *Envis Fores. Bull.*, **9**, 8–12.

Sudheendrakumar, V.V., Jalali, S.K. and Singh, S.P. (1995) Acceptance of the teak defoliator Cramer (Lepidoptera, Hyblaeidae) by two exotic species of *Trichogramma* (Hymenoptera : Trichogrammatidae). *J. Biol. Control*, **9**, 43–44.

Sudheendrakumar, V.V., Mohamed Ali, M.I., Verma, R.V. (1988) Nuclear polyhedrosis virus of the teak defoliator *Hyblaea puera*. *J. Invertebr. Pathol.*, **51**, 307–308.

Sudheendrakumar, V.V.; Varma, R.V. and Sajeev, T.V. (2006) Demonstration of Mass Production, Formulation and Application of a Baculovirus for Management of the Teak Defoliator, Hyblaea puera. Kerala Forest Research Institute, Peechi, Kerala.

Tewari, D.N. (1992) *A Monograph on Teak* (*Tectona grandis* Linn. f.). International Book Distributors, Dehradun.

Thakur, M.L. (1983). Insect factor in forest tree improvement programme. *J. Indian Acad.Wood Sci.*, **14**, 63–69.

Thakur, M.L. (2000) *Forest Entomology*. Sai Publishers, Dehradun.

Troup, R.S. (1921) *The Silviculture of Indian Trees*. Clarendon Press, Oxford.

Weiser, J. (1955) *Neoaplectana carpocapsae* n. sp. (Anguillulata: Steinernematinae) novy cizopasnik housenek obalece jablecného, *Carpocapsa pomonella* L. *Ve¢stník C¡eskoslovenské Spole¡cnosti Zoologické,* **19**, 44–52.

Wright, R.J., Villani, M.G. and Agudelo-Silva, F. (1988) *Steinernematidae* and *Heterorhabditis* nematodes for control of larval European chafers and Japanese beetles (Coleoptera: Scarabaeidae) in potted yew. *J. Econ. Entomol.*, **81**, 152–157.

Yousuf, M., Joshi, K.C. and Kulkarni, N. (2004) *Trichogramma*: the noble egg parasitoids and their potential in biological control of insect pests. In: G.T. Gujar (ed.), *Contemporary Trends in Insect Science*, Campus Books International, New Delhi, pp. 307–369.

15

Role of Biopesticides in Tea from a Japanese Perspective: Viral Control of Tea Pests in Japan and the Effects of Virus Infection on Domestic Endoparastioids

MADOKA NAKAI

Division of Bioregulation and Biointeraction, Institute of Agriculture, Tokyo University of Agriculture and Technology, Saiwai, Fuchu, Tokyo 183-8509, Japan.
E-mail: madoka@cc.tuat.ac.jp

INTRODUCTION

In Japan, the food sufficiency rate has been decreasing in recent years and is currently approximately 40% (calory-based). As compared to other agricultural products, tea production is relatively high in Japan: the food sufficiency rate for green tea was 88% in 2002. Japan is the world's eighth tea producer (Fig. 15.1).

Tea is a perennial and evergreen plant which has a unique structure and provides a habitat for various pests and natural enemies. The major insects pests of tea are Kanzawa spider mite (*Tetranychus kanzawai*), tea leaf roller (*Caloptilia theivora*), oriental tea tortrix (*Homona magnanima*), smaller tea tortrix (*Adoxophyes honmai*), mulberry scale (*Pseudaulacaspis pentagona*), black citrus aphid (*Toxoptera aurantii*), yellow tea thrips (*Scirtothrips dorsalis*), tea green leafhopper (*Empoasca onukii*), mugwort looper (*Ascotis selenaria*), and chafer (*Heptophylla picea*) (Kawai, 1997). Tea cultivation in Japan requires a high chemical pesticide and fertilizer input, with conventional management involving 7-14 chemical applications per year. Because of consumers' safety consciousness, however, excessive use of chemical pesticides is undesirable. Prominent biological control agents in Japanese tea fields are a pesticide-resistant predatory mite (*Amblyseius womersleyi*) against the Kanzawa spider mite, granuloviruses (GVs) (Baculoviridae) and sex pheromones against *H. magnanima* and *A. honmai* (Tortricidae: Lepidoptera) (Kodomari, 1991). Biological control agents that are sensitive to chemical pesticides can not be used in tea fields as various pests still need to be controlled primarily by chemical pesticides. Details of biopesticides in Japanese tea system are discussed earlier by Nakai (2009).

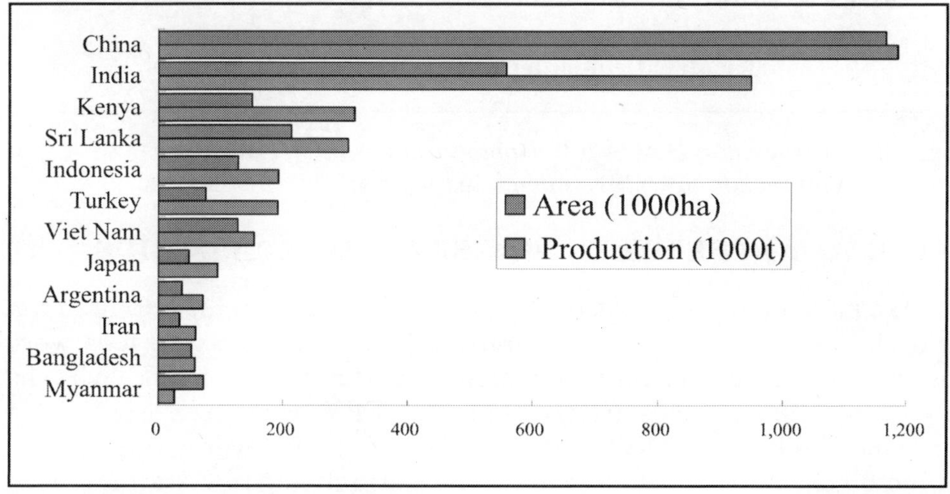

Fig. 15.1. Annual production of tea in the world. Area (× 1000 ha) and production (× 1000 t). *Source*: FAOSTAT (2008).

A. honmai and *H. magnanima* are two of the most important pests for tea plants in Japan (Fig. 15.2). Larvae of both tortricid species roll tea leaves, causing damage that severely reduces the quality of tea and the harvest. While conventional control of these tortricids has depended on the use of chemical pesticides, the development of resistance by both species to several pyrethroid pesticides, and the persistence of pesticide residues on harvested tea leaves, have stimulated interest in microbial control for these pests. GVs, double-stranded DNA viruses belonging to the family Baculoviridae, have successfully controlled these two species in Japan. As well as GVs and other viral pathogens including entomopoxviruses (EPVs), *A. honmai* and *H. magnanima* also have various natural enemies, particularly hymenopteran parasitoids.

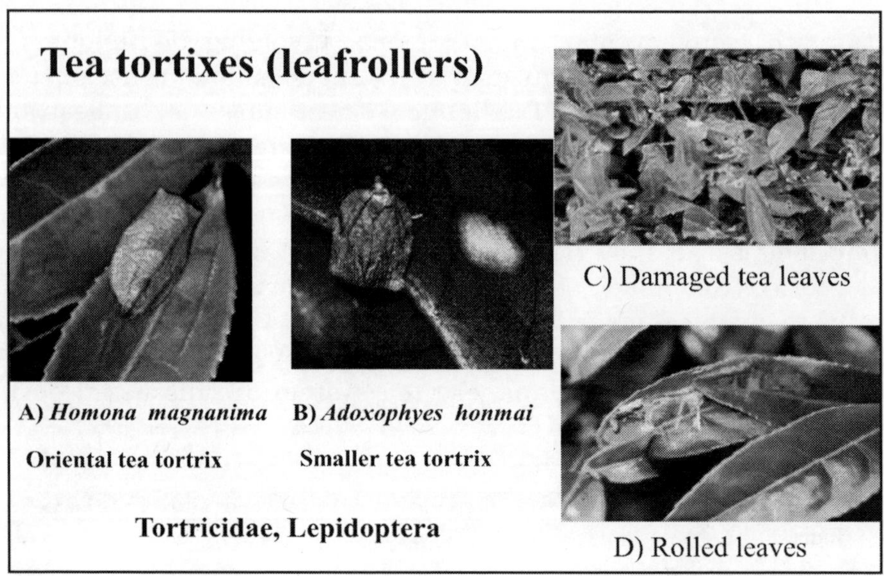

Fig. 15.2. Target pests. A) Adult *Homona magnanima.* B) Adult and egg of *Adoxophyes honmai.* C) Damaged tea leaves. D) Rolled leaves.

INTERACTION BETWEEN VIRUSES AND PARASITOIDS

We have monitored natural populations of *A. honmai* larvae in tea fields in Tsukuba, Ibaraki prefecture (Fig. 15.3). Insect tissues were observed under phase contrast microscopy for infection with viruses having occlusion bodies (OBs) including EPV, cypovirus (CPV) and nucleopolyhedrovirus (NPV). OBs of NPV and CPV were stained with amido black. NPV and CPV OBs were distinguished by staining with bromophenol blue (BPB), as OBs of CPV are stained by BPB but those of NPVs are not. GV-infected insects were diagnosed by ELISA using an antibody against capsules (OBs of GV) of *A. honmai* GV. Surveys in Tsukuba revealed that many *A. honmai* larvae were killed by parasitic natural enemies including insect viruses and endoparasitoids (Nakai *et al.*, 1997). The most common cause of death was parasitism by *Ascogaster reticulatus* (38%) (Fig. 15.4) and the second commonest was infection with EPV (31%). Dissection of larval cadavers revealed that 12% of larvae that had been killed by viral infection also harbored parasitoid larvae. These parasitoids died inside the host cadaver, however, suggesting that epizootics of insect viruses affect parasitoid survival deleteriously (Nakai *et al.*, 1997). Since insect viruses do not infect the natural enemies of their hosts, virus application should have no direct impact on natural enemies when a biological control program is started.

Fig. 15.3. Tea field in Tsukuba, Ibaraki prefecture.

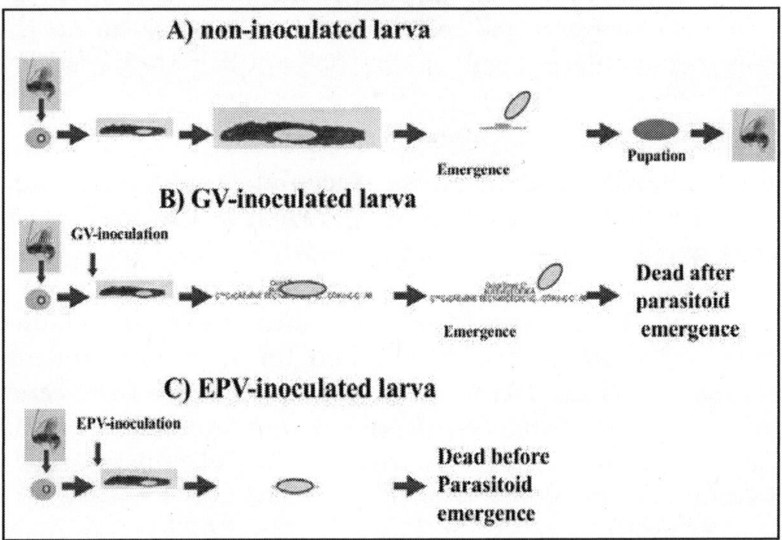

Fig. 15.4. Consequences of parasitization by *Ascogaster reticulatus* on *A. honmai* larvae. A) Parasitization of a healthy host larva. An adult female parasitoid oviposits an egg into a host egg. The parasitoid larva develops in the host, emerges from the host larva and develops externally to adulthood. B) A parasitized host was inoculated as a neonate with AdhoGV. Most parasitoid larvae emerged from the host but died prior to pupation. C) A parasitized host was inoculated as a neonate with AdhoEPV. Most parasitoid larvae did not emerge and died within the host.

GRANULOVIRUS APPLICATION

Granulovirus application in tea fields in Kagoshima prefecture was highly effective because *A. honmai* and *H. magnanima* larvae were killed not only directly by the GVs but also by several species of hymenopteran parasitoids (Kodomari, 1987). However, if more than one species share a common resource as their host, interspecies competition may occur; or, more specifically, "interkingdom competition" in the case of a virus and an insect parasitoid (Hochberg and Lawton, 1990). Thus, resident natural enemy populations should be evaluated in the field, and biological control agents should be used in a way that is compatible with their beneficial effects. We studied the effect of *A. honmai* GV (AdhoGV) infection on the development and survival of *A. reticulatus* in *A. honmai. A. reticulatus* (Braconidae: Hymenoptera) is a solitary egg-larval endoparasitoid; adult females oviposit into the egg, and the parasitoid larva hatches in the host egg and develops in the host larva. The parasitoid larva emerges from 4th (penultimate) instar larvae of the moribund host. As described below in detail, just before emergence of the *A. reticulatus* larva, the host initiates precocious metamorphosis in the penultimate instar, which is similar to the onset of normal molting in prepupae of non-parasitized final-instar larvae (Fig. 15.5, Fig. 15.8, upper).

Purified AdhoGV suspension was inoculated onto discs of artificial diet (Insecta LF, Nihon Nosanko), and parasitized neonate *A. honmai* larvae were exposed to the contaminated food disc for 24 hours and then transferred to untreated diet and reared individually. As a control, distilled water was inoculated onto the artificial diet. The larvae were observed daily and checked for mortality and survival time or pupation (Fig. 15.6). Approximately 90% of *A. reticulatus* larvae emerged from AdhoGV-infected *A. honmai*, but most of them died before pupation, whereas approximately 100% parasitoid larvae emerged from the non-infected host, pupated, and emerged as adults (Nakai and Kunimi, 1997). Why they failed to pupate is unknown, but we speculated that insufficient nutrition was available to parasitoid larvae in a virus-infected host. *A. reticulates* larvae normally feed externally on the host cadaver after they emerge; however, larvae that had developed in an AdhoGV-infected host did not do so. When parasitoid larvae emerging from an AdhoGV-infected host were transferred to a non-infected host cadaver, they did not feed externally. However, when parasitoid larvae emerging from a non-infected host were transferred to an AdhoGV-infected host, they did feed externally (Fig. 15.6). We also found that parasitoid larvae developing in an AdhoGV-infected host accumulated undegraded virus capsules in their gut lumen. These results suggest that parasitoid larvae which develop in an AdhoGV-infected host cannot acquire sufficient nutrition; they develop until emergence, but are not strong enough to feed externally or to proceed to pupation.

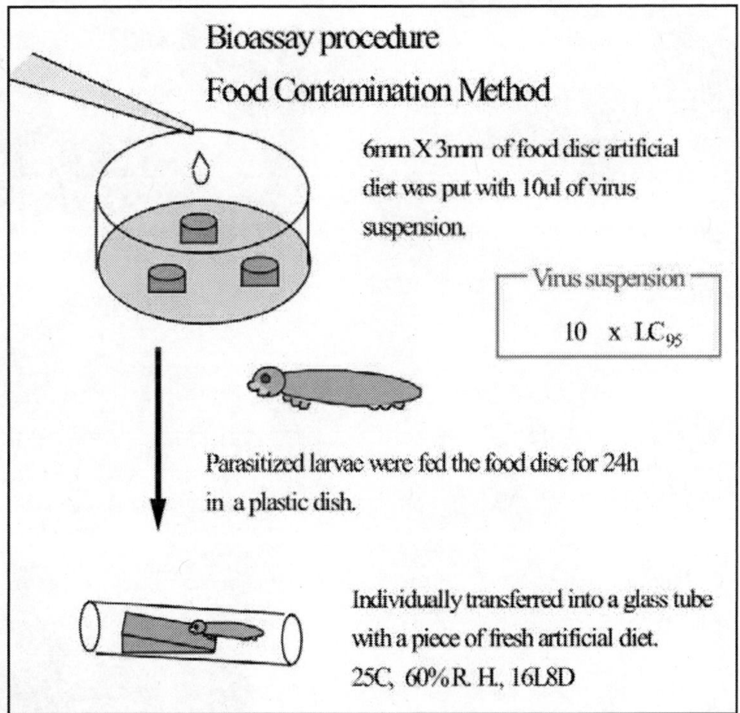

Fig. 15.5. Bioassay procedure for diet contamination method. Purified virus occlusion bodies prepared as 10 × 95% lethal concentration (LC95) were applied to a food disc. Parasitized larvae were fed the food discs for 24 h in a plastic dish. The larvae were then transferred to tubes containing fresh artificial diet.

EPV CASE STUDY

We also conducted a similar study for *A. honmai* EPV (AdhoEPV) and *A. reticulatus*. We expected that the impact of AdhoEPV on *A. reticulatus* would be similar to that of AdhoGV, because both viruses prevent pupation of the host and kill it slowly (Nakai *et al.*, 1998). In fact, the results were not the same: the emergence rate of parasitoids was lower from AdhoEPV-infected hosts than from AdhoGV-infected hosts (Fig. 15.7). This low emergence rate is due to early death of the host, which is caused by abnormal molting (Fig. 15.8, lower; Nakai and Kunimi, 1998). Dually AdhoGV-infected and parasitized *A. honmai* larvae, in contrast, do not show abnormal metamorphosis but are killed by parasitoid emergence.

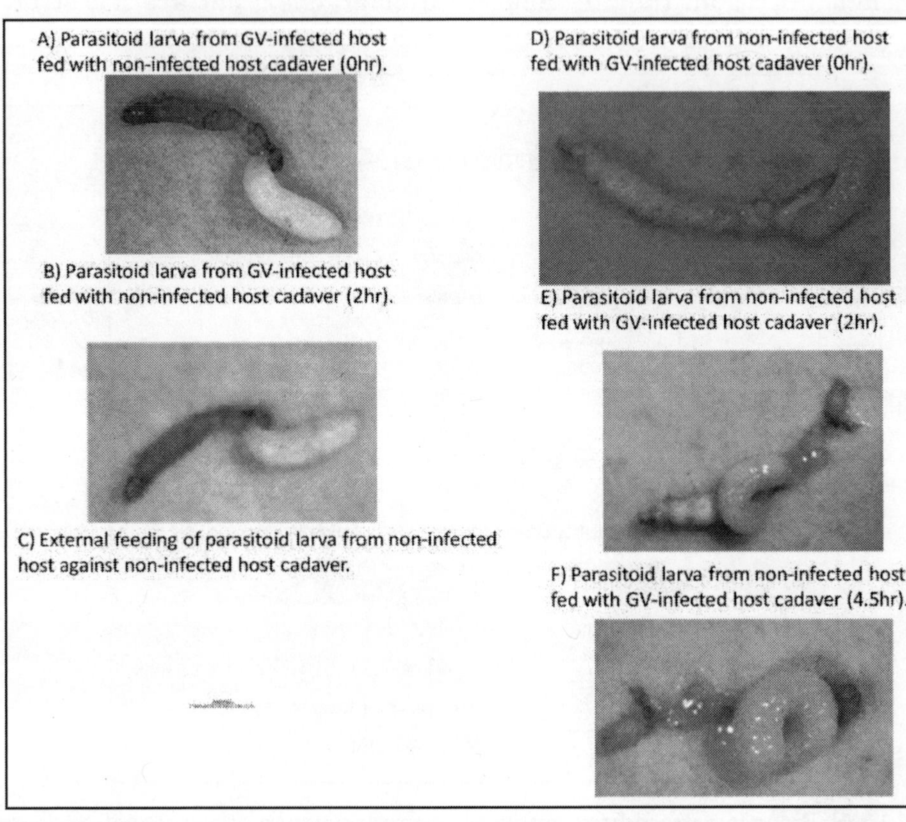

Fig. 15.6. External feeding of parasitoid larvae developed in non-infected or GV-infected host to switched non-infected or GV-infected host cadaver. A) Parasitoid larva emerged from GV-infected host was fed with cadaver of non-infected host at 0 h B) Parasitoid larva emerged from GV-infected host was fed with cadaver of non-infected host at 2 hr. No external feeding was observed during the two hours. C) Normal external feeding. Parasitoid larva from non-infected host feeds the non-infected host cadaver externally. Parasitoid completely consumes the host cadaver before spinning cocoon and pupation. D) Parasitoid larva from non-infected host fed with GV-infected host cadaver at 0 hr. E) Parasitoid larva from non-infected host fed with GV-infected host cadaver at 2 hr. The parasitoid larva had enough viability to feed the GV-infected cadaver externally. F) Parasitoid larva from non-infected host fed with GV-infected host cadaver at 4.5 hr. The parasitoid larva complete to feed the GV-infected cadaver externally. However, even the parasitoid consumed GV-infected host cadaver completely, the parasitoid did not pupate successfully.

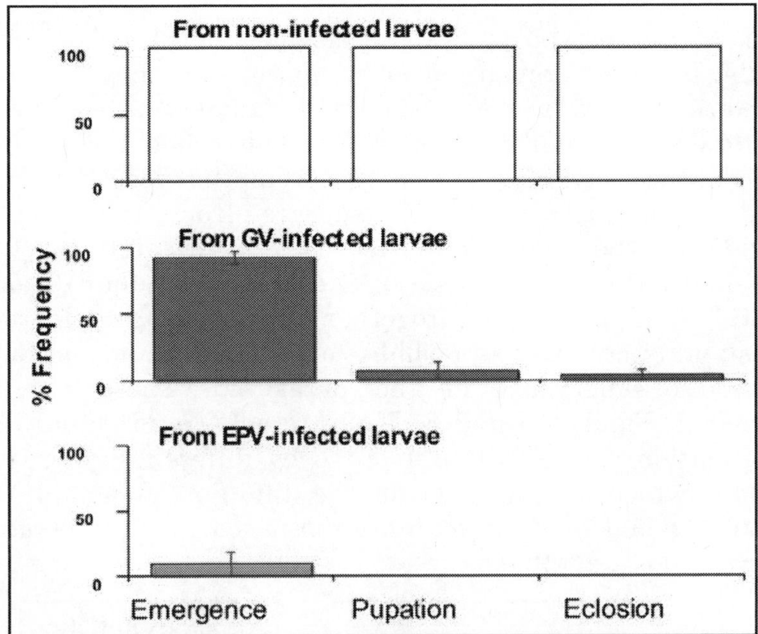

Fig. 15.7. Percent emergence, pupation and adult emergence of *A. reticulatus*. Neonate *A. honmai* larvae were not inoculated (control; top), inoculated with AdhoGV (middle), or inoculated with AdhoEPV (bottom).

These observations prompted us to study the hormonal status of AdhoGV- and AdhoEPV-infected *A. honmai* larvae. Insect molting is regulated by fluctuations of ecdysone and juvenile hormone (JH). Parasitization by *A. reticulatus* also regulates host endocrinology, causing a precocious initiation of metamorphosis in the penultimate instar (Fig. 15.8 upper). Brown *et al.* (1993) demonstrated that *A. reticulatus* releases 20 OH-ecdysone into the hemolymph of *A. honmai* before it emerges from the penultimate-instar host. Infection with AdhoGV alone prevents pupation of *A. honmai* larvae, because JH esterase activity is suppressed and ecdysteroids are inactivated (Nakai *et al.*, 2002). JH esterase is released from fat body cells, in which AdhoGV replicates, and this may be the reason that JH esterase activity is suppressed in AdhoGV-infected hosts. On the other hand, AdhoGV is known to encode ecdysteroid UDP-glucosyltransferase (EGT), an enzyme that inactivates ecdysteroids in the hemolymph of AdhoGV-infected hosts. AdhoEPV also prevents pupation of *A. honmai* larvae when it is inoculated alone; in this case, however, AdhoEPV suppresses the release of ecdysteroids in infected *A. honmai* larvae (Nakai *et al.*, 2004). Like AdhoGV, AdhoEPV also infects fat body tissue, and JH esterase activity is suppressed in the hemolymph of AdhoEPV-infected hosts, so an elevated level of JH may be responsible for the prevention of pupation in these insects. In parasitized *A. honmai* larvae infected with AdhoEPV, the JH esterase level was low and a small peak of ecdysteroids was observed. In these larvae, ecdysone may be released not from the host but from the parasitoid larva just before it emerges from the host. Such ecdysteroids in the hemolymph are not inactivated by EGT in AdhoEPV-infected hosts, because, unlike AdhoGV, AdhoEPV does not encode EGT. A high JH level, together with ecdysone released from the parasitoid, may be responsible for extraordinary larval-larval molting before emergence of the parasitoid. These results are summarized in Figures 5 and 8. Thus, AdhoGV and AdhoEPV both prevent pupation of the host but do so by different host-regulation mechanisms, which may account for the different impacts of the two viruses on survival of *A. reticulatus* discussed above (Nakai *et al.*, 2005; Burand *et al.*, 2009).

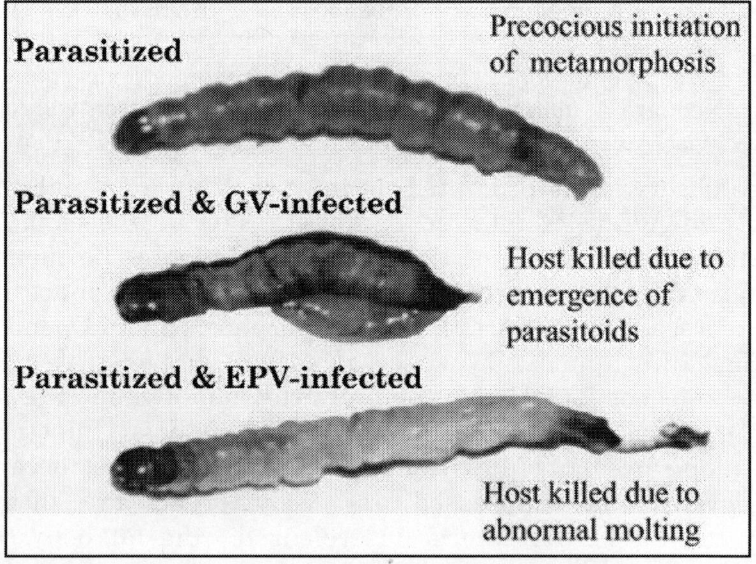

Fig. 15.8. Gross pathology of *A. honmai* larvae parasitized by *A. reticulatus* and inoculated with AdhoGV or AdhoEPV. Non-inoculated, parasitized larva showed precocious initiation of metamorphosis. Parasitized and AdhoGV-inoculated hosts were killed by parasitoid emergence. Parasitized and AdhoEPV-inoculated hosts were killed by abnormal molting.

CONCLUSIONS

Tea is an important crop for Japanese farming. The morphological features of tea, as a perennial and evergreen plant having dense leaves on its canopy, provide a habitat for various insects including pests and their natural enemies. Excessive input of chemical pesticides has caused resistance to develop in insect pests, and the combination of pesticide resistance and the persistence of pesticide residues on the harvested tea leaves has stimulated the development of microbial control strategies for these pests. *Adoxophyes honmai* and *Homona magnanima* are important pests of tea plants. These two species have been managed by biological control using granuloviruses (GVs) in Japan. According to field survey data, an entomopoxvirus (EPV) and the braconid parasitoid *Ascogaster reticulatus* are the major natural enemies of *A. honmai* larvae. However, the interactions between parasitoids and insect viruses are not well understood. The effect of viral infection on the development of the parasitoid was studied in the laboratory. Bioassays in which *A. honmai* larvae parasitized by *A. reticulatus* were then inoculated with GV or EPV revealed that both viruses deleteriously affect the survival of *A. reticulatus*. The GV and the EPV both retarded development of the host, although the fate of parasitoid larvae differed because host endocrinological regulation by the two viruses was different. In this study, we revealed that (1) AdhoGV and AdhoEPV infection of the host deleteriously affect the survival of *A. reticulatus*; (2) AdhoGV and AdhoEPV both retard development of the host; however, (3) the fate of parasitized hosts differs because the endocrinological regulation of host larvae by AdhoGV and AdhoEPV is different. Insect viruses generally have narrow host ranges, but this does not mean that they have no impact on non-target organisms. While viruses do not kill adult parasitoids (as chemical pesticides normally do), they can nonetheless have a deleterious impact on the survival of parasitoid larvae that are developing in virus-infected hosts.

REFERENCES

Brown, J.J., Kiuchi, M., Kainoh, Y. and Takeda, S. (1993) *In vitro* release of ecdysteroids by an endoparasitoid, *Ascogaster reticulatus* Watanabe. *J. Insect Physiol.*, **39**, 229–234.

Burand, J.P., Nakai, M. and Smith, I. (2009) Host-Virus Interactions. In: S.P. Stock, J. Vandenberg, I. Glazer and N. Boemare (eds.), *Insect Pathogens: Molecular Approaches and Techniques*, CAB International, UK, 417 pp.

FAOSTAT (2008) http://faostat.fao.org/

Hochberg, M.E. and Lawton, J.H. (1990) Competition between Kingdoms. *Trends Ecol. Evol.*, **5**, 367–371.

Kawai, A. (1997) Prospect for integrated pest management in tea cultivation in Japan. *Jpn. Agr. Res. Q.*, **31**, 213–217.

Kodomari, S. (1987) Control of leafrollers with granulosis viruses in tea field. 1. Effectiveness of application of GV mixture of oriental tea tortrix and smaller tea tortrix. *Bull. Shizuoka Tea Exp. Stn.,* **13**, 39–48.

Kodomari S. (1991) Pests and diseases of tea and their control. Science of tea (Cha no kagaku), Asakura Publishing Co. Ltd., Japan, 42–51 (in Japanese).

Nakai, M., Takeda M. and Kunimi Y. (1997) Seasonal changes in prevalence of viral disease and parasitism by parasitic insects in a larval population of the smaller tea tortrix, *Adoxophyes* sp. (Lepidoptera: Tortricidae) in a tea field. *Appl. Entomol. Zool.*, **32**, 609–615.

Nakai, M., Sakai, T. and Kunimi, Y. (1997) Effect of entomopoxvirus infection of the smaller tea tortrix, *Adoxophyes* sp. (Lepidoptera: Tortricidae) on the development of the endoparasitoid, *Ascogaster reticulatus. Entomol. Exp. Appl.*, **84**, 27–32.

Nakai, M. and Kunimi Y. (1998) Effects of the timing of entomopoxvirus administration to the smaller tea tortrix, *Adoxophyes* sp. (Lepidoptera: Tortricidae) on the survival of the endoparasitoid, *Ascogaster reticulatus* (Hymenoptera: Braconidae). *Biol. Control,* **13**, 63–69.

Nakai, M., Goto, C., Shiotsuki, T. and Kunimi, Y. (2002) Granulovirus prevents pupation and retards development of *Adoxophyes honmai* larvae. *Physiol. Entomol.,* **27**, 157–164.

Nakai, M., Shiotsuki, T. and Kunimi, Y. (2004) An entomopoxvirus and a granulovirus use different mechanisms to prevent pupation of *Adoxophyes honmai. Virus Res.*, **101**, 185–191.

Nakai, M., Shiotsuki, T. and Kunimi, Y. (2005) An entomopoxvirus and a granulovirus have different impacts on *Adoxophyes honmai* (Tortricidae: Lepidoptera) simultaneously parasitized by *Ascogaster reticulatus* (Braconidae: Hymenoptera). *Biol. Control,* **33**, 230–237.

Nakai, M. (2009) Biological control of Tortricidae in tea fields in Japan using insect viruses and parasitoids. *Virologica Sinica,* **24**, 323–332.

16

Current Status and Potential of Biopesticides in Pest Management of Stored Products

C. DURAIRAJ*, M. KALYANASUNDARAM AND T. BHARATHIMEENA

Department of Agricultural Entomology, Centre for Plant Protection Studies, Tamil Nadu Agricultural University, Coimbatore-641 003, India
*E-mail: tcdurairaj@gmail.com

INTRODUCTION

Post harvest preservation of grains is a critical factor in food production. Globally, almost one third of the yearly harvest is destroyed by pests. This accounts for about 10–20% of the entire produce in India which is about 20 million tonnes annually (Dhuri, 2006). The stored food grains suffer quantitative as well as qualitative losses due to insect pests which are mostly from order Coleoptera and several from order Lepidoptera (Brower, 1990). Stored grain pests have a remarkably high rate of multiplication, destroying almost 10–15% of the produce (Naik and Geetanjali, 2010) and contaminating the rest with offensive odours and secretions. Primary storage pests consume whole grains whereas secondary pests infest only milled or processed food grains. Industrialised countries suffer only 1% storage losses whereas developing countries it is up to10–30% of their produce in storage (Adam *et al.*, 2006). Often, chemical pesticides are illicitly used in order to reduce the population of insect pests in storage godowns (Kellouche and Soltani, 2006). There have been several cases of resistance development in stored insect pests towards the commonly used insecticides including phosphine (Champ and Dyte, 1976).

Certain fumigants like methyl bromide are known to threaten the health and well-being of workers involved in application. This necessitates the consolidation of subtle but unfailing management strategies for protecting the harvested produce in stores. Biopesticides, including plant products, parasitoids, predators and entomopathogens form key elements in the sustainable management of stored product pests. They are less toxic and more consumer-friendly since they leave least residual effects. There is no demonstrated resistance in many cases and safety to non targets is most assured. In the recent wake of consumer awareness towards chemical pesticide residues in stored grains, biopesticides are likely to assume greater importance in storage pest management.

BOTANICALS

Botanicals are the most appealing candidates of all biopesticides (Maheshwari and Dwivedi, 1996) as they are easily accepted by consumers and also pose very meagre chances for insect resistance. Neem, *Azadirachta indica* had shown considerable antifeedant and repellent potential against several stored product pests (Table 16.1). Egwurube *et al.* (2010) found that seed powder of neem when applied @ 10% w/w killed 100% of the larval population of khapra beetle, *Trogoderma granarium* (Everts) infesting stored groundnut and recorded only 6.5% seed damage. Grain damage by rice weevil, *Sitophilus oryzae* L. was significantly minimised in rice grains treated with neem leaf powder@ 2% (w/w). Volatiles, specifically the di-n-propylsulfides from neem have also toxic to stored grain pests (Koul, 2004). Neem oil (1% w/v) is also toxic to stored grain pests inducing 100% mortality, three days after release in stored chick pea (Pokharkar and Chauhan, 2010). Significant deterrence on oviposition when treated with neem leaf powder has also been observed. Govindan and Nelson (2009) found that plant powders of *Vitex negundo* L. proved highly effective against the pest followed by, *Nelumbium speciosum* Wild, *Alpinina officinarum* Hance and *Gloriosa superba* L. Some *Labiatae* spp. oils have also been shown to have significant insecticidal as well as repellence potential against the lesser grain borer, *Rhyzopertha dominica* (F.), saw toothed beetle *Oryzaephilus surianamensis* (L.) and the pulse beetle, *Callosobruchus chinensis* L. Bruchids in stored mung bean could be effectively managed by *Chenopodium ambrosioides* L. (0.4%) leaf powder which killed about 60% of the population. Root extracts of *Withania somnifera* Dunal were toxic to *C. chinensis* causing 63.33% mortality

(Gupta and Srivastava, 2008). Sweet flag, *Acorus calamus* is known as anti-insect substance for a long time now (Saxena *et al.*, 1976; Koul *et al.*, 1977) and Nadaf *et al.* (2010) recently studied the effect of sweet flag on a bruchid pest of groundnut and found that the rhizome powder @ 10 g/kg disrupted the development of the pest. Further there was no adult emergence and the pods were undamaged for two months. Essential oils from aromatic plants also play a significant role in the control of stored grain pests (Shaaya *et al.*, 1997; Koul *et al.*, 2007, 2008). Turmerone and *ar*-turmerone (dehydroturmerone), the major constituents of turmeric rhizome powder oil are strong repellents to stored grain pests. The turmeric oil has been reported to provide protection to wheat grains against red flour beetle, *T. castaneum* (Herbst) (Chahal *et al.*, 2005). The fruit oil of *Piper retrofractum* has also shown high repellency (52–90%) against *T. castaneum* at 0.5–2% concentration (Koul *et al.*, 2008). Similalrly, effects of volatile oil constituents of *Mentha species* against stored grain pests, *Callosobruchus maculatus* and *Tribolium castaneum*; toxicity, feeding deterrence, and effect of activity of 1, 8-cineole from *Artemisia annua* on progeny production of *Tribolium castaneum* and bioactivity of l-carvone, d-carvone and dihydrocarvone towards stored product beetles are well known (Tripathi *et al.*, 2000, 2001, 2003). The efficacy of these plant based materials falls short when compared to synthetic pesticides although there are specific pest contexts where control equivalent to that with conventional products has been observed. Essential oils also require somewhat greater application rates (as high as 1% active ingredient) and may require frequent reapplication when used out-of-doors (Koul *et al.*, 2008). However, it is obvious that pesticides derived from plant essential oils do have several important benefits. Due to their volatile nature, there is a much lower level of risk to the environment than with current synthetic pesticides.

PARASITOIDS AND PREDATORS

Insect predators and parasitoids are being used to control insects for a long time and in fact, first parasitic wasp was seen in 1911 in a flour mill in London. For the control of stored product moths, egg parasitoids combined with a larval parasitoid is very effective proposition, like larval parasitoid *Habrobracon hebetor* complements the egg parasitoid *Trichogramma* spp. Because one attacks the larva and other attacks the eggs (Grieshop *et al.*, 2006). According to Brower and Press (1990), such a combination when released in peanut ware houses to control Indianmeal moths and almond moths, the population

Table 16.1. Efficacy of some botanicals against major stored grain pests

Family and Plant	Extract	Bioactivity against	References
Apocyanaceae			
Tabernamontana coronaria	Alcohol (leaf)	*Trogoderma granarium*	Pal *et al.* (1996)
Araceae			
Murraya paniculata	Alcohol (leaf)	*T. granarium*	Pal *et al.* (1996)
Meliceae			
Azadirachta indica	Leaf extract volatiles	*T. granarium* *S. oryaze, T. castaneum*	Pal *et al.* (1996) Koul (2004)
Fabaceae			
Crotolaria ochroleuca	Volatile constituents	*Sitophilus zeamais*	Tiwari and Dixit (1995)
Lauraceae			
Cinnamomum zeylanicum	Volatile constituents	*Tribolium casteneum*	Tiwari and Dixit (1995)
Caesalpiniaceae			
Cassia occidentalis	Leaf powder	*T. castaneum*	Maheshwary and Dwivedi (1996)
Verbinaceae			
Vitex negundo	Methanol (leaf)	*T. confusum*	Khanam *et al.* (1990)

was reduced by 84.3% in combination. In USA three species of *Trichogramma* are used to suppress *Plodia interpunctella* populations (Grieshop *et al.*, 2007). Similarly, *T. evanescens* alone and in combination with *H. hebetor* has been released in commercial facilities in Germany to control stored product moths and it has also been recommended that such parasitoids should be released for about 9 weeks (Flinn and Schöller, 2012). For stored product beetle control, *Anisopteromalus calandre* has been studied (Smith, 1992; Wen and Brower, 1994) specifically against rice weevil which is suppressed by about 76% in the loose grain (Cline *et al.*, 1985). Laripphagus distinguendus has been shown to disperse at least 4 months horizontally and vertically in bulk grain (Steidle and Schöller, 2001). *Trichogramma pretiosum* and *T. evanescens* Westwood attack eggs of *Cadra cautella* Walker, *Ephestia* spp. and *Plodia interpunctella* (Hubner), suppressing 42–57% of population in shell peanuts in small buildings. Several braconid, ichneumonid, pteromalid and bethylid parasitoids are also associated with storage pests.

The ware house pirate bug, *Xylocoris flavipes* (Reuter) consumes the red flour beetle, *Tribolium castaneum* (Herbst), the confused flour beetle, *Tribolium confusum* Duval, *R. dominica* and *T. granarium* in corn bins. Releases of *X. flavipes* @ 50 pairs resulted in more than 78% suppression of moth populations, *O. surinamensis* and *Tribolium* spp. in a simulated peanut warehouse (Rahman *et al.*, 2009). The larger pirate bug, *Lyctocoris campestris* (F.) and the reduviid bug, *Allaeocranum biannulipes* (Montr. et Sign.) are also potential predators of warehouse moths.

From commercial point of view the use of parasitoids and predators has not made much impact. The reason for this is the constraints in the release of live insects into areas where food is kept or packaged and needs a specialized approach for industrial applications. In Germany *T. evanescens* has been released in milling areas and grocery stores to prevent the stored food from moths, however, the number of parasitoids required to suppress pest populations is very high. For example, in a factory producing 1.5 tonnes of bread and cereals/year requires about 3 million parasitoids to control the stored grain pests (Prozell and Schöller, 1998).

ENTOMOPATHOGENS

Entomopathogenic viruses, bacteria, fungi and protozoa are capable of initiating natural epizootics in storage structures. These organisms could be formulated as biopesticides. Being benign with reduced risk factors, microbial biopesticides form a key element in insect pest management of stored products. They are in several ways advantageous compared to chemical insecticides used in storage godowns. Since most microbial pathogens are self replicating, they provide long term control.

Viruses

Viruses are obligate pathogens which offer great promise in biological control of stored insect pest (Burges, 1964). They are highly virulent, cause acute lethal infections and persist in the storage structures for long periods. Several isolates, including nucleo polyhedro virus (NPV), granulo virus (GV), cytoplasmic polyhedro virus (CPV), and a number of non-occluded viruses have been reported from order Lepidoptera and Coleoptera. NPVs and GVs have been widely experimented against stored insect pests. The Indian meal moth, *P. interpunctella* a major pest of stored food products throughout most of the world can be controlled by using *P. interpunctella* granulo virus (Pi GV) as a biopesticide. Young larvae of *P. interpunctella, C. cautella* and *C. cephalonica are* known to be killed by nucleopolyhedrosis and granulosis infections. Both infections of *C. cautella* are known to severely affect *P. interpunctella.* However, granulo virus of *P. interpunctella* did not cross infect *C. cautella.* A nucleo-polyhedro virus of almond moth, *C. cautella* @ 64 × 10^3 polyhedra/g of bran diet was found effective in killing 80% of the test larvae on the 9th and 10th day after exposure at 27°C. The same virus also proved infectious to *P. interpunctella.*

Granulovirus formulated as suspension or dust containing 3.2 × 10^3 virus capsules/mg @ 1.9 mg/kg effectively suppressed *P. interpunctella* when applied in the surface layers of bulk stored wheat or maize. The application of an aqueous suspension of the Granulovirus of *P. interpunctella* was found to be an effective protectant for stored raisins and almonds. The only negative aspect was that the high storage temperature was not conducive for maintaining the viability of the virus. The virulence was best retained at 27°C even for 5–6 months and the pest population was reduced even up to 93%. At 32°C, only 16% control was observed after 6 months. In Indonesia, a Granulovirus, Po GV was found to be an effective biocontrol agent against potato tuber moth, *Pthorimaea operculella* (Zeller). The LC_{50} value ranged between 15.6–18.5 LE/100 ml (Zeddam *et al.*, 1999). A granulosis virus formulation was patented (Vail *et al.*, 1991) and was registered for control of Indianmeal moth larvae on dried fruit and nuts in USA in 2001. However, the formulation is no more commercially available.

Bacteria

Bacillus thuringiensis Berliner, the most common bacteria used for the control of stored insect pests was isolated from diseased larvae of the Mediterranean flour moth, *Ephestia kuehniella* Zeller in 1911. It has been proven as an effective control agent against a wide range of ware house moths, *P. interpunctella, Ephestia elutella* (Hubner), *Ephestia cautella* (Walker), and *Corcyra cephalonica* (St.). The Angoumois grain moth, *Sitotroga cereallella* (Oliver) was not susceptible to *B. thuringiensis* as the pest did not come to sufficient contact with the pathogen, which may be due to its internal feeding behaviour and spending most of its life cycle within a single grain. Treating the surface layers of bulk stored wheat and corn with dust or aqueous suspension of *B. thuringiensis* @ 125 mg/kg offered good control of *P. interpunctella* and *E. cautella* even up to 92%. The recommended depth of treatment was 100 mm at least and the pathogen has to be well mixed to ensure an even treatment with the top layer (McGaughey, 1976). Admixture to surface layers of bulk grain during grain transfer and after filling of grains were found easily effective. The pathogen culture should be re-inoculated after one year because *B. thuringiensis* loses its viability by then. *B. thuringiensis* var *kurstaki* have been found infectious to *E. kuehniella* (Jacobs, 1951). Tounsi *et. al.* 2005 found that the toxin from the parental strain BNS3 was more toxic than the individual toxins Cry1Aa, Cry1Ac, Cry2Aa or toxins used as mixtures against the larvae of *E. kuehiniella.* Gamma irradiation (1.0 krad) coupled with 400 ppm commercial formulation of *B. thuringiensis kurstaki* was found to have synergistic effect in increasing the mortality and extending the developmental period of almond moth *C. cautella* (Faruki and Khan, 2001). The cigarette beetle, *Lasioderma serricorne* (F.) succumbed to toxic isolates of four known serovars of *B. thuringiensis* viz., *thuringiensis, kurstaki, kenyae* and *dermstadiensis* (Kaelin *et al.*, 1999; Tsuchiya *et al.*, 2002). The *B. thuringiensis* isolates PME 3904, PME 1401 and PME 2173 appeared very similar to the anticoleopteran *B. thuringiensis tenebrionis* with 81% mortality against this pest. The commercial formulation of *B. thuringiensis kurstaki* and fine sand dust with quartz controlled *P. operculella* in potato tubers for even three months. The treatment was 95% effective against adult development and hence would best meet the demands of pest control in unsorted, infested seed potatoes (Kroschel and Koch, 1996). Overall, it is apparent that current strains are mostly effective against moths and not stored grain beetles. Resistance also has been reported (McGaughey and Beeman, 1985)

Fungi

Entomopathogenic fungi are generally facultative pathogens of insects. Several fungi attack stored product insects and most notable is Beauveria bassiana (Ferron and Robert, 1975; Hluchy and Samsinakova, 1989). A soil isolate of this white muscardine fungus, *Beauveria bassiana* (Bals.-Criv.) Vuill. was evaluated as a potential microbial control agent of *O. surianamensis*. However, the humidity requirement was well past 90%. In storage of wheat grains, conidial suspension of *B. bassiana* @ 10^7 conidia/ml reduced the percentage weight loss due to insect pests by 81.5% (Padin *et al.*, 2002). Lord

(2007) reported a synergism between *B. bassiana* and diatomaceous earth formulation against adults of the lesser grain borer, *R. dominica* and *O. surianamensis*. *B. bassiana* infection of *R. dominica* was enhanced by reduced moisture and hence could have great potential for control of this pest under storage conditions. The green muscardine fungus, *Metarhizium anisopliae* (Metsch.) has been found effective against *R. dominica*, rice weevil, *Sitophilus oryzae* (L.) and *T. confusum*. However, the effectiveness was highly influenced by exposure interval, target species, dosage and characteristics of fungal preparation. A combined mix of the isolates *B. bassiana* ARSEF 5500 and *M. anisopliae* ARSEF 2974 could cause a mortality of 51.66% in *S. oryzae*. Kavallieratos *et al.* (2006) observed 9-fold increase in mortality after 7 days of exposure with *M. anisopliae* conidial powder @ 8×10^{10} conidia/kg of wheat. Batta (2005) observed 93.3% mortality of *R. dominica* adults in chick pea treated with *M. anisopliae* @ 8×10^{10} conidia/kg of produce.

Cherry *et al.* (2005) found that indigenous isolates of *B. bassiana* and *M. anisopliae* have been found highly infectious to *Callosobruchus maculatus* (F.) in stored cowpea with LC_{50} values ranging from 1.15×10^7 and 4.44×10^7 conidia/g of grain, respectively. *B. bassiana* significantly reduced the counts of the beetle, despite the initial high population of 50 adults/kg of grain. *M. anisopliae* combined with diatomaceous earth could cause 100% mortality in *R. dominica* adults after 14 days of exposure. Also, the progeny production of *R. dominica* on wheat treated with conidial powder of *M. anisopliae* @ 8×10^{10} conidia/kg of grain was significantly lowered and the progeny of *T. confusum* was completely absent. Conidial powder of *M. anisopliae* proved more effective than conidial suspension in *S. oryzae* and *T. confusum*. However for *R. dominica* conidial suspension was more effective than conidial powder. Hence, insect species should be seriously taken into account when planning control strategies which include the use of entomopathogenic fungi preparations against stored product pests.

Mahdneshin *et al.* (2011) tested the virulence and LT_{50} and LT_{90} values of different Iranian isolates of Entomopathogenic fungi such as *B. Bassiana* and *M. anisopliae* against adults of the cowpea weevil, *Callosobruchus maculates* (Table 16.2). The lowest LC_{50} value of 9.7×10^6 conidia/ml was observed in the treatment IRAN 429C (*B. bassiana*). Based on the LC_{50} values, the isolate IRAN 429C (*B. bassiana*) was the most effective to cowpea weevil adults followed by IRAN 429C (*B. bassiana*), DEM I001 (*M. anisopliae*), IRAN 441C (*B.

bassiana), IRAN 187C (*B. bassiana*) and IRAN 715C (*M. anisopliae*). The LC_{50} values showed that isolates of *B. bassiana* in the current research were more virulent than isolates of *M. anisopliae*. Comparison among the estimated LT_{50} values of *B. bassiana* and *M. anisopliae* isolates indicated that IRAN 429C had the lowest LT_{50} values. The different Iranian isolates of *B. bassiana* and *M. anisopliae* were infective to adults of *C. maculatus* and required very high concentrations to caused any mortality to adults that too after 11 days of exposure. The IRAN 441C of *B. bassiana* had the highest virulence against adults of cowpea weevil because it had lower LC_{50} (2.4×10^7) and LT_{50} (7.2 days) and caused the highest mortality (76%) at 1×10^8 conidia/ml concentration as compare with other isolates. So, it is clear that *B. bassiana* IRAN 441C could significantly reduce *C. maculatus* adults in stored cowpea (Mahdneshin *et al.*, 2011). As of today there are no fungi registered for use on stored product insects at least in the United States or India.

Nematodes

They are lethal endoparasites of insects. They enter the host through natural body openings, penetrate into the haemocoel, release symbiotic bacteria in their gut that kill the host within 24–48 h and make the environment inside the insect suitable for nematode development. The infective juvenile, the only free living stage leaves the depleted host and searches for a new one. Since, entomopathogenic nematodes need a thin film of moisture for their mobility; they appear poorly fit for the relatively dry stored product environment. However, they do have a considerable potential as treatment for refugia, outside spillage and product accumulations which are potential sources of infestation. Ramos-Rodriguez *et al.* (2006) reported that *Steinernema riobrave* @ 10 IJs caused 80% mortality against larvae of *E. kuehniella, T. castaneum, O. surinamensis* and pupae of *T. castaneum*. In *P. interpunctella,* larval mortality up to 99% was observed under laboratory conditions (Ramos- Rodriguez *et al.* (2007).

Protozoans

Protozoa are one of the most extensively studied pathogen groups of storage pests as they are capable of causing natural epizootics under favourable conditions (Henry, 1981). Unlike, bacteria and virus, many protozoa have been isolated from coleopterans. Protozoans pose debilitating and weakening effect on their hosts causing subtle and chronic infection rather than immediate kill. Symptoms and

Table 16.2. Virulence, LT_{50} and LT_{90} of different Iranian isolates of *B. bassiana* and *M. anisopliae* against adults of *Callosobruchus maculates*

Fungi	Isolate	Day	Slope ± SE	Chi square value	Lethal concentrations (conidia/ml)	
					LC_{50} (95% FL)	LC_{90} (95% FL)
B. bassiana	IRAN 187C	11	0.45±0.05	72.2	1.9×10^{7} (9.3×10^{6}–3.5×10^{7})	1.2×10^{10} (3.7×10^{9}–7.6×10^{10})
	IRAN 429C	11	0.65±0.07	88.2	9.7×10^{6} (5.7×10^{6}–1.6×10^{7})	8.5×10^{8} (3.5×10^{8}–2.9×10^{9})
	IRAN 441C	11	0.63±0.07	79.0	2.4×10^{7} (1.4×10^{7}–1×10^{7})	2.5×10^{9} (1×10^{9}–1×10^{10})
M. anisopliae	DEM 1001	11	0.35±0.03	84.3	2.6×10^{8} (1.2×10^{8}–6.2×10^{8})	9.9×10^{11} (1.9×10^{11}–1.1×10^{13})
	IRAN 715C	11	0.44±0.05	81.3	1.2×10^{8} (6.3×10^{7}–2.3×10^{8})	9×10^{10} (2.5×10^{10}–5.9×10^{11})
B. bassiana	IRAN 429C	2.7×10^{9}	5.2 ± 1.04	24.2	7.8 (6.5–9.4)	13.8 (10.9–28.7)
	IRAN 187C	3.9×10^{8}	4.3 ± 0.4	94.0	6.7 (6.1–7.2)	13 (11.5–15.5)
	IRAN 441C	7.8×10^{9}	7.1 ± 1.3	30.3	7.2 (6.1–8.3)	11 (9.4–16.3)
M. anisopliae	DEM 1001	5.2×10^{10}	5.9 ± 0.6	89.0	7.7 (7.3–8.1)	12.7 (11.5–14.5)
	IRAN 715C	1.4×10^{10}	6.2 ± 0.65	89.0	7.8 (7.4–8.3)	12.5 (11.5–14.35)

(*Source*: Mahdneshin *et al.*, 2011)

pathological effects also vary considerably. Microsporidians like *Nosema parasiticum* which encompass a wide host range provides some insight into the potential host range of protozoans. *T. castaneum* infected by *Farinocystis tribolii* Weiser were found to be more susceptible to malathion, chlorpyrifos, fenvalerate and cypermethrin. The susceptibility increased by 1.75–7.35-fold depending on the insecticide. One another protozoan infection rendered the pest hypersensitive to radiations.

Nosema whitei (Weiser) has been found effective against *T. castaneum per os* when the larvae feed on medium contaminated with spores (Blaser *et al.*, 2005). However, the success of *N. whitei* to infect its host and its virulence expressed was strongly dependent on age of host larva such that only young larvae were susceptible. A distinct level of synergism was observed in extending the larval period of the pest when *N. whitei* spores were fed along with an insecticide. Subtle interactions are more pronounced in protozoan infections than direct lethal effects. There is a definite possibility that these interactions could also be used to complement other new technologies that would gradually regulate the population dynamics of stored product pests and might reduce the need or frequency of specific control measures. Transovarial and veneral transmission of *Nosema plodiae* Kellen and Lindegren was observed in *P. interpunctella* apart from *per os* route of infection. Infection of ovaries in diseased females was reflected in lowered fecundity and progeny reduction by 27%.

BIORATIONALS

There are some biorational insect controlling agents that affect stored product insects. Spinosad @ 1 mg/kg on corn was found to be very effective against six out of the seven major stored grain insects including *C. ferrugineus*, *R. dominica*, *O. surinamensis*, *S. oryzae*, *S. zeamensis*, and *P. interpunctella.* Spinosad @ 1 mg/kg provided complete suppression of progeny production in *T. castaneum* (Huang and Subramanyam, 2007) (Table 16.3). Impact of hexaflumuron (0–2 µg/µl), a chitin synthesis inhibitor, on growth, development and reproductive performance of the progeny in *Callosobruchus maculatus* after adult treatments has been observed and duration of the embryonic and post embryonic development of eggs laid by treated females of *Callosobruchuus chinensis* was increased (Kellouche and Soltani, 2006).

The external morphology, mortality in the various developmental stages, life span, numbers and percentage reduction of F_1 adults of susceptible (S) and actellic resistant (R) strains of *T. castaneum* were analysed by Kostyukovsky *et al.* (2000) using IGRs such as methoprene and pyriproxifen. The results showed that methoprene @ 0.1 ppm prevented the development of F_1 S-strain adults and most larvae were abnormal and died in the larval stage except for a small number which died during pupation (Table 16.4). As the concentration increased, the life span prolonged from 105 days at 0.1 ppm to 139 days at 3.0 ppm, while that of control was only 67 days. When the F_0

adults were allowed to breed to F_1, the R-strain was less sensitive to methoprene and only at a concentration of 3 ppm was a 100% reduction of F_1 adults obtained. Insect mortality occurred in the pupal stage and during adult emergence but the life span was lengthened only to 82 days compared with that of the S-strain. With pyriproxyfen, all larvae of both S-strain and R-strain appeared abnormal in their external morphology and died in this stage at all tested concentrations and life span was extended to 158 and 300 days in S strain and R strain, respectively.

Table 16.3. Efficacy of some biorationals against storage pests

Biorationals	Target insects	Dose	Reference
Spinosad	*Tribolium castaneum, Cryptolestes ferrugineus, Rhyzopertha dominica, Sitophilus oryzae, S. zeae Oryzaephilus surinamensis*	0.125%	Mutambuki (2012)
Juvenile hormone agonists	*C. maculates, T. castaneum, R . dominica, O. surinam-ensis, P. interpunctella*	-	Oberlander *et al.*, 1997
Ecdysteroid agonists	*P. interpunciella*	-	Oberlander *et al.*, 1997
Methoprene, Pyriproxyfen	*T. castaneum*	0.5 to 3 ppm	Kostyukovsky *et al.* (2000); Arthur (2006)
Hexaflumuron	*C. chinensis*	0.5 μg	Kellouche and Soltani (2006)

Similarly, the biological activity of the four IGRs on F_1 susceptible laboratory strain of *S. oryzae* was analysed by Kostyukovsky *et al.* (2000) as shown in Table 16.5. The maximum mortality recorded for parental adults, including the control was 7%. The life span of the F_1 generation in all treatments including control was about 42 days. Pyriproxyfen showed very high insecticidal effect on *S. oryzae* and reduced the number of F_1 adults by 97% at 2.5 ppm, and its LC_{50} and LC_{90} were only 0.3 and 1.2 ppm, respectively. However, methoprene, RH- 5849 and tebufenozide were much less effective for the control of *S. oryzae*. Methoprene at 10 ppm, and RH- 5849 and tebufenozide at 20 ppm caused F_1 adult mortality of 29, 44 and 40%, respectively. The LC_{50} and LC_{90} were 13 and 23 ppm for methoprene, 21 and 46 ppm for RH-5849 and 22 and 32 ppm for tebufenozide, respectively.

Table 16.4. The biological activity of JH agonists Methoprene and Pyriproxifen on development of F_1 adults of susceptible (S) and actellic-resistant (R) strains of *T. castaneum*

IGR	Strain	Conc. (ppm)	Stage of abnormal death	Life span (days ± SE)	Adults in F_1 (No)*	Reduction of adults in F_1 (%)
Methoprene	S	3.0	Larva	139 (4.3)[i]	0	100
		1.0	Larva	117 (5.4)[h]	0	100
		0.5	Larva (few) and pupa	105 (3.6)[g]	0	100
		0.1	Larva (few) and pupa	105(5.4)[g]	0	100
	R	3.0	Larva	82 (2.9)[f]	13.0 (0.4)[i]	81.9
		1.0	Larva	82 (3.2)[f]	56.7 (1.3)[h]	21.2
		0.5	Larva (few) and pupa	82 (2.5)[f]	109 (6.8)[f]	–51.4
		0.1	Larva (few) and pupa	82 (3.3)[g]	0	100
Pyripro-xyfen	S	2.0	Larva	158 (6.4)[g]	0	100
		1.0	Larva	158 (7.5)[g]	0	100
		0.5	Larva	158 (5.7)[g]	0	100
		0.1	Larva	158 (6.2)[g]	0	100
	R	2.0	Larva	300 (7.9)[k]	0	100
		1.0	Larva	300 (10.2)[k]	0	100
		0.5	Larva	300 (9.8)[k]	0	100
		0.1	Larva	300 (12.3)[k]	0	100
Control	S	0.0	None	67.3 (2.3)[e]	69.3 (0.8)[e]	0
	R	0.0	None	67 (3.5)[e]	72.0 (0.8)[e]	0

*The data are the average of three experiments, each with three replicates. *Number of normal adults obtained at the end of F_1 generation period. Different letters indicate statistically significant difference of averages in the same column with 95% confidence.

Reecently, Mutambuki *et al.* (2012) have evaluated the effect of spinosad along with the Acetylic dust (pirimiphos-methyl 1.6% and permethrin 0.3% dust) against the three major storage pests, *S. zeamensis, T. castaneum* and *R. dominica* ( ). They reported that spinosad at 0.35 ppm was effective up to 12 weeks. However, the Actellic super dust at 10.5 ppm and spinosad 0.125% at 0.7 and 1.4

ppm produced total control of *S. zeamais* up to 24 weeks (Table 16.6). There were no significant differences (P = 0.05) in insect mortality between the two dosage rates and Actellic super (10.5 ppm) over the same period. Spinosad at all levels controlled *R. dominica* with 99.3% mortality over the 24 week trial period. Actellic super dust at 10.5 ppm killed 22.8 (76%) of *T. Castaneum* at 24 weeks, but spinosad 0.125% dust was not effective even at the highest rate of 1.4 ppm, which killed only 3.3 insects (11%) over the same period. So, *T. castaneum* was less susceptible to spinosad than the other insects. There were significant differences between spinosad and Actellic super throughout the trial period for *T. castaneum*. A progressive decrease was apparent by the 24th week for all the treatments. So application of spinosad is advisable before infestation to control *S. zeamais*, *P. truncatus* and *R. dominica* that would also protect the products against opportunistic secondary pests that attack damaged grains.

Table 16.5. The biological activity of various IGRs on development of F_1 adults of susceptible laboratory strain of *S. oryzae*

IGRs	Concentration (ppm)	Adults F_1 (No. ± SEm)	Reduction of F_1 adults
Methoprene	10	163.3 (4.9)[i]	29.3
	7.5	203.7 (6.4)[h]	11.8
	5	226.0 (11.4)[e]	2.2
	2.5	250.7 (10.8)[g]	-8.5
Pyriproxyfen	2.5	6.9 (0.8)[o]	97.0
	2	9.0 (0.9)[n]	96.1
	1	27.0 (1.6)[m]	88.3
	0.5	72.1 (3.0)[l]	68.8
RH – 5849	20	128.9 (7.9)[k]	44.2
	15	152.9 (8.1)[i]	33.8
	10	207.9 (10.7)[h]	10.0
	5	267.3 (13.8)[f]	-15.7
Tebufenozide	20	138.7 (12.3)[j]	40.0
	15	204.3 (9.3)[h]	11.6
	10	230.1 (15.8)[e]	0.4
Control	0	231.0(13.2)[e]	-

(*Source*: Kostyukovsky *et al.*, 2000)

Table 16.6. Mean mortality of adults (n = 30) of stored product pests exposed to grains admixed with spinosad compared to actellic super dust (permethrin)

Treatment	Conc. (%)	Rate (g/90kg)	Dose (ppm)	Post treatment period (weeks)						
				0	4	8	12	16	20	24
S. zeamensis										
Spinosad	0.125	25	0.35	20.5^{b}	19.5^{b}	28.0^{b}	15.5^{c}	22.5^{a}	24.5^{b}	25.5^{b}
Spinosad	0.125	50	0.7	28.5^{a}	28.5^{a}	29.5^{a}	25.5^{b}	27.3^{a}	28.0^{ab}	29.5^{a}
Spinosad	0.125	100	1.4	29.8^{a}	29.8^{a}	30.0^{a}	28.8^{ab}	22.8^{a}	29.5^{ab}	29.8^{a}
P- methyl/ Permethrin	1.6/0.3	50	10.5	30.0^{a}	30.0^{a}	30.0^{a}	30.0^{a}	29.8^{a}	30.0^{a}	30.0^{a}
Control	0.0	-	-	0.5^{c}	0.0^{c}	0.0^{c}	0.0^{c}	0.0^{c}	0.5^{c}	0.0^{c}
LSD ($P = 0.05$)	-	-	-	4.35	2.65	1.09	3.73	9.67	5.16	1.96
R. dominica										
Spinosad	0.125	25	0.35	30.0^{a}	30.0^{a}	30.0^{a}	30.0^{a}	30.0^{a}	30.0^{a}	29.8^{a}
Spinosad	0.125	50	0.7	30.0^{a}	30.0^{a}	30.0^{a}	30.0^{a}	30.0^{a}	30.0^{a}	30.0^{a}
Spinosad	0.125	100	1.4	30.0^{a}	30.0^{a}	30.0^{a}	30.0^{a}	30.0^{a}	30.0^{a}	30.0^{a}
P- methyl/ Permethrin	1.6/0.3	50	10.5	30.0^{a}	30.0^{a}	29.3^{a}	28.3^{a}	30.0^{a}	26.0^{b}	30.0^{a}
Control	0.0		-	0.5^{b}	1.5^{b}	0.0^{b}	0.0^{b}	0.0b	0.0^{b}	0.0^{b}
LSD ($P = 0.05$)	-	-	-	0.40	1.19	0.0	1.92	0.0	0.3	0.0
T. castaneum										
Spinosad	0.125	25	0.35	6.0^{c}	3.5^{b}	8.0^{c}	1.5^{bc}	5.0^{bc}	11.5^{b}	1.3^{b}
Spinosad	0.125	50	0.7	15.8^{b}	2.3^{bc}	12.0^{bc}	1.8^{bc}	6.8^{b}	11.8^{b}	1.3^{b}
Spinosad	0.125	100	1.4	17.5^{b}	1.5^{cd}	16.0^{b}	2.5^{b}	11.3^{b}	18.0^{b}	3.3^{b}
P- methyl/ Permethrin	1.6/0.3	50	10.5	30.0^{a}	30.0^{a}	30.0^{a}	30.0^{a}	25.0^{a}	28.0^{a}	22.8^{a}
Control	0.0	-	-	1.5^{c}	0.0^{c}	1.3^{d}	0.0^{c}	0.0^{c}	0.0^{c}	0.0^{c}
LSD ($P - 0.05$)	-	-	-	6.34	1.88	5.09	2.21	6.49	7.80	4.60

(*Source*: Mutambuki *et al*., 2012)

CONCLUSIONS

Even though there is a good scope for employing biopesticides such as, botanicals, microbials, predators, parasitoids and IGRs against stored insect pests, their use is often limited by factors like slow kill, limited success, high specificity, sensitivity to abiotic factors and failure to check insect pests in deep layers/mixed infestations. Since, microbial insecticides are effective against only a narrow range of pests so, they are vulnerable to rapid degradation. Their availability is not consistent and they lack competence with newer synthetics which are more cost effective. Hence, there is a need to explore the possibilities of expanding the host range of entomopathogens of stored insect pests through molecular engineering. Novel production technologies of microbial biopesticides should be developed with a view to enhance their viability and virulence. Biopesticides offer effective alternatives for the control of storage insect pests. Their greatest strength is their safety, as they are essentially nontoxic and nonpathogenic to animals and humans. So, more studies have to be conducted to screen various strains of native isolates of different microbials.

Among different botanicals, much works have been carried out on neem only. Research focus is to be given on the identification of active principles and fumigant toxicity of many unexplored plant oils. Regulatory barriers on biopesticides should be addressed by extensive studies and corresponding well proven results on their safety. However, strict regulatory aspects need to be imposed on quality control standards of biopesticides so that they gain consumer acceptance and preference. Effectiveness of microbial derived insecticides like spinosad and avermectins are already proven to be effective against various crop pests. Hence, research is needed to use these novel insecticides against stored grain insect pests.

Due to various harmful effects of conventional insecticides, the insect growth regulators (IGRs) are receiving more practical attention to provide for safer foods and a cleaner environment. Since the introduction of diflubenzuran, a number of other BPU derivatives have been developed such as flucycloxuron, hexaflumuron and triflumuron (Soltani *et al.,* 1999) suggesting that IGRs could be very effective under controlled conditions, which can severely influence the growth and development of insects. However, strategy based research is required to make biopesticides potential agents for control of wide array of storage insect pests.

REFERENCES

Adam, B.D., Phillips, T.W. and Flinn, P.W. (2006) The economics of IPM in stored grain: Why don't more grain handlers use IPM? In *9th International Working Conference on Stored Product Protection,* 15-18 Oct., 2006, Campinas, Sao Paulo, Brazil, 12 p.

Blaser, M. and Schmid-Hempel, P. (2005) Determinants of virulence for the parasite *Nosema whitei* in its host *Tribolium castaneum. J. Invertebr. Pathol.,* **89**, 251–257

Brower, J.H. (1990) Pests of stored products. In D.H. Habeck F.D. Bennet and Frank, J.H. (eds.), *Classical Biological Control in the Southern United States,* Southern Cooperative Series, Bulletin No. 355, pp. 113–122

Brower, J.H. and Press, J.W. (1990) Interaction of *Bracon hebetor* (Hymenoptera: Braconidae) and *Trichogramma pretiosum* (Hymenoptera: Trichogrammatidae) in suppressing stored-product moth populations in small inshell peanut storage. *J. Econ. Entomol.*, **83**, 1096–1101.

Burges, H.D. (1964) Possibilities of biological control of stored products insects. In: *Proceedings of XII International Congress of Entomology*, London, England, Ind. Stored Products Entomology Section 9b, 659

Chahal, K.K., Arora, M., Joia, B.S. and Chhabra, B.R. (2005) Bioefficacy of turmeric oil against *Tribolium castaneum* (Herbst) under laboratory conditions. In: V.K. Dilawari, G.S. Deol, B.S. Joia and P.K. Chuneja (eds.), *Proc. 1st Congress on Insect Science*: *Contributed Papers*, PAU Ludhiana, pp. 147–148.

Champ, B.R. and Dyte, C.E. (1976) Report of the FAO Global Survey of Pesticide Susceptibility of Stored Pests. FAO, Rome.

Cherry, A.J., Abalo, P. and Hell, K. (2005) A laboratory assessment of the potential of different strains of the entomopathogenic fungi *Beauveria bassiana* (Balsamo) Vuiellemin and *Metarhizium anisopliae* (Metsch.) to control *Callosobruchus maculatus* (F.) (Coleoptera: Bruchidae) in stored cowpea. *J. Stored Prod. Res.,* **41**, 295–309

Cline, L.D., Press, J.W. and Flaherty, B.R. (1985) Supression of the rice weevil, *Sitophilus oryzae* (Coleoptera : Curculionidae) inside and outside of burlap, woven polypropylene, and cotton bags by the parasitic wasp, *Anisopteromalus calandrae* (Hymenoptera : Pteromalidae). *J. Econ. Entomol.*, **78**, 835–838.

Dhuri, A.V. (2006) Fumicover: an effort in reducing losses in stored grains at farm levels. In *9th International Working Conference on Stored Product Protection,* 15-18 Oct., 2006, Campinas, Sao Paulo, Brazil, pp. 612–617

Egwurube, E., Magaji, B.T. and Lawal, Z. (2010) Laboratory evaluation of neem (*Azadirachta indica*) seed and leaf powders for the control of khapra beetle, *Trogoderma granarium* (Coleoptera : Dermestidae) infesting groundnut. *Int. J. Agric. Biol.,* **12**, 638–640

Faruki, S.I. and Khan, A.R. (2001) The pathogenicity of *Bacillus thuringiensis* ssp. *kurstaki* to γ – irradiated *Cadra cautella* (Walker) (Lepidoptera: Phycitidae). *J. Invertebr. Pathol.*, **77**, 92–98

Flinn, P.W. and Schöller, M. (2012) Biological control: Insect pathogens, parasitoids and predators. In: D.W. Hangstrum, T.W. Phillips and G.

Cuperus (eds.), *Stored Product Protection*, Kansas State University, Manhattan, KS, pp. 203–212.

Govindan, K. and Nelson, S.J. (2009) Insecticidal activity of twenty plant powders on mortality, adult emergence of *Sitophilus oryzae* L. and grain weight loss in paddy. *J. Biopestic.,* **2**, 169–172

Grieshop, M.J., Flinn, P.W. and Nechols, J.R. (2006) Biological control of Indianmeal moth (Lepidoptera: Pyralidae) on finished stored products using egg and larval parasitoids. *J. Econ. Entomol.*, **99**, 1080–1084.

Grieshop, M.J., Flinn, P.W. and Nechols, J.R. (2007) Host-foraging success of three species of *Trichogramma* (Hymenoptera: Trichogrmmatidae) in a simulated retail environment. *J. Econ. Entomol.*, **100**, 591–598.

Gupta, L. and Srivastava, M. (2008) Effect of *Withania somnifera* extracts on the mortality of *Callosobruchus chinensis* L. *J. Biopestic.,* **1**, 190–192

Henry, J.E. (1981) Natural and applied control of insects by protozoa. *Annu. Rev. Entomol.,* **26**, 49–73

Huang, F. and Subramanyam, B. (2007) Effectiveness of spinosad against seven major stored grain insects on corn. *Insect Sci.,* **14**, 225–230

Jacobs, S.E. (1951) Bacteriological control of the flour moth, *Ephestia kuehniella Z. Proc. Soc. Appl. Bacteriol.,* **1**, 83–91

Kaelin, P., Zaugg, L., Albertini, A.M. and Gadani, F. (1999) Activity of *Bacillus thuringiensis* isolates on *Lasioderma serricorne* (F.) (Coleoptera: Anobiidae). *J. Stored Prod. Res.,* **35**, 145–158

Kavallieratos, N.G., Athanassiou, C.G., Michalaki, M.P., Batta, Y.A., Rigatos, H.A., Pashalidou, F.G., Balotis, G.N., Tomanovie, Z. and Vayaias, B.J. (2006) Effect of the combined use of *Metarhizium anisopliae* (Metsch.) Sorokin and diatomaceous earth for the control of three stored-product beelte species. *Crop Prot.*, **25**, 1087–1094

Kellouche, A. and Soltani, N. (2006) Impact of hexaflumuron, a chitin synthesis inhibitor, on growth, development and reproductive performance of the progeny in *Callosobruchus maculatus* after adult treatments. *Afr. J. Agric. Res.*, **1**, 57–64

Khanam, L.A.M, Khan, A.R. and Talukder, D. (1990) Insecticidal properties of royna, *Aphanamixis polystachea* Wall (Parker) (Meliaceae) against *Tribolium confusum* Duval. *J. Asiatic Soc. Bangladesh Sci.,* **16**, 71–74

Kostyukovsky, M., Chen, B., Atsmi, S. and Shaaya, E. (2000) Biological activity of two juvenoids and two ecdysteroids against three stored product insects. *Insect Biochem. Mol. Biol.*, **30**, 891–897

Koul, O. (2004) Bioactivity of volatile di-n-propyl disulfide from seeds of neem, *Azadirachta indica* (Meliaceae) to two species of stored grain pests, *Sitophilus oryzae* (L.) and *Tribolium castaneum* (Herbst.). *J. Econ. Entomol.*, **97**, 1142–1147.

Koul, O., Tikku, K. and Saxena, B.P. (1977) Follicular regression in *Trogoderma granarium* due to sterilizing vapours of *Acorus calamus* L. oil. *Curr. Sci.*, **46**, 724–725.

Koul, O., Walia, S. and Dhaliwal, G.S. (2008) Essential oils as green pesticides: Potential and constraints. *Biopestic. Int.*, **4**, 63–84.

Koul, O., Gurmeet Singh, Rajwinder Singh and Jasbir Singh (2007) Mortality and reproductive performance of *Tribolium castaneum* exposed to anethole vapours at high temperature. *Biopestic. Int.*, **3**, 126–137.

Kroschel, J. and Koch, W. (1996) Studies on the use of chemicals, botanicals and *Bacillus thuringiensis* in the management of the potato tuber moth in potato stores. *Crop Prot.* **15**, 197–203

Lord, J.C. (2007) Desiccation increases the efficacy of *Beauveria bassiana* for stored-grain pest insect control. *J. Stored Prod. Res.,* **43**, 535–539

Mahdneshin, Z., Vojoudi, S., Ghosta, Y. Safaralizadae, M.H. and Saber, M. (2011) Laboratory evaluation of the entomopathogenic fungi, Iranian isolates of *Beauveria bassiana* (Balsamo) Vuillemin and *Metarhizium anisopliae* (Metsch) Sorokin against the control of the cowpea weevil, *Callosobruchus maculatus* F. (Coleoptera: Bruchidae). *Afr. J. Microbiol. Res.,* **5**, 5215–5220

Maheshwari, H.K. and Dwivedi, S.C. (1996) Evaluation of botanicals for the management of *Tribolium castaneum* (Coleoptera: Tenebrionidae). *Insect Environ.,* **2,** 72–73

McGaughey, W.H. (1976) *Bacillus thuringiensis* for controlling three species of moths in stored grain. *Can. Entomol.*, **108**, 105–112.

McGaughey, W.H. and Beeman, R.W. (1985) Insect resistance to biological insecticide bacillus thuringiensis. *Science*, **229**, 193–195.

Mutambuki, K., Ngatia, C.M., Mbugua, J.N. and Likhayo, P. (2012) Evaluation on the efficacy of spinosad dust against major storage insect pests. *J. Stored Prod. Postharvest Res.* **3**, 19–23

Nadaf, H.A., Rao, S.R.K., Verma, N.R.G. and Murthy, V.R.K. (2010) Bio-efficacy of certain grain protectants against groundnut bruchid, *Caryedon serratus* (Olivier). *Int. J. Plant Prot.,* **3,** 60–64

Naik, S.N. and Geetanjali, K. (2010) Grain Storage in India: An Overview. Available: http://www.vigyanprasar.gov.in/pdf (online).

Oberlander, H., Silhacek, L.D., Shaaya, E. and Ishaaya, I. (1997) Current status and future perspectives of the use of insect growth regulators for the control of stored product insects. *J. Stored Prod. Res.,* **33**, 1–6

Padin, S., Bello, G.D. and Fabrizio, M. (2002) Grain loss caused by *Tribolium castaneum, Sitophilus oryzae* and *Acanthoscelides obtectus* in stored durum wheat and beans treated with *Beauveria bassiana. J. Stored Prod. Res.,* **38,** 69–74

Pal, R., Tripathi, R., Prasad, T. and Rameshwar, P. (1996) Relative toxicity of certain plant extracts to khapra beetle, *Trogoderma granarium*. *Ann. Plant Prot. Sci.,* **4**, 35–37

Pokharkar, P.K. and Chauhan, N.R. (2010) Bioefficacy of botanicals material against *Callosobruchus chinensis* (L.) in stored chickpea. *Int. J. Plant Prot.,* **3,** 72–76

Pozell, S. and Schöller, M. (1998) Insect fauna of a bakery, processing organic grain and applying Trichogramma evanescens Westwood. In: C. Adler and M. Schöller (eds.), *Integrated Protection of Stored Products,* IOBC wprs Bulletin 21, pp. 39–44

Rahman, M.M., Islam, W. and Ahmed, K.N. (2009) Functional response of the predator *Xylocoris flavipes* to three stored product insect pests. *Int. J. Agric. Biol.,* **11,** 316–320

Ramos-Rodriguez, O., Campbell, J.F. and Ramaswamy, S.B. (2006) Pathogenicity of three species of entomopathogenic nematodes to some major stored-product insect pests. *J. Stored Prod. Res.,* **42,** 241–252

Ramos-Rodriguez, O., Campbell, J.F. and Ramaswamy, S.B. (2007) Efficacy of the entomopathogenic nematode, *Steinernema riobrave* against the stored-product insect pests *Tribolium castaneum* and *Plodia interpunctella. Biol. Control,* **40,** 15–21

Saxena, B.P., Koul, O. and Tikku, K. (1976) Nontoxic protectant against the stored grain insect pests. *Bull. Grain Technol.,* **14**, 190–193.

Schöller, M., Prozell, S., Al-Krishi, A.G. and Reichmuth, C.H. (1997) Towards Biological Control as a major component of integrated pest management in stored product protection. *J. Stored Prod. Res.,* **33,** 81–97

Shaaya, E., Kostjukovski, M., Eilberg, J. and Sukprakarn, C. (1997) Plant oils as fumigants and contact insecticides for the control of stored-product insects. *J. Stored Prod. Res.,* **33,** 7–15.

Smith, L. (1992) Effect of temperature on life history characteristics of *Anisopteromalus calandrae* (Hymenoptera: Pteromalidae) parasitizing maize weevil larvae in corn kernels. *Environ. Entomol.,* **21**, 872–876.

Soltani, N., Soltani-Mazouni, N. and Delachambre, J. (1999) Evaluation of triflumuron, a benzoylphenyl urea derivative, on *Tenebrio molitor* pupae: effects on cuticle. *J. Appl. Entomol.,* **120**, 627–629

Steidle, J.L.M. and Schöller, M. (2001) Fecundity and ability of the parasitoid *Lariophagus distinguendus* (Hymenoptera: Pteromalidae) to parasitize larvae of the granary weevil, *Sitophilus granarius* (Coleoptera: Curculionidae) in bulk grain. *J. Stored Prod. Res.,* **38**, 43–53.

Tiwari, R. and Dixit, V. (1995) Organoleptic properties of rice treated with the essential oil of *Cinnamomum zeylanicum. Indian Perf.,* **39**, 67–69

Tounsi, S., Dammak, M., Rebai, and Jaoua, S. (2005) Response of larval *Ephestia kuehiniella* (Lepidoptera: Pyralidae) to individual *Bacillus thuringiensis* var *kurstaki* toxins and toxin mixtures. *Biol. Control*, **35**, 27–31

Tripathi, A.K., Prajapati, V. and Kumar, S. (2003) Bioactivity of l-carvone, d-carvone and dihydrocarvone towards three stored product beetles. *J. Econ. Entomol.*, **96**, 1594–1601.

Tripathi, A.K., Prajapati, V., Aggarwal, K.K., Sushil Kumar, Prajapti, V., Kumar, S. Kukreja, A.K. Dwivedi, S. and Singh, A.K. (2000) Effects of volatile oil constituents of *Mentha species* against stored grain pests, *Callosobrunchus maculatus* and *Tribolium castanum*. *J. Med. Arom. Plant Sci.*, **22**, 549–556.

Tripathi, A.K., Prajanpati, V., Aggarwal, K.K. and Kumar, S. (2001) Toxicity, feeding deterrence, and effect of activity of 1, 8-cineole from *Artemisia annua* on progeny production of *Tribolium castanaeum* (Coleoptera : Tenebrionidae). *J. Econ. Entomol.*, **94**, 979–983.

Tsuchiya, S. Kasaishi, Y., Harada, H., Ichimatsu, T., Saitoh, H., Mizuki, E. and Ohba, M. (2002) Assessment of the efficacy of Japanese *Bacillus thuringiensis* isolates against the cigarette beetle, *Lasioderma serricorne* (Coleoptera: Anobiidae). *J. Invertebr. Pathol.*, **81**, 122–126

Vail, P.V., Tebbets, J.S., Cowan, D.C. and Jenner, K.E. (1991) Efficacy and persistence of a granulosis virus infestations of *Plodia interpunctella* (Hubner) (Lepidoptera: Pyralidae) on raisins. *J. Stored Prod. Res.*, **27**, 103–107

Wen, B. and Brower, J.H. (1994) Competition between *Anisopteromalus calandrae* and *Choetospila elegans* at different parasitoid densities on immature maize weevils in corn. *Environ, Entomol.*, **23**, 367–373.

Zeddam, J.L., Pollet, A., Mangoendiharjo, S., Ramadhan, T.H. and Ferber, M.L. (1999) Occurence and virulence of a granulosis virus in *Pthorimaea operculella* (Lepidoptera: Gelechiidae) populations in Indonesia. *J. Invertebr. Pathol.*, **74**, 48–54.

Subject Index